QUANTUM FLUIDS AND SOLIDS—1989

Symposium on Quantum Fluids and Solids -1989

University of Florida
Gainesville, Florida
24-28 April 1989

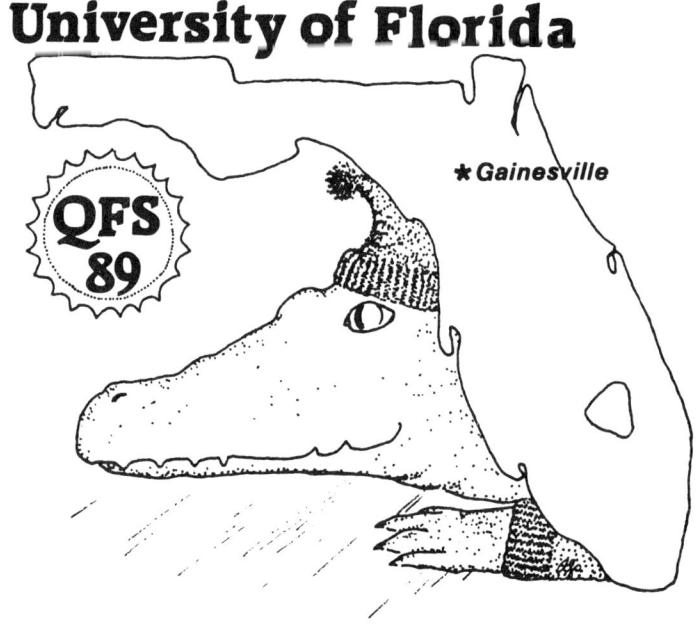

AIP CONFERENCE PROCEEDINGS 194

RITA G. LERNER
SERIES EDITOR

QUANTUM FLUIDS AND SOLIDS—1989

GAINESVILLE, FL 1989

EDITORS:
GARY G. IHAS & YASUMASA TAKANO
UNIVERSITY OF FLORIDA

American Institute of Physics   New York

To the memory of William M. Fairbank
1917-1989

Authorization to photocopy items for internal or personal use, beyond the free copying permitted under the 1978 US Copyright Law (see statement below), is granted by the American Insitute of Physics for users registered with the Copyright Clearance Center (CCC) Transactional Reporting Service, provided that the base fee of $3.00 per copy is paid directly to CCC, 27 Congress St., Salem, MA 01970. For those organizations that have been granted a photocopy license by CCC, a separate system of payment has been arranged. The fee code for users of the Transactional Reporting Service is: 0094-243X/87 $3.00.

Copyright 1989 American Institute of Physics.

Individual readers of this volume and non-profit libraries, acting for them, are permitted to make fair use of the material in it, such as copying an article for use in teaching or research. Permission is granted to quote from this volume in scientific work with the customary acknowledgment of the source. To reprint a figure, table or other excerpt requires the consent of one of the original authors and notification to AIP. Republication or systematic or multiple reproduction of any material in this volume is permitted only under license from AIP. Address inquiries to Series Editor, AIP Conference Proceedings, AIP, 335 E. 45th St., New York, NY 10017.

L.C. Catalog Card No. 89-081079
ISBN 0-88318-395-1
DOE CONF-8904257

Printed in the United States of America.

Contents

Preface xi

Organizing Committee xiii

I. TUNNELING AND PHASE SLIPPAGE IN SUPERFLUIDS

Phase Slippage Experiments in Superfluid ^{3}He–A 3
    O. Avenel and E. Varoquaux
Magnetic Relaxation in Superfluid ^{3}He 15
    A. S. Borovik-Romanov, Yu. M. Bunkov, V. V. Dmitriev, and Yu. M. Mukharskiy
Josephson Effect in Spin Supercurrent in ^{3}He–B 27
    A. S. Borovik-Romanov, Yu. M. Bunkov, V. V. Dmitriev, Yu. M. Mukharskiy, and D. A. Sergatskov
Escape from a Metastable Potential in Dissipative Quantum
and Classical Systems 39
    D. A. Browne, V. Ambegaokar, and K. S. Chow
Superfluid Phase Slippage in ^{3}He Flow Through a Narrow Channel 51
    M. M. Salomaa and N. B. Kopnin

II. SUPERFLUID ^{3}He

Scattering of Ballistic Quasiparticles at the A–B Phase Boundary
of Superfluid ^{3}He 55
    N. Schopohl and D. Waxman
Ultrasonic Investigation of Rotating Superfluid ^{3}He 63
    J. P. Pekola, R. H. Salmelin, A. J. Manninen, K. Torizuka, J. M. Kyynäräinen, and G. K. Tvalashvili
Five-Fold Splitting of the Squashing Mode of Superfluid ^{3}He–B
in a Magnetic Field 75
    R. Movshovich, N. Kim, D. M. Lee, and E. Varoquaux
Nonlinear Acoustics in Superfluid ^{3}He–B 87
    R. H. McKenzie and J. A. Sauls
Fermi Parameter Estimates from Non-Resonant Measurements of the Real
Squashing Mode Frequency in ^{3}He–B 96
    P. N. Fraenkel
Collective Mode Resonances in Superfluid ^{3}He–A 103
    E. R. Dobbs, R. Ling, J. Saunders, and W. Wojtanowski
Anisotropic Attenuation of Zero Sound in Superfluid ^{3}He–A 105
    J. Saunders, W. Wojtanowski, R. Ling, and E. R. Dobbs
Pair Breaking Edge in Superfluid ^{3}He 107
    S. Adenwalla, Z. Zhao, J. B. Ketterson, and B. K. Sarma
Observation of a Doublet Splitting of the Squashing Mode
in Superfluid ^{3}He–B 109
    Z. Zhao, S. Adenwalla, P. N. Brusov, J. B. Ketterson, and B. K. Sarma

The Linear Zeeman Effect for Clapping and Pairbreaking Modes in ^{3}He–A 111
    P. N. Brusov, M. Y. Nasten'ka, T. V. Filatova-Novoselova, M. V. Lomakov,
    and V. N. Popov
New Sound and NMR Experiments in ^{3}He–B and Common Approach to
Their Theoretical Description 113
    M. Y. Nasten'ka and P. N. Brusov
Gyrosonic Effect in Rotating Superfluid ^{3}He–B 115
    M. M. Salomaa and G. E. Volovik
Finite Amplitude Acoustic Waves in Liquid ^{3}He and ^{4}He 117
    M. Chapellier, J. Joffrin, M. W. Meisel, and A. Schuhl
Intrinsic Magnus Effect in Superfluid ^{3}He–A 119
    M. M. Salomaa and R. H. Salmelin

III. QUANTUM INTERACTIONS

Specific Heat of Liquid ^{3}He Bubbles in Solid Matrices 123
    E. Syskakis, Y. Fujii, M. Gebhardt, and F. Pobell
Fractional Statistics and Analogs of Quantum Hall Effect
in Superfluid ^{3}He Films 136
    G. E. Volovik
Interaction of Vortices with the Homogeneously Precessing Resonance Mode
in Superfluid ^{3}He–B 147
    J. S. Korhonen, Z. Janú, Y. Kondo, M. Krusius, Yu. M. Bunkov, V. Dmitriev,
    and Yu. M. Mukharskiy
Vortex Nucleation in Superfluid ^{4}He 149
    R. M. Bowley
Anomalous ^{3}He Solubility in Superfluid ^{4}He at Low Temperatures 156
    H. Fukuyama, I. Kurikawa, H. Ishimoto, and S. Ogawa

IV. LOW DIMENSIONAL HELIUM

The Superfluid Transition in Porous Media 161
    J. Machta and R. A. Guyer
Superfluid Transition of ^{4}He Confined in Porous Glasses 170
    M. H. W. Chan
Boson Localization and the Superfluid–Insulator Transition 179
    M. P. A. Fisher
Ultrasonic Studies of Helium in Porous Glasses 182
    J. R. Beamish and N. Mulders
Attenuation Mechanisms in Porous Media Containing Helium 191
    K. Warner, N. Mulders, and J. R. Beamish
Vortex String Dissipation in ^{4}He Films Adsorbed on Porous Materials 193
    H. Cho, F. Gallet, and G. A. Williams
Pore-Size Dependence of the Superfluid Transition in ^{4}He Films Adsorbed
on Porous Glasses 195
    K. Shirahama, N. Wada, M. Kubota, S. Ogawa, T. Watanabe, and K. Eguchi
The Singular Boundary Resistance of Superfluid ^{4}He 197
    F. Zhong, J. Tuttle, and H. Meyer

Wetting of Helium Films to Silver Substrates 199
    R. J. Dionne and R. B. Hallock
Evidence for Magnetic Ordering in the Boundary Layers
of ^{3}He on Grafoil 201
    L. J. Friedman, A. L. Thomson, C. M. Gould, H. M. Bozler, P. B. Weichman,
    and M. C. Cross
Nuclear Spin Heat Capacity of ^{3}He Adsorbed on Graphite 213
    D. S. Greywall
Field Dependence of the ^{3}He–Substrate Spin Coupling 217
    T. J. Gramila, F. Van Keuls, Y. Hu, and R. C. Richardson
Oscillations of Critical Parameters in Thin Films of ^{3}He 219
    L. S. Borkowski, G. Harań, and L. Jacak
Superfluid ^{3}He Film Flow—A Phase Transition and a Substrate Effect 221
    J. P. Harrison, A. Sachrajda, S. C. Steel, and P. Zawadzki
Magnetic Relaxation of Normal ^{3}He on a Surface Coated with ^{4}He 223
    Yu. M. Bunkov, V. V. Dmitriev, Yu. M. Mukharskiy, and D. A. Sergatskov
Structure in the Magnetization of Thin ^{3}He–^{4}He Mixture Films 225
    R. H. Higley, D. T. Sprague, and R. B. Hallock
Quasiparticle Interaction of ^{3}He Impurities on ^{4}He Films 227
    M. Saarela, E. Krotscheck, and J. L. Epstein

V. POLARIZED QUANTUM SYSTEMS

Kinetic Phenomena in Spin-Polarized Quantum Systems 231
    A. E. Meyerovich
Numerical Study of Nonlinear Spin Waves in a Fermi Liquid 241
    D. Candela
Transverse Spin Diffusion in Polarized Fermi Gases 243
    J. W. Jeon and W. J. Mullin
Viscosity Change in Liquid ^{3}He on Polarization 245
    G. A. Vermeulen, A. Schuhl, F. B. Rasmussen, J. Joffrin, and M. Chapellier
Bose Einstein Condensation: Compress or Expand? 247
    I. F. Silvera
The Role of Magnetic Field Gradient in the Spin Wave Spectrum of Spin
Polarized Atomic Hydrogen 257
    N. P. Bigelow, J. H. Freed, and D. M. Lee

VI. SOLID HELIUM

Measurements of Magnetic Susceptibility of hcp Solid ^{3}He 261
    T. Mamiya, H. Yano, T. Kato, Y. Minamide, Y. Miura, S. Inoue, and T. Uchiyama
Pressure Study of hcp Solid ^{3}He in Magnetic Fields 267
    Y. Miura, S. Abe, S. Sugiyama, Y. Mamiya, and R. C. Richardson
Charge Motion in Solid Helium 273
    A. F. Andreev
Nuclear Magnetism of bcc Solid ^{3}He in a High Magnetic Field 281
    H. Ishimoto, H. Fukuyama, T. Fukuda, T. Okamoto, T. Tazaki, K. Sakayori,
    and S. Ogawa

Domain Structure of uudd ^{3}He Studied by NMR 286
    Y. Sasaki, Y. Hara, T. Mizusaki, and A. Hirai
Towards a First Principles Theory of ^{3}He Magnetism 288
    J. H. Hetherington and M. Roger
Low Temperature Susceptibility of Large Molar Volume bcc Solid ^{3}He 290
    M. E. R. Bernier, M. Bassou, M. Chapellier, and M. Rotter
Vacancies in bcc Solid ^{3}He 292
    M. E. R. Bernier and J. H. Hetherington
Solid ^{4}He: Search for Superfluidity 294
    G. Bonfait, H. Godfrin, and B. Castaing
Roughening and Wetting Transitions in Dilute ^{3}He–^{4}He Mixture Crystals 299
    Y. Carmi, S. Lipson, and E. Polturak
The Dynamics of the Isotopic Phase Separation in Solid Helium 301
    N. Alikacem and M. Richards

VII. NUCLEAR MAGNETIC ORDERING

Nuclear Magnetism in Copper at Nanokelvin Temperatures 305
    H. E. Viertiö and A. S. Oja
Nonlinear Spin Dynamics and Nuclear Ordering: Tl Metal
as an Example 316
    G. Eska
Antiferromagnetic Resonance of Hyperfine-Enhanced Nuclear
Antiferromagnet HoVO$_4$ 328
    H. Suzuki, M. Ono, and N. Mizutani

VIII. EXOTIC STUDIES AT LOW TEMPERATURES

Hunting for Most of the Universe—Dark Matter 333
    D. O Caldwell
Superconducting Detectors for Laboratory Dark Matter Searches 342
    B. Cabrera
Search for Dark Matters Axions 352
    P. Sikivie
Bolometric Detection of Rare Events: A Challenge for Low Temperature
Physics 364
    M. Chapellier, G. Chardin, H. Ji, and J. Joffrin
A Cosmic Axion Detector 366
    C. Hagmann, P. Sikivie, N. S. Sullivan, and D. B. Tanner

IX. LOW TEMPERATURE TECHNIQUES

Quasiparticle Spectrometry, the Engineering and the Physics 371
    A. M. Guénault and G. R. Pickett
Some Design Features, Non-Features, and Ex-Non-Features of the Cornell
Microkelvin Cryostat 382
    E. N. Smith
New Design for a Copper Demagnetization Stage 393
    J. D. Kilian, P. B. Chilson, G. G. Ihas, and E. D. Adams

Vibration Analysis of an Ultra-Low Temperature Cryostat 395
  G. J. Labbe, H. Royce, Y. Takano, and G. G. Ihas
Dilution Refrigerator with Internal Cryogenic ^{3}He Cycle.................... 397
  Yu. M. Bunkov, D. A. Sergatskov, J. Nyéki, and I. N. Ivoilov
A High-Resolution, Low-Power Ultrasonic Spectrometer 399
  N. Mulders and J. R. Beamish
Acoustic Study in Liquid and Solid ^{3}He: Using Novel and Sophisticated Ultrasonic Techniques 401
  F. Guillon
Field-Resolved NQR Spectroscopy for Ultra-Low Temperature Thermometry 403
  N. S. Sullivan, M. Rall, and D. Zhou
Magnetic-Field-Induced Metal–Insulator Transition in Degenerately Doped n-Type Germanium 405
  M. J. Burns

EPILOGUE

Quantum Fluids and Solids: Where We Have Come, Where We Are Going 409
  D. M. Lee

List of Participants 419

Author Index 427

Symposium logo designed by Anna-Lisa Paul

Preface

These are the proceedings for a conference on quantum fluids and solids (QFS-89) which has become a regular event held approximately midway between the tri-annual LT Conferences. Since 1977 they have alternated between Florida and other sites (Itacha, New York—1980, and Banff, Alberta—1986). A tacit arrangement has therefore developed so that the Low Temperature Group at the University of Florida holds this conference every other time, which we are happy to do.

The usefulness of a series of conferences such as this must continually be examined. In a time of many conferences and shrinking research dollars, great care must be taken by would-be organizers so that tradition is not the sole justification for the meetings to continue. In our case, planning began at the 1987 March American Physical Society meeting in New York with an opinion questionnaire regarding the continuance of this particular series. That and later polls done by mail and at other meetings indicated strong support for QFS-89.

In fact, quantum fluids and solids has continued to be a proving ground for fundamental physics, as this conference and these proceedings attest. With 130 participants from 11 countries, most work in the field both experimental and theoretical is presented. (Because of the unusual circumstance of an Aspen Conference on Low Temperature Physics being held in the same year, the organizers attempted to not duplicate the work presented at that conference. Unfortunately, there are no proceedings from the Aspen Conference.) Also, with the LT Conference having left the North American Continent in 1981, not to return before 1993 at the earliest, the QFS Conferences have become particularly important to graduate students and postdoctoral fellows in the United States and Canada.

The program for the conference was set by the International Organizing Committee through suggestions to the Local Organizing Committee. These two large committees worked well, thanks to the superior leadership of Mark Meisel, the Conference Chairman. His ability to both initiate and carry out all phases of the work associated with the conference was the cornerstone of success. After him, we owe our gratitude to the Organizing Committee, and then to all the speakers and poster preparers whose presentations were uniformly outstanding. Finally, the gator gophers, who "go for this and go for that," kept the conference organizers sane (or less crazy) throughout the week.

This conference was a true international gathering and was the most costly we have ever run. Conference fees defrayed only about one-third of the expenses. In order of their level of support, we wish to thank the University of Florida Low Temperature Group, the corporate sponsors, the National Science Foundation, the University of Florida Department of Physics, and the University of Florida Division of Sponsored Research.

These proceedings, although they represent 88% of the work presented at QFS-89, cannot substitute for conference attendance. Only participants could become involved in discussion and share in the passion and enthusiasm that exists in low temperature physics today. And nowhere is the future brighter than where the next QFS participants may meet in 1992—Moscow. We sincerely hope these proceedings serve you well in the interim.

Gary G. Ihas
Yasumasa Takano

Organizing Committee

M. W. Meisle, Chair

E. D. Adams
P. J. Hirschfeld
G. G. Ihas
P. Kumar
K. Muttalib

C. J. Stanton
N. S. Sullivan
Y. Takano
P. Wölfle

Advisory Committee

A. F. Andreev, Moscow
R. M. Bowley, Nottingham
J. S. Brooks, Boston
M. C. Cross, CalTech
G. Frossati, Leiden
C. M. Gould, Los Angeles-USC
R. A. Guyer, Massachusetts
J. P. Harrison, Ontario
T. L. Ho, Ohio
H. Ishimoto, Tokyo

J. M. Kurkijärvi, Abo
A. J. Leggett, Illinois
H. Meyer, Duke
F. D. M. Pobell, Bayreuth
R. C. Richardson, Cornell
J. A. Sauls, Northwestern
H. Smith, Copenhagen
T. Tsuneto, Kyoto
E. Varoquaux, Saclay
R. A. Webb, IBM

Sponsors

Center for Ultralow Temperature Research, University of Florida
Division of Materials Research, National Science Foundation
Division of Sponsored Research, University of Florida
EG & G PARC, Princeton, New Jersey
ENI Power Systems, Rochester, New York
Gainesville Avis Rent A Car, Gainesville, Florida
Harra Technical Sales, Clearwater, Florida
Holiday Inn University Center, Gainesville, Florida
Joe's Deli, Gainseville, Florida
Lectromagnetics, Inc., Los Angeles, California
Linde, Union Carbide Industrial Gases, Danbury, Connecticut
Oxford Instruments, Inc., Bedford, Massachusetts
Precise Power Corporation, Bradenton, Florida
Schmidt, Garden & Ericson, Inc., Sarasota, Florida
Scientific Instruments, Inc., West Palm Beach, Florida

Gofers

J. P. Brison
M. J. Burns
P. B. Chilson
J. Clark
G. J. Labbe

S. Mishra
W. Puttika
M. Rall
H. Royce
D. Zhou

TUNNELLING AND PHASE SLIPPAGE

IN SUPERFLUIDS

CHAIRMAN

Richard E. Packard
Department of Physics
University of California, Berkeley

This session was made possible, in part, through
a generous donation provided by

A Member of the Oxford Instruments Group plc

Oxford Instruments North America Inc.
3A Alfred Circle, Bedford, MA 01730, USA
Tel: (617) 275-4350

Oxford Instruments Limited
Osney Mead, Oxford OX2 ODX, England
Tel: (0865) 241456

PHASE SLIPPAGE EXPERIMENTS IN SUPERFLUID ^{3}He–A

O. Avenel
DPhG/PSRM, CEN–Saclay, 91191 Gif–sur–Yvette Cedex, France

E. Varoquaux
Laboratoire de Physique des Solides, 91405 Orsay, France

ABSTRACT

Phase slippage experiments similar to those already performed in ^{4}He first, and then in low pressure ^{3}He–B, have been extended to the A and B phases at 28.4 bar. The observed staircase patterns show very analogous features in both phases. The state of the superfluid in the micro-orifice is thus likely to be independent of which phase prevails in the bulk liquid. Critical currents in the micro-orifice are large and in broad agreement with Kurkijärvi's calculations. A new low-level periodic pattern has also been discovered, which we attribute tentatively to textural effects.

INTRODUCTION

The concept of phase slippage was introduced in 1966 by Anderson [1] to account for dissipation in superfluid ^{4}He flow. When a vortex crosses streamlines of potential (irrotational) flow, it collects energy from the stream and slows the stream down. If it crosses all streamlines going through an orifice connecting two superfluid baths, the difference in the quantum mechanical phases between these baths, $\delta\varphi$, changes by 2π and the total kinetic energy taken away from the stream in the process is expressed in terms of the total mass flow rate J and the quantum of circulation κ_i ($i = 3, 4$) by:

$$\Delta E = \kappa_i J . \qquad (1)$$

Direct experimental evidence for the existence of such 2π phase slips was obtained first in ^{4}He [2] with the help of a miniature hydromechanical resonator in which the main superfluid passage between the inner chamber and the main body of the cell was nominally a 0.3 × 5 μm rectangular slit bored in a thin metal foil 0.2 μm in thickness.[3] Individual slips could be observed, and eq.(1) was found to be verified. This equation is a direct consequence of the ac Josephson relation between the rate of change of the phase $\partial\varphi/\partial t$ and the chemical potential μ

$$\hbar\, \partial\varphi/\partial t = -\mu . \qquad (2)$$

After this discovery, the effect of temperature, hydrostatic pressure, ^{3}He impurity concentration and rate of production on these phase slips and their onset threshold, i.e. the critical superflow velocity, was investigated [4]. It was found that the dissipative jumps obey eq.(1) in all cases, including when as many as 300 jumps are produced in rapid succession in a total lapse of time of less than 4.5 msec. This body of experiments puts on quite firm a footing the interpretation of the observed dissipative events in ^{4}He as being phase slips in Anderson's sense [5-6].

The application of the phase slippage concept to the superfluid phases of ^{3}He requires caution, as reviewed by Cross [7] and by Hall and Hook [8]. In the A-phase, the motion of the l vector contributes to the variation of the chemical potential so that the superfluid velocity v_s is not simply related to the gradient of the phase. The hydrodynamic circulation is no longer necessarily quantized if l is inhomogeneous in space. These difficulties do not arise in the B—phase which behaves as a conventional superfluid.

In order to probe the superfluid phases of ^{3}He, the next series of experiments was carried out in the same cell, with the same micro-orifice, but with another larger, longer channel of fairly well specified geometry connecting the inner chamber to the main cell. This second channel now provides the main passage for superfluid flow, with mass flow rates which are larger by a factor of 5 to 20 than through the micro-orifice, but with a flow velocity which remains much smaller. The presence of the large parallel hole enables cooling of the inner chamber to superfluid ^{3}He temperatures.

The operation of the double-hole resonator, first used by Zimmermann *et al.* [9-10-11] in ^{4}He, is similar to that of the well-known rf-SQUID [12]. The response in amplitude of the hydromechanical resonator to a linearly increasing input excitation was found, both in the case of ^{4}He and ^{3}He, to develop a highly characteristic staircase pattern quite similar to that of rf-SQUIDs [13-14-15].

These experiments show that phase coherence is preserved between the two superfluid baths when the critical velocity is reached in the micro-orifice. They give access to the current-phase relation. If the micro-orifice acts as an ideal weak link between the two baths, this relation is the dc Josephson relation:

$$J = J_c \sin\theta \qquad (3)$$

In most instances in superconducting weak links [16-17], the current-phase relation is not a pure sine as in (3) but takes a slanted shape which can even become multiple-valued but which always remains periodic with period 2π in θ. One possible representation for such a non-ideal relation is given by

$$J = J_c \sin\zeta, \qquad \theta = \zeta + \alpha \sin\zeta, \qquad (4)$$

which has been proposed for superconducting junctions [16] but is also quite appropriate for superfluid weak links ‡. The pure Josephson case is recovered for $\alpha = 0$. Eq.(4) becomes multiple-valued for $\alpha > 1$ and describes a hysteretic behaviour of the weak link. For $\alpha \gg 1$, a regime leading to eq.(1) is attained.

Experiments with the double-hole resonator in ^{3}He—B at low pressure (0.8 bar), where the coherence length ξ_0 is large (~ 600 Å [18]), lead to observations of periodic current-phase relations which are single-valued and close to sinusoïdal at temperatures of the order of $0.9\ T_c$ [13-14]. Close to T_c, the temperature dependent coherence length, $\xi(T) \simeq \xi_0/\sqrt{1 - T/T_c}$, becomes comparable to or larger than the size of the orifice and the behaviour becomes close to ideal.

---

‡ Wayne Saslow (private communication).

These experiments constitute the first observation of a Josephson effect in a superfluid. They called forth a number of theoretical studies aiming at a better understanding of phase slip events in superfluid ^{3}He, and in particular of the departures from ideality in a finite-size orifice [19,20,21,22,23,24,25].

In this paper, we discuss new phase slippage experiments in the B phase at high pressure, and in the A phase. These experiments have not been conducted in an exhaustive and systematic manner, owing to the difficulty of obtaining noise-free high quality data in the present setup. Nonetheless, they already reveal a complex behaviour of the anisotropic superfluid and show a number of features which can only partly be accounted for by the simple weak link description of eq.(4) or by the theoretical studies referred to above.

DOUBLE HOLE RESONATOR OPERATION

The experimental cell, sketched in fig. 1, and its operation have been described in detail [2,3,5,6,13,14,26]. Its two main parts are first, the two channels already mentionned, the micro-orifice acting as weak link (w) and the main channel (m) in which flow follows ideal fluid hydrodynamics, and, secondly, the flexible membrane which is used both for driving the resonator and detecting the fluid displacement. This membrane, of stiffness σ and cross-section S, provides the elastic restoring force while the superfluid flow in the channels yields the inertia in the resonator equation of motion. For ideal fluid motion in both channels, this flexible wall hydromechanical resonator shows a Helmholtz-like resonance at a frequency given by

$$\Omega^2 = \omega_w^2 + \omega_m^2 = \left\{\frac{s_w}{l_w} + \frac{s_m}{l_m}\right\} \frac{\sigma}{S^2} \frac{\rho_s}{\rho^2}, \qquad (5)$$

where ω_w and ω_m are the resonance frequencies corresponding to each channel acting alone, s_w/l_w and s_m/l_m the hydraulic aspect ratios (cross-section over length) of the two holes, ρ the fluid density and ρ_s/ρ the superfluid fraction.

The motion of the resonator occurs at the driving frequency ω and is tracked by the ultra-sensitive displacement gauge. We record at each half-cycle the peak amplitude of this motion (defined by the mean deflection d of the diaphragm).

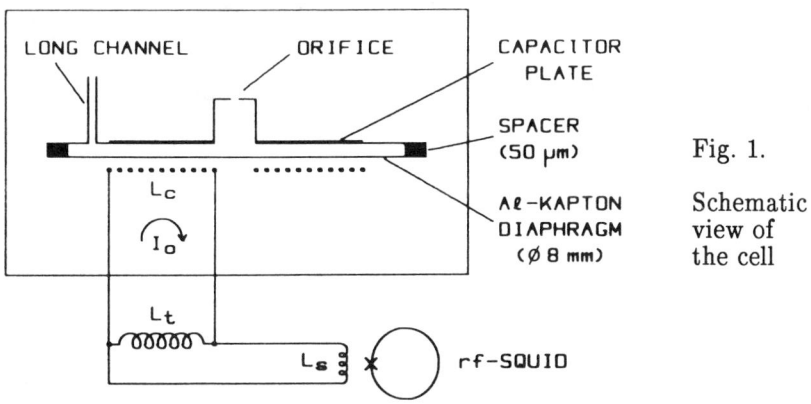

Fig. 1. Schematic view of the cell

For a nonlinear current–phase relation such as (4), the resonator behaviour departs from that of a simple harmonic oscillator. It is given, as discussed in [13], by the following equations:

$$a \frac{\partial d}{\partial t} = \omega_m^2 \left\{ \theta - \theta_b + \frac{1+\alpha}{R} \sin \zeta \right\}, \qquad (6)$$

$$\frac{\partial \theta}{\partial t} = 2\ ab\ \mathcal{V}_0\ \mathcal{V}_1 \cos \omega t - ad - (a/\omega Q) \frac{\partial d}{\partial t}, \qquad (7)$$

in which a and b are the signal and drive amplitude normalizing constants, R is the ratio of the flow rate in the main channel to that in the weak link (determined at small oscillation amplitudes with $\theta_b = 0$), Q is the quality factor of the resonator and α and ζ are defined by eq.(4). The bias phase θ_b describes the state of the junction when the resonator is at rest, *i.e.* it leaves open the possibility that a persistent current be trapped in the loop between the two holes or that rotation be applied to the cell. The bias phase is usually trapped when cooling through T_c and remains constant afterwards (modulo 2π) unless the resonator is very strongly perturbed. The first term of the r.h.s. of (7) is the externally applied electrostatic drive at frequency ω, with amplitude $\mathcal{V}_1$, $\mathcal{V}_0$ being a *dc* polarisation potential.

Eqs. (6) and (7) can be shown to be formally equivalent to those describing the behaviour of an *rf*–SQUID with a non-ideal junction: they have the same types of solutions and lead to the existence of staircase patterns. We note in eq.(6) that, when slips occur for a given value of θ, increasing θ by 2π does not change ζ so that the r.h.s. increases by the constant quantity $2\pi\omega_m^2$. To the extent that the resonator motion is well approximated by a sine function so that $\partial d/\partial t \simeq i\omega d$, the peak amplitude increases by a constant quantity

$$\Delta d = 2\pi \omega_m^2 / a\omega, \qquad (8)$$

which is the approximate height of the staircase steps. The validity of (8) requires a high Q value and a large flow rate ratio R.

The detailed solutions of the set of nonlinear eqs. (6) and (7) are obtained by numerical integration. They yield complex peak amplitude versus drive level curves which, among other things, depend greatly on the difference between ω and Ω. The outcome of this numerical simulation can be directly compared to experiment. Examples of such comparisons are given in ref.(13) and pertain to ^{3}He–B at low pressure and at temperatures from 0.9 to 0.7 T_c. In this case, where the coherence length is large, the agreement is found to be excellent and gives confidence in the interpretation of the observations and in the validity of the model.

In the present paper, which deals with higher pressure experiments carried in both the A and B phases, the overall description of the data does not fall in line as readily with the model calculation, but the same concepts of phase slips and staircase steps apply and are well supported.

The experiments were conducted at a pressure of 28.4 bar in the cell pictured in fig. 1. In this cell and in the main channel, where the typical dimensions are a few tens of μm (50 μm close to the membrane, 15 μm in the main channel) except in the duct connecting to the weak link (0.6 mm), the superfluid is mildly confined, that is, the order parameter is not significantly depressed in the bulk of the fluid but l is pinned. In all the graphs of staircase patterns presented below, we have plotted the peak amplitude of the resonator oscillation recorded at each half-cycle as the drive level (i.e. $\mathcal{V}_1$) at constant excitation frequency ω is linearly ramped up in time. Time is an implicit parameter. The sweep time Θ plays a role: if it is too fast, the pattern rounds up and some features may disappear altogether. The model simulations are performed in conditions as close as possible to those of the experiments and show the same features in most instances, as discussed below.

THE A PHASE CLOSE TO T_c

Although phase slippage in the A phase may occur with non-integral values of 2π because the motion of l is coupled to that of the overall phase, the A-phase was nonetheless observed to sustain well defined staircase patterns in the various situations that we have studied.

The patterns observed close to T_c (0.92 T_c), where only the A phase is stable, are shown in fig. 2 (upper curves) at frequencies larger and smaller than ω_m (1.783 Hz). The numerical simulation yields the lower curves with the following parameter values: $R = 10$, $Q = 80$. The nonlinearity parameter α ($\simeq 5$) is fairly large, that is, phase slips are dissipative. The critical phase difference θ_c is of the order of 2π and $\theta_b \simeq 1.1\,\pi$. The model accounts well for this situation where dissipative phase slips occur and the current-phase relation is strongly hysteretic. These dissipative slips can be seen directly on the first few plateaus of the left staircase, obtained at 1.85 Hz, both in the

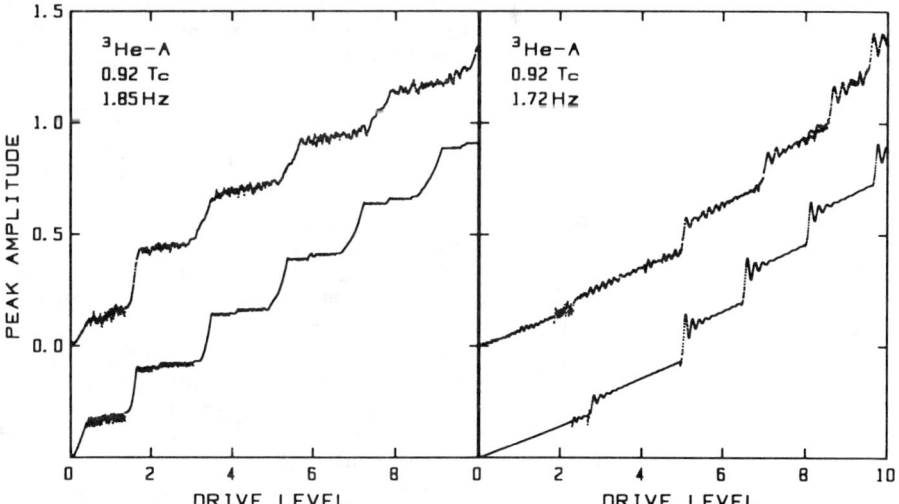

Fig. 2. Observed (top) and computed (bottom) staircase patterns close to T_c with $\alpha = 5$, $R = 10$, $Q = 80$, $\omega_m = 1.783$ Hz, $\theta_b = 1.1\pi$, $\Theta = 10$ min.

experimental data and in the computed curves. The quality factor of the resonator is about three times larger than at low pressures. Dissipation is due mainly to the compression of the normal component by the membrane and second viscosity damping ‡.

For temperatures closer to T_c, we expected that, as in the low pressure B–phase experiments, the current-phase relation would evolve toward a non-hysteretic, ideal Josephson-type sinusoïdal relation. What was observed in the A phase at 28.4 bar was indeed a decrease of θ_c and a trend toward a non-hysteretic current-phase relation. However, the recorded patterns could not be fit by computer simulations and no value could be firmly assigned to the nonideality parameter α. This may be taken as an indication that the A phase displays a somewhat less ideal Josephson-type behaviour than the low pressure B–phase. But, we wish to point out that the experiments at 28.4 bar are more difficult, and less precise, than at 0.8 bar because, as the zero temperature coherence length is shorter, a comparable value of $\xi(T)$ is obtained at temperatures much closer to T_c. This implies lower values of ρ_s/ρ and, by eq.(5), lower frequencies and makes the resonator even more sensitive to external perturbations. Thermal drifts also become quite troublesome. The likelihood of spurious response increases, preventing firm conclusions on the behaviour of the high pressure A phase very close to T_c to be drawn.

Let us at this point discuss the frequency dependence of the patterns shown in fig. 2, also observed at low pressure in the B–phase [13-14]. Just on resonance, the steps are largely smeared out. This indicates that dispersion effects, *i.e.* a frequency-dependent response, play a large role in shaping the pattern.

On the high frequency side (left part of fig. 2), the steps and plateaus are well defined. Phase slips start to occur at the beginning of the first (lowermost) plateau and, since θ_b is close to π, occur at the rate of precisely 2 per cycle at its end. The slip rate increases gradually from 2 to 6 per cycle along the second plateau, from 6 to 10 along the next, and so on. The response levels off and plateaus form because dissipation increases with slip rate.

When the excitation frequency is lower than the Helmholtz frequency due to the main channel, ω_m, the processes taking place are markedly different, as seen on the right of fig. 2. The basic reason behind this lies in the fact that phase slips are accompanied not only by a slowing down of the membrane motion but also by an effective phase lag (although the slips themselves are assumed to be instantaneous). This phase lag shifts the effective resonance frequency of the hydromechanical device downward, closer to the drive frequency. The improved frequency matching enhances the power transfer between the external drive and the resonator. The enhancement takes place rather suddenly as soon as the first pair of slips occurs and gives the steep first rise in the pattern. The system reaches a steady state after some overshoot at a slip rate of 2 slips per cycle while the amplitude continues to grow as the drive level increases. The next steep rise is attained when the amplitude becomes such that more slips (6 slips) per cycle may take place.

---

‡ The observed Q values in the B phase correspond well to those estimated by John Hook (private communication).

The process goes on repeating itself (10 slips, 14 slips, etc...) with slowly up–sloping segments at constant slip rate followed by steep rises as soon as additional pairs of slips at each half-cycle can be created. This process clearly lends itself to hysteresis if the drive level is ramped up and down. The hysteretic behaviour is illustrated in fig. 3 by computer simulations for the same conditions as in fig. 2. This digression on the pattern shapes shows that, although the resonator behaviour may look fairly complex, it is quite satisfactorily described by eqs.(6) and (7).

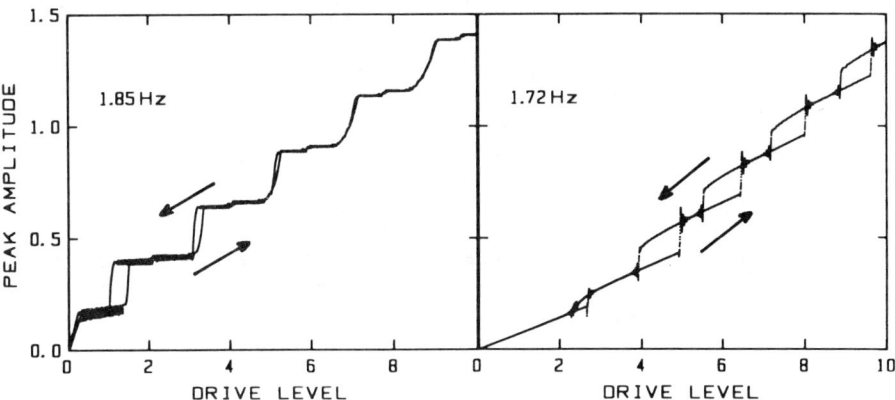

Fig. 3. Simulated staircases in the same conditions as in fig. 2 showing the hysteretic behaviour when ramping up and down the drive level, with $\Theta = 30$ min.

When the drive level is increased further, the regularity of the staircase pattern breaks up, as shown in fig. 4, and a more dissipative regime sets in which we interpret as the critical velocity regime in the main channel. Quite remarkably, rather well defined steps can also be seen in this regime as well as quite regular and reproducible giant phase slips, shown in the inset of fig. 4 (which may be compared to fig. 6 of ref.(5) in order to better see the stroboscopic effect between slip rate and drive amplitude).

A comment on the bias current is in order here. The fact of exceeding the critical velocity in the main channel (as well as in the weak link) does not affect θ_b which is, in the A phase, very robustly pinned down to the value 1.1 π. This particular value seems always to be obtained when cooling into the superfluid phase. It is usually not altered by severe overdriving of the resonator, even at a level nearly two orders of magnitude higher than the drive level required to reach the critical velocity in the main channel. It was observed to change only once in about ten attempts. In contrast, it proved in the B–phase at low pressure easy to change θ_b to any value at random with the same overdriving procedure.

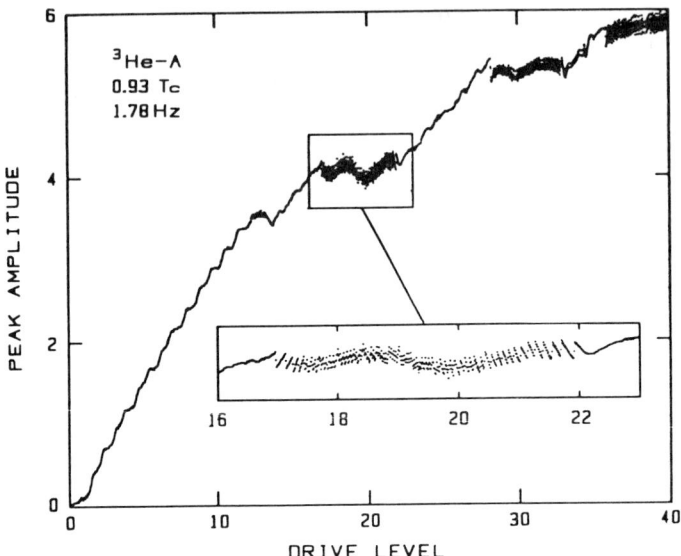

Fig. 4. Critical regime in the main channel, signaled by the lowest kink in the staircase. The inset, magnified by 4 horizontally, shows individual phase slips taking place in the main channel.

A DIRECT COMPARISON BETWEEN THE A AND B PHASES

At 28.4 bar, the A to B transition occurs at a temperature of about 0.81 T_c. It is signaled by an abrupt drop of the resonator frequency from ~ 3.30 Hz to ~ 3.05 Hz. The B to A transition on warming up, at about 0.85 T_c, is accompanied by a smooth N-shaped variation of the resonance frequency. The supercooling of the A phase allows both phases to be studied at the same temperature, or rather, for our purpose here, at the same frequency (*i.e.* the same ρ_s/ρ).

Staircase patterns obtained at nearly identical frequencies in both phases are shown in fig. 5. Apart for a difference in bias, they are seen to be very similar, yielding the same step heights (as measured on fig. 8 below) and nearly the same critical threshold. We take this similarity to mean that the state of the superfluid in the weak link depends little on whether the bulk liquid is in the A or B phases.

A striking new feature appears on the low-level, low-dissipation portion of the response in the form of a periodic structure superposed on that part of the response curve. Magnified plots of these periodic structures in the A and B phases are shown in fig. 6. Both phases again display a very similar behaviour.

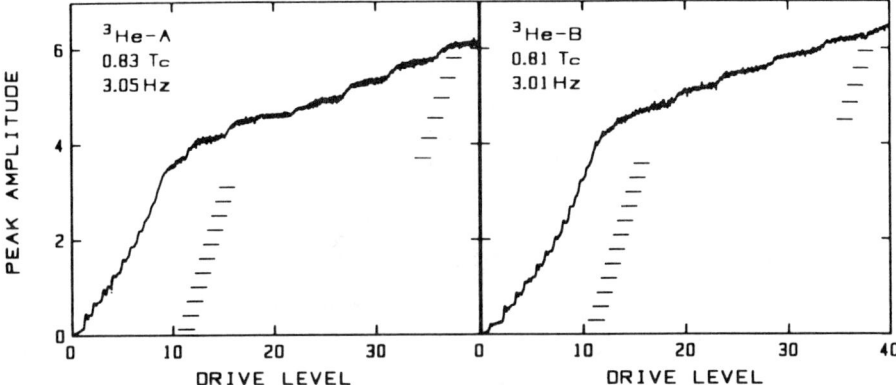

Fig. 5. Comparison of the A and B phases at nearly identical frequencies. The horizontal ticks mark the periodicity of the staircase pattern (top) and of the low-level structure (bottom).

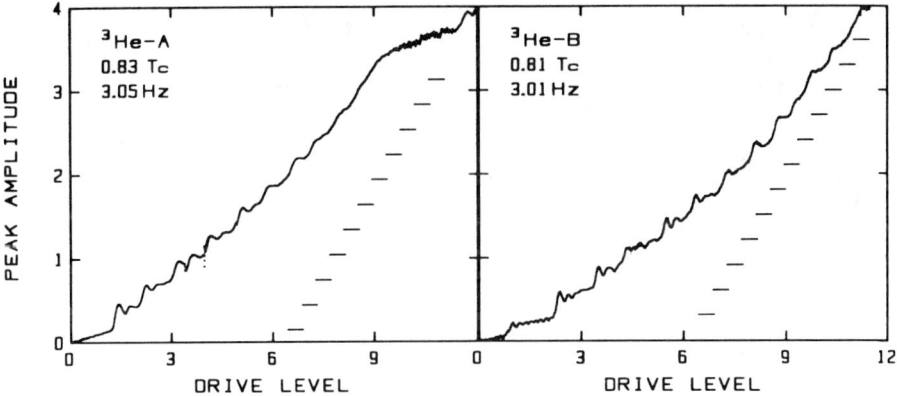

Fig. 6. The low-level periodic structures, as in fig.5.

The periodicity of the new pattern, which looks like a low-level staircase pattern, can also be measured by the height difference between steps and is found to be highly regular and roughly 25 % smaller than that of the supercritical staircase. The low level structure is slightly dissipative, at least in the B–phase, as can be seen in fig. 7, which shows free ringing decays. Decay times are comparable in both phases, but the resonator amplitude decreases abruptly to zero in the B–phase. This peculiar decay shape is characteristic of 'solid friction' which we certainly do not expect to find in the bulk B–phase. The occurence of this apparent 'solid' friction acting on the membrane was also confirmed in *cw* measurements which exhibited the corresponding history-dependent behaviour.

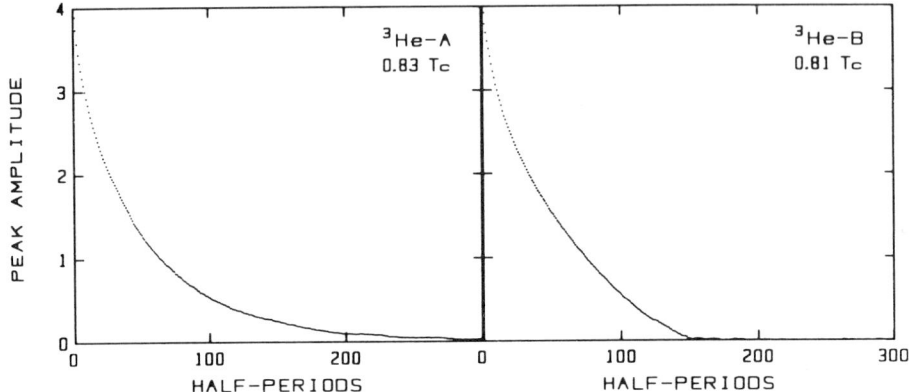

Fig. 7. Free ringing decays (with zero drive).

Finally, the response of both phases to high drive levels is shown in fig. 8. The critical regime in the main channel can be reached in the A phase at this temperature also, as it was at 0.93 T_c (fig. 4). It exceeds the dynamical range of the displacement gauge in the B–phase, although up to 70 steps can be tracked (only about 25 are shown in fig. 8). Thus, a clear-cut difference in the critical velocity in the main channel exists between the two phases. The step height of the supercritical staircase measured in fig. 8 is the same in both phases, and is also in good agreement with that observed at higher temperature when the frequency change is taken into account by eq. (8). The height of the few first steps just after the onset on the critical regime shown in fig. 5 is however slightly different in both phases, and different from the value higher up in the staircase by $\sim$ 10 %.

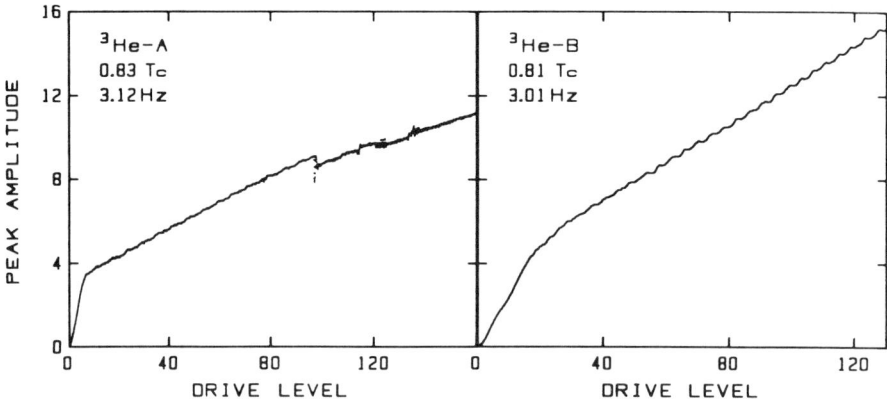

Fig.8. Staircase patterns for high drive levels. A critical regime is obtained in the A phase (left) but not in the B phase at about the same temperature.

DISCUSSION

A detailed discussion of these preliminary results might be premature at this stage. Nonetheless, some definite conclusions can already be drawn.

Since most of the characteristics of the supercritical regime do not depend on which phase is stable in the bulk, it seems highly probable that the microscopic state of the superfluid in the weak link is fixed locally by the presence of the walls.

It has been predicted by Li and Ho [27], along with other authors [28-29], that an A–polar phase is favoured in narrow slabs in the strongly confined limit, while a deformed B state takes over when typical dimensions become larger than a few $\xi(T)$. The exact transition from one state to the other depends heavily on the nature of the reflection (specular or diffuse) of quasiparticles at the walls.

The case of a point-like orifice in an infinitely thin wall has been discussed by Kurkijärvi [24] in the A phase with specular reflection at the walls. This author finds a skewed sine current-phase relation with a critical current J_c which is about one order of magnitude larger than the bulk depairing critical velocity times ρ_s and which possesses the same temperature dependence as ρ_s. This critical Josephson current decreases with increasing surface roughness, and also with the effective length of the junction.

We estimate the critical current in the weak link from the critical phase difference ($\theta_c \simeq 2\pi$ at $0.93\ T_c$, $\simeq 8.9 \times 2\pi$ at $0.83\ T_c$) to which corresponds a well-defined velocity v_s in the main channel (of length 0.5 cm). Using the values of ρ_s of Hook et al. [30], and the known cross-section of the annular channel (a gap of $\sim 15\ \mu$m on a perimeter of $\pi \times 0.23$ mm), we obtain the corresponding mass flow rate. The fitted value of R leads to J_c (6.9×10^{-11} gs^{-1} at $0.93\ T_c$, 1.9×10^{-9} gs^{-1} at $0.83\ T_c$). The value of ref.(24) at the lower temperature is 10^{-8} gs^{-1} with the nominal area of the micro-orifice (which could well be sizably reduced by contamination [26]) and the agreement is satisfactory. The experimental critical current is however found to decrease more rapidly than ρ_s as $T \to T_c$.

Thus, the theoretical picture and the experimental findings support one another and one can safely infer that the micro-orifice, at 28.4 bar and down to at least $0.83\ T_c$, is filled with an A-like phase irrespectively of the bulk equilibrium phase.

From the data shown in figs. 4 and 8, we can estimate the critical velocity in the main channel and find it to be 0.19 mm/s at $0.93\ T_c$ and 0.31 mm/s at $0.83\ T_c$. These values are significantly lower, by one order of magnitude, than those found for persistent currents in the A phase at 29 bar in a 20 μm powder by Gammel et al. [31]. The origin of this discrepancy is unclear, although geometry dependent and textural effects very likely play an important, if undetermined, role.

A final comment is in order on the extraordinary behaviour shown in figs. 6 and 7 which display a nonlinear response at vanishing membrane amplitude. We discard first of all the possibility of actual solid friction in the B phase: hydrodynamic flow must prevail in the main channel. Also, we consider as remote the possibility of Josephson-type effects in some minute partition of the system: the very small, very short hole necessary for such an effect is totally unlikely to form (but very easy to plug) spontaneously. We are forced, once more, to invoke the interplay of flow and textures as a plausible cause for these low-level nonlinearities. The experimental study of these minute features is extremely difficult and will require a much improved vibration isolation that we are currently attempting to achieve.

The help of Mark Meisel at various stages of the double-hole resonator experiments is gratefully acknowledged. We also express our thanks to Bill Zimmermann for suggesting many improvements to the manuscript.

REFERENCES

1. P.W. Anderson, Rev. Mod. Phys. 38, 298 (1966).
2. O. Avenel, E. Varoquaux, Phys. Rev. Lett. 55, 2704 (1985).
3. P. Sudraud, P. Ballongue, E. Varoquaux, O. Avenel, J. Appl. Phys. 62, 2163 (1987).
4. E. Varoquaux, M.W. Meisel, O. Avenel, Phys. Rev. Lett. 57, 2291 (1986).
5. E. Varoquaux, O. Avenel, M.W. Meisel, Can. J. Phys. 65, 1377 (1987).
6. E. Varoquaux, O. Avenel, Physica Scripta 19, 445 (1987).
7. M.C. Cross, in 'Quantum Fluids and Solids – 1983', ed. E.D. Adams, G.G. Ihas (American Institute of Physics, N.Y., 1983), p. 325.
8. H.E. Hall, J.R. Hook, in 'Prog. in Low Temp. Physics', vol. IX ed. D.F. Brewer (Elsevier, Amsterdam, 1986), p. 143.
9. F.H. Wirth, W. Zimmermann, Jr., Physica B&C 107, 579 (1981).
10. B.J. Anderson, B.P. Beecken, W. Zimmermann, Jr., in 'Proc. XVIIth Int. Low Temp. Conf.', ed. U. Eckern, A. Schmid, W. Weber, H. Wühl (North–Holland, Amsterdam, 1984), p. 313.
11. B.P. Beecken, W. Zimmermann, Jr., Phys. Rev. 35, 74 (1987).
12. J.E. Zimmerman, P. Thiene, J. Harding, J. Appl. Phys. 41, 1572 (1970).
13. O. Avenel, E. Varoquaux, Jpn. J. Appl.Phys. 26–3, 1798 (1987)
14. O. Avenel, E. Varoquaux, Phys. Rev. Lett. 60, 416 (1988).
15. W. Zimmermann (private communication) has reported the observation of a staircase pattern at the Washington APS meeting in April 1987.
16. B.S. Deaver, J.M. Pierce, Phys. Lett. 38A, 81 (1972).
17. K.K. Likharev, Rev. Mod. Phys. 51, 101 (1979).
18. K. Ichikawa, S. Yamasaki, H. Akimoto, T. Kodama, T. Shigi, H. Kojima, Phys. Rev. Lett. 58, 1949 (1987).
19. H. Monien, L. Tewordt, J. Low Temp. Phys. 62, 277 (1986), Can. J. Phys. 65, 1388 (1987).
20. D. Rainer, P.A. Lee, Phys. Rev. B 35, 3181 (1987).
21. D. Rainer, J.A. Sauls, Jpn. J. Appl. Phys. 26–3, 1804 (1987).
22. J.R. Hook, Jpn. J. Appl. Phys. 26–3, 159 (1987).
23. E.V. Thuneberg, Europhys. Lett. 7, 441 (1988).
24. J. Kurkijärvi, Phys. Rev. B 38, 11184 (1988).
25. N.B. Kopnin, M.M. Salomaa, preprint (Feb. 1989).
26. O. Avenel, E. Varoquaux, in 'Proc. XIth Int. Cryogenic Eng. Conf.', ed. G. Klipping, I. Klipping (Butterworths, Guildford, 1986), p. 587.
27. Y–H. Li, T–L. Ho, Phys. Rev. B 38, 2362 (1988).
28. A.L. Fetter, S. Ullah, J. Low Temp. Phys. 70, 515 (1988).
29. N.B. Kopnin, J. Low Temp. Phys. 65, 433 (1986).
30. J.R. Hook, E. Faraj, S.G. Gould, H.E. Hall, J. Low Temp. Phys. 74, 45 (1989).
31. P.L. Gammel, T–L. Ho, J.D. Reppy, Phys. Rev. Lett. 55, 2708 (1985).

MAGNETIC RELAXATION IN SUPERFLUID ^{3}He

A.S. Borovik-Romanov, Yu.M. Bunkov, V.V. Dmitriev, Yu.M. Mukharskiy
Institute for Physical Problems, 117334, Kosygin st. 2, Moscow, USSR

ABSTRACT

A review of recent investigations of magnetic relaxation in ^{3}He-B is presented. A new method for the measurement of relaxation is developed which makes use of the properties of the homogeneously processing domain (HPD). Terms corresponding to intrinsic and spin diffusion relaxation are distinguished, as well as surface relaxation and possible additional relaxation. At temperatures below 0.45 T_c a very fast process of relaxation is observed, which may be a manifestation of the collisionless regime.

Some investigations of magnetic relaxation processes in ^{3}He-A are also presented.

INTRODUCTION

The recent investigations of the Moscow experimental group in connection with theoretical work by Fomin have shown that the transport of magnetization by spin supercurrents plays an important role in the NMR of the superfluid phases of ^{3}He. The flow of spin supercurrents after an NMR excitations pulse leads to a nontrivial spatial structure of precessing magnetization. Magnetic relaxation processes take place in such structures. This structure can also be excited using nonlinear continuous wave NMR regime.

In ^{3}He-B the spatial structure arises in an inhomogeneous magnetic field and consists of two domains. In one domain situated in the higher magnetic field the magnetization is stationary. In the second domain it precesses spatially uniformly in spite of the inhomogeneity in the magnetic field. This is referred to as a homogeneously precessing domain (HPD). The main processes of magnetic relaxation in this structure are intrinsic relaxation inside the HPD and relaxation due to spin diffusion through the domain boundary. The relaxation leads to a decrease in the size of the HPD and an increase in the size of the stationary domain.

In ^{3}He-A the uniform precession of magnetization is unstable. Spin supercurrents redistribute the magnetization to a structure where both the phase of precession and the tipping angle of magnetization become spatially modulated. This leads to very effective magnetic relaxation due to spin diffusion and the spin-orbit interaction.

SPIN SUPERCURRENT

The existence of the spin supercurrent in superfluid ^{3}He has been suspected for a long time, as described by Leggett[1]. Cooper pairs of ^{3}He have a spin equal to 1 and as a result can transport magnetization. Superfluid ^{3}He may be imagined as a mixture of superfluid liquids with different projections of spin on their quantization axes. The order parameter can be written in the form (for references see ref. 2)

$$\psi = \begin{bmatrix} \psi_{\uparrow\uparrow} & \psi_{\downarrow\uparrow} \\ \psi_{\uparrow\downarrow} & \psi_{\downarrow\downarrow} \end{bmatrix} = \begin{bmatrix} -\mathbf{d}_x(\mathbf{k}) + i\mathbf{d}_y(\mathbf{k}) & \mathbf{d}_z(\mathbf{k}) \\ \mathbf{d}_z(\mathbf{k}) & \mathbf{d}_x(\mathbf{k}) + i\mathbf{d}_y(\mathbf{k}) \end{bmatrix} \qquad (1)$$

© 1989 American Institute of Physics

where the vector **d** characterizes the spin part of the order parameter, and the components $\psi_{\downarrow\downarrow}, \psi_{\downarrow\uparrow}, \psi_{\uparrow\uparrow}$ represent wave functions of three superfluid liquids with different spin projections. A rotation of **d** around the z axis changes the phases of $\psi_{\downarrow\downarrow}$ and $\psi_{\uparrow\uparrow}$ oppositely. The gradients of phase rotation **d** around z gives the gradients of phases of these components in opposite directions and as a result there will be a counterflow of these two superfluid liquids. In the transport of mass they compensate each other (if one does not take into consideration the different density of these components in a magnetic field). This simplified scheme demonstrates the main difference between spin supercurrents in superfluid ^{3}He and solid-like spin transport that can be observed in more usual magnetically ordered materials.

The vector **d** of the order parameter is not an experimental observable value, but it is strongly connected with the magnetic and orbital moments of ^{3}He. In the case of NMR, the gradients of the magnetic moment excite gradients in **d**, and as a result, a spin supercurrent. From the experimental point of view it is more convenient to characterize the spin supercurrent by the gradients of the magnetization. For the transverse NMR mode one uses the gradients $\nabla\alpha$ and $\nabla\beta$ where α is the phase of precession and β the deflection of the magnetization from the field direction. The equation for NMR, which takes into account the spin supercurrent, has been given by Fomlin[3] in the form:

$$\frac{du}{dt} - \frac{1}{\omega}\frac{\gamma^2}{\chi}\frac{\partial}{\partial z}\left(\frac{\partial F_\nabla}{\partial \nabla\alpha}\right) = 0$$

$$\frac{d\alpha}{dt} + \omega - \frac{1}{\omega}\frac{\gamma^2}{\chi}\left(\frac{\partial F_\nabla}{\partial u} - \frac{\partial}{\partial z}\frac{\partial F_\nabla}{\partial \nabla u}\right) = o \quad (2)$$

where $u = \cos\beta$ and F_∇ is the gradient energy.

In ^{3}He-B the gradient energy, if all gradients are directed along the magnetic field, has the form:

$$F_\nabla = \frac{\chi}{\gamma^2}(1-u)c^2(u)(\nabla\alpha)^2 + \frac{1}{2}\frac{c^2(1)}{1-u^2}(\nabla u)^2 - \omega u(z-z_o)\nabla\omega$$
$$+ \left\{\frac{1}{2}\frac{3c^2(-1)}{(1+4u)(1+u)^2}\nabla u \pm (1-u)\frac{c^2(-1)}{1+u}\sqrt{\frac{3}{1+4u}}\nabla\alpha\right\}\nabla u \quad (3)$$

where $c^2(u) = uc_\parallel^2 + (1-u)c_\perp^2$. $c_\parallel$ and $c_\perp$ are the spin wave velocities parallel and transverse to the magnetic field, and $\alpha - \nabla\omega/3\gamma$ is the gradient of the magnetic field. (Remember that γ for ^{3}He is negative.) The term in the curly brackets represents the minimum of the dipole-dipole energy with $\cos\beta > -1/4$. When $\cos\beta < -1/4$ this term is equal to 0.

It is clear from the first equation is (2) that the spin supercurrent is equal to $\gamma \partial F_\nabla/\partial \nabla\alpha$ and in this case

$$J_{M_z} = -\frac{\chi}{\gamma}\left\{2\left[(u-u^2)c_\parallel^2 + (1-u)^2 c_\perp^2\right]\nabla\alpha\right.$$
$$\left.\pm (2c_\perp^2 - c_\parallel^2)\frac{1-u}{1+u}(1+4u)^{-1/2}\sqrt{3}\nabla u\right\} \quad (4)$$

for $\cos\beta > -1/4$. In the case of $\cos\beta < -1/4$ the term with ∇u is equal to 0.

These equations do not take into consideration the relaxation processes, but as was shown experimentally[4], there is a wide region of experimental conditions where spin transport is much faster than relaxation. The numerical solutions of these equations for pulsed and CW NMR for ^{3}He-B in a closed chamber with walls parallel to the magnetic field are shown in Fig. 1 and 2. Here the spin diffusion has been taken into account by including the appropriate terms in Eqs. (2).

In Figs. 1 and 2 we have used $c_\parallel = 800$ cm/s, $\nabla H = 0{,}2$ Oe/cm, $D = 0.1$ cm/s, and a length of the chamber 2.5 mm. Fig. 1 shows the redistribution of the longitudinal component of magnetization ($\cos\beta$) with time after a 90° RF pulse.

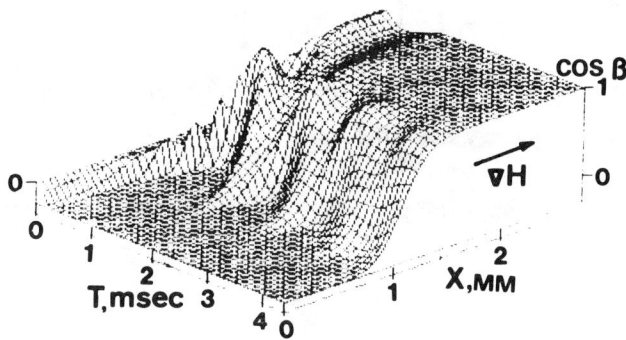

Fig. 1. Formation of the HPD after a 90° pulse.

It is clear that the formation of the two-domain structure takes place after transient processes consisting of the flow of spin supercurrents have finished. In the high field domain $\beta = 0$, while in the low field domain $\beta \geq 104°$. The magnetization in this domain precesses homogeneously with a frequency equal to the Larmor frequency at the domain boundary. The gradient in the Larmor frequency inside the HPD is compensated by a frequency shift that takes place in ^{3}He-B at $\beta > 104°$. The two domains are divided by a domain wall. The spin current through the wall vanishes because terms with ∇u in (4) are compensated by the $\nabla\alpha$ term.

The same structure was calculated for CW NMR. Fig. 2 shows how the HPD is formed after turning on an RF field of 0.005 Oe. The location of the domain boundary corresponds to the place where the frequency of the RF field equals the Larmor frequency. The formation of the HPD was studied by the Moscow experimental group[4,5] and has been reviewed in ref. 6. It was shown that in the conditions of pulsed NMR the HPD generates the so-called long lived induction decay signal which has been observed[7,8]. In the conditions of CW NMR one obtains a large signal with hysteresis which also has been observed[9,10]. In pulsed NMR the relaxation leads to a decrease in the dimensions of the HPD and as a result to a decay of the induction signal. In CW NMR the energy absorbed from the RF field equals the energy dissipated in the HPD.

The properties of the HPD have been used by our group for measurements on magnetic relaxation in ^{3}He-B.

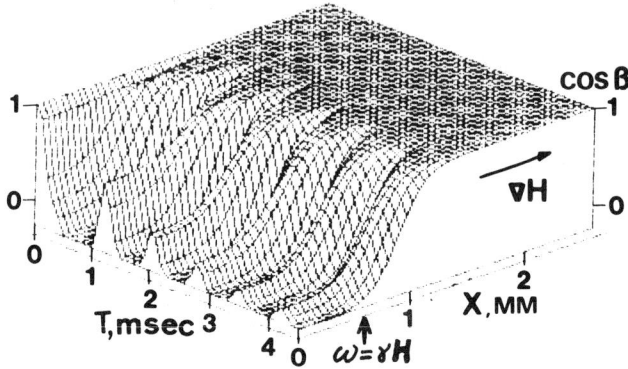

Fig. 2. Formation of the HPD after turning on the CW RF field.

In ^{3}He-A the gradient energy has the form

$$F_\nabla = \frac{\chi}{\gamma^2} 2c^2 \left[\frac{1}{2}(1 - \cos\beta)(3 - \cos\beta)(\nabla\alpha)^2 - 2(1 - \cos\beta)\nabla\alpha\nabla\Phi + (\nabla\Phi)^2 \right]. \tag{5}$$

Where Φ is a phase that characterizes the orientation of the order parameter. Consequently, the spin supercurrent is given by

$$J_{M_z} = \frac{\chi}{2\gamma} c^2 (1 - \cos\beta)(3 - \cos\beta)\nabla\alpha. \tag{6}$$

If we take into consideration the dependence of the frequency on the angle β, then $\omega = -\gamma H + \nabla\omega(1 + 3\cos\beta)$ and one can see that $\nabla\beta$ leads to an increase in $\nabla\alpha$. As a result, spin currents transport magnetization in the direction which leads to increases of $\nabla\beta$. This describes the instability of the homogeneous precession in ^{3}He-A. Any perturbations in the homogeneity of the precessing magnetization grow with a time constant that was calculated by Fomlin[11] and measured in our experiments[12].

MAGNETIC RELAXATION IN ^{3}He-B

The relaxation processes in superfluid ^{3}He are associated with the fact that there exists a normal component of ^{3}He whose magnetic moment interacts with the magnetic moments of the superfluid components. In the case of disequilibrium between the quasiparticles and the condensate contributions to the total spin polarization of the system, the so-called intrinsic spin relaxation mechanism takes place. It was introduced by Leggett and Takagi[13] and is in agreement with the experimental results on the relaxation of the wall pinned ringing mode[14] and the time dependent frequency shift in spin tipping experiments[15]. But in general the experimental results on pulsed NMR were puzzling for a long time[2]. A second mechanism of relaxation is associated with the spin diffusion of the normal component of ^{3}He and takes place in spatially inhomogeneous structures. A third mechanism of relaxation was proposed for ^{3}He-B in ref. 16 and is connected with the distortion of the bulk liquid precession mode near the walls (surface relaxation). This mechanism was experimentally estimated in ref. 17. Below we shall demonstrate that all these mechanisms of relaxation can be measured by using the properties of the HPD.

The pulsed NMR experiments of our group have shown that in a closed chamber the magnetic relaxation develops according to the following scenario. First, the HPD structure is established. Next, the dimensions of the HPD start decreasing due to magnetic relaxation, and as a result, the amplitude and frequency of the induction signal diminish. By measuring the time dependence of the signal amplitude and frequency, it is possible to calculate the velocity of the moving domain boundary and to separate the intrinsic and spin diffusion terms of relaxation by their different dependencies on the HPD dimensions. The details of this method are described in Refs. 4 and 6. Our experiments show a qualitative agreement of these two relaxation mechanisms with theoretical calculations as will be demonstrated below.

CW NMR studies of the HPD have rich possibilities for the investigation of different relaxation mechanisms. The HPD is formed by CW NMR when the RF field is sufficiently strong and the magnetic field is swept down. At the field Ha, when $-\omega rf/\gamma$ is equal to the lowest field in the chamber, the HPD is created. During the field sweep it grows, such that the domain wall is always where $H = -\omega rf/\gamma$. As a result, the transverse component of the magnetic moment of a sample also grows in Fig. 3, where we show the phase portrait of the transverse component of the magnetic moment of the sample measured with a lock-in amplifier. At the field Hb, when $-\omega rf/\gamma$ equals the highest field in the chamber, the HPD fills all of the chamber. If one continues to sweep down the field, the angle β in the HPD grows in order to be in resonance with the RF field. At the same time the energy dissipation grows and the phase of the precession changes. The details of CW NMR excitation of the HPD can be found in ref. 5. It is important to note that the energy absorbed from the RF field is equal to the energy dissipated in the HPD. Fig. 4 shows the absorption signal in the same experimental conditions as in Fig. 3 (P = 11 bar, ω = 460 kHz, T = 0.65 T_c).

The relaxation rate that leads to energy dissipation in the HPD is

$$\dot{W} = \chi H^2 S \left[\omega_{rf}^{1/3} \frac{D\sigma(\nabla\omega)^{1/3}}{c^{2/3}} + \frac{5}{16} \tau_{LT}(\nabla\omega)^2 F(L) \right] + W_s 2\pi RL \qquad (7)$$

where S is the cross sectional area of the chamber, $2\pi R$ its perimeter, σ a parameter characterizing the shape of the domain boundary, τ_{LT} the time constant characterizing the intrinsic relaxation, W_s the value of the surface relaxation rate and L the length of the HPD. If the domain boundary is in the chamber, then $F(L) = L^3$. When $-\omega rf/\gamma$ is larger than the highest field inside the chamber, the domain boundary and the first term in (7) vanish, and L becomes equal to L_o (the length of the chamber) and $F(L) = (L_o + 1)^3 - l^3$, where $l = \Delta H/\nabla H$ and ΔH is the difference between the highest field in the chamber and $\omega rf/\gamma$.

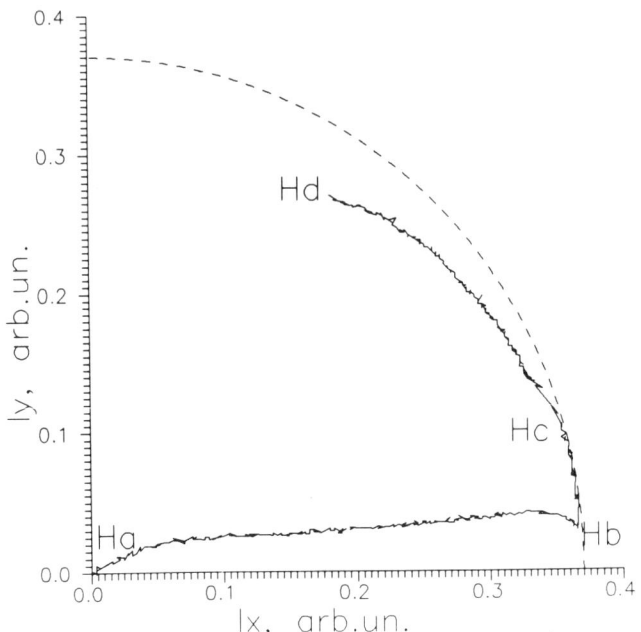

Fig. 3. Phase portrait of the transverse component of the magnetic moment of the sample.

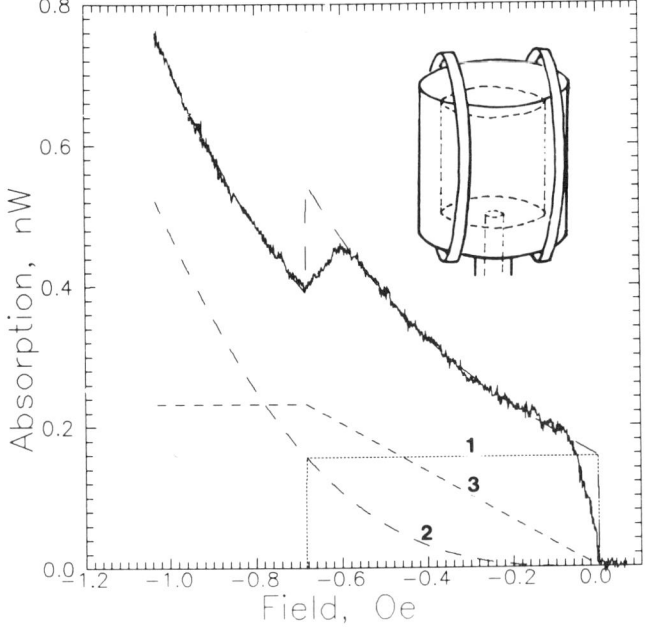

Fig. 4. Absorption signal and its theoretical fit.

One can see that all three terms have a completely different dependence from the dimensions of the HPD. In Fig. 4 are shown the theoretical curves corresponding to the three terms in Eq. (7). A good agreement is found between the theoretical and experimental signal shapes. The term with L^3 (curve 2) is determined with good accuracy at low fields, term L^0 (curve 1) is determined by the jumps at Ha and Hb. The linear term (curve 3) can be mixed with additional signal from the large dispersion signal.

In Figs. 5 and 6 are shown temperature dependencies of the time constants of the intrinsic relaxation and of the spin diffusion coefficient calculated by fitting the curves. In Fig. 6 is also shown the spin diffusion coefficient calculated from pulsed NMR experiments. Clearly, both of these constants drop rapidly in value with decreasing temperature which is associated with the increasing effective collision time when $\omega\tau \gg 1$ (analog of Leggett-Rice phenomena).

Figure 5 shows the theoretical curve which has been calculated from data of ref. 15 by scaling for 11 bar and multiplying by $1/(1+(\omega\tau)^2)$. Experimental points are systematically shifted to lower temperature or higher value. This may be caused by overheating of the cell by absorption of energy from the HPD. By our estimation the overheating can not be more than 0.1 T_c.

The surface relaxation term of the fitted curves is unexpectedly large. It may be the result from mixing absorption and dispersion signals at very large amplitude of the transverse magnetic moment. We have made additional measurements in a cell with increased surface area using mylar foils. The surface area in this cell was 3 times larger but otherwise the cell had the same dimensions. At the field H_b the surface relaxation term in these cells was 0.10 nW and 0.12 nW respectively. If the difference corresponds to the difference in surface relaxation due to the increased surface area, W_s in Eq. (7) is equal to 0.01 nW/cm^2 and agrees with theoretical estimates. Of course one needs to develop this method in order to measure surface relaxation directly. It may be that the large value of this term is connected to some bulk liquid relaxation.

CATASTROPHIC RELAXATION

We have observed a crucial change in the NMR properties of ^{3}He-B at temperatures below 0.45 T_c. At this temperature the lifetime of the HPD in pulsed NMR experiments rapidly decreases from 100-1000 msec to 1 msec with a short temperature region. The dissipation in the HPD in CW NMR drastically grows in the same temperature region. At the lowest temperatures the process of relaxation is so fast that the HPD cannot be formed. We have named this new relaxation process **catastrophic relaxation**. This phenomenon does not depend on pressure, magnetic field or the presence of ^{4}He covering the surface of the chamber. The first description of this phenomenon is published in ref. 18. Recently we have performed pulsed NMR experiments in homogeneous magnetic fields which show that the decay time of the precession signals decreases at temperatures below 0.45 T_c. In Fig. 7 are shown the signals at different temperatures and tipping pulses. The characteristic shape of the signals shows that this mode of precession is unstable below 0.45 T_c. The time constant that signifies this instability is shorter as the temperature is lowered and the tipping angle increased.

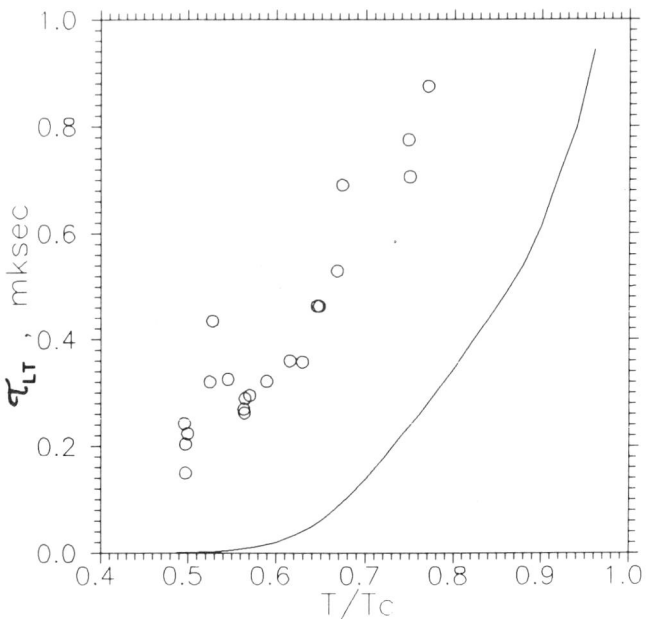

Fig. 5. The time constants of intrinsic relaxation at 11 bar and 460 kHz.

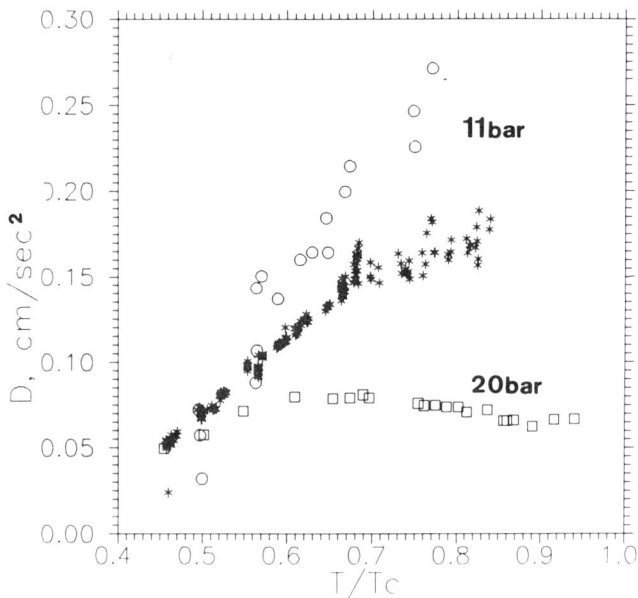

Fig. 6. The spin diffusion constant at 460 kHz.

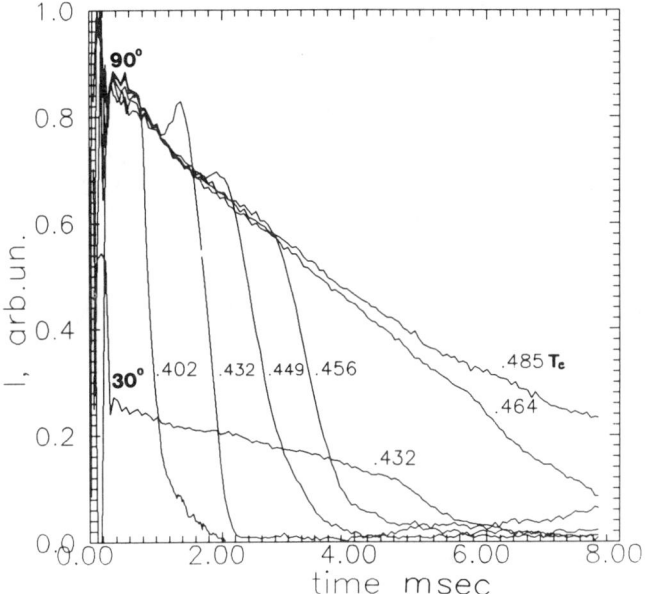

Fig. 7. The decay of the induction signals at different temperatures for two tipping angles.

A possible explanation of this phenomenon has been suggested recently by Markelov[19]. He shows that when $\omega\tau \gg 1$ and $T < 0.5\,T_c$, a new mode of oscillations exists. This is the relative oscillation of the magnetization of the superfluid component and the quasiparticles which are connected with the fermi liquid interactions. In pulsed NMR the homogeneous precession excites this mode of oscillations. As a result, the difference in the directions of the magnetizations of the condensate and quasiparticles grows and intrinsic relaxation sharply increases. Let us name this process the intrinsic instability of phase precession. Possibly the same mechanism is broadening the line in longitudinal NMR at $0.2\,T_c$ as was observed in ref. 20.

MAGNETIC RELAXATION IN ^{3}He-A

As was shown, the spin supercurrents promote the growth of spatial inhomogeneity in the free precession. In contrast to ^{3}He-B, where spin currents tend to minimize the gradient energy, in ^{3}He-A they attempt to minimize the dipole-dipole energy. As was shown by Fomin[11], a spatial modulation of the phase of the precession with a period of 10^{-1}-10^{-3} mm will be established. This phenomenon was studied experimentally by observing the shape of induction decay signal in a rather homogeneous magnetic field[12]. In agreement with theoretical predictions[11], the shape of the signals are best described by the function: $I = I_0(1-A\exp(t/T_f))$. A comparison of the rate $1/T_f$ with theory is shown in Fig. 8. It is quite clear that experimental and theoretical vales are quantitatively close to each other.

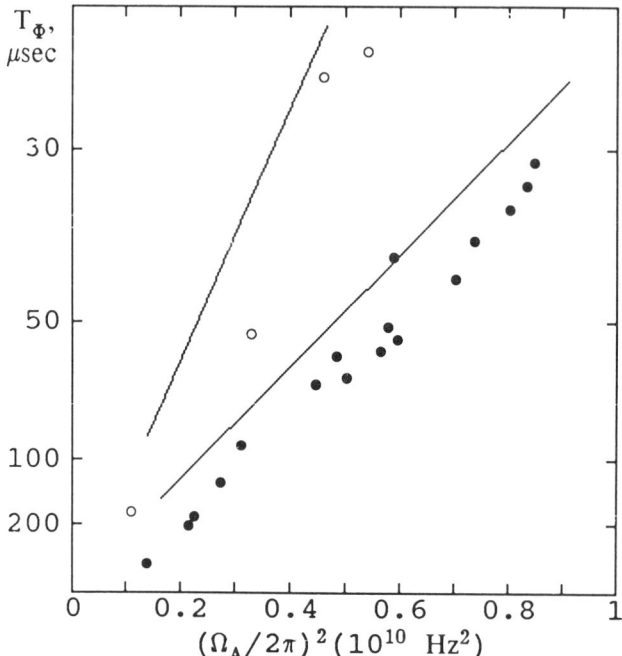

Fig. 8. The rate of instability of the homogeneous precession in ^{3}He-A versus Ω_A for 250 (o) and 500 (·) kHz.

We are sure that the enhancement[2] of relaxation in ^{3}He-A is to be attributed to spin diffusion and spin-orbit interaction in an inhomogeneous magnetic texture. To test this we have compared the decay of the homogeneous precession with the recovery of the longitudinal magnetization measured by a two pulse method in a magnetic field gradient. As shown in Fig. 9, the time scale for these two processes is the same. The good quantitative agreement between the experimental results[12] and the theory of Fomin, which depends on the rate of instability of temperature, magnetic field and tipping angle, demonstrates that homogeneous precession in ^{3}He-A splits into an inhomogeneously precessed texture due to magnetization transport by spin supercurrents. The main difference between theory and experiment is a rather large value of initial fluctuations of order of 10^{-3} to 10^{-5}, whereas thermal fluctuations give the value 10^{-7}. The main source of these fluctuations is a stationary texture due to the dimensions of the chamber. A similar behavior for the induction decay signal is observed in solid antiferromagnetic ^{3}He.[21] The main difference is that the value of the initial fluctuation is of the order of 1. This can be attributed to the small size of the magnetic domains.

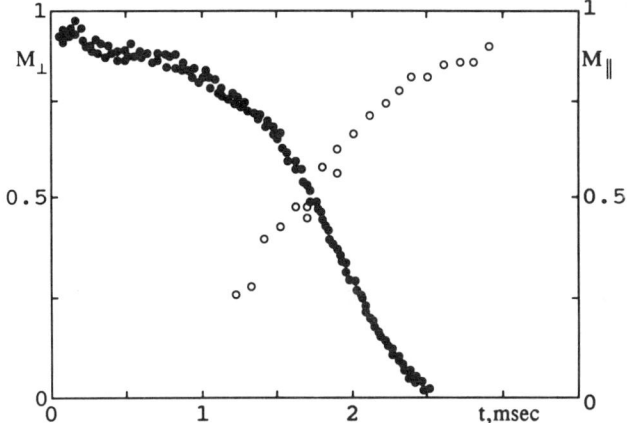

Fig. 9. The comparison of the decay of the induction signal (·) and the recovery of the longitudinal magnetization (○).

CONCLUSION

Now that there is almost complete understanding of spin relaxation processes in superfluid ^{3}He, it looks very interesting to investigate the properties of nonlinear NMR phenomena in the collisionless region. It is clear that at very low temperatures the intrinsic instability of homogeneous precession in ^{3}He-B must disappear. Down to a temperature of 0.2 T_c, however, no decrease was observed in the catastrophic relaxation.

The second set of investigations presented in this review concerns the method of maintaining the HPD for indefinite time by use of CW NMR. By use of this method we have studied the spin current in a channel and in collaboration with the Helsinki experimental group, the properties of vortices. Both these topics are discussed elsewhere in these proceedings.

Finally, we wish to express our deep gratitude to I.A. Fromin, A. de-Waard, A.V. Markelov, J. Nyeki, D.A. Sergatskov and G.K. Tvalashvily for their very helpful collaboration, and the organizing committee of this symposium for financial support.

REFERENCES

1. A. Leggett, Rev. Mod. Phys. **47**, 331 (1975).
2. D.M. Lee, R.C. Richardson, In "The Physics of Liquid and Solid He" part 2, ed. Benneman, Ketterson (Wiley, NY, 1978).
3. I.A. Fomin, Sov. Phys. JETP **61**, 1207 (1985).
4. A.S. Borovik-Romanov, Yu.M. Bunkov, V.V. Dmitriev, Yu.M. Mukharskiy, K. Flachbart, Sov. Phys. JETP **61**, 1199 (1985).
5. A.S. Borovik-Romanov, Yu.M. Bunkov, V.V. Dmitriev, Yu.M. Mukharskiy, K. Poddyakova, O.D. Timofeevskaya, Sov. Phys. JETP, to be published.

6. Yu.M. Bunkov, Proc. 18th Int. Conf. on Low Temperature Physics, Kyoto, 1987, Jap. J. Appl. Phys., v. 26, p. 1809.
7. R.W. Giannetta, E.N. Smith, D.M. Lee, J. Low Temp. Phys. **45**, 295 (1981).
8. L.R. Corruccini, D.D. Osheroff, Phys. Rev. **B17**, 12 (1978).
9. D.D. Osheroff, in "Quantum Fluids and Solids" ed. by Trickey, Adams, Dufty (Plenum Press, NY, 1977), p. 164.
10. R.A. Webb, Phys. Rev. Lett. **39**, 1008 (1977).
11. I.A. Fomin, JETP Lett. **39**, 466 (1984), J. Low Temp. Phys. **31**, 509 (1977).
12. Yu.M. Bunkov, V.V. Dmitriev, Yu.M. Mukharskiy, Sov. Phys. JETP **61**, 719 (1985).
13. A.J. Leggett, S. Takagi, Ann. Phys. **106**, 79 (1977).
14. R.A. Webb, R.E. Sager, J.C. Wheatley, Phys. Rev. Lett. **35**, 1164 (1975).
15. G. Eska, K. Neumaier, W. Shoepe, K. Uhlig, W. Wiedman, Phys. Lett. **87A**, 311 (1982).
16. T. Ohmi, M. Tsubota, T. Tsuneto, Proc. 18th Int. Conf. on Low Temperature Physics, Kyoto, Jap. J. Appl. Phys. **26**, 169 (1987).
17. O. Ishikawa, Y. Sasaki, K. Sasayama, T. Mizusaki, A. Hirai, Jap. J. Appl. Phys. **26**, 171 (1987).
18. Yu.M. Bunkov, V.V. Dmitriev, Yu.M. Muharskiy, J. Nyeki, D.A. Sergatskov, Europhys. Lett. (1989), to be published.
19. A.V. Markelov, JETP Letters (1989), to be published.
20. D. Candela, D.O. Edwards, A. Heft, N. Masuhara, Y. Oda, D.S. Sherrill, Phys. Rev. Lett. **61**, 420 (1988).
21. T. Kusumoto, O. Ishikawa, T. Mizusaki, A. Hirai, J. Low Temp. Phys. **59**, 269 (1985).

JOSEPHSON EFFECT IN SPIN SUPERCURRENT IN ^{3}He-B

A.S. Borovik-Romanov, Yu.M. Bunkov, V.V. Dmitriev,
Yu.M. Mukharskii, D.A. Sergatskov
Institute for Physical Problems
Kosygin st. 2, Moscow 117334, USSR

ABSTRACT

Investigations of spin supercurrent flow in narrow channels are described. Zeeman energy transfer, connected with spin superflow, is observed. The critical spin current is reached when phase slips take place. In a channel with an aperature, a transition from phase slips to a non-hysteretic current-phase relationship is observed. Creation of spin vortices and possible future experiments are discussed.

1. INTRODUCTION

The nonzero spin of Cooper pairs of ^{3}He atoms in the superfluid phases of ^{3}He opens up the possibility of spin (and, consequently, magnetic moment) transfer by supercurrents. The transport of spin may occur independently of mass transport. The spin supercurrent has been considered in a number of theoretical papers[1-3]. The essential progress in the theory is connected with recent works by I. Fomin[4-6]. He has shown that the analog of the phase of the order parameter for the spin superfluidity are Euler angles α, β and γ, which define the orientation of the spin part of the order parameter. For NMR experiments α and β have a clear meaning: α is the phase of the precession of the magnetization, and β is the tipping angle. There is an important difference between spin and other types of supercurrents. The point is that in the superfluid phases of ^{3}He a spin-orbital (dipole) interaction exists. This interaction results in local nonconservation of spin: equations of spin dynamic now contain sources and sinks. Therefore, in the A-phase it is impossible to make direct investigations of spin supercurrents[7]. Fortunately, in the B-phase, the dipole energy is degenerate with respect to α. The corresponding conserved value is longitudinal magnetization (i.e., the projection of spin on the direction of the external magnetic field H). The flow of the longitudinal magnetization in ^{3}He-B for the case of $\nabla\alpha, \nabla\beta \perp H$ is defined by the following equation[5]:

$$j_s = \frac{\chi}{\gamma}(1-u)c_\parallel^2(A(u)\nabla\alpha + B(u)\nabla u) \qquad (1)$$

where $A(u) = (1-u) + (1+u)c_\perp^2/c_\parallel^2$, $B(u) = (3/(1+4u))^{1/2}(1+u)^{-1}\nabla u$, $u = \cos\beta$, and $c_\perp$ and $c_\parallel$ are spin wave velocities. In the case $\nabla\alpha, \nabla\beta \perp H$ the formula for the spin flow is similar to (1) (see Eq. (4) in Ref. 7). It is interesting to note that these formulae have the common multiplier $\chi/\gamma(1-u)c^2$ which may be considered as the "spin superfluid density."

Recently our group in Moscow has carried out experiments on spin supercurrents in ^{3}He-B [8-11] using NMR methods. The ^{3}He sample is located in a magnetic frequency ($\nabla\omega = \gamma\nabla H$) along the sample causes, after the RF tipping pulse has been applied, a growth in the gradient of the precession phase

α, i.e., a spin supercurrent. This current redistributes the magnetization of the sample and a two-domain structure may arise[4,12]. In one of these domains the magnetization is in equilibrium, and in the other it is deflected by an angle slightly larger than θ = arccos (-1/4) = 104° and precesses with a common phase (homogeneously precessing domain, HPD). The width of the wall between the domains in the absence of the RF-field may be determined from the following equation[5]:

$$\lambda = \left(\frac{c_\parallel^2}{\omega \nabla \omega}\right)^{1/3} \quad (2)$$

For our experimental conditions ($c_\parallel \sim 10^3$cm/s, $\nabla\omega \sim 10^2 - 10^4$) $\lambda \sim 0.1 - 1$mm.

The phase of the precession of the HPD equals the local Larmor frequency in the middle of the domain wall. The HPD may be created also in CM NMR experiments. In this case it can be maintained as long as necessary: magnetic relaxation in the HPD is compensated by RF-power absorbed from the RF-field. The frequency of the precession (and, consequently, the position of the domain wall) is defined now by the frequency of the RF-field (ω_{rf}). If ω_{rf} is fixed then the position of the domain wall may be changed by sweeping the external magnetic field. The absorbed power is proportional to the product of the RF-field amplitude and $\sin(\alpha - \varphi)$, where α is the phase of precession and φ is the phase of the RF-field and is dissipated in the HPD. If the amplitude of the RF-field is high enough, the $\alpha - \varphi$ is small, i.e., the phase of precession coincides with the phase of the RF-field. Thus, in the HPD β=104° and $\alpha \approx \varphi$; one of the two angles defining the spin supercurrent is constant and the other is controlled by the RF-field. For this reason the HPD, maintained by the RF-field, is a very convenient "tool" for the study of the spin supercurrent.

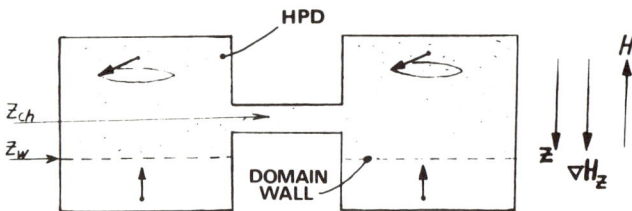

Fig. 1. The scheme of the experiments

The idea of all experiments described below is to connect two HPDs with a channel (Fig. 1). Each HPD may be considered a massive "electrode" in which the order parameter is homogeneous. Let both HPDs fill the channel. The change of the phase of precession of one HPD with respect to the phase of the other (reference) HPD creates an increasing gradient of α in the channel, i.e., a spin supercurrent. This current carries the longitudinal magnetization and, consequently, the Zeeman energy, from one HPD to the other. The power, absorbed by the HPDs from the RF-fields should change: in one HPD the absorption grows and in the other - it decreases.

2. SPIN SUPERFLOW ALONG A NARROW CHANNEL

In order to observe and investigate the spin supercurrent we have used a number of experimental cells with different geometries. Much of the work was done in a cell, the upper part of which is shown in Fig. 2. The experimental volumes filled with ^{3}He (5,6) are cylinders with horizontal axes. The length of these cylinders is 6 mm and the diameter equals 4.5 mm. The volumes are connected by channel 7, which has a constriction of 5.5 mm length and 0.6 mm diameter. The necessary temperature was obtained by a nuclear demagnetization cryostat and was measured with a PLM-3 NMR-thermometer (not shown in Fig. 2). To create and maintain HPDs two NMR coils (1,2) are used. Two miniature coils (3,4) wound on the channel allow us to follow the behavior of the magnetization in different parts of the channel. Copper shields (8) (one of them has cylindric form and is placed around the channel) prevent RF-fields from penetrating in the channel and diminish the mutual inductance of RF-coils 1 and 2.

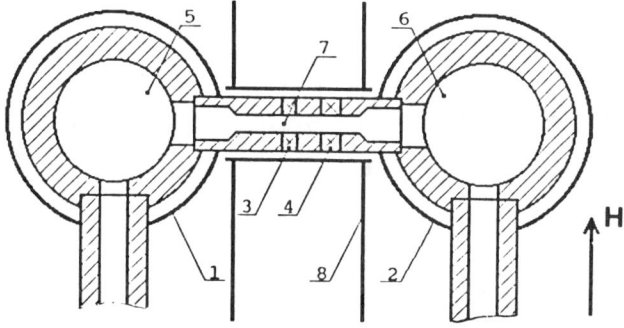

Fig. 2. The cell for the study of the spin superflow in a long channel

Experiments were carried out in fields of 71, 142 and 284 Oe (corresponding NMR frequencies are 230, 460 and 920 KHz) and pressures from 0 to 29.3 bar. Special gradient coils are used to apply and control the gradient of the magnetic field. Standard electronics for CW NMR are used. Two high stability RF-generators are used for creating the HPDs and for controlling their phases of precession. The NMR signal from any of the coils is first fed to a preamplifier and a lock-in detector, and then recorded with a computer or with an XY-recorder.

The experiment proceeds as follows. At first, both RF-generators are tuned to the same frequency (say 460 KHz) and the RF-power is applied to NMR coils 1 and 2. Then two HPDs are created and "grown" by sweeping the external magnetic field downward. The applied RF-power is maintained at a level sufficient such that $\alpha = \varphi$ (the corresponding RF-fields are of the order of 10^{-2} Oe); therefore we shall not distinguish between α and φ in the text below. After the HPDs have penetrated into the channel, the field sweep is stopped and the field is kept constant. If we neglect magnetic relaxation in the channel then, at this moment, HPDs in both experimental volumes and the magnetization in the channel should precess in phase. The next step in the experiment is to change slightly (by 0.01-1 Hz) the frequency of one of the RF-generators and to monitor the absorbed RF-power in the second HPD. The result from such an experiment is shown in Fig. 3. It is seen from the figure that the increase in $\nabla\varphi$ causes the absorption to grow. At some phase difference $\Delta\varphi_c$, the absorbed power jumps to some initial value (B→C). With a constant frequency difference between the two RF-generators after the jump, the absorption continues to rise until it reaches the critical value again, and so on. For the opposite sign of the frequency difference the absorption decreases, reaches point C and then continues to decrease until the critical value at point B' is reached. Thus, in this experiment we directly observe the transfer of Zeeman energy from one HPD to another. This transfer is surely connected with the spin supercurrent - the change of the absorption is accompanied by a corresponding change of the gradient in the phase of the precession in the channel (this is clearly seen by the miniature NMR- coils[8]). We believe that jumps in the absorption occur when the critical value of the supercurrent is reached. At this moment phase slips (by $2\pi n$) take place. For example, in Fig. 3 phase slips by 4π occur. The state which exists in the channel with constant frequency difference between the two RF-generators is analogous to a resistive state in a superconducting weak link and the frequency difference corresponds to the voltage across the weak link. It should be mentioned that if at some moment we return the frequency difference to zero then the absorbed power, i.e., the spin supercurrent, remains at the corresponding value.

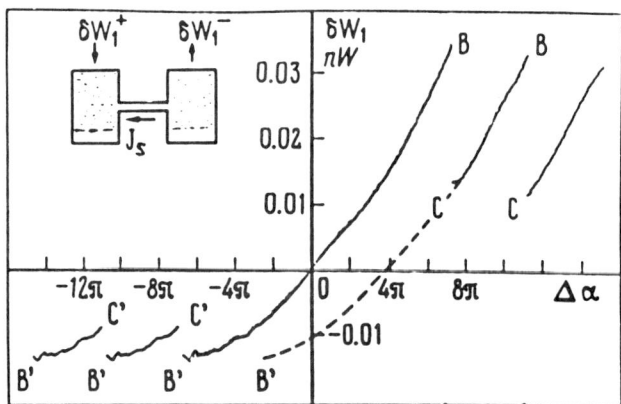

Fig. 3. The change of the absorption in one of the experimental volumes versus the phase difference between two HPDs. P = 29.3 bar, T = 0.48 T_c, H = 142 Oe.

3. CRITICAL SPIN CURRENT

The flow of the spin supercurrent in a long channel has been considered by I. Fomin[6]. Unlike the other types of supercurrents, for the phase slippage to occur in the spin supercurrent, it is not necessary to destroy the superfluid state in a phase slip center. The phase slip may occur if the magnetization in some place in the channel is rotated to lie along the direction of the external magnetic field. The phase of precession in this place is undefined and may slip (note that in the phase slip center $\beta = 0$, i.e., the "spin superfluid density", mentioned in Sec. 1, equals 0). The creation of a phase slip center is connected with a change in the Zeeman energy of the system δE. The total longitudinal magnetization of the sample has to be conserved. Therefore:

$$\delta E = \chi \, \omega \, \Delta \omega \, V / \gamma^2 \qquad (3)$$

where V is the volume of the phase slip center, ω is the frequency of precession in the channel, $\Delta\omega$ is the frequency shift in the channel ($\Delta\omega = \gamma\nabla H(z_w - z_{ch})$, z_{ch}, and z_w are z-coordinates of the channel and the domain wall (see Fig. 1). This energy plays the role of the energy gap and should be compared with the energy due to the gradient of α in the channel: $E \sim c_\perp^2 (\nabla\alpha)^2$. By comparison[6]:

$$\nabla\alpha_c = 1/\xi_s \tag{4}$$

where

$$\xi_s = c_\perp/(\omega\nabla\omega)^{1/2} \tag{5}$$

and ξ_s may be considered to be an analog of the Ginzburg-Landau correlation length in the theory of superconductivity.

From Fig. 3 we may obtain the value for the critical phase difference between the ends of the channel. To evaluate $\nabla\alpha_c$ this phase difference should be divided by the length of the channel. Values of $\nabla\alpha_c$ obtained by such means agree qualitatively with (4). In particular, the measured critical gradient of the phase of precession increases with the frequency shift in the channel[13]. The quantitative disagreement (experimental values are usually 1.5-2 times smaller than theoretical ones) may be connected with the magnetic relaxation in the channel. Part of the Zeeman energy transferred by the supercurrent is spent to compensate for this relaxation. Therefore, the values of the spin current at the ends of the channel are different. That is, the gradient of the phase is not constant along the channel. The fact that the dependence in Fig. 3 is not quite antisymmetric may also be explained in this way.

4. JOSEPHSON EFFECT IN THE SPIN SUPERCURRENT

Nowadays, the name "Josephson effect" is commonly used for the phenomenon of flow of electrical current across a superconducting weak link[13] or of mass superflow[14] in ^{3}He and ^{4}He when the current-phase relationship is non-hysteretic (single-valued). To observe the Josephson effect in these cases the dimensions of the weak link should be of the order of the coherence length. If this length is much larger than the dimensions of the link then the Josephson current obeys the following relationship:

$$J = J_c \sin(\Delta\Phi) \tag{6}$$

where J_c is a critical current of the weak link, $\Delta\Phi$ is the difference in the phases of the wave functions of the superfluid (superconducting) states on the two sides of the link. An increase in the dimensions of the link result in a more complex dependence of the current on $\Delta\Phi$ (but the current is still periodic with respect to $\Delta\Phi$ and the period equals 2π). As it has been mentioned above, the analog of the coherence length for the spin supercurrent is the spin coherence length ξ_s. In our experiments we may easily change ξ_s by moving the domain wall closer to the channel and by decreasing the field gradient. The residual inhomogeneity of the external magnetic field in our experiments was about 0.05 Oe/cm. This values gives the minimum for the gradient which must be applied in order to have a definite two-domain structure in each experimental volume. It is small enough such that ξ_s may be of the order of the channel dimensions if $z_w - z_{ch} \sim$ 1 mm. The spin superflow along a channel with a length less than or of the order of ξ_s was considered by A. Markelov[15]. He has obtained an analytic solution for the current-phase dependence for the case when $\beta = \theta$ at the ends of the channel. This solution and also results of numerical calculations for $\beta < \theta$ are shown in Fig. 4.

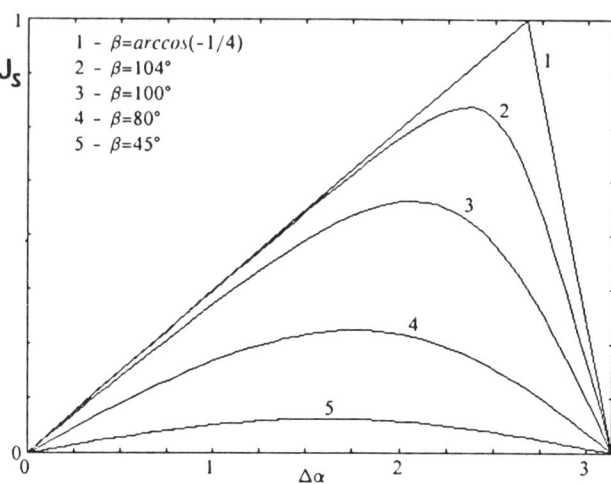

Fig. 4. The spin supercurrent vs. the phase difference across the weak link for different values of β[15].

At first we attempted to observe the non-hysteretic current-phase relationship in the cell shown in Fig. 2. For this purpose we measured the absorption in one of the HPDs vs. the phase difference between RF-fields for different lengths of HPDs. We found that the critical phase difference decreases when the length of the HPD is reduced, reaches the value of the order of a few π (minimum observed value was equal to 1.3π) and then both HPDs are "pushed" from the channel. The angle β in the channel, in this case, equals 0 and, consequently, the spin supercurrent vanishes. The absence of the HPD in the channel is clearly seen with the miniature NMR-coils. This phenomenon may be explained if one supposes that the HPD does not "wet" the walls of the cell. The interaction between the HPD and the wall may be evaluated[16] from the following equation:

$$E_i = -\Delta\chi \, \xi \, H^2 R_{xy} S/2 \qquad (7)$$

where $\Delta\chi$ is the difference between the susceptibilities of normal and B-liquid, ξ is the "real" coherence length, S a surface area, and R_{xy} is a coefficient of the order of unity, depending on the orientation of the order parameter near the wall. It is clear that in equilibrium R_{xy} has a minimum value, i.e., the HPD should not wet the wall.

The energy (7) should be compared with the change in the Zeeman energy connected with the shift of the domain boundary near the wall of the cell. If we take into account that the total magnetization of the sample should not change, then the formula which coincides with (3) is obtained ($\Delta\omega = \gamma\nabla H \, \Delta x$, where Δx is the displacement of the domain wall near the surface). A comparison of (3) and (7) gives the estimate for Δx for our experimental conditions:

$$\Delta x = \left(\frac{\Delta \chi \, \xi \, \omega}{\chi \nabla \omega}\right)^{1/2} \simeq 0.1 \text{ to } 1 \text{mm} \tag{8}$$

Thus, the fact that the HPD is pushed from the channel prevents one from achieving the Josephson regime in the cell shown in Fig. 2. Therefore, we have prepared another cell which is shown in Fig. 5. Two cylindrical experimental volumes (diameter 4.5 mm, length 6 mm) are connected with a wide channel (diameter 1.4 mm). In the middle of the channel is a circular orifice (diameter 0.48 mm) which is shown in more detail in the inset. The cross section of the orifice is much less that that of any other part of the channel so that this aperature may be considered to be the weak link. Two NMR-coils (4.5) are used to maintain the HPDs and to measure the change in absorption connected with the spin superflow. The spin supercurrent of the expected "ideal" Josephson effect is proportional to $\sin(\Delta\varphi)$. The pick-up signal from one RF-coil to another has the same dependence. For this reason we have put special effort into RF-shielding. The NMR-coils are placed inside copper "barrels". Such a construction has essentially decreased the mutual inductance between coils and prevented RF-fields from penetrating the weak link. Special electronics are used to compensate the small residual mutual inductance between the RF-coils.

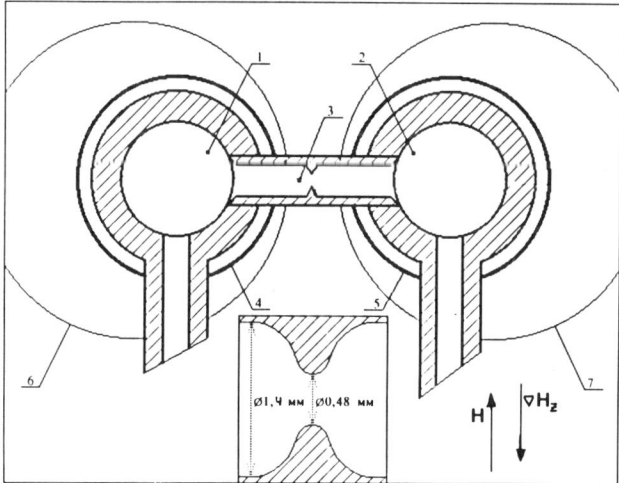

Fig. 5. The cell with the orifice in the channel.

This circuit is tuned in the absence of the HPDs in the cell. The experiments were carried out in the same way as is described in Sec. 2 and under the same experimental conditions. The frequency shift in the channel is changed by changing the external magnetic field (i.e., the lengths of the HPDs). The dependence of the absorption on the difference in the phase of precession of the two HPDs (measured at constant frequency difference between RF-generators) is shown in Fig. 6. For large values of frequency shift in the weak link phase slips were observed (Figs. 6a,b).

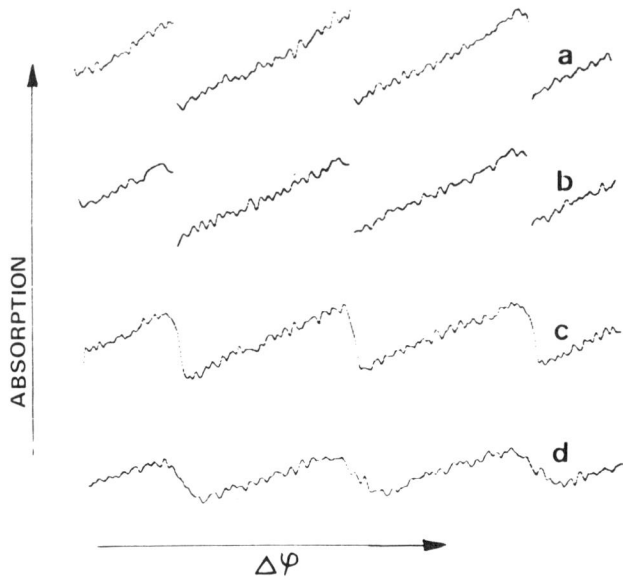

Fig. 6. The absorption in one of the HPDs vs. the phase difference between HPDs for different values of the frequency shift in the weak link. P = 0 bar, T = 0.47 T_c, H = 71 Oe, ∇H = 0.15 Oe/cm.
a: $\Delta\omega/2\pi$ = 53 Hz, $\xi_s \sim$ 0.8 mm; b: $\Delta\omega/2\pi$ = 36 Hz, $\xi_s \sim$ 1 mm; c: $\Delta\omega/2\pi$ = 31 Hz, $\xi_s \sim$ 1.1 mm; d: $\Delta\omega/2\pi$ = 20 Hz, $\xi_s \sim$ 1.3 mm

The decrease in lengths of the HPDs results in a reduction of the critical phase difference down to π. After that a continuous dependence was observed (Figs. 6c,d). Three other examples, recorded both for increasing and diminishing phase difference, are shown in Fig. 7. Figure 7a clearly represents phase slips while Figs. 7b,c illustrate non-hysteretic continuous current-phase relationship. Thus, in these experiments we observe the Josephson effect in the spin supercurrent system. Figures 6c,d correspond to the AC Josephson current with the frequency difference between RF-fields the analog of the voltage. If we keep some phase difference between the HPDs, then the spin supercurrent also is constant, i.e., we obtain in this case the analog of the DC Josephson effect.

The values of ξ_s shown in Figs. 6 and 7 should be considered as approximate. The point is that we cannot measure the value of $z_w - z_{ch}$ with good accuracy. An additional difficulty is that the width of the domain wall is roughly equal to the diameter of the aperture. To calibrate the location of the HPD with respect to the channel we have measured the slope A of the linear part of the current-phase relationship for small phase differences. As seen from Eq. (1), this slope should not depend on the frequency shift in the channel when the HPD fills the weak link but decreases when the domain wall reaches the link. In Fig. 8 the dependence of A on $\Delta H = \Delta\omega/\gamma$ is shown for ∇H = 0.9 Oe/cm (circles) and 0.15 Oe/cm (diamonds). The origin of ΔH is chosen such that the slope A is equal to zero at the same point for both gradient values. The closed symbols represent points where a non-hysteretic current-phase relationship is observed. These points correspond to the situation that at least part of the domain wall is located in the aperture. The field change where A drops from 1 to 0 corresponds to 0.1 mm ($\nabla H = 0.15$ Oe/cm) and 0.4 mm ($\nabla H = 0.9$ Oe/cm) which is less than the diameter of the narrowing. We believe that this fact is also connected with the interaction between the domain and channel walls. Unfortunately, it is impossible to compare our results with the theory[15] because when the domain wall is close to or inside the link the problem is not one-dimensional: ξ_s in the channel varies along the z-axis.

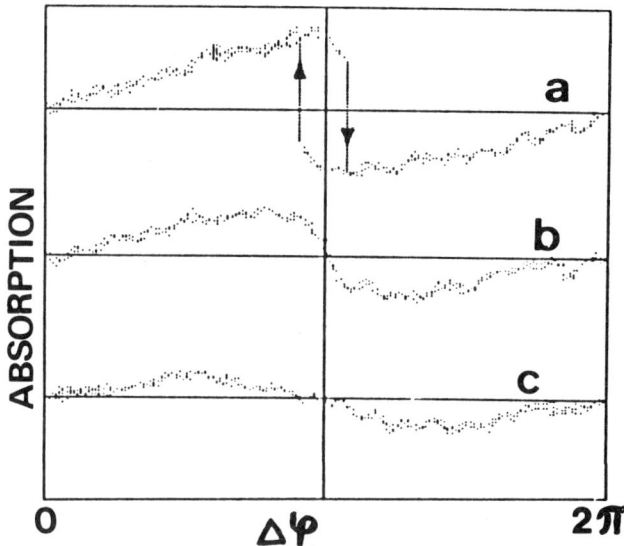

Fig. 7. The absorption in one HPD recorded both by increasing and diminishing the phase difference between HPDs: P = 0 bar, H = 71 Oe, T = 0.47 T_c, ∇H = 0.9 Oe/cm.
a: $\xi_s \sim 0.7$ mm; b: $\xi_s \sim 0.8$ mm; c: $\xi_s \sim 1.3$ mm.

In the Josephson effect case, the angle β in the link continuously changes from 104° (at $\Delta\varphi = 0$ or $2\pi n$) to 0° (at $\Delta\varphi = \pi$ or $\pi(2n+1)$). For $\beta < \theta$ the dipole frequency shift is zero. Nevertheless, the magnetization in the link precesses with the frequency of HPDs which differs from the local Larmor frequency in the link. The point is that the frequency shift in the weak link is connected mainly with the gradient energy[4].

5. EFFECTS CONNECTED WITH INHOMOGENEITY OF ξ_s

We observed the single-valued current-phase dependencies only in magnetic fields of 71 and 142 Oe. In a field of 284 Oe the smaller value of ξ_s (for otherwise similar conditions) possibly prevents one from obtaining the Josephson effect. However, at this field we observed another interesting effect. The decrease of the length of the HPD resulted in jumps of the absorption being changed to double jumps. The transition from the usual jumps to double jumps is illustrated in Fig. 9. For a constant frequency difference these double jumps occur with a period of 2π, so the double jumps also correspond to phase slips by 2π. After the first jump, the absorption may be kept at a constant level if we stop changing the phase difference at this point. If we change the sign of the frequency difference then a hysteresis after the jump is observed. A possible explanation of double jumps is the creation of a spin vortex with a core perpendicular to the external magnetic field.

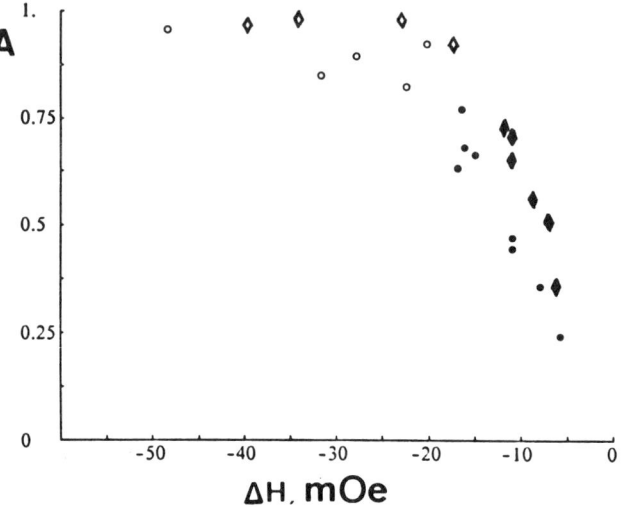

Fig. 8. The slope of the initial part ($\Delta\varphi \approx 0$) of the current-phase relationships of the type shown in Figs. 6 and 7 versus ΔH.

Actually, if the distance from the domain wall is of the order of the diameter of the weak link, then ξ_s is sufficiently inhomogeneous along the z-axis. For this reason the critical current is achieved first in the lower part of the weak link. If ξ_s is much less than the diameter of the link, then the creation of the spin vortex may be more likely than the creation of the phase slip center. If, in the upper part of the link, the critical current is not achieved then the vortex will not move to the upper wall of the channel and disappear there. Therefore, the vortex stops at some level to minimize the gradient energy connected with it. The suggested distribution of the magnetization (in the rotating frame) near the vortex in a rectangular channel is shown in Fig. 10. For simplicity the angle β far away from the vortex is taken to be 90° (not 104° as in the real case). The possibility of creating a spin vortex with the core axis parallel to the external magnetic field has been predicted earlier by I. Fomin[17].

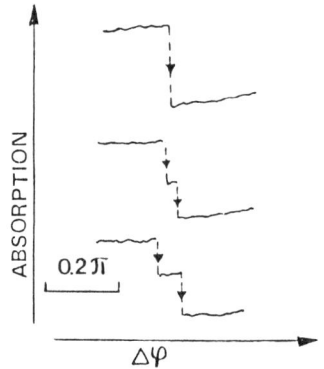

Fig. 9. Jumps of the absorption for different domain lengths. H = 284 Oe, P = 0 bar, T = 0.52 T_c, $\nabla H \simeq 0.1$ Oe/cm. a: $\Delta\omega/2\pi \sim 400$ Hz; b: $\Delta\omega/2\pi \sim 300$ Hz; c: $\Delta\omega/2\pi \sim 160$ Hz

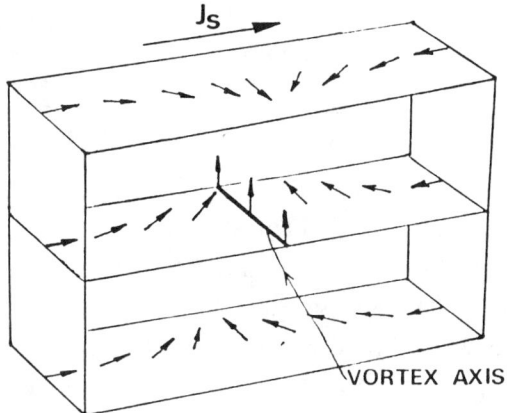

Fig. 10. The supposed distribution of the precessing magnetization near the "perpendicular spin vortex."

6. DISCUSSIONS

Thus, the spin supercurrent exhibits many of the properties of its mass and charge analogs. The difference is that at nonzero temperature, the magnetic relaxation exists. The analog of a chemical potential for the spin superflow is the precession frequency and the longitudinal magnetization plays the role of the three particle density. If the precession frequencies of both HPDs are equal, then the difference in chemical potentials equals zero. Nevertheless, the phase difference and, consequently, the spin supercurrent may not equal 0. So, the energy dissipation is due to a decrease in "the number of the particles" and not to the difference in the potentials. The dissipation does not change the main features of the superfluid state: the current is caused by gradients of the order parameter phases but not by the gradient of the chemical potential.

The analogy between the spin and other types of supercurrents permits us to predict new effects. We are preparing not an experiment where we hope to follow the motion of spin vortices. Another experiment being discussed now is the study of the influence of an electric field on the spin supercurrent. If we consider the spin supercurrent as the counterflow of magnetic moments then the electric field should change the gradient of the phase of precession. For the current between two HPDs ($\nabla \beta = 0$) we may write:

$$j_s \propto \left(\nabla \alpha - \frac{2}{\hbar c} [\vec{\mu} \times \vec{E}] \right) \qquad (9)$$

where $\vec{\mu}$ is the magnetic moment of the ^{3}He atom, $\vec{E}$ is the electric field and c is the speed of light. An estimate for reasonable values of channel length ($\sim$1 cm) and the electric field ($\sim$30 kV/cm) gives the change of the phase difference between two HPDs of the order of 0.1°. This change is large enough to be measured.

ACKNOWLEDGEMENTS

We acknowledge the help from A. de Waard and V. Makroczyove, who participated in part of the described experiments. We also express our gratitude to I. Fomin for helpful and stimulating discussions and to M. Krusius for useful comments.

REFERENCES

1. M. Vuorio, J. Phys. C7, L5 (1974).
2. A.J. Leggett, Rev. Mod. Phys. 47, 331 (1975).
3. E.B. Sonin, JETP Lett. 30, 662 (1979).
4. I.A. Fomin, Sov. Phys. JETP 61, 1207 (1985).
5. I.A. Fomin, Can. J. Phys. 65, 1510 (1987); Sov. Phys. JETP 66, 1142 (1987).
6. I.A. Fomin, JETP Lett. 45, 135 (1987).
7. A.S. Borovik-Romanov, Yu.M. Bunkov, V.V. Dmitriev and Yu.M. Mukharskii, this conference.
8. A.S. Borovik-Romanov, Yu.M. Bunkov, V.V. Dmitriev and Yu.M. Mukharskii, JETP Lett. 45, 124 (1987); Jap. J. Appl. Phys. 26, Suppl. 26-3, 175 (1987).
9. Yu.M. Bunkov, Jap. J. Appl. Phys. 26, Suppl. 26-3, 1809 (1987).
10. A.S. Borovik-Romanov et al., JETP Lett. 47, 478 (1989).
11. A.S. Borovik-Romanov, Yu.M. Bunkov, V.V. Dmitriev, Yu.M. Mukharskii and D.A. Sergatskov, Phys. Rev. Lett. 62, 1631 (1989).
12. A.S. Borovik-Romanov, Yu.M. Bunkov, V.V. Dmitriev, Yu.M. Mukharskii and K. Flachbart, Sov. Phys. JETP 61, 1199 (1985).
13. K.K. Likharev, Rev. Mod. Phys. 51, 101 (1979).
14. O. Avenel and E. Varoquaux, Phys. Rev. Lett. 60, 416 (1988).
15. A.V. Markelov, Sov. Phys. JETP 67, 520 (1988).
16. T. Ohmi, M. Tsubota and T. Tsuneto, Jap. J. Appl. Phys. 26, Suppl. 26-3, 169 (1987).
17. I. Fomin, Sov. Phys. JETP 67, 1148 (1988).

ESCAPE FROM A METASTABLE POTENTIAL IN DISSIPATIVE QUANTUM AND CLASSICAL SYSTEMS

Dana A. Browne
Louisiana State University, Baton Rouge, LA 70803

Vinay Ambegaokar and Ken S. Chow
Cornell University, Ithaca, NY 14853

ABSTRACT

The resistively shunted Josephson Junction with an applied DC bias current has become a paradigm for studying the tunneling of a quantum object out of a metastable well in the presence of a dissipative environment. We use a simple kinetic theory approach to discuss the results of several experiments in the classical and the quantum regime, as well as the crossover between them. We also describe the enhancement of the tunneling in an applied microwave field.

INTRODUCTION

The escape of a particle out of a confining region through thermal activation over a barrier or quantum mechanical tunneling is a process that occurs in a variety of physical situations. Examples are alpha-particle decay of nuclei, nucleation and growth phenomena, hopping transport in disordered systems, and, of most importance in this paper, tunneling of a Josephson junction from the zero voltage state to a finite voltage state. Many of these phenomena occur in the purely classical regime, while others, for example the last two, can be purely quantum in nature. What will concern us here is the crossover from the classical to the quantum regime, and the effect of a time-dependent driving term on the escape rate. While this subject is quite old[1], the recent work by Caldeira and Leggett[2] revived interest in this field, particularly concerning the quantum regime and the effect of dissipation on the quantum tunneling process. For the most part, we will discuss only the low-damping regime, because that is where we have the best understanding of the physical processes.

The theoretical effort in this subject has been stimulated by the ability to produce ultrasmall Josephson junctions where the quantum fluctuations induced by the Coulomb charging of the junction become appreciable. This has led to a number of fascinating experiments on tunneling processes in quantum systems in the presence of dissipation. Let us recall briefly how a Josephson junction can act as a quantum particle in a tilted cosine potential,[3] in particular how the resistively shunted junction (RSJ model) can act quantum mechanically.

© 1989 American Institute of Physics

Consider a junction driven by an external current I_{ext} with a resistance produced by quasiparticle tunneling or an actual resistive shunt. The current in the junction is divided into the Josephson tunneling current, the current through the resistance and the displacement current. The equations of motion of the junction are therefore

$$C\frac{dV}{dt} + \frac{V}{R} + I_c \sin(\phi) = I_{ext}$$
$$\frac{\hbar}{2e}\frac{d\phi}{dt} = V \quad (1)$$

where C is the junction capacitance, R is the resistance, I_c is the critical current of the junction, V is the voltage, and ϕ is the phase difference of the superconducting order parameter across the junction.

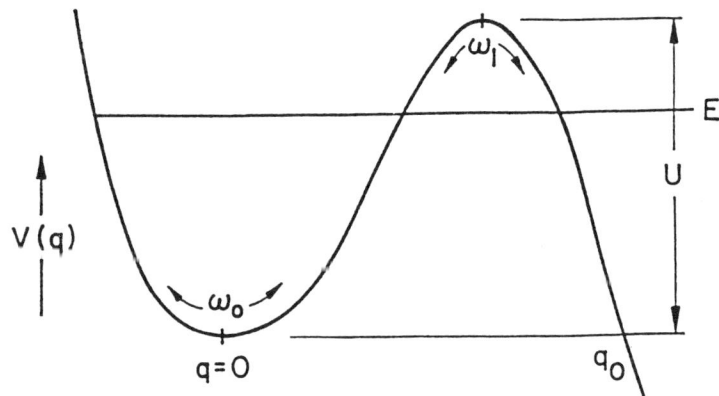

Fig. 1. Generic metastable potential with barrier height U and curvature ω_o and ω_1 and the bottom and top respectively.

In order for a definite phase difference to appear across a junction, the Cooper pairs must tunnel through the barrier. So the difference in the numbers of Cooper pairs on each side must be indefinite. Therefore the pair difference, or equivalently the charge on the junction, is a conjugate variable to the phase and has a Heisenberg uncertainty relation with the it.

In the absence of the damping term, the equations of motion of the junction can be derived from the Hamiltonian given by

$$H_o = \frac{p^2}{2m} + m\omega_J^2(1 - \cos(\phi)) - \frac{\hbar}{2e}I_o\phi \quad (2)$$

where $m = (\hbar/2e)^2 C$ is the "mass" of the quantum particle, $p = m\dot{\phi}$ is the canonical momentum and $\omega_J^2 = 2eI_c/\hbar C$ is the square of the Josephson plasma frequency. Since typically I_o is nearly equal to I_c, the potential in Eqn. (2) can be approximated by the "quadratic plus cubic" potential $V(q) = \frac{1}{2} m\omega_o^2 q^2 (1 - q/q_o)$, where the coordinate q is given by $q = \phi - \sin^{-1}(s)$, $\omega_o^2 = \omega_J^2 \sqrt{1-s^2}$, and $q_o^2 = 27(1 - \cos^{-1}(s)/\sqrt{1-s^2})$ with $s = I_o/I_c$. This potential is depicted in Fig. 1. So we see that the as the junction capacitance is lowered, the particle gets "lighter" and the actual phase across the junction becomes less well defined due to the energy cost of charging up the junction.

CLASSICAL AND QUANTUM TUNNELING

The escape of a classical particle from a metastable minimum, originally discussed by Kramers[1], is controlled by two processes: (1) acquiring sufficient energy via thermal fluctuations to mount the free energy barrier and then avoiding being driven back into the metastable well by those same fluctuations. Let us suppose the dynamics of a particle of mass m is given by a Langevin equation of the form

$$m\frac{d^2x}{dt^2} + m\gamma\frac{dx}{dt} + \frac{dV(x)}{dx} = f(t) \tag{3}$$

where γ is the damping rate, $V(x)$ is the potential shown schematically in Fig. 1, and the noise $f(t)$ obeys the fluctuation dissipation relation $\langle f(t)f(t')\rangle = 2m\gamma k_B T \delta(t-t')$. The classical escape rate is then given by an expression of the form[1]

$$\Gamma_{classical} = \frac{\omega_o}{2\pi} a_c \exp(-\frac{U}{k_B T}) \quad . \tag{4}$$

where U is the barrier height and ω_o is the frequency of oscillation in the bottom of the well, the so-called attempt frequency. The correction factor a_c, originally derived by Kramers[1], depends weakly on the damping and would be unity in a transition state model where the distribution at the top of the barrier is nearly the equilibrium distribution and every particle crossing the barrier never returns. Note that in the classical regime the dissipation affects only the attempt rate, while the exponential factor remains unchanged. Buttiker et. al.[4] have re-examined the Kramers result using a more detailed description of the behavior near the top of the barrier and found that his result remains essentially correct.

At very low temperatures, escape from the well via thermal activation becomes rare and we reach the regime of quantum tunneling. We now have a particle trapped in the lowest metastable state of the potential and we need to calculate the rate for it to tunnel out. This can be done for an isolated particle via WKB techniques. However, this quantum degree of freedom is interacting with a heat bath, and we must include the effect of the bath on the particle in such a way as to obey all the rules of quantum mechanics. Caldeira and Leggett[2], drawing on earlier work by Feynman and others[5], showed that the correct way to proceed is to treat the combined system of particle and bath quantum mechanically, and then to sum over all the bath degrees of freedom to find the particle dynamics.

We will follow the approach of Caldeira and Leggett[2] and assume that the combined system of particle and bath is governed by the Hamiltonian

$$H = H_o + \sum_i \hbar\omega_i \, a_i^\dagger \, a_i - \sum_i C_i x_i q + \sum_i \frac{C_i^2}{2m_i\omega_i} q^2 \tag{5}$$

where H_o is the Hamiltonian of the isolated particle, the second term describes the bath as a collection of harmonic oscillators, and the last two terms describe the coupling between the particle and the bath. Here C_i is the strength of the coupling, $x_i = \sqrt{\hbar/(2m_i\omega_i)} \, (a_i^\dagger + a_i)$ and q are the bath and particle coordinates respectively, and the last term is a counter term[2] which ensures that the static potential seen by the particle is the same with the bath present as when it is absent.

The dynamics of the bath is not arbitrary, because we want to ensure that the classical behavior of this system is described by the Langevin equation (3). To reproduce Eqn. (3) we must require that the spectral density of bath oscillators $J(\omega)$, defined by

$$J(\omega) = \pi \sum_i \frac{C_i^2}{2m_i\omega_i} \, \delta(\omega - \omega_i), \tag{6}$$

be given by $J(\omega) = m\gamma\omega$.

Caldeira and Leggett[2] showed in this model that the quantum tunneling rate is always smaller in the presence of dissipation. For the potential shown in Fig. 1, the quantum decay rate is given by

$$\Gamma_{quantum} = \frac{\omega_o}{2\pi} a_q \exp\left(-\frac{36}{5} \frac{U}{\hbar\omega_o}(1 + \frac{\gamma}{\omega_o}(0.87 - \alpha(\frac{k_BT}{\hbar\omega_o})^2))\right) \tag{7}$$

In the quantum case a_q is not unity but depends on the details of the barrier shape as well as the damping. The new feature is the sharp (exponential) suppression of the tunneling with damping. The last term in the exponential is the low temperature enhancement of the tunneling rate found by Grabert et. al.[6] for ohmic damping, with α a constant of order unity.

These two expressions (3) and (7) are quite different, and one subject that has been studied is the crossover behavior from one to the other. Most of the theoretical work has used an imaginary time path integral approach and has computed the tunneling via instanton[7] methods for evaluating the dominant contribution to the path integral. The interpretation of this work can be subtle since it requires an analytic continuation[7] from imaginary time to real time to produce an imaginary part of the free energy which is interpreted as the decay rate. The upshot of this work is that there is a critical temperature $T_o = (\hbar\omega_o)/(2\pi k_B)$, first noticed by Affleck[8], where the dominant path in the integral suddenly changes from a motion starting near the bottom of the well and going under the barrier, which yields Eqn. (7), to a motion localized at the top of the barrier, which produces Eqn. (3). Later calculations[9] implied that there is no real critical point but rather a rapid crossover from one behavior to the other as the temperature is changed.

KINETIC EQUATION

As useful as the imaginary time approach has been, there are a number of experiments that directly address the real-time dynamics of a metastable state that are hard to untangle from the imaginary time formalism. Therefore we will describe a real-time approach[10] that treats the dynamics by deriving an equation of motion for to the density matrix of the particle. In fact, this equation is nothing more than a generalization of the Bloch-Wangness-Redfield equations[11] used so extensively in studying magnetic resonance phenomena. There has been other work,[12] using a Kadanoff-Baym formulation[13] of transport, aimed at developing equations for the real time response. In that work, the potential felt by the particle is broken into a piece for which the path integral including the bath can be explicitly evaluated, and a remainder whose effect is accounted for in perturbation theory. We take the complementary approach of assuming that the problem of the quantum motion of the isolated particle has been solved, and study the effect of dissipation on the particle motion via an expansion in the strength of the particle-bath coupling.

The density matrix ρ of the combined system is given in the interaction representation by

$$\rho(t) = U(t,0)\rho(0)U^\dagger(t,0) \qquad (8)$$

where $U(t,t')$ is the time evolution operator. Since we are interested solely in the properties of the particle, we define the reduced density matrix $\hat{\rho}$ by $\hat{\rho}(t) = \text{Tr}[\rho(t)]$ where the Tr denotes a trace over the bath variables. In general, it is not possible to directly write a kinetic equation for $\hat{\rho}$ in terms of itself because the bath does not respond instantaneously to the motion of the particle. We therefore define a two time Green function by

$$\hat{G}(t_1,t_2) = \text{Tr}[U(t_1,0)\rho(0)U^\dagger(t_2,0)] \qquad (9)$$

which for $t_1 = t_2$ is the density matrix. We also introduce the retarded and advanced Green functions

$$g^r(t_1,t_2) = -i\Theta(t_1-t_2)\text{Tr}[U(t_1,t_2)\rho_{eq}^{bath}]$$
$$g^a(t_1,t_2) = +i\Theta(t_2-t_1)\text{Tr}[U(t_1,t_2)\rho_{eq}^{bath}] \qquad (10)$$

which describe the modification of the time evolution of the quantum particle due to its coupling to the bath. The propagators for the decoupled particle are denoted $g_o^{r,a}$.

Let us now expand U in Eqns. (9) and (10) in powers of V_I and perform the trace over the bath. The n-th term in the expansion requires the evaluation of a correlation function of n bath coordinates $x_i(t)$. Since we assume the initial density matrix $\rho(0)$ describes the bath as being in thermal equilibrium, this correlation function can be reduced to a sum over all distinct products of the two time pair correlation functions of the bath variables given by

$$\alpha(t,t') = \sum_i C_i^2 <x_i(t)x_i(t')>$$
$$= \hbar \int_{-\infty}^{\infty} \frac{d\omega}{\pi} J(\omega)(1+n(\omega))e^{-i\omega(t-t')} \qquad (11)$$

where $n(\omega) = 1/(\exp(\beta\hbar\omega)-1)$ is the Bose distribution function and $J(\omega)$ is defined in Eqn. (6). We find that $g^{r,a}$ and $\hat{G}$ obey the Dyson equations

$$g^{r,a}(t_1,t_2) = g_o^{r,a}(t_1,t_2)$$
$$+ \int_{-\infty}^{\infty} dt_1' \int_{-\infty}^{\infty} dt_2' \, g_o^{r,a}(t_1,t_1') \, \sigma^{r,a}(t_1',t_2') \, g^{r,a}(t_2',t_2)$$
$$\hat{G}(t_1,t_2) = g^r(t_1,0)\hat{\rho}(0) \, g^a(0,t_2) \qquad (12)$$
$$+ \int_0^{\infty} dt_1' \int_0^{\infty} dt_2' \, g^r(t_1,t_1') \, \hat{\sigma}(t_1',t_2') \, g^a(t_2',t_2)$$

The above equations have the same structure as those used in the quantum kinetic theory of many-body transport[13], the principal difference being that the self-energies $\sigma^{r,a}$ and $\hat{\sigma}$ in our formalism have no terms corresponding to "exchange" or "particle-hole annihilation" processes because we are treating one particle, not several, and there is no way to impose quantum statistics upon it. With this one difference, it is obvious that the Feynman diagrammatic approach can be applied here as well.

In order to express the above equations in the form of a kinetic equation, we differentiate Eqn. (12) with respect to t_1 and t_2 and add to find a kinetic equation

$$i(\frac{\partial}{\partial t_1}+\frac{\partial}{\partial t_2})\hat{G}(t_1,t_2) - [H_o + \text{Re }\sigma, \hat{G}] - [\hat{\sigma}, \text{Re }g] = \frac{i}{2}\{A,\hat{\sigma}\} - \frac{i}{2}\{\Gamma,\hat{G}\} \qquad (13)$$

where the brackets denote a commutator and the braces an anticommutator. We have defined $g^{r,a} = \text{Re } g \mp iA/2$, and $\sigma^{r,a} = \text{Re } \sigma \mp i\Gamma/2$.

While the above results are formally exact, they do not present us with a convenient calculational scheme. To compare with experiments, we go only to leading order in the damping, which is equivalent a "Born approximation" in the scattering of the particle by the bath, where the self energies $\sigma^{r,a}$ and $\hat{\sigma}$ are given by

$$\sigma^{r,a}(t_1,t_2) = \alpha(t_1,t_2)\, q_{op}\, g^{r,a}(t_1,t_2)\, q_{op} + \delta(t_1-t_2) \int_0^\infty \frac{d\omega}{\pi\omega} J(\omega)\, (q_{op})^2$$

$$\hat{\sigma}(t_1,t_2) = \alpha(t_2,t_1)\, q_{op}\, \hat{G}(t_1,t_2)\, q_{op} \tag{14}$$

where q_{op} is the position operator of the particle. With a fair amount of labor similar to that used to derive[13] quantum Boltzman equations for transport, we can reduce the above equations to evolution equations for the density matrix itself. The interested reader is referred to Ref. 10 for the precise details. The form of the evolution equation is

$$\frac{\partial}{\partial T}\hat{\rho}_{mn}[T] = -2\pi i \sum_k \{\sigma^r_{mk}[T]\,\hat{\rho}_{kn}[T] - \hat{\rho}_{mk}[T]\,\sigma^a_{kn}[T]$$
$$+ i\hat{\sigma}_{mn}[T]\} \tag{15}$$

where the labels denote quantum states and where the self energy terms are given by

$$\sigma^r_{nm}[T] = \int \frac{d\omega'}{2\pi^2} J(\omega')(1+n(\omega')) \frac{<n|q_{op}|k><k|q_{op}|m>}{E_n - \omega' - E_k + i0^+}$$
$$+ \int_0^\infty \frac{d\omega'}{\pi\omega'} J(\omega') <n|(q_{op})^2|m> e^{i(E_n-E_m)T} \tag{16}$$
$$+ <n|H_{ext}(T)|m> e^{i(E_n-E_m)T}$$

$$\hat{\sigma}_{nm}[T] = \frac{1}{\pi} J(E_n - E_k) n(E_n - E_k) <n|q_{op}|k>\, \hat{\rho}_{kl}[T]\, <l|q_{op}|m>$$

Equations (15) and (16) constitute a kinetic equation akin to the master equation. They resemble the Bloch-Wangness-Redfield[11] equations of magnetic resonance theory, but are more inclusive because they do not discard nonresonant couplings between levels as is usually done. These nonresonant terms are crucial in deriving the classical Fokker-Planck[1] equation from this quantum kinetic equation. To see the similarity to the master equation, suppose we ignore the self-consistency in the self-energies. Then the real part of σ_{nn} is given by

$$2\pi \text{Re } \sigma_{nn} = \sum_k |\langle n|q_{op}|k\rangle|^2 \int_0^\infty \frac{d\omega}{\pi} J(\omega) \{\frac{1+n(\omega)}{E_n - E_k - \omega}$$
$$+ \frac{n(\omega)}{E_n - E_k + \omega} + \frac{1}{\omega}\} \tag{17}$$

which is nothing but the energy shift δE_n caused by the bath as calculated in second-order perturbation theory. Furthermore, let us define the usual Golden Rule transition rates from state n to state m as

$$W_{n\to m} = 2|<n|q|m>|^2 J(|\omega_{nm}|) \times \left[(1+n(\omega_{nm}))\Theta(\omega_{nm})+n(\omega_{mn})\Theta(\omega_{mn})\right]. \tag{18}$$

These transition rates satisfy the detailed balance condition $W_{n\to m}/W_{m\to n} = \exp(-(E_m - E_n)/k_B T)$. We find from the kinetic equation (15) that the diagonal elements of the density matrix obey

$$\frac{d\hat{\rho}_{nn}(T)}{dT} = \sum_m W_{m\to n}\hat{\rho}_{mm}(T) - (\sum_m W_{n\to m})\hat{\rho}_{nn}(T) \\ - i\langle n|[H_{ext}(T),\hat{\rho}(T)]|n\rangle \tag{19}$$

where the scattering in and scattering out terms are clearly seen. The off-diagonal components of $\hat{\rho}$ evolve according to

$$\frac{d\hat{\rho}_{nm}(T)}{dT} = -\frac{1}{2}\sum_k (W_{m\to k} + W_{n\to k})\hat{\rho}_{nm}(T) - i(\delta E_n - \delta E_m)\hat{\rho}_{nm}(T) \\ - i\langle n|[H_{ext}(T),\hat{\rho}(T)]|m\rangle. \tag{20}$$

The net effect of the self-consistency is to alter the actual value of the relaxation rates and energy shifts, but the form of equations (19) and (20) is not changed significantly. These equations are simply the master equation description of the quantum system, and have also been examined by Larkin and Ovchinnikov[14]. In deriving these final equations (19) and (20) we have assumed that the eigenvalues do not have any regularity in their spacing. and used a "rotating wave" approximation. For the harmonic oscillator, or for a nearly harmonic system, this is not true and we find additional terms for the off-diagonal elements which represent the transfer of coherence between one pair of levels and another pair. These terms turn out only to be important in the classical limit where many levels are nearly in resonance. In that case, the full form of equation (16) for $\hat{\sigma}_{mn}$ must be used in the equation for the off-diagonal elements.

To describe the dynamics of the particle in the well and its escape we use as quantum levels the metastable levels in the well. To these equations we must also add terms $-\frac{1}{2}(\gamma_j+\gamma_k)\hat{\rho}_{jk}$, where γ_j is the tunneling rate of the $j-th$ metastable level. We will find below that in order to describe the behavior completely, we need to also include one or two "autoionizing" resonant levels *above* the barrier. The matrix elements are calculated semiclassically and checked against exact calculations from the solution of Schrodinger's equation for the uncoupled particle.

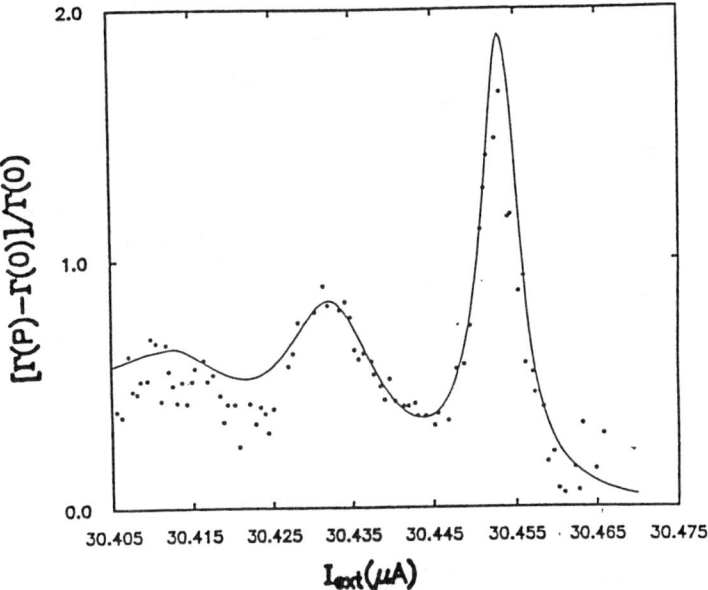

Fig. 2. Fit to the microwave enhancement data of Martinis et. al. The junction parameters are taken from the experiment: $I_c = 30.572\mu A, C = 47pF, R = 134.7\Omega, \omega_o/2\pi = 2.0 GHz$ and $T = 28mK$.

RESULTS

We have used Eqns. (19 20) to describe the results of recent experiments by Martinis, Devoret, Esteve and Clarke[15] on the enhancement of the tunneling rate in a microwave field. In those experiments they observed an enhancement of the tunneling rate whenever the level spacing of the metastable levels equalled the applied microwave power. We solved the equations numerically, which required keeping typically 6 or fewer levels in the well for their experimental parameters. The results are shown in Fig. 2. In the temperature regime of the experiments, the tunneling occurs essentially only at the top of the well, because they are higher in temperature than the critical temperature discussed above. However this is not in the classical regime, since typically there are only 4 levels in the well. Hence we see that in this regime we are seeing a quantized version of the classical escape process, where the microwaves disturb the population of the low-lying states in resonance, and the coupling to the bath provides the mechanism to communicate the presence of this disturbed distribution to the upper levels where the decay occurs. We do not see the direct tunneling from the upper resonant level as might be expected. So the rate would become exponentially small as the bath coupling decreased. We have used the experimental parameters given by Martinis et.

al.[15] to calculate the observed enhancement of the decay rate. The only free parameter is magnitude of the applied power, although we did need to use a slightly different I_c than they quoted to get good fits, although the difference was within the error they quoted.

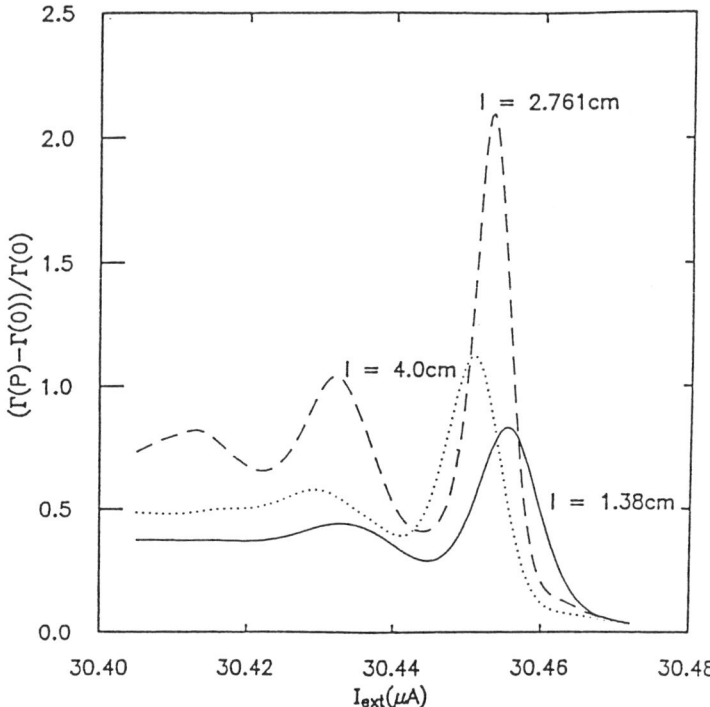

Fig. 3. Microwave enhancement of the escape rate vs. bias current for a transmission line. ℓ is the length of the transmission line.

In the actual experimental setup[15], the load across the junction was actually supplied by a resistively terminated transmission line. As a result, there is now a frequency-dependent damping in the equation of motion for the junction. The effect of this in the classical regime has been discussed by Grabert and Linkwitz[16], and the dependence of the decay without the microwave field is discussed by Devoret et. al.[17]. In the quantum regime, the only effect a change in the form of $J(\omega)$ to

$$J(\omega) = \left(\frac{\hbar}{2e}\right)^2 \omega \text{Re}\left(\frac{1}{Z_{load}(\omega)}\right) \qquad (21)$$

where Z is the complex impedance of the transmission line. As a result, the energy levels are now shifted by the bath as well as broadened. We show this result in Fig. 3.

Finally, we have examined[18] the classical limit by keeping a large number of levels (110) and simply following the dynamics on the computer. We have also found that by starting with these generalized Redfield equations (15-16), expanding the Bose factor in powers of $\hbar$ and using sum rules to evaluate the moments of the self energy, that one can produce the low damping limit of the classical Fokker-Planck equation[1]. In this derivation it is essential to start with Eqn (16) and not the nonresonant version Eqns. (19,20). It is reassuring that our results also agree well with the resonant activation data of Devoret et. al.[19] shown in Fig 4.

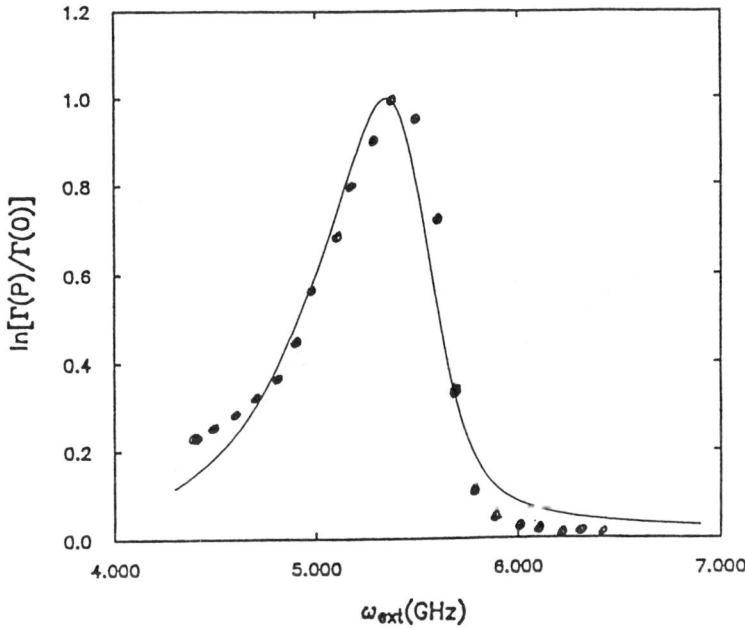

Fig. 4. The microwave enhancement of the classical tunneling rate vs. frequency ω_{ext} at fixed bias current I_b. The dots are data from Ref. 19 and the solid curve is our result. The junction parameters are $I_c = 3.42\mu A, I_b = 1.94\mu A, C = 6.75pF, R = 95\Omega$ and $T = 4.2K$.

So we see that the real time approach is very versatile even if it requires extensive numerical work to implement at present. There are a number of interesting dynamical properties that it can predict with good accuracy, and we see that it can reproduce the information gathered in the imaginary time approach, while also rendering dynamical information more easily than the imaginary time formalism. We am deeply indebted to John Martinis and Michel Devoret for their interest and for supplying their data in advance of publication. This research was supported in part by the National Science Foundation under Grant No. DMR 84-17555.

REFERENCES

1. H. A. Kramers, *Physica (Utrecht)* **7**, 284 (1940).
2. A. O. Caldeira and A. J. Leggett, *Ann. Phys. (New York)* **149**, 374 (1983).
3. W. C. Stewart, *Appl. Phys. Lett.* **12**, 277 (1968); D. E. McCumber, *J. Appl. Phys.* **39**, 3133 (1968).
4. M. Buttiker, E. P. Harris and R. Landauer, *Phys. Rev. B* **28**, 1268 (1983).
5. R.P. Feynman, F.L. Vernon and R.W. Hellwarth, *J. Appl. Phys.* **28**, 49 (1957).
6. H. Grabert, U. Weiss and P. Hanggi, *Phys. Rev. Lett.* **52**, 2193 (1984).
7. J. S. Langer, *Ann. Phys. (New York)* **41**, 108 (1967); S. Coleman in *The Whys of Subnuclear Physics*, ed. A. Zichichi, Plenum Press, 1979.
8. I. Affleck, *Phys. Rev. Lett.* **46**, 388 (1981).
9. A. I. Larkin and Yu. Ovchinnikov, *Sov. Phys. JETP* **58**, 876 (1983); H. Grabert and U. Weiss, *Phys. Rev. Lett.* **53**, 1787 (1984); H. Grabert, P. Oschowski and U. Weiss, *Phys. Rev. B* **32**, 3348 (1985).
10. D. A. Browne, K. S. Chow and V. Ambegaokar, *Phys. Rev. B* **35**, 7105 (1987); K. S. Chow, D. A. Browne and V. Ambegaokar, *Phys. Rev. B* **37**, 1624 (1988); K. S. Chow, thesis (Cornell University, 1988).
11. C. P. Schlichter, *Principles of Magnetic Resonance*, Springer, New York, 1985, Ch. 6.
12. A. Schmid, *J. Low Temp. Phys.* **49**, 609 (1982); U. Eckern and F. Pelzer, *Europhys. Lett.* **3**, 131 (1987).
13. D.C. Langreth, in *1975 NATO Advanced Study Institute on Linear and Nonlinear Electron Transport in Solids*, edited by J.T. de Vreese and E. van Boren (Plenum, New York, 1976), p. 3.
14. A.I. Larkin and Yu. N. Ovchinnikov, *J. Low Temp. Phys.* **63**, 317 (1986); *Sov. Phys. JETP Lett.* **64**, 185 (1986). See also W. Bialek, S. Chakravarty and S. Kivelson, *Phys. Rev. B* **35**, 120 (1987).
15. J.M. Martinis, M.H. Devoret and J. Clarke, *Phys. Rev. Lett.* **55**, 1543 (1985).
16. H. Grabert and S. Linkwitz, preprint.
17. M. Devoret, E. Turlot, D. Esteve, C. Urbina and J. M. Martinis, *Bull. Am. Phys. Soc.* **34**, 561 (1989).
18. K. S. Chow and V. Ambegaokar, submitted to Phys. Rev. B; D. A. Browne and V. Ambegaokar, unpublished.
19. M. H. Devoret, D. Esteve, J. M. Martinis, A. Cleland and J. Clarke, *Phys. Rev. B* **36**, 58 (1987).

SUPERFLUID PHASE SLIPPAGE IN ^{3}He FLOW THROUGH A NARROW CHANNEL

M.M. Salomaa and N.B. Kopnin

Low Temperature Laboratory, Helsinki University of Technology, Finland

L.D. Landau Institute for Theoretical Physics, 117334 Moscow, USSR

In this contribution, we present an investigation of phase slippage[1] of superfluid ^{3}He flowing through narrow channels. Other workers[2-6] have also made attempts to theoretically understand this system. Our work is based on the numerical integration of time-dependent Ginzburg-Landau equations. The quantum-dynamical phase-slip processes in ^{3}He are found to be associated with superfluid-core phase-slip centers (PSC) - in contrast with the resistive state for superconductors[7] where phase-slip centers have normal cores. Just like quantized vortices[8], PSC structures in ^{3}He could prove more diverse than those in superconductors.

We find states of superfluid ^{3}He displaying phase slippage[9] by direct solution of the dynamic equations for the order-parameter field $A_{\alpha i}$, where α and i denote the spin and orbital indices, respectively. We employ time-dependent Ginzburg-Landau (TDGL) equations[6], which can be derived microscopically for gapless superfluid ^{3}He; one can use them to model the order-parameter dynamics in superfluid ^{3}He. We consider all quantities as functions of only one spatial coordinate z and of time t.

The order parameter in the channel can differ significantly from that in the bulk liquid. Therefore, the particular geometry is crucial, see Fig. 1. The width of the channel (along x) is assumed large: $a \gg \xi_0$, hence $A_{\alpha x}$ and $A_{\alpha z}$ are both finite. This is one of the simplest restricted geometries, yet preserving the fundamental multi-component order-parameter structure for ^{3}He.

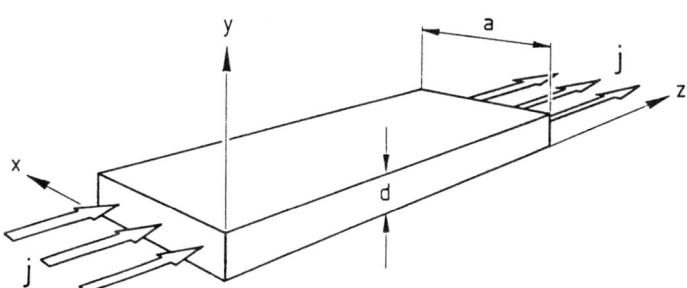

Figure 1. The flow-channel geometry considered.

© 1989 American Institute of Physics

We assume the total current j_z through the channel to have a known constant value, i.e., the fluid is incompressible. For low currents, the system has stationary homogeneous nondissipative solutions with constant p_x and p_z, constant $\partial \varphi_x / \partial z$ and $\partial \varphi_z / \partial z$, and chemical potential $\mu = 0$. Two classes of solutions exist: (i) **Class 1** having $\varphi_z = \varphi_x + \pi/2 + \pi k$ with k an integer. For $j_z = 0$, this is an A phase with critical current $j_{cr} = \frac{2}{9}$. (ii) **Class 2** is polar phase with two subclasses (a) $p_x = 0, p_z \neq 0$, with $j_{cr} = \frac{2}{9}$, and (b) $p_z = 0, p_x \neq 0$, with $j_{cr} = \frac{2}{9\sqrt{3}}$. For currents $0 \leq j \leq j_{cr} = \frac{2}{9}$, the lowest free energy occurs for the A phase of Class 1. The highest free energy corresponds to Class 2(b).

We specify a constant value for j_z and an initial distribution for the complex variables $a_x(z,0)$ and $a_z(z,0)$, compatible with the boundary conditions, and we evaluate the functions $\mu(z,t), a_x(z,t)$, and $a_z(z,t)$. We calculate the distribution of $\mu(z)$, which then is input for the integration of the order-parameter field, $A_{\alpha i}$. Iterations continue until transients decay and a certain periodic limit-cycle solution is attained. Stability of the one-dimensional solutions with respect to the formation of vortices under dc currents deserves further investigation, as well as a theoretical study of the ac Josephson effect[10] in superfluid ^{3}He.

The most interesting new features of the phase slips in superfluid ^{3}He are their superfluid cores, which are in contrast to PSCs in usual superconductors. This result is due to the multi-component ^{3}He order parameter, and may prove relevant also for heavy-fermion and high-T_c superconductors.

We thank G.E. Volovik for useful discussions. This research has been supported through funds for the Advancement of European Science by the Körber Foundation (Hamburg, FRG) and by the Soviet Academy of Sciences and the Academy of Finland through the project ROTA.

REFERENCES

1. P.W. Anderson, *Rev. Mod. Phys.* **38**, 298 (1966).
2. M. Monien and J. Tewordt, *J. Low Temp. Phys.* **62**, 277 (1986); *Can. J. Phys.* **65**, 1388 (1987).
3. E.V. Thuneberg, *Europhys. Lett.*, to be published.
4. J.R. Hook, *Jpn. J. Appl. Phys.* **26**, Suppl. 26-3, 159 (1987).
5. D. Rainer and J.A. Sauls, *Jpn. J. Appl. Phys.* **26**, Suppl. 26-3, 1804 (1987).
6. N.B. Kopnin, *Pis'ma Zh. Eksp. Teor. Fiz.* **43**, 451 (1986); *J. Low Temp. Phys.* **65**, 433 (1986).
7. B.I. Ivlev and N.B. Kopnin, *Adv. Phys.* **33**, 47 (1984).
8. M.M. Salomaa and G.E. Volovik, *Rev. Mod. Phys.* **59**, 533 (1987).
9. N.B. Kopnin and M.M. Salomaa, to be published.
10. O. Avenel and E. Varoquaux, *Phys. Rev. Lett.* **60**, 416 (1988).

SUPERFLUID ^{3}He

CHAIRMAN

Douglas F. Brewer
School of Mathematics and Physical Sciences
University of Sussex

This session was made possible, in part, through
a generous donation provided by

 EG&G PRINCETON APPLIED RESEARCH
P.O. BOX 2565, PRINCETON, NEW JERSEY 08540

The advanced design line
of RF power amplifiers

100 Highpower Road
Rochester, NY 14623
Telephone (716) 427-8300
Telex 6711542 ENI UW
Telfax (716) 427-7839

SCATTERING OF BALLISTIC QUASIPARTICLES AT THE A-B PHASE BOUNDARY OF SUPERFLUID ^{3}He

N. Schopohl and D. Waxman[a]
Institut Laue-Langevin, 156X, 38042 Grenoble Cedex, France

ABSTRACT

We present quasiclassical calculations of the transmission and Andreev reflection of ballistic wave packets of excitations incident onto the A-B phase boundary.

Superfluid ^{3}He is known to exist in two quite distinct phases: the A and the B phase. In the absence of a magnetic field the two superfluid phases coexist at a particular temperature T_{AB}, which is a function of pressure[1,2].

In each phase the total spin S and the orbital angular momentum L of the Cooper pairs is unity. However in the A phase the pairs are oriented along a common angular momentum direction $\hat{\ell}$. The result is that the A phase is highly anisotropic, in contrast to the B phase which is isotropic in the absence of orienting fields[3].

In the present work we consider a planar A-B phase boundary lying in the x=0 plane. We model the exact profile of the order parameter[4] by a piecewise constant complex vector

$$\underline{\Delta}(x,\underline{\hat{k}}) = \begin{cases} \Delta_A \underline{\hat{k}} \cdot (\underline{\hat{\phi}}_I + i\underline{\hat{\phi}}_{II})\hat{w} & \text{for } x < 0 \\ \Delta_B \underline{\hat{k}} & \text{for } x > 0 \end{cases} \qquad (1)$$

where $\underline{\hat{\phi}}_I + i\underline{\hat{\phi}}_{II} = \hat{x} + i\hat{z}$, $\hat{w} = \hat{x}$, $\hat{\ell} = \underline{\hat{\phi}}_I \times \underline{\hat{\phi}}_{II} = -\hat{y}$ as illustrated in Fig. 1.

This correctly describes the known asymptotic orientation of the order parameter in equilibrium and provides an adequate description of phenomena occurring on length scales large compared to the phase boundary thickness $\Lambda \simeq 3\xi$.

In what follows we determine waves of low energy excitations moving in the order parameter Eq. (1). We work to quasiclassical accuracy, in the limit of a low excitation density and neglect quasiparticle collisions[5,8]. With these restrictions the advanced, retarded and Keldysh propagators $\{\hat{g}^A, \hat{g}^R, \hat{g}^K\}$ of the quasiclassical theory all obey the same transport-like differential equation[5,6]:

$$iv_F \underline{\hat{k}} \cdot \underline{\nabla} \hat{g}(\underline{R}, t; \underline{\hat{k}}, \epsilon) + [\hat{\mathcal{H}}, \hat{g}]_- + \frac{i}{2}[\hat{\tau}_3, \partial_t \hat{g}]_+ = \hat{0} \qquad (2)$$

Apart from satisfying the above equation $\hat{g}^{R(A)}$ is subject to the necessary condition of analyticity in the upper (lower) complex ϵ plane. In equilibrium ($\partial_t \hat{g} = \hat{0}$) the propagators are related by

$$\hat{g}^K_{eq} = \tanh\frac{\epsilon}{2T}[\hat{g}^R_{eq} - \hat{g}^A_{eq}], \qquad (3a)$$

---

[a] School of Mathematical and Physical Sciences, University of Sussex, Brighton BN1 9QH, Sussex, U.K.

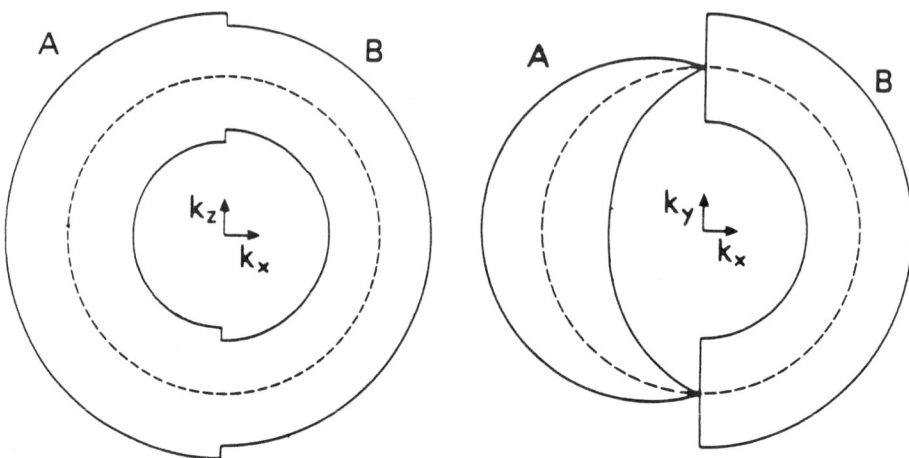

Fig. 1. Schematic cross sections of the Fermi sphere (dashed lines) indicating the maximum and minimum bulk gaps on either side of the A-B interface.

where T is the absolute temperature. Furthermore[6], the propagators $\hat{g}^{R(A)}$ are constrained by

$$\hat{g}^R_{eq} \cdot \hat{g}^R_{eq} = -\pi^2 = \hat{g}^A_{eq} \cdot \hat{g}^A_{eq}. \tag{3b}$$

The elementary excitations described by $\hat{g}$ are characterized by 4 labels: p↑, p↓, h↑, h↓: p, h stand for particle or hole and ↑, ↓ for spin up or down. A complete specification furthermore requires ϵ, a small deviation from the Fermi energy and $\hat{k}$, a unit vector giving the direction of the excitation momentum at position $\underline{R}$ and time t. (Further details and a beautiful physical discussion are given in Ref. 6.) It follows that $\hat{g}, \hat{\mathcal{H}}, \hat{\tau}_3$ are 4 × 4 matrices given in 2 × 2 subblock form as

$$\hat{\mathcal{H}} = \begin{pmatrix} \epsilon & \Delta \\ -\Delta^+ & -\epsilon \end{pmatrix}, \quad \hat{\tau}_3 = \begin{pmatrix} 1 & 0 \\ 0 & -1 \end{pmatrix} \tag{4}$$

with $\Delta = \underline{\Delta}(x, \hat{k}) \cdot \vec{\sigma} i \sigma^y$ and $\{\sigma^x, \sigma^y, \sigma^z, 1\}$ the Pauli matrices.

We look for solutions of (2) describing small deviations from equilibrium of the form

$$\hat{g}(\underline{R}, t; \hat{\underline{k}}, \epsilon) = \int_{-\infty}^{\infty} \frac{d\omega}{2\pi} e^{-i\omega t} \hat{G}(\underline{R}, \omega; \hat{\underline{k}}, \epsilon) \tag{5a}$$

Then $\hat{G}(\underline{R}, \omega; \hat{\underline{k}}, \epsilon)$ satisfies

$$iv_F \hat{\underline{k}} \cdot \underline{\nabla} \hat{G} + \hat{\mathcal{H}}(\omega) \hat{G} - \hat{G} \hat{\mathcal{H}}(-\omega) = \hat{0} \tag{5b}$$

with

$$\hat{\mathcal{H}}(\omega) = \hat{\mathcal{H}} + \frac{\omega}{2} \cdot \hat{\tau}_3. \tag{6a}$$

In our case, $\underline{\Delta} \times i\underline{\Delta}^* = \underline{0}$, hence $[\hat{\mathcal{H}}(\omega)]^2 = [E(\omega)]^2 \hat{1}$ with

$$E(\omega) = [(\epsilon + \frac{\omega}{2})^2 - \underline{\Delta} \cdot \underline{\Delta}^*]^{1/2} \tag{6b}$$

an eigenvalue of $\hat{\mathcal{H}}(\omega)$. Defining

$$\hat{E}(\omega) = E(\omega)\hat{\tau}_3, \quad \hat{K}(\omega) = (\hat{E}(\omega) + \hat{\mathcal{H}}(\omega))/\{2E(\omega)[E(\omega) + \epsilon + \frac{\omega}{2}]\}^{1/2} \tag{6c}$$

it is found that $\hat{K}(\omega)$ diagonalizes $\hat{\mathcal{H}}(\omega)$:

$$\hat{\mathcal{H}}(\omega)\hat{K}(\omega) = \hat{K}(\omega)\hat{E}(\omega) \quad \text{and} \quad \hat{K}(\omega)\hat{K}(\omega) = \hat{1}. \tag{6d}$$

Because we work with a piecewise constant order parameter the general solution of Eq. (5b) may be derived in closed form; on either side of the interface it is given by

$$\hat{G}(\underline{R}, \omega; \underline{k}, \epsilon) = \hat{K}(\omega) \exp\{i\hat{\tau}_3 \underline{k}(\omega) \cdot \underline{R}\} \hat{C} \exp\{-i\hat{\tau}_3 \underline{k}(-\omega) \cdot \underline{R}(-\omega)\} \hat{K}(-\omega) \tag{7}$$

Here $\hat{C}$ is an arbitrary 4×4 matrix $\hat{C} = \begin{pmatrix} \tilde{C}_{11} \tilde{C}_{12} \\ \tilde{C}_{21} \tilde{C}_{22} \end{pmatrix}$ whose $\underline{R}$-dependence is of the form $\underline{R} \times \hat{\underline{k}}$ since $\hat{\underline{k}} \cdot \underline{\nabla} \hat{C} = \hat{0}$, and $\underline{k}(\omega) = \hat{\underline{k}} \cdot E(\omega)/v_F$ is a wave vector of the excitations. Insight can be gained into the nature of the solutions of (2) by expressing $\hat{C}$ in 2×2 block form and inserting (7) into (5a):

$$\hat{g}(R, \underline{t}; k, \epsilon) =$$

$$\int_{-\infty}^{\infty} \frac{d\omega}{2\pi} \hat{K}(\omega) \begin{bmatrix} \tilde{C}_{11} e^{i[(\underline{k}(\omega) - \underline{k}(-\omega)) \cdot \underline{R} - \omega t]}, & \tilde{C}_{12} e^{i[(\underline{k}(\omega) + \underline{k}(-\omega)) \cdot \underline{R} - \omega t]} \\ \tilde{C}_{21} e^{-i[(\underline{k}(\omega) + \underline{k}(-\omega)) \cdot \underline{R} + \omega t]}, & \tilde{C}_{22} e^{-i[(\underline{k}(\omega) - \underline{k}(-\omega)) \cdot \underline{R} + \omega t]} \end{bmatrix} \hat{K}(-\omega)$$
$$\tag{8}$$

The phases $[\underline{k}(\omega) - \underline{k}(-\omega)] \cdot \underline{R} \pm \omega t$ describe waves of excitations propagating along $\mp \hat{\underline{k}}$ while the phases $[\underline{k}(\omega) + \underline{k}(-\omega)] \cdot \underline{R} \pm \omega t$ describe nonpropagating waves due to their different symmetry under $\omega \to -\omega$.

If we consider propagating waves of low ω, incident from the B phase onto the interface ($\hat{\underline{k}} \cdot \hat{\underline{x}} < 0$), then in the A-phase these can only lead to waves propagating away from the interface, provided $|\epsilon| > \max(\Delta_B, \Delta_A | \hat{\underline{k}} \times \hat{\underline{\ell}} |)$. This implies that in the A-phase $\hat{C}$ must be of the form[5]

$$\hat{C}^A(\underline{R} \times \hat{\underline{k}}) = \begin{bmatrix} C_{11}^A e^{-i(\hat{\underline{k}} \times b_{11}^A) \cdot \underline{R}} & , & 0 \\ 0 & , & 0 \end{bmatrix} \tag{9a}$$

and correspondingly in the B-phase

$$\hat{C}^B(\underline{R} \times \hat{\underline{k}}) = \begin{bmatrix} C_{11}^B & , & C_{12}^B \cdot e^{-i(\hat{\underline{k}} \times b_{12}^B) \cdot \underline{R}} \\ C_{21}^B \cdot e^{i(\hat{\underline{k}} \times b_{21}^B) \cdot \underline{R}} & , & C_{22}^B \cdot e^{i(\hat{\underline{k}} \times b_{22}^B) \cdot \underline{R}} \end{bmatrix} \tag{9b}$$

Here all the 2×2 blocks C_{11}^A, C_{11}^B, etc. are independent of $\underline{R}$. The continuity of the scattering solutions (7) at the interface where $\underline{R} = \underline{R}_\perp = (o,y,z)$ requires (the notation is explained in Ref. (10)):

$$\hat{K}^A(\omega)\exp\{i\hat{\tau}_3\underline{k}^A(\omega)\cdot\underline{R}_\perp\}\hat{C}^A(\hat{\underline{R}}_\perp\times\hat{\underline{k}})\exp\{-i\hat{\tau}_3\underline{k}^A(-\omega)\cdot\underline{R}_\perp\}\hat{K}^A(-\omega)$$
$$= \hat{K}^B(\omega)\exp\{i\hat{\tau}_3\underline{k}^B(\omega)\cdot\underline{R}_\perp\}\hat{C}^B(\hat{\underline{R}}_\perp\times\hat{\underline{k}})\exp\{-i\hat{\tau}_3\underline{k}^B(-\omega)\cdot\underline{R}_\perp\}\hat{K}^B(-\omega)$$
(10)

This fixes for arbitrary $\underline{R}_\perp$ the b's appearing in the phase factors of Eq. (9):

$$\begin{aligned}\underline{b}_{11}^A &= \hat{\underline{x}}\times[(\underline{k}^B(\omega)-\underline{k}^B(-\omega))-(\underline{k}^A(\omega)-\underline{k}^A(-\omega))]/(\hat{\underline{k}}\cdot\hat{\underline{x}})\\ \underline{b}_{12}^B &= -2\hat{\underline{x}}\times\underline{k}^B(-\omega)/(\hat{\underline{k}}\cdot\hat{\underline{x}})\\ \underline{b}_{21}^B &= -2\hat{\underline{x}}\times\underline{k}^B(\omega)/(\hat{\underline{k}}\cdot\hat{\underline{x}})\\ \underline{b}_{22}^B &= -2\hat{\underline{x}}\times[\underline{k}^B(\omega)-\underline{k}^B(-\omega)]/(\hat{\underline{k}}\cdot\hat{\underline{x}}).\end{aligned} \quad (11)$$

Additionally, we can express all the 2×2 blocks $C_{11}^A, C_{12}^B, C_{21}^B, C_{22}^B$ in terms of the incoming weight C_{11}^B by using elementary operations on the 2×2 blocks in the following equation which is equivalent to Eq. (10).

$$\begin{pmatrix} C_{11}^B & C_{12}^B \\ C_{21}^B & C_{22}^B \end{pmatrix} = \hat{K}^B(\omega)\hat{K}^A(\omega)\cdot\begin{pmatrix} C_{11}^A & 0 \\ 0 & 0 \end{pmatrix}\cdot\hat{K}^A(-\omega)\hat{K}^B(-\omega) \quad (12)$$

The results for C_{11}^A, C_{12}^B, etc. are given below in terms of the overlaps

$$\hat{K}^B(\omega)\hat{K}^A(\omega) = \begin{pmatrix} U_{11} & U_{12} \\ U_{21} & U_{22} \end{pmatrix} \text{ and } \hat{K}^A(-\omega)\hat{K}^B(-\omega) = \begin{pmatrix} V_{11} & V_{12} \\ V_{21} & V_{22} \end{pmatrix};$$

$$\begin{aligned}C_{11}^A &= (U_{11})^{-1}C_{11}^B(V_{11})^{-1}\\ C_{12}^B &= C_{11}^B(V_{11})^{-1}V_{12}\\ C_{21}^B &= U_{21}(U_{11})^{-1}C_{11}^B\\ C_{22}^B &= U_{21}(U_{11})^{-1}C_{11}^B(V_{11}^{-1})V_{12}\end{aligned} \quad (13)$$

where the explicit forms for the 2×2 blocks $U_{11}, U_{21}, V_{11}, V_{12}$ follow from the defining equations (6).

The precise choice of the weight $C_{11}^B(\omega)$ is obtained from the initial conditions imposed on the incoming waves.

In the steady state limit $|\omega|\ll|\epsilon|$ the elements of C_{11}^B determine the number and spin of the incident quasiparticles. As an example we present in Fig. 2 the conversion probability $T_{oo}^A = \text{tr}(C_{11}^A)/\text{tr}(C_{11}^B)$ and the spin conversion probability $\vec{T}_{OS}^A = \text{tr}(C_{11}^A\vec{\sigma})/\text{tr}(C_{11}^B)$ for the transmission of excitations from an unpolarized beam in the B-phase into the A-phase.

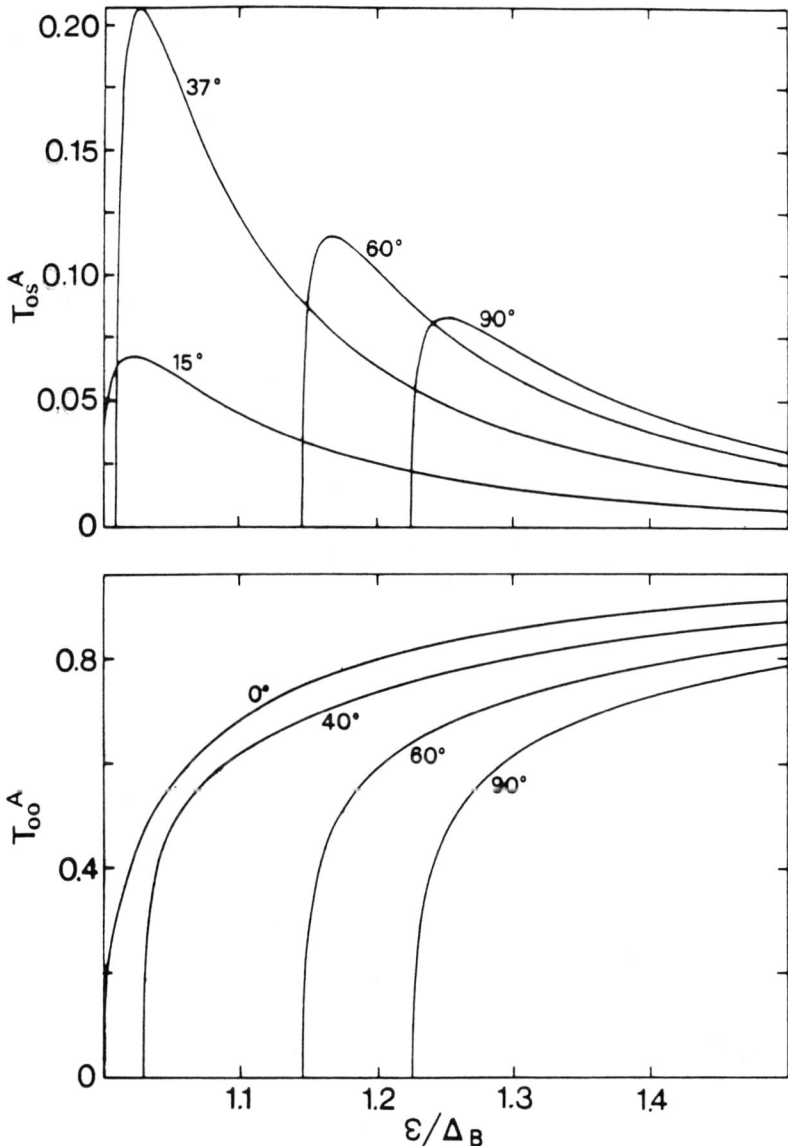

Fig. 2. Transmission of excitations from B into A for an unpolarized incoming beam: $\text{tr}(C_{11}^B \vec{\sigma}) = \underline{0}$. Conversion probability T_{oo}^A and magnitude T_{OS}^A of spin-conversion probability ($\vec{T}_{OS}^A = T_{OS}^A \underline{\hat{k}} \times \underline{\hat{x}} / |\underline{\hat{k}} \times \underline{\hat{x}}|$) are plotted as functions of scaled energy ϵ/Δ_B. The orientation of the incident $\underline{\hat{k}}$ vector is given in non-standard polar coordinates by $\underline{\hat{k}} = \cos\theta_k \underline{\hat{x}} + \sin\theta_k (\cos\phi_k \underline{\hat{y}} + \sin\phi_k \underline{\hat{z}})$. Curves are shown with $\theta_k = 135°$, $\phi_k = 0°, 40°, 60°, 90°$ for T_{oo}^A and $\theta_k = 135°$, $\phi_k = 15°, 37°, 60°, 90°$ for T_{OS}^A.

In a related way, the elements of C_{22}^B determine the number and spin of the Andreev reflected excitations. Their conversion and spin conversion probabilities are

$$R_{oo}^B = -\text{tr}(C_{22}^B)/\text{tr}(C_{11}^B) \text{ and } \vec{R}_{OS}^B = -\text{tr}(C_{22}^B \vec{\sigma})/\text{tr}(C_{11}^B).$$

From the trace of (12) we derive the sum rule $R_{oo}^B + T_{oo}^A = 1$ in the steady state limit.

The group velocity $\underline{\nabla}_q \omega(q)$ of the waves follow from the ω-dependence of their wave vectors $\underline{q}(\omega)$. It is essential to note that the full wave vectors $\underline{q}(\omega)$ of the blocks in Eq. (8) have the form following from (9) and (11):

$$\begin{aligned}
\underline{q}_{11}^B &= \underline{k}^B(\omega) - \underline{k}^B(-\omega) \\
\underline{q}_{12}^B &= \underline{k}^B(\omega) + \underline{k}^B(-\omega) - \hat{\underline{k}} \times \underline{b}_{12}^B \\
\underline{q}_{21}^B &= -(\underline{k}^B(\omega) + \underline{k}^B(-\omega)) + \hat{\underline{k}} \times \underline{b}_{21}^B \\
\underline{q}_{22}^B &= -(\underline{k}^B(\omega) - \underline{k}^B(-\omega)) + \hat{\underline{k}} \times \underline{b}_{22}^B \\
\underline{q}_{11}^A &= \underline{k}^A(\omega) - \underline{k}^A(-\omega) - \hat{\underline{k}} \times \underline{b}_{11}^A \; .
\end{aligned} \qquad (14)$$

The additional parts $\hat{\underline{k}} \times \underline{b}$ are necessary to preserve the wavevector components $\underline{q} \times \hat{\underline{x}}$ parallel to the interface[9]

$$\underline{q}_{11}^B \times \hat{\underline{x}} = \underline{q}_{12}^B \times \hat{\underline{x}} = \underline{q}_{21}^B \times \hat{\underline{x}} = \underline{q}_{22}^B \times \hat{\underline{x}} - \underline{q}_{11}^A \times \hat{\underline{x}} \; . \qquad (15)$$

A consequence of Eqs. (14) is that only the incident wave is along $\hat{\underline{k}}$ whereas the transmitted and reflected waves deviate from this direction. We interpret Eq. (15) as an analogue of Snell's law that describes the scattering of a ballistic wave packet of 1-dim nature being initially modulated along $\hat{\underline{k}}$. This wave packet splits at the A-B phase boundary into two wave packets of 1-dim nature. The Andreev reflected part[7] is modulated along $\underline{q}_{22}^B$ and the transmitted part is modulated along $\underline{q}_{11}^A$. Along the directions of $\underline{q}_{12}^B$ and $\underline{q}_{21}^B$ there are non-propagating waves (Tomasch oscillations) due to the interference of the incoming quasiparticles with the Andreev reflected quasiholes[5].

The angle of the transmitted beam θ^A vs. the incident angle θ^B may be obtained from (15). It differs qualitatively for ϵ in the range $\Delta_B \leq |\epsilon| \leq \Delta_A$ and $|\epsilon| > \Delta_A$ as shown in Fig. 3. In fact, for $|\epsilon| > \Delta_B$ transmission of waves in the A-phase is impossible for $\hat{\underline{k}}$- orientations such that $|\epsilon| < \Delta_A \cdot |\hat{\underline{k}} \times \hat{\underline{l}}|$. In this case the waves incident from the B-side undergo complete Andreev reflection[5,8]. The angle θ^R of the Andreev reflected beam is independent on ϵ and depends only on the incident angle θ^B:

$$\sin \theta^R = \sin \theta^B \cdot \cos \theta^B / (4 - 3\cos^2 \theta^B)^{1/2} \qquad (16)$$

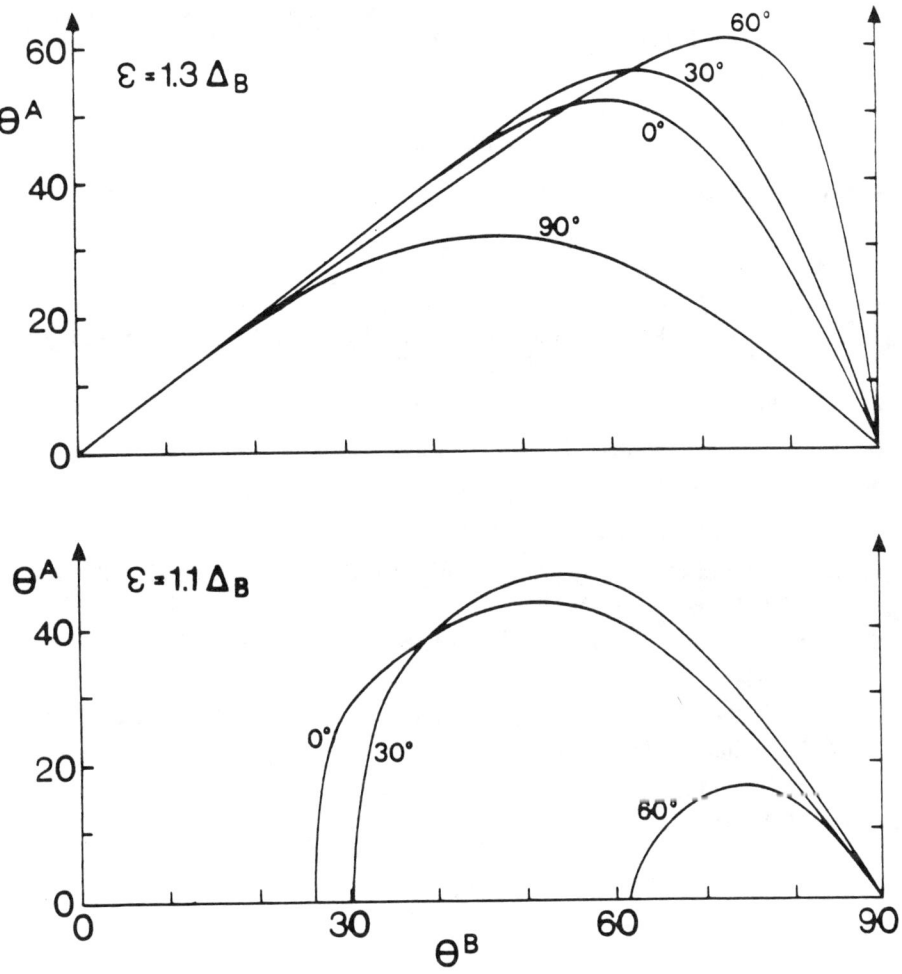

Fig. 3 Snell's law for waves incident from B into A. Curves are shown for two values of scaled energy $\frac{\epsilon}{\Delta_B} = 1.1$, 1.3 and $\phi_k = 0°$, 30°, 60°, 90°. The incident angle $\theta^B = \arcsin(|\,\hat{\underline{k}} \times \hat{\underline{x}}\,|)$ and the angle of transmission $\theta^A = \arcsin(|\,\underline{q}_{11}^A \times \hat{\underline{x}}\,|\,/\,|\,\underline{q}_{11}^A\,|)$ are measured in degrees.

We acknowledge helpful discussions with G. Barton, G. Eilenberger, J. Kurkijärvi, A.J. Leggett, P. Nozières, P. Thalmeier, E. Thuneberg and W. Zhang.

REFERENCES

1. D.D. Osheroff and M.C. Cross, Phys. Rev. Lett. $\underline{38}$, 905 (1977).
2. D.S. Buchanan, G.W. Swift and J.C. Wheatley, Phys. Rev. Lett. $\underline{57}$, 341 (1986).
3. For example: D.M. Lee and R.C. Richardson, in "The Physics of Liquid and Solid Helium, Part II" edited by K.H. Bennemann and J.B. Ketterson, John Wiley and Sons, 1978.
4. N. Schopohl, Phys. Rev. Lett. $\underline{58}$, 1664 (1987).
5. G. Kieselmann and D. Rainer, Z. Phys. B$\underline{52}$, 267 (1983).
6. J. Serene and D. Rainer, Phys. Rep. $\underline{101}$, 221 (1983).
7. A.F. Andreev, Sov. Phys. JETP $\underline{19}$, 1228 (1964).
8. S. Yip, Phys. Rev. B$\underline{32}$, 2915 (1985); Ph.D. thesis, University of Illinois at Urbana-Champaign, 1986.
9. Conservation of $\underline{q} \times \hat{\underline{z}}$ holds since the excitation momentum parallel to the interface is conserved.
10. $\hat{K}^B(\omega)$ diagonalizes $\hat{\mathcal{H}}(\omega)$ on the B-side (x > 0) of the phase boundary, $\hat{K}^A(\omega)$ diagonalizes $\hat{\mathcal{H}}(\omega)$ on the A-side (x < 0); $\pm E^{B(A)}(\omega)$ are the corresponding eigenvalues of $\hat{\mathcal{H}}(\omega)$ (see Eqs. (6a-d)) and $\underline{k}^{B(A)}(\omega) = \hat{\underline{k}} E^{B(A)}(\omega)/v_F$.

ULTRASONIC INVESTIGATION OF ROTATING SUPERFLUID ³He

J.P. Pekola, R.H. Salmelin, A.J. Manninen, K. Torizuka, J.M. Kyynäräinen, and G.K. Tvalashvili[#]

Low Temperature Laboratory, Helsinki University of Technology, 02150 Espoo, Finland
[#]Institute of Physics of the Georgian Academy of Sciences, 380077 Tbilisi, USSR

ABSTRACT

The coupling of the real squashing collective mode to the zero sound in rotating superfluid ³He-B has been investigated. In a magnetic field rotation enhances the coupling of the non-zero m_J substates, and the five or sometimes six-fold line splitting becomes observable even when $\vec{H}$ is parallel to $\vec{\Omega}$ and to the direction of sound propagation. Equilibrium vortex lattices and vortex free states can be distinguished by their characteristic absorption spectra. The data are compared with microscopic calculations and a recent phenomenological theory. In the A-phase the anisotropy of the sound attenuation serves as a powerful means to detect the $\hat{l}$ texture rearrangement when a vortex lattice is formed; measurements have been made in zero and in weak (1.7 mT) magnetic fields.

INTRODUCTION

Propagation of zero sound has recently proven to be a sensitive probe of textural changes due to counterflow and vortices in rotating ³He-B.[1] In ³He-A, our experiments show that vortex formation and vortex free flow substantially change attenuation and velocity of ultrasound, but the interpretation of the data is by no means straightforward. Measurements in both phases, with emphasis on the B-phase results, will be discussed below.

THE REAL SQUASHING MODE IN ROTATING ³He

Ultrasound in ³He-B probes several collective modes which arise from temporal oscillations of the 3 x 3 complex order parameter matrix. The possible modes can be classified according to their total angular momentum quantum number J, which may have values 0, 1, or 2. We have investigated the real J = 2 mode, also called the real squashing (rsq) mode.[2]

© 1989 American Institute of Physics

When no magnetic field is applied to a sample of ^{3}He-B, the rsq mode gives rise to one intense attenuation peak, located at $\hbar\omega = \sqrt{(8/5)}\Delta$; here Δ is the isotropic energy gap in the B-phase and $f = \omega_0/2\pi$ is the frequency of the zero sound. The rsq mode is made of substates whose resonance frequencies are given[3] in the high field limit (H > 50 mT) by

$$\omega^2 = \omega_0^2 + 2\omega_0 gHm_J + c_1^2 q^2 + (1/2)c_2^2 q^2[(1 + m_J^2/2)/3 + (\hat{q}\cdot\hat{h})^2(1 - m_J^2/2)], \tag{1}$$

where $\hat{q} = \vec{q}/q$ denotes the direction of sound propagation and $\hat{h}$, in this case, gives the local direction of the quantization axis for the rsq mode in the texture. The first of the correction terms on the right, $2\omega_0 gHm_J$, determines[2] the linear Zeeman splitting. Dispersion constants c_1^2 and c_2^2 are on the order of Fermi velocity squared; the value of c_2^2 from our experiments will be[4] compared to the theoretical weak coupling value $c_2^2 = 0.33v_F^2$).

The coupling of each of the m_J substates depends on the angle between h and q. In weak and intermediate magnetic fields (H < 50 mT) one has to solve[5] either phenomenological or fully microscopic equations numerically in order to calculate the spacings and intensities of the rsq modes.

In our experimental chamber we have two X-cut quartz-crystal transducers with the fundamental frequency of f = 8.9 MHz, L = 4 mm apart. The sound propagates along Ω, the axis of rotation. The cylindrical ultrasonic chamber, with a radius R = 3 mm, has quartz surfaces, and it is connected to the rest of the ^{3}He sample only through tiny holes in its side wall. The transmitting crystal was excited by a pulse whose duration was typically 12 μs. A dual-line phase sensitive detector decomposed the received signal into its 0- and 90-degree components, from which the attenuation and the phase of the signal were extracted.

Our main observations on the coupling of zero sound to the rsq modes in rotating ^{3}He-B have been described elsewhere.[1] Further analysis of the experimental data will be given below.

Figure 1 shows the attenuation data taken at three different pressures and frequencies around the rsq mode. The common features in each of the sets of results are: i) $m_J = 0$ peak loses intensity with increasing Ω, ii) intensities of the $m_J = \pm 1$ peaks behave nonmonotonically, iii) the $m_J = \pm 2$ peaks gain intensity when increasing Ω, and iv) a clear asymmetry between positive and negative m_J substates exists. An additional splitting of the $m_J = 0$ peak can be observed at large angular velocities (eg. at $\Omega = 4.0$ rad/s when p = 6.5 bar).

The spectra were fitted into a set of five or six Lorentzian

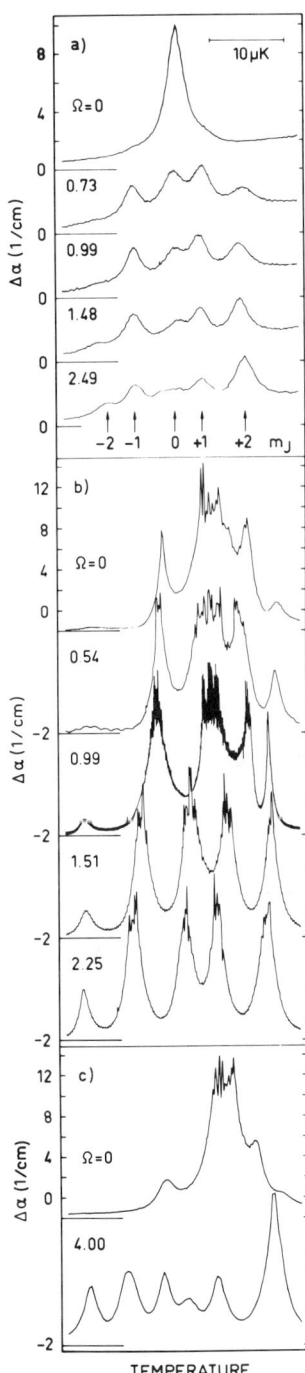

Fig. 1. Effect of rotation on the attenuation of the rsq mode $\vec{H} \parallel q \parallel \Omega$. a) Rsq spectra at p = 3.2 bar, f = 44.7 MHz and H = 32 mT. b) p = 11.5 bar, f = 80.6 MHz and H = 25 mT. c) Two spectra at p = 6.5 bar, f = 62.6 MHz and H = 25 mT demonstrating the textural splitting of the m_J=0 peak. The horizontal scale is reliable in the p = 3.2 bar data in (a); in (b) and (c) it is arbitrary. (From Ref. 1)

lineshapes

$$\Delta\alpha = \sum_{i=1}^{5,6} (A_i \delta^2)/[(t-t_i)^2 + \delta^2]$$

during a temperature sweep through the rsq-modes; t in this equation denotes real time, and $\Delta\alpha$ is the change of attenuation with respect to the attenuation at T_c. An example of an rsq spectrum and the corresponding Lorenzian fit is seen as an inset in Fig. 2.; the width δ has been forced to be equal for all 6 peaks. Figure 2 shows the intensities of the Lorentzian lines for each substate as a function of Ω at pressures p = 3.2 bar and p = 11.5 bar.

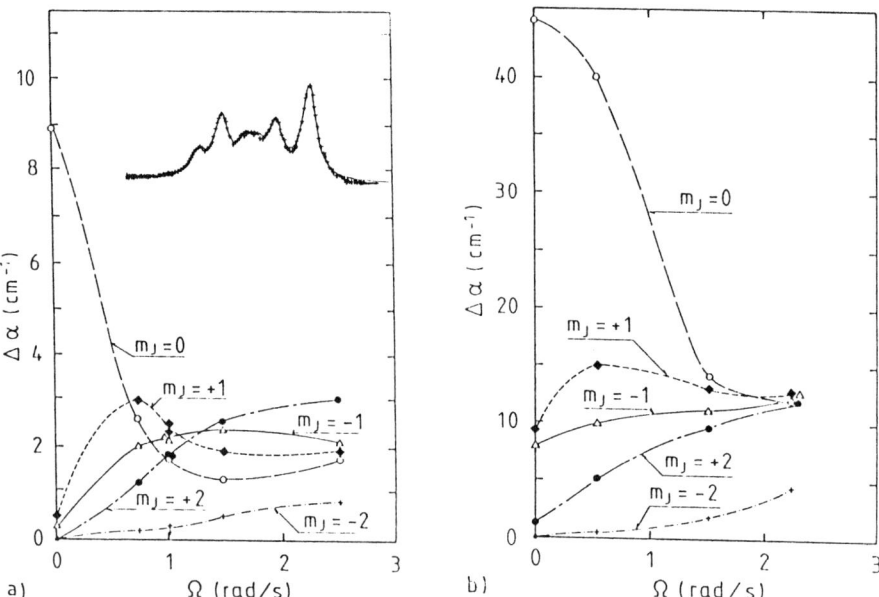

Fig. 2. Intensities of the various substate peaks as a function of Ω. a) p = 3.2 bar, f = 44.7 MHz, H = 32 mT; the inset shows the Lorentzian fit to the experimental data points used to extract the intensity of each peak. b) p = 11.5 bar, f = 80.6 MHz, H = 25 mT.

Our data allows comparisons with the existing theories on the rsq mode in weak magnetic fields and under rotation, even without a thorough textural calculation for the cylindrical chamber. When $\vec{H} \parallel q$ is applied, the external field tends to align $\hat{h}$ in the same direction. Therefore, in the absence of rotation, $\hat{h} \parallel \vec{H} \parallel q$ holds everywhere except near the walls of the experimental volume. When the sample is set into rotation, nonzero counterflow $\vec{w} = \vec{v}_s - \vec{v}_n$ is created, which tends to orient $\hat{h}$ parallel to $\vec{w}$. In the notation of Salomaa and Volovik[3] the angle γ between q and $\hat{h}$ increases in the vortex free counterflow state monotonically from $\gamma = 0$ towards $\gamma = \pi/2$ with increasing Ω.

In addition to the vortex free counterflow states, we have created vortex states with both equilibrium and non-equilibrium number of vortices. The former states are believed to be generated by rotating the sample from a temperature $T > T_c$ down to the temperature where the rsq mode is reached in the B-phase (eg. $T/T_c = 0.8$ at $p = 3.2$ bar and $f = 44.7$ MHz). Spectra of the three types are seen in Fig. 3. The intensities of the various m_J substates in the vortex states are very different from those in the vortex free flow states. This means that the effective value of γ is different in the various flow regimes at the same angular velocity of the cryostat.

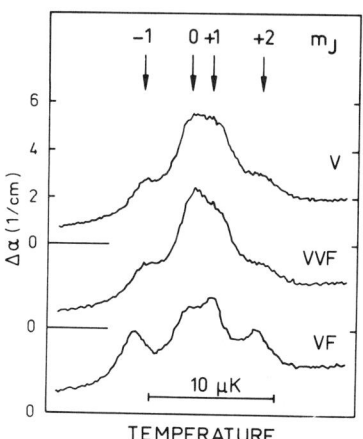

Fig. 3. Rsq mode absorption spectra of ^{3}He-B rotating at $\Omega = 0.87$ rad/s, $p = 3.2$ bar, $f = 44.7$ MHz, $H = 25$ mT. Type V spectrum represents a vortex state, whereas type VF was obtained in a vortex-free counterflow state. The VVF spectrum indicates an intermediate non-equilibrium texture. (From Ref. 1)

Besides the direct textural coupling due to the particle hole asymmetry, suggested by Koch and Wölfle (K-W),[6] there exists another mechanism, originally suggested by Sauls and Serene (S-S),[7] to couple the rsq modes to w. This effect has a different angular dependence on γ and it is proportional to w^2. The relative strength of the S-S to the K-W coupling is on the order of $\Delta^2/\epsilon_F p_F$, but the numerical value is not known accurately; a measurement thus provides new information on Fermi liquid parameters.

Figure 4 shows a collection of results from Ref. 3 which correspond to our experimental conditions. The value for H_o was obtained from the low Ω ($\gamma \simeq 0$) or high Ω ($\gamma \simeq \pi/2$) spectra where

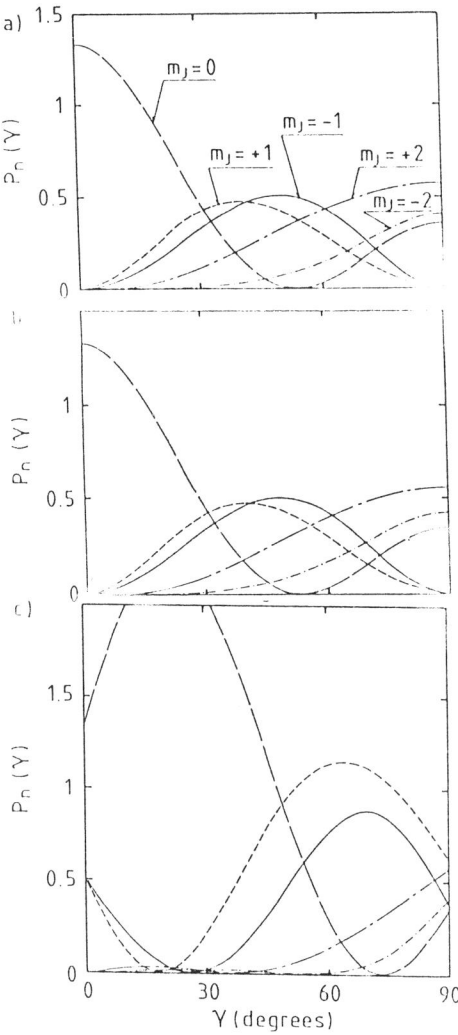

Fig. 4. Calculated curves by Salomaa and Volovik:[3] a) Relative intensities of the m_J substates at $H = 2H_o$, $w = 0$. b) $H = 2.5H_o$, $w = 0$ (M. Salomaa, unpublished). c) $H = 2H_o$, $w = w_o$.

the unequal spacing of the m_J substates yields

$$H_0/H = 2[(\omega_{-1} - \omega_{-2}) - (\omega_{+2} - \omega_{+1})]/(\omega_{+2} - \omega_{-2}), \gamma = 0$$

$$H/H_0 = [(\omega_{+2} - \omega_{+1}) + (\omega_{-1} - \omega_{-2})]/[2(\omega_{+2} - \omega_{+1}) - (\omega_{-1} - \omega_{-2})], \gamma = \pi/2$$

where ω_J is the position of the center of the m_J peak. These equations yield H_0 = 15.5 mT, 12.5 mT, and 13.0 mT at pressures p = 3.2 bar, 6.5 bar, and 11.5 bar and frequencies f = 44.7 MHz, 62.6 MHz, and 80.6 MHz, respectively. Thus, according to our data, H = 25 mT (32 mT) corresponds to $H/H_0 \simeq 2$ (2.5) at all pressures and frequencies employed. The relative intensities shown in Fig. 4a at H/H_0 = 2 and in Fig. 4b at H/H_0 = 2.5 do agree with our Ω-dependence of Fig. 2. There is, however, one inconsistency: the maximum intensity of the m_J = +1 peak exceeds the maximum intensity of the m_J = -1 peak at all experimental fields and pressures analyzed. A possible explanation is the non-zero S-S coupling, illustrated in Fig. 4c where the relative strength w/w_0 in Salomaa and Volovik's notation is set equal to 1. In this case the maximum of m_J = +1 exceeds that of m_J = -1.

A direct indication of the counterflow coupling is the "gyrosonic effect".[3] In the counterflow state $\vec{w}$ and q are at certain directions with respect to the texture in the cell. The distributions of these angles determine the attenuation due to the S-S coupling in the experiment. For a counterflow state the texture looks different if either $\vec{w}$ or q is reversed; thus by reversing Ω or q one should be able to detect a difference in the attenuation spectrum. The former method is less convenient for probing the gyrosonic effect because the texture is not completely reproducible from one rotation to another.

Our observations are very tentative so far. In the experiment sound was propagated alternately in ± Ω directions and a comparison of the two attenuation spectra was made. Figure 5 shows a clean vortex free state at Ω=0.90 rad/s; a difference is evident in the center section of the spectrum, which is composed of the m_J = 0 and m_J = +1 peaks. The effect should vanish in the vortex states because the flow field is "random" on the scale of the experimental chamber.

Using Figs. 4c and 5, we can deduce the rough estimate $w_0 \simeq 10$ mm/s for the S-S coupling parameter, which is 10 times larger than the one given in Ref. 7.

Our experiment provides a value for the parameter c_2^2 quoted in Equation (1). In the phenomenological model $c_2^2 = (H_0/H)(\Delta\omega/\omega_0)$ where $\Delta\omega$ is the spacing between the m_J = +2 and m_J = -2 peaks. The values of c_2^2/v_F^2 from our experiment are listed in Table 1. It is also possible to derive c_2^2 from the experiment by Shivaram et al.[8]; the accuracy of determining c_2^2 from their data is, however, poor because we do not know the effective value of γ in low magnetic fields in their case.

Table I: c_2^2 from our experiment

p(bar)	T/T$_c$	f(MHz)	c_2^2 (m^2/s^2)	c_2^2/v_F^2
3.2	0.81	44.7	679	0.24
6.5	0.68	62.6	643	0.26
11.5	0.60	80.6	753	0.38

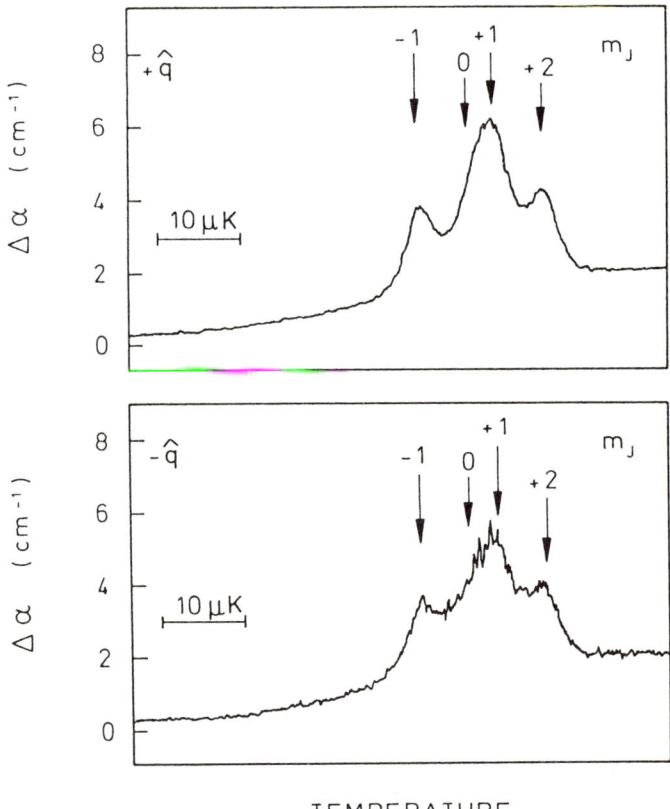

Fig. 5. Rsq mode probed by sound pulses propagating alternately parallel and antiparallel to Ω. p = 3.0 bar, H = 25 mT, f = 44.7 MHz. The spectrum is of vortex free type.

ZERO SOUND IN ROTATING ³He—A

We have performed a few experimental runs in the rotating A-phase in low magnetic fields.

In ^{3}He-A ultrasound probes, due to the anisotropic sound attenuation $\alpha = \alpha(\theta)$, the distribution of the l-texture. θ is the angle between q and l. More explicitly,

$$\alpha(\theta) = \alpha_\perp \sin^4\theta + 2\alpha_c \sin^2\theta \cos^2\theta + \alpha_{||} \cos^4\theta \quad (2)$$

The three coefficients $\alpha_\perp$, α_c and $\alpha_{||}$ are strongly temperature dependent and thoroughly investigated both experimentally[9,10] and theoretically.[11] A consequence of this anisotropy is that in the vortex texture the attenuation of sound is different from that of the stationary texture where it is determined by the presence of the magnetic field and the walls of the experimental chamber. Two examples of the attenuation in a start/stop experiment are shown in Fig. 6. The data were taken at (a) p = 29.3 bar, f = 26.8 MHz, T/T_c = 0.95, and H = 0 and (b) H = 1.7 mT. After stopping the cryostat, one can observe orbital oscillations at H = 0 (Fig. 6a) which were detected in a stationary experiment by Paulson et al.[12]; these oscillations are probably induced by heat flow. They do not exist when H is nonzero or during rotation since the l-texture is more rigid under these conditions. The sharp decrease in the attenuation at the beginning of rotation at H = 0 is the fingerprint of the rapidly decaying vortex free state before the vortex lattice forms.

An axial magnetic field ($\vec{H} \,||\, \vec{\Omega}$) turns $\hat{l}$ perpendicular to itself almost everywhere in the sample except near the walls. Even a field of H ≈ 1 mT is enough to orient the texture in the stationary sample; thus at H = 1.7 mT, which we employed, the stationary attenuation equals $\alpha_\perp$. Therefore $\alpha_\perp$ provides a reference level in the experiment. Figure 7 shows the dependence of α in a start/stop experiment (Ω: 0 → 1.5 rad/s → 0) as a function of T/T_c. The temperature scale is by no means accurate.

It is evident from the results and by estimating the responce of the various possible vortex textures to zero sound that our data does not have enough resolution to distinguish between different vortex types, unless a transition between them would be observed.

Our results illustrate the increased size of the soft vortex core in ^{3}He-A in low magnetic fields (Fig. 8). The change in the attenuation should increase linearly as a function of Ω if the sample were threaded with vortices completely isolated from each

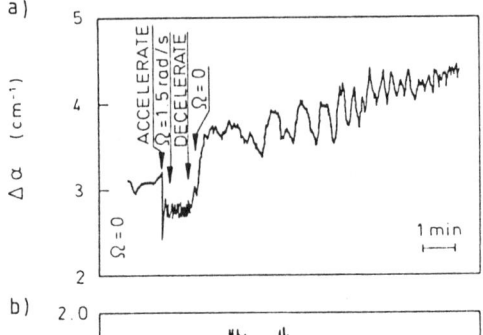

Fig. 6. Examples of start/stop experiments (0 → 0.15 rad/s^2 acceleration → 1.50 rad/s → 0.15 rad/s^2 deceleration → 0) in ^{3}He-A at $T/T_c \simeq 0.95$. a) H = 0, b) H = 1.7 mT.

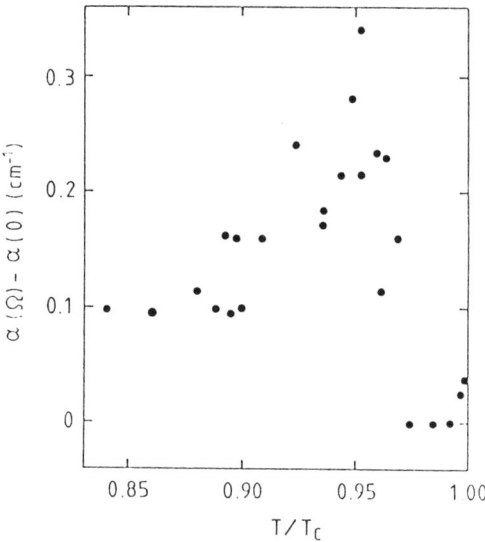

Fig. 7. $\Delta\alpha$ vs. T/T_c at H = 1.7 mT, p = 29.3 bar, and f = 26.8 MHz in start/stop experiments (0 → 1.5 rad/s → 0). Temperature scale is tentative.

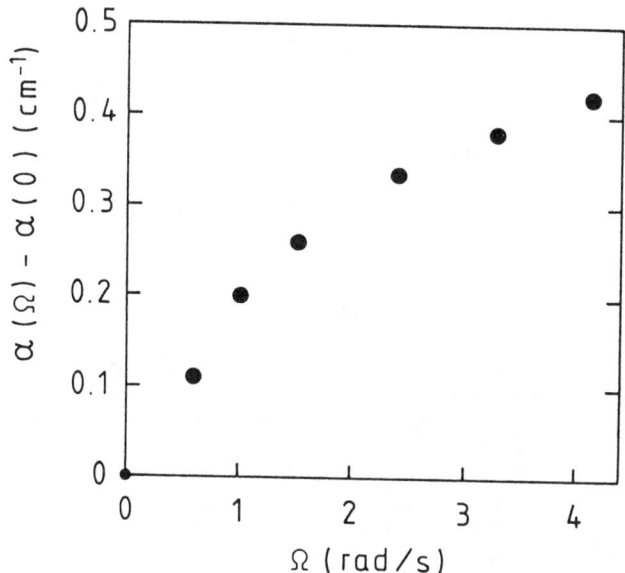

Fig. 8. Δα vs. Ω at H = 1.7 mT, p = 29.3 bar, f = 26.8 MHz, and T/T_c = 0.95.

other. One sees, however, that the soft cores, ie. regions where the l-texture is rearranged around the center of a vortex, start to overlap already at Ω ≃ 1 rad/s at 1.7 mT. Our experience at H = 0 is that even when Ω≈20 mrad/s (the slowest even angular velocity used) was enough to saturate the change in the attenuation with respect to Ω = 0 state. These angular velocities correspond to lattice constants for singly quatized vortices, of 0.1 mm and 0.7 mm, yielding an order of magnitude estimate of the soft core sizes at H = 1.7 mT and H = 0.

To our knowledge there exists no rigorous calculation on the attenuation in a vortex lattice for the A-phase. It seems to us that such an estimate should take into account that: i) the wavelength of the sound ($\lambda = 2\pi/q \simeq 10$ μm) is on the order of the soft vortex core size at least in low magnetic fields, ii) the velocity of zero sound is anisotropic leading to focusing/defocusing effects in the vicinity of the soft cores, and that iii) the spectrometer output is not proportional to the average attenuation in a unit cell of a vortex lattice but to the average sound amplitude reaching the quartz crystal; the phase of the signal has to be taken into account, too.

We thank O. Avenel, P. Berglund, M. Krusius, O. Lounasmaa, O. Magradze, V. Mineev, M. Salomaa, J. Simola, E. Varoquaux, and G. Volovik for their contributions and discussions. This work was done under the auspices of the ROTA-project, a joint undertaking between the Academy of Finland and the USSR Academy of Sciences. Our work was partly supported by the Körber-Stiftung of Hamburg. Scholarships from the Finnish Cultural Foundation (R.H.S. and J.M.K.) and from the Magnus Ehrnrooth Foundation (R.H.S., J.M.K., and J.P.P) are gratefully acknowledged.

REFERENCES

1. R.H. Salmelin, J.P. Pekola, A.J. Manninen, K. Torizuka, M.P. Berglund, J.M. Kyynäräinen, O.V. Lounasmaa, G.K. Tvalashvili, O.V. Magradze, E. Varoquaux, O. Avenel, and V.P. Mineev, to be published.
2. W.P. Halperin, Physica 109 & 110B, 596 (1982).
3. M.M. Salomaa and G.E. Volovik, to be published in J. Low Temp. Phys. (1989), and this Proceedings.

4. Yu.A. Vdovin, in Primenie metodov kvantovoi teorii polya k zadacham mnogikh tel (Application of Methods of Quantum Field Theory to Many Body Problems), Gosatomizdat Moscow, 1963, p.65 (Proceedings of the Moscow Engineering Physics Institute).
5. R.S. Fishman and J.A. Sauls, Phys. Rev. Lett. 61, 2871 (1988).
6. V.E. Koch and P. Wölfle, Phys. Rev. Lett. 46, 486 (1981).
7. J.A. Sauls and J.W. Serene, in Proc. 17th Internat. Conf. on Low Temp. Phys., edited by U. Eckern, A. Schmid, W. Weber, and H. Wühl (Elsevier, Amsterdam, 1984), p.775.
8. B.S. Shivaram, M.W. Meisel, B. K. Sarma, W.P. Halperin, and J.B. Ketterson, J. Low Temp. Phys. 63, 57 (1986).
9. D.N. Paulson, M. Krusius, and J.C. Wheatley, J. Low Temp. Phys. 26, 73 (1977).
10. J.B. Ketterson, P.R. Roach, B.M. Abraham, and P.D. Roach, in Quantum Statistics and the Many Body Problem, edited by S.B. Trickey, W.P. Kirk, and J.W. Dufty (Plenum Press, New York and London, 1975), p. 35.
11. P. Wolfle and V.E. Koch, J. Low Temp. Phys. 30, 61 (1978).
12. D.N. Paulson, M. Krusius, and J.C. Wheatley, Phys. Rev. Lett. 37, 599 (1976).

FIVE-FOLD SPLITTING OF THE SQUASHING MODE OF SUPERFLUID ^{3}He-B IN A MAGNETIC FIELD

Roman Movshovich[a], Nam Kim and David M. Lee
Laboratory of Atomic and Solid State Physics
Cornell University, Ithaca, NY 14853, USA

Eric Varoquaux
Laboratoire de Physique des Solides
Université de Paris-Sud, F-91405, Orsay, France

ABSTRACT

We have conducted an ultrasonic pulse-time-of-flight investigation of the squashing mode of the superfluid ^{3}He -B in a magnetic field. All five of the expected Zeeman sublevels for this $J = 2^-$ mode were observed. The investigation was done at a very low temperature ($T/T_c < 0.35$) to reduce the quasi-particle broadening that tends to smear the collective modes and makes resolution of the individual sublevels impossible. Variation of the pressure was used to sweep the collective modes' frequencies through the operating frequency, 137.6 MHz of the resonant sound transducers. Both group velocity spectroscopy and pulse-shape analysis were used for identification of different sublevels. Performing experiment at different orientations of the sound propagation with respect to magnetic field direction (parallel, perpendicular, 55 and 13 degrees) was the key to observation of all 5 modes and their identification. The Landé g factor was obtained from the linear dependence of the $J_z=+2$ mode on magnetic field and compared to theory.

INTRODUCTION

The collective modes of superfluid ^{3}He-B,[1] which correspond to the Cooper pair excited states, can be classified according to their total angular momentum quantum number, J, which can take on the values 0, 1, and 2 for $\ell = 1$ pairing.[2] Three of these modes can be excited by passing compressional sound waves[3] through the superfluid allowing ultrasonic investigations of these modes. With these methods, studies were performed on a J=1$^-$ mode[4] and two J=2 modes.[5,6] The $J_z = 2$ modes are the real squashing and the imaginary squashing modes. The latter mode is usually called the squashing mode. The real squashing mode was discovered at about the same time by the Cornell[5] and the Northwestern[6] groups. It was manifested by a very narrow sound attenuation peak observed as the temperature was swept through the mode position. The real squashing mode is only weakly coupled to sound due to particle-hole asymmetry.[7]

In contrast with the real squashing mode of superfluid ^{3}He-B, the squashing mode is coupled much more strongly to density fluctuations. As a result, the attenuation feature usually can not be resolved completely. Calder et al.[8] used the "group velocity spectroscopy" technique to study the position of the squashing mode and its coupling to sound.

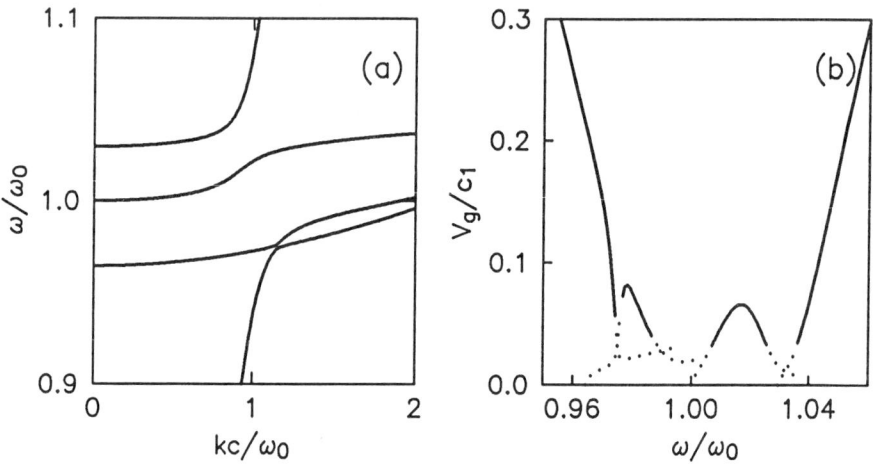

Fig. 1. *(a) Model dispersion curves[9] for the crossing of zero sound with two strongly coupled modes and one weakly coupled mode.*

(b) Group velocity spectrum calculated from the dispersion curves in (a). The mode frequencies are 1.03, 1 and .964 and their coupling constants are[9] .01, .01 and .00002 respectively. The squashing mode velocity is given by $c_{sq}/c_1 = .125$. The dotted line represents the regions of high attenuation and hence no received signal.

The coupled sound and propagating collective modes can be expressed by a pair of coupled differential equations[9]:

$$\frac{\partial^2 \rho}{\partial t^2} + \frac{2}{\tau_o}\frac{\partial \rho}{\partial t} - c_o^2 \frac{\partial^2 \rho}{\partial x^2} = \gamma_o \frac{\partial^2 \delta}{\partial x^2}$$

$$\frac{\partial^2 \delta}{\partial t^2} + \frac{2}{\tau}\frac{\partial \delta}{\partial t} + \omega_m^2 \delta - c_m^2 \frac{\partial^2 \delta}{\partial x^2} = \gamma_m \frac{\partial^2 \rho}{\partial x^2}$$

where c_m = the collective mode velocity, ω_m the mode frequency, δ is the instantaneous disturbance of the order parameter and the γ_i are coupling constants. For the squashing mode at zero magnetic field these equations give rise to an anti-crossing in the $\omega - k$ plane at a finite (non-zero) value of k. The size of the gap in ω at this value of k is determined by the strength of the coupling. Figure 1a portrays the ω vs. k spectrum of two strongly coupled modes and a weakly coupled mode. Figure 1b shows the group velocities obtained from figure 1a as a function of ω.

Tewordt and Schopohl[10] predicted that both of the J=2 modes (real and imaginary squashing modes) should exhibit a five-fold Zeeman splitting in the presence of a magnetic field. Shortly thereafter, Avenel, Ebisawa and Varoquaux[11] were able to use a conventional pulse-time-of-flight technique to observe Zeeman splitting for the real squashing mode. These investigators also performed experiments to obtain the directional dependence of the coupling strengths of different components by rotating the magnetic field relative to the direction of sound propagation. The results were in good agreement with the theoretically predicted $|Y_2^m(\theta)|^2$ dependence where θ is the angle between sound propagation direction and applied field.

There are several factors that combine to make observation of the splitting of the squashing mode rather more difficult. One was that the predicted[10] effective g value for this mode should be smaller than that for the real squashing mode.

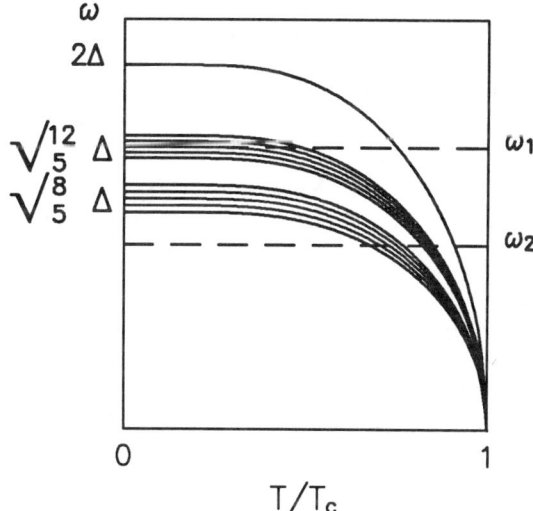

Fig. 2. *A schematic plot in ω vs. T space for pairbreaking, the squashing mode and the real squashing modes. The Zeeman splitting (not to scale) of the collective modes in an applied magnetic field is shown in the plot. The dashed lines labelled ω_1 and ω_2 correspond to two sound frequencies.*

Two additional factors virtually rule out the method of sweeping temperature in a pulse-time-of-flight apparatus. The first of these is quasi-particle collision broadening of the mode. As the temperature is reduced below T_c this effect becomes less important. Since we already expect to be dealing with Zeeman components having the broad evanescent regions associated with the squashing mode, any further broadening mechanism might completely destroy any chance of resolving these components. The decision was made therefore to operate at the lowest possible reduced temperature (T/T_c) in order to maximize the chance of observing Zeeman splitting.

The second factor involves the temperature dependence of the squashing mode frequency which, according to weak-coupling theory, varies as $\omega(T) = (12/5)^{\frac{1}{2}}\Delta(T)$. The gap $\Delta(T)$ becomes almost completely independent of temperature as $T \to 0$. Consequently, since the experiment requires the lowest possible temperature, a temperature sweep would cover only a small fraction of the Zeeman spectrum. This difficulty is illustrated in figure 2, where two different fixed frequencies, corresponding to the frequencies of two different sets of quartz transducers in a pulse-time-of-flight experiment, are represented as horizontal lines. Sound attenuation signatures occur where these horizontal lines, which correspond to a sweep of the temperature, cross the collective mode frequencies.

EXPERIMENTAL PROCEDURE

Instead of varying the temperature, the sweeping of the collective modes frequencies through the operating frequency of the quartz sound transducer was accomplished by varying the pressure, since Δ is proportional to T_c via the weak coupling BCS formula $\Delta_{BCS}(T=0) = 1.764 k_B T_c$ and T_c is a rather strong function of pressure.[1] For this work the Greywall temperature scale[12] has been used to specify values of T_c for the various pressures.

The experimental cell was cooled to temperatures well below 1mK by a $PrNi_5$ demagnetization stage precooled by a dilution refrigerator. A ^{3}He melting curve thermometer mounted in the zero field region and thermally linked to the cell by annealed silver rods provided the temperature measurement. Below 1mK where most of our data was taken, the melting curve obtained by Osheroff and Yu[13] was employed. A more detailed description of the cryostat and its performance has been given elsewhere[14].

Two 15 MHz quartz sound transducers separated by a 4.5 mm precision quartz spacer were mounted in the experimental cell. Since the magnetic field applied to induce Zeeman splitting was provided by a fixed, vertically oriented 0-9 tesla superconducting solenoid, changes of orientation of the sound path relative to the applied field could be accomplished only by dismantling the cell at room temperature. In these experiments the sound cell was mainly operated at 137.6 MHz which was the ninth harmonic of the quartz transducers.

It is the highest odd harmonic of the transducers at which we could operate the sound cell and still have access to the squashing mode. The higher frequency in turn required higher pressure of ^{3}He and resulted in lower T/T_c thus giving a smaller amount of quasiparticle broadening. Low temperature operation had the added advantage of minimizing the temperature dependence of the collective mode frequencies, so that temperature corrections were minimal.

In a typical group velocity run, the cell was cooled to its minimum temperature by demagnetization of .3 mole of PrNi$_5$ from 6T to a few hundred Gauss and then the pressure was reduced slowly (over period of a few days) in order to avoid viscous heating. Sound pulses were repetitively applied to the transmitting quartz crystal and were received by the receiving crystal after traversing the 4.5 mm path through the superfluid. The group velocity was determined from the time of arrival of the peak of the propagating sound pulse. The attenuation of the received pulse as well as its detailed shape also provided important information in these experiments. Most of the data was taken in the pressure range extending from 17 to 22 bars.

DATA

The results of measurements of the group velocity of sound propagated in parallel and perpendicular orientation with respect to magnetic field as the pressure was swept are shown in figure 3 and were reported previously.[15,16] The horizontal axes in figures 3a and 3b correspond to the variable ω/Δ_{BCS} where ω is the applied sound frequency and Δ_{BCS} is the pressure dependent weak-coupling energy gap. The splitting of the group velocity spectrum of the squashing mode is clearly evident in figure 3a which displays results obtained for the $q \perp H$ orientation, where q is the propagation vector for the applied sound wave. The data shown in figure 3a can be related to the group velocity obtained from the model calculation for collective modes coupled to sound shown in figure 1b. The strongly coupled collective modes shown in figure 1 are located at the positions of vanishing group velocity, according to the model. In figure 3a, we therefore identify the position of the collective modes with the points of intersection of linear extrapolations of the group velocity and the ω/Δ_{BCS} axis. The intersection occurring at the higher value of ω/Δ_{BCS} for each evanescent region is empirically chosen as the point corresponding to location of the collective mode. A rather striking feature of the group velocity data is the extremely low values of group velocity in the neighborhood of the collective modes. Group velocities as low as 6 m/s were recorded corresponding to 750 microsecond flight times in the 4.5 mm sound cell. As a result, the data points extend down nearly to the ω/Δ_{BCS} axis, so that the points of intersection between linear extrapolations of the group velocity data and the ω/Δ_{BCS} axis are known with rather high precision.

The $q \perp H$ data of figure 3a contains points corresponding to three different values of the magnetic field because it was not possible to sweep through the entire Zeeman spectrum for a single value of the magnetic field. The sharp dip on the left hand side of the spectrum only emerged from the adjacent evanescent region for fields greater than 4.0 kilogauss. On the other hand, the far right hand side of the spectrum was only observable for fields less than 4.0 kilogauss. Above this field, the onset of pair breaking absorption associated with the magnetic field led to an extremely strong sound attenuation which completely obscured the signal.

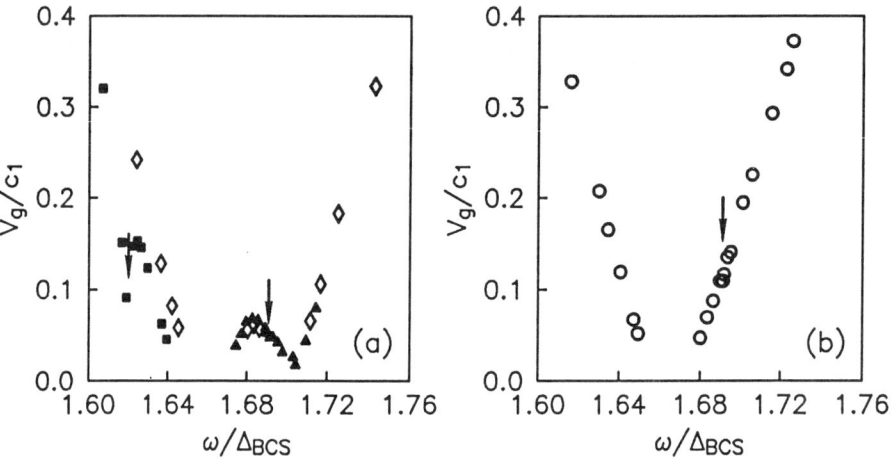

Fig. 3. *(a) Observed group velocity spectra for $q \perp H$: $\Diamond$, 3.6kG; , 3.8kG; , 4.6kG. Arrows indicate locations of $J_z = \pm 1$ as explained in the text.*

(b) Group velocity spectrum for $q \| H$, for $H=3.8kG$. Only the $J_z = 0$ mode is strongly coupled. The position of $J_z = +1$ for $H=3.8kG$ is indicated by the arrow.

The orientation dependence of the coupling strength of the individual Zeeman components has been discussed earlier for the real squashing mode. A similar proportionality to $|Y_2^m(\theta, \phi)|^2$ has been theoretically predicted for the squashing mode[10] corresponding to coupling strengths proportional to $(3\cos^2\theta - 1)^2$, $\sin^2\theta\cos^2\theta$ and $\sin^4\theta$ for the $J_z = 0$, $J_z = \pm 1$ and $J_z = \pm 2$ modes respectively. Thus, for $\theta = 90°(q \perp H)$, the $J_z = 0$ and $J_z = \pm 2$ sublevels should be strongly coupled to sound. On the other hand the $J_z = \pm 1$ sublevels should only be coupled to the sound either because of textural distortions or a deviation from perfect perpendicularity between q and H. Thus, the strongly coupled $J_z = 0$ and ± 2 components should, according to figure 1b, have group velocity spectra with broad evanescent regions for this orientation. In figure 3a we see two components with broad evanescent regions rather than three. The broad evanescent region on the far right is identified as the $J_z = +2$ component. The other broad feature is interpreted to be a combination of the strongly coupled $J_z = 0$ and -2 components, which are too close together (due to non-linear behavior) to be resolved.

The last prominent feature in the group velocity spectrum of figure 3a is the narrow dip in the group velocity data for 4.6 kilogauss at $\omega/\Delta_{BCS}=1.620$. This narrow feature is identified with the $J_z = -1$ Zeeman component which must then cross the $J_z = -2$ mode at somewhat lower fields. The very narrowness of the evanescent region means that this Zeeman component is quite weakly coupled to the sound as expected from our previous discussion since $|Y_2^1(\theta, \phi)|^2$ is proportional to $sin^2\theta cos^2\theta$ which vanishes for $\theta = 90°$. Finally, a very narrow feature at $\omega/\Delta_{BCS} = 1.691$ and 3.8 kilogauss was observed. This feature has a very small group velocity signature and is mainly manifested by a sharp drop in pulse amplitude. It falls between the $J_z = 0$ and the $J_z = +2$ evanescent regions, and is taken to be the $J_z = +1$ component.

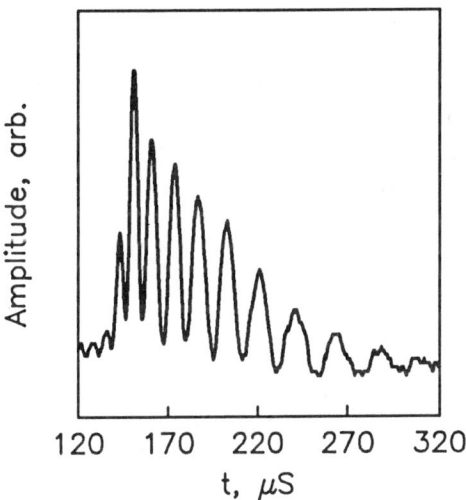

Fig. 4. *Pulse break up corresponding to the precursory behavior for weakly coupled Zeeman components. This behavior enabled us to identify these Zeeman components.*

Figure 3b shows data obtained for the parallel orientation ($\theta = 0$ and $q \| H$). As a result of the θ dependence of the spherical harmonics, we expect $J_z = 0$ to be very strong, $J_z = \pm 1$ to be weak because of the $sin^2\theta$ coupling, and $J_z = \pm 2$ to be very weak due to the $sin^4\theta$ coupling. The broad evanescent region associated with the $J_z = 0$ Zeeman component is the most prominent feature in figure 3b. Although they exhibited no pronounced group velocity effects, the $J_z = \pm 1$ components were also observed for this parallel orientation. They were identified by a dramatic break-up of the received pulse into a sequence of closely spaced spikes as shown in figure 4. This phenomenon has been discussed previously in terms of precursory sound pulse propagation[9] for the crossing of the weakly coupled real squashing mode and sound. The $J_z = \pm 1$ signatures were so narrow for the parallel orientation that they could be traversed completely by sweeping the frequency of the signal from the oscillator through the narrow quartz resonance. Finally, no evidence was found for the $J_z = \pm 2$ components in this orientation because of the very weak coupling for these modes, mentioned above.

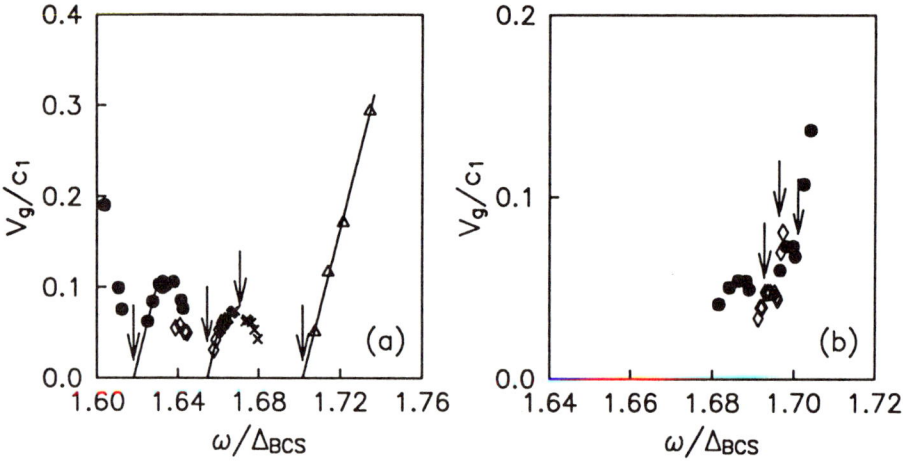

Fig. 5. *(a) Observed group velocity spectra for $\theta = 55°$:* •, *4.6kG;* ◊, *4.2kG;* ×, *4.1kG;* △, *3.2kG. Arrows indicate positions of the $J_z = -1, -2, 0, 2$ modes from left to right.*

(b) Group velocity spectrum for $\theta = 13°$: ◊, *3.0kG;* •, *3.6kG. Arrows indicate positions of the $J_z = 1(3.6kG), 2(3.0kG)$ and $2(3.6kG)$ from left to right.*

To observe the fifth mode ($J_z = -2$) we performed an experiment with sound propagated at an angle $\theta = cos^{-1}(1/\sqrt{3}) \approx 55°$ to the direction of a magnetic field. This is the only orientation at which the $J_z = 0$ mode was expected not to couple to sound, and so the $J_z = -2$ could be observed on its own. Figure 5a displays the data obtained in this orientation. All $J_z = \pm 1, \pm 2$ modes couple strongly to sound and their positions are indicated by arrows for the appropriate fields. Note that this time the $J_z = 1$ and 2 modes could not be separated. For the run at $H = 4.1kG$ the narrow high attenuation region (where there was no received signal) falls at the exact location of the $J_z = 0$

mode obtained from previous orientations and identified as such. The strongly coupled mode to the left of $J_z = 0$ for the $H = 4.2 kG$ run then is the $J_z = -2$ mode and we obtain its frequency by the group velocity spectroscopy technique.

The last experiment was done at the $\theta = 13°$ orientation. It was performed to bring out the $J_z = 2$ mode which did not couple to sound in the parallel orientation. The data is displayed on figure 5b where the narrow high attenuation regions are the weakly coupled $J_z = 2$ modes, and are denoted by arrows. Also the position of the $J_z = 1$ mode for the $H = 3.0 kG$ run is show by an arrow. Note the expanded scales of both axes.

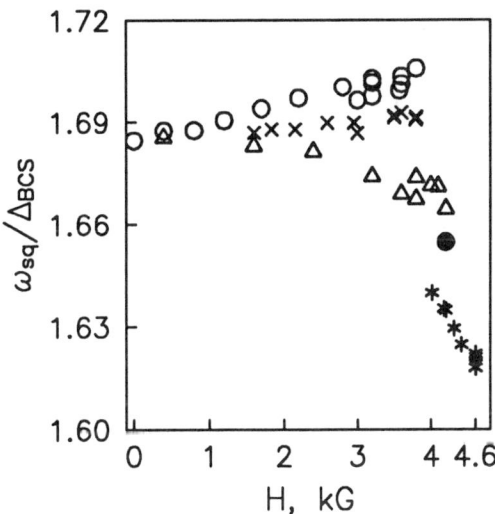

Fig. 6. *Zeeman sub-levels as a function of magnetic field.* o, $J_z = +2$; ×, $J_z = +1$; △, $J_z = 0$; *, $J_z = -1$; •, $J_z = -2$.

All of the raw data for the various squashing mode Zeeman components in terms of the reduced frequency ω/Δ_{BCS} for both perpendicular and parallel orientations are displayed in figure 6 as a function of applied field. The data for strong coupled components were obtained by extrapolating the group velocity spectra as explained earlier. The mode frequency data for the weakly coupled components were obtained from pulse break up for the parallel orientation and from attenuation peaks for the perpendicular orientation. The uppermost curve in figure 6, corresponding to the $J_z = +2$ Zeeman component, is nearly a straight line for the data obtained in the perpendicular orientation. The Landé g factor, defined by the equation $\Delta\omega = gJ_z\Omega$, can be obtained from the slope of this line. In the above equation $\Delta\omega$ is the frequency shift away from the zero field squashing mode frequency for a given pressure and Ω is the renormalized Larmor frequency given by $\Omega = \gamma H[1 + F_0^a \chi_{BW}(\Omega)/\chi_n]^{-1}$. Since the pressure rather than the frequency was swept in these experiments, it was necessary to correct for the pressure variation. Measurements of the zero field squashing mode were made at the 7th harmonic (107 MHz) and the 9th harmonic (137.6 MHz) corresponding to two different values of the pressure. The zero field squashing mode frequency as a function of pressure was obtained by interpolation. The difference between the zero field squashing mode frequency so obtained and the frequency of the experimental point (which was always 137 MHz) gave the value of $\Delta\omega$.

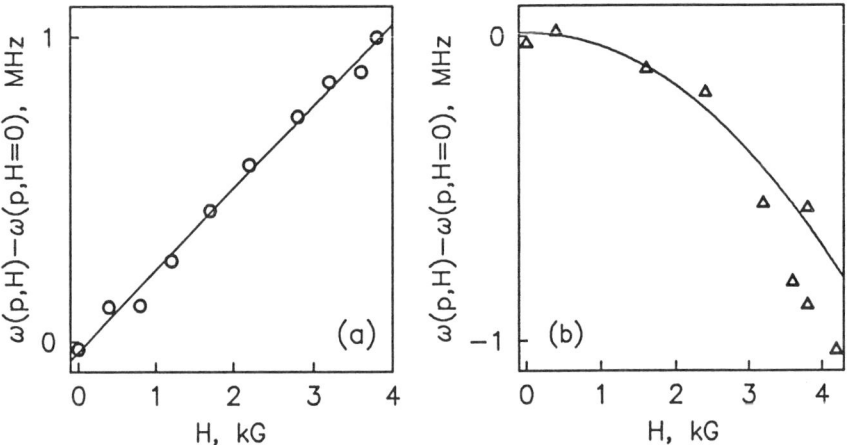

Fig. 7. *Pressure corrected data showing the Zeeman splitting as a function of magnetic field:*

(a) $J_z = +2$ data which permitted calculation of the g value from $\Delta\omega = gJ_z\Omega$.

(b) $J_z = 0$ data which permitted calculation of the quadratic splitting factor Γ from $\Delta\omega = -\Gamma H^2$.

In this analysis it is assumed that g is only a weak function of the pressure, an approximation which is valid provided that the total pressure variation is kept small. Since the pressure varied only between 17 and 22 bars, this criterion should be obeyed. A plot of the pressure corrected $\Delta\omega$ vs. the applied magnetic field for the $J_z = +2$ mode is given in figure 7a. From the slope of the linear fit, a g value of $0.042 \pm .002$ is obtained, in good agreement with the theoretical value of g=0.0435 calculated using expressions of Sauls and Serene[16] where we use the weak coupling plus energy gap of Serene and Rainer[19] and $X_3^{-1} = 0$ (X_3^{-1} is a parameter indicating the strength of f-wave pairing).

The $J_z = 0$ component shows quadratic dependence on field in figure 6. Figure 7b is obtained by applying the same pressure correction procedure to $J_z = 0$ as for the case of the $J_z = +2$ mode. The resulting plot can be fitted to the quadratic function $\Delta\omega = \omega - \omega_o = -\Gamma H^2$ with $\Gamma = 45 \pm 15$ kilohertz per kilogauss squared. This result is consistent with the acoustic impedance studies of Shivaram et al.[17] and the prediction of Schopohl et al.[20].

SUMMARY

In summary, we have been able to show by means of pulse-time-of-flight sound spectroscopy and by sweeping the pressure applied to the ^{3}He sample that Zeeman splitting occurs for the squashing mode of superfluid ^{3}He, in accordance with the 1979 theoretical studies of Schopohl and Tewordt[10]. All five Zeeman components have been identified. The Landé g factor has been measured and compares favorably with theoretical value. Strong non-linear behavior of some of the modes has been observed, with the $J_z = -2, -1$ modes crossing at a field $H < 4kG$.

We express our gratitude to the National Science foundation for supporting this research via Grant DMR-841605 and through the Cornell Materials Science Center via Grant DMR-8516616. One of us (R.M.) thanks AT&T Bell Laboratories for fellowship support, and another of us (D.M.L.) gratefully acknowledges the hospitality of the Aspen Center for Physics. We are also most grateful to Peter Wölfle, Tilo Kopp, Mark Meisel, Jim Sauls and Bill Halperin for stimulating and informative discussions. Finally, we owe a debt of gratitude to David Sagan, Larry Friedman, Emil Polturak, Paul de Vegvar and Eric Ziercher for designing and constructing some of the apparatus used in these experiments.

REFERENCES

(a) Invited paper given by R. Movshovich.

1. D. M. Lee and R. C. Richardson, *The Physics of Liquid and Solid Helium, Part II*, edited by K. H. Benneman and J. B. Ketterson, (Wiley, 1978, pp. 287-496).

2. P. Wölfle, Physica **90 B+C**, 96 (1977).

3. P. Wölfle, *Progress in Low Temperature Physics, VIIa*, edited by D. F. Brewer, (North Holland, 1978), pp. 191-281.

4. R. Ling, J. Saunders and E. R. Dobbs, Phys. Rev. Lett. **59**, 461 (1987).

5. R. W. Giannetta, A. Ahonen, E. Polturak, J. Saunders, E. K. Zeise, R. C. Richardson and D. M. Lee, Phys. Rev. Lett. **45**, 262, (1980).

6. D. B. Mast, B. K. Sarma, J. R. Ower-Bradley, I. D. Calder, J. B. Ketterson and W. P. Halperin, Phys. Rev. Lett. **45**, 266, (1980).

7. V. E. Koch and P. Wölfle, Phys. Rev. Lett. **46**, 486 (1981).

8. I. D. Calder, D. B. Mast, B. K. Sarma, J. R. Owers-Bradley, J. B. Ketterson and W. P. Halperin, Phys. Rev. Lett. **45**, 1866 (1980).

9. E. Varoquaux, G. A. Williams and O. Avenel, Phys. Rev. B **34**, 7617 (1986).

10. N. Schopohl and L. Tewordt, J. Low Temp. Phys. **45**, 67 (1981); L. Tewordt and N. Schopohl, J. Low Temp. Phys. **37**, 421 (1979).

11. O. Avenel, E. Varoquaux, and H. Ebisawa, Phys. Rev. Lett. **45**, 1952 (1980).

12. D. S. Greywall, Phys. Rev. B **33**, 7520 (1986).

13. D. D. Osheroff and C. Yu, Phys. Lett. **77A**, 458 (1980).

14. D. C. Sagan, Ph.D. thesis, Cornell University, 1984.

15. R. Movshovich, E. Varoquaux, N. Kim, D. M. Lee, Phys. Rev. Lett. **61**, 1732 (1988).

16. D. M. Lee, R. Movshovich, E. Varoquaux, Spin Polarized Quantum Systems, World Scientific Publishing Co., New Jersey, 1989, ed. S. Stringari (Procedures of Torino Conference, Torino, Italy, June 1988), p. 142-154

17. B. S. Shivaram, M. W. Meisel, B. K. Sarma, W. P. Halperin, and J. B. Ketterson, J. Low Temp. Phys. **63**, 57 (1986).

18. J. A. Sauls and J. W. Serene, Phys. Rev. Lett. **49**, 1183 (1982); J. A. Sauls and J. W. Serene, Phys. Rev. B **23**, 4798 (1981).

19. J. W. Serene and D. Rainer, Phys. Rep. **101**, 221 (1983).

20. N. Schopohl, M. Warnke, and L. Tewordt, Phys. Rev. Lett. **50**, 1066 (1983).

NONLINEAR ACOUSTICS IN SUPERFLUID ^{3}He-B

Ross H. McKenzie and J.A. Sauls
Department of Physics and Astronomy
Northwestern University
Evanston, Illinois 60208

ABSTRACT

The nonlinear interaction of zero sound with the order parameter collective modes in superfluid ^{3}He-B is considered within perturbation theory in the amplitude of the sound field. Selection rules for nonlinear excitation of the order parameter modes are determined by the approximate particle-hole symmetry of the ^{3}He Fermi liquid. A diagrammatic algorithm, based on the quasiclassical theory of superfluid ^{3}He, is used to calculate nonlinear coupling constants. These nonlinearities are sufficiently large that it should be possible to observe two phonon absorption and stimulated Raman scattering of zero sound by the real squashing ($J = 2^+$) mode. Finally, we discuss the possibility of using these nonlinearities to produce zero sound with 'squeezed' noise.

INTRODUCTION

The order parameter collective modes of superfluid ^{3}He have been studied extensively with zero sound. Most of these studies have been devoted to the linear response of the superfluid.[1] Interesting phenomena should also be observable in the nonlinear acoustic response.

The order parameter for the superfluid phases of ^{3}He is the Cooper pair amplitude $\langle\psi_\alpha\psi_\beta\rangle$ which for a p-wave ($\ell = 1$), spin triplet (s = 1) state can be written in terms of a 3x3 complex matrix A_{ij}.[2] The order parameter collective modes in the B phase are oscillations A_{ij} about is equilibrium value $\Delta(T)\delta_{ij}$ (where $\Delta(T)$ is the temperature dependent energy gap), and can be classified[3] by the quantum numbers J^ζ and M where $J = 0,1,2$ is the total angular momentum, $M = \{-J,\cdots,0,\cdots J\}$ is the magnetic quantum number, and $\zeta = +, -$ is the 'parity' under particle-hole symmetry (discussed below) for the real and imaginary parts of the order parameter, respectively.

The $J = 2^+$ and $J = 2^-$ modes, which are also known as the real and imaginary squashing modes, respectively, are of particular interest because they couple to zero sound. These modes lie below the pair breaking edge $2\Delta(T)$ ($\omega_{2+}\approx1.1\Delta(T)$ and $\omega_{2-}\approx1.5\Delta(T)$) and are

weakly damped by quasiparticle collisions. Consequently, the
J = 2^+, 2^- modes result in sharp features in the zero sound
attenuation, phase velocity and group velocity whenever the
frequency and wave vector of sound is equal to that of one of the
collective modes. In a magnetic field, H, a five-fold Zeeman
splitting of the J = 2^+ mode is observed,[4] the splitting becomes
nonlinear as a function of H for large magnetic fields[5] as a result
of gap distortion and level repulsion.[6] Recently, the Zeeman
splitting of the J = 2^- modes has been observed in the group
velocity spectrum.[7] In general, the acoustic spectroscopy of ^{3}He-B
involves phenomena similar to those seen in the optical spectroscopy
of atoms, molecules and solids. Here we show that it should also be
possible to observe nonlinear acoustic processes in ^{3}He-B analogous
to the well known nonlinear optical effects of two photon absorption
and stimulated Raman scattering. Both of these processes in ^{3}He
involve quanta of the real squashing mode (*real squashons*
hereafter). The first process is the excitation of a real squashon
by two zero sound phonons. This occurs if the frequencies ω_1 and ω_2
and wavevectors $\vec{q}_1$ and $\vec{q}_2$ of the phonons satisfy the conditions

$$\omega_1 \pm \omega_2 = \omega_M$$
$$\vec{q}_1 \pm \vec{q}_2 = \vec{q}_M \qquad (1)$$

with positive signs, where ω_M and $\vec{q}_M$ are the frequency and
wavevector respectively of the real squashon. The second process is
the decay of a zero sound phonon into a real squashon and a second
zero sound phonon, which occurs if eq. (1) is satisfied with the
negative sign. Whether or not these processes are observable
depends on the answers to two questions. Are the processes allowed
by the selection rules that are implied by the symmetries of ^{3}He?
And, if so, what acoustic energy density is needed to detect them?

Liquid ^{3}He at low temperatures has an approximate symmetry
under the interchange of quasiparticle and quasihole states near
the Fermi surface. An important selection rule is imposed by
exact particle-hole symmetry which is represented by a unitary
operator C that maps quasiparticle states just above the Fermi
surface into quasihole states just below the Fermi surface (and
vice-versa).[9,10] Particle-hole symmetry determines the selection
rules for the coupling of zero-sound to the order parameter
collective modes because the real (imaginary) components of the
order parameter are even (odd) under C, whereas the density
fluctuations are odd under C.[9] Thus, with exact particle-hole
symmetry the J = 2^+ modes do not couple linearly to sound.
However, particle-hole symmetry is weakly broken in ^{3}He because
the density of states just above and just below the Fermi

surface differ slightly. Consequently, there is a weak linear coupling between the $J = 2^+$ modes and sound. In the linear response limit the dynamical equations for the $J = 2^+$ modes are

$$\tilde{\lambda}(\omega) \, [(\omega+i\Gamma)^2 - \omega_{M+}(q)^2] \, D^M_+(\omega,\vec{q}) = \frac{6}{1+F_0^s} \, \beta_M \, \delta n(\omega,\vec{q}), \qquad (2)$$

where $D^M_+(\omega,\vec{q})$ is the amplitude of the mode with magnetic quantum number M, $\frac{1}{\Gamma}$ is the lifetime of the mode due to quasiparticle collisions, $\tilde{\lambda}(\omega)$ is the Tsuneto function and $\delta n(\omega,\vec{q})$ is the density fluctuation. The coupling constant β_M is small, of order $\eta = N'(E_F)\Delta/N(E_F)$, where $N(E_F)$ and $N'(E_F)$ are the density of states and its' slope at the Fermi surface.[11]

The propagation of sound in superfluid ^{3}He is described by a wave equation

$$\left[\frac{\partial^2}{\partial t^2} - c_1^2 \nabla^2 \right] \delta n = 2c_1^2 \nabla^2 \delta\Pi, \qquad (3)$$

where $\delta\Pi(\vec{R},t)$ is related to the stress tensor of the superfluid, and c_1 is the hydrodynamic sound velocity. Equation (3) is a consequence of the mass and momentum conservation laws. It is important to note that although this equation is linear in δn and $\delta\Pi$ it describes nonlinear sound propagation because the longitudinal stress $\delta\Pi$ is in general a nonlinear functional of the density fluctuation and, in general, the amplitudes of the collective modes of the system which couple to zero-sound. The relationship between the fluctuating stress $\delta\Pi$ and the density fluctuation δn must be obtained from a more microscopic theory than hydrodynamics. Under certain conditions the constitutive relation is of the form

$$\delta\Pi \approx \chi^{(1)} \delta n + \chi^{(2)} (\delta n)^2 + \chi^{(3)} (\delta n)^3 + \cdots \qquad (4)$$

In the linear response limit the frequency dependent attenuation $\alpha(\omega)$ and the phase velocity $c(\omega)$ of sound are given by

$$\alpha = -q\,\text{Im}\,\chi^{(1)}, \qquad \frac{c(\omega)-c_1}{c_1} = \text{Re}\,\chi^{(1)}. \qquad (5)$$

For exact particle-hole symmetry the second order susceptibility $\chi^{(2)}$ vanishes. Analogues between the above situation and that in optics[8] have been pointed out recently.[12]

Although the $J = 2^+$ modes can only be excited by a single zero sound phonon via the small intrinsic particle-hole asymmetry in ^{3}He, the excitation of the $J = 2^+$ mode by two phonon processes is not forbidden by particle-hole symmetry selection rules. Thus, at higher sound amplitudes the right-hand side of eq. (2) contains a driving term which is second order in the density; and the stress tensor has a term which is bilinear in δn and the amplitude D_+^M for the $J = 2^+$ mode. These nonlinear couplings have been calculated from microscopic theory.[12] The results are

$$\delta \Pi(\omega) = \frac{1}{(1+F_0^s)\Delta} \sum_M \int d\nu \, A^M(\omega,\nu,\omega-\nu) \, \delta n(\nu) \, D_+^M(\omega-\nu), \quad (6)$$

$$\tilde{\lambda}(\omega)\left[(\omega+i\Gamma)^2 - \omega_{M+}(q)^2\right] D_+^M(\omega,\vec{q}) = \frac{6}{(1+F_0^s)^2\Delta} \times$$

$$\int d\nu \, A^M(\nu-\omega,\nu,-\omega)^* \, \delta n(\nu) \, \delta n(\omega-\nu), \quad (7)$$

where A^M is a dimensionless function of order one. These are the central equations describing the interaction of the $J = 2^+$ modes with two zero sound waves.

TWO PHONON PROCESSES

Equations (2) and (3) are now applied to the nonlinear interaction of two zero sound waves with the $J = 2^+$ mode. The density fluctuation is written in the form

$$\delta n(\vec{R},t) = \text{Re}[N_1(\vec{R},t) + N_2(\vec{R},t)] \quad (8)$$

where $N_j(\vec{R},t) = \tilde{N}_j(\vec{R},t)e^{i[\omega_j t - \vec{q}_j \cdot \vec{R}]}$, $j = 1,2$ and it is assumed that the wave amplitudes $\tilde{N}_j(\vec{R},t)$ vary slowly on the time scale of the mode lifetime $1/\Gamma$. In this quasi-steady-state approximation eq. (6) can be solved for $D_+^M(\vec{R},t)$. It is straightforward to show that the $J = 2^+$ mode amplitudes contains terms oscillating with frequencies, 0, $2\omega_1$, $2\omega_2$, $\omega_1 + \omega_2$, and $\omega_1 - \omega_2$. If these solutions, together with eq. (8), are substituted in eq. (6) it is found that $\delta\Pi(\vec{R},t)$ contains terms oscillating with frequencies ω_1, ω_2, $3\omega_1$, $3\omega_2$, $2\omega_1 \pm \omega_2$, and $2\omega_2 \pm \omega_1$.

The term with frequency ω_1 is written as

$$\frac{\delta\Pi_1(\vec{R},t)}{N_1} = [\chi^{(3)}(\omega_1,-\omega_2,\omega_1+\omega_2) + \chi^{(3)}(\omega_1,\omega_2,\omega_1-\omega_2)]|N_2|^2, \quad (9)$$

where the nonlinear susceptibility is

$$\chi^{(3)}(\omega,\nu,\omega-\nu) = \frac{6}{5\Delta^2(1+F_0^s)^3} \sum_M \frac{|A^M(\omega,\nu,\omega-\nu)|^2}{\bar{\lambda}(\omega-\nu)[(\omega-\nu+i\Gamma)^2 - \omega_{M+}^2]}. \quad (10)$$

If the wave with frequency ω_2 is of much higher intensity than the wave with frequency ω_1, then the intensity $|N_2|^2$ can be treated as a constant. The attenuation and shift in phase velocity of the sound wave with frequency ω_1, due to the nonlinear interaction with the sound wave with frequency ω_2 and the $J = 2^+$ collective modes is then calculated from eqs. (3), (9) and (10). Well defined features in the spectrum occur whenever one of the resonance conditions $\omega_1 \pm \omega_2 = \omega_{M+}$ is satisfied. These resonance features correspond to two phonon absorption (+) and stimulated Raman scattering (-) of phonons by the $J = 2^+$ modes.

Figures 1 (a) and (b) show the change in the phase velocity in a zero sound wave of frequency ω_1 due to its linear and nonlinear interaction with the $J = 2^+$ modes in the presence of a second wave of high intensity and frequency ω_2. The features on the left ($T_+ \simeq 0.62 T_c$) are due to two-phonon absorption by the $J = 2^+$ mode and occur at a temperature such that $\omega_1 + \omega_2 = \sqrt{8/5}\,\Delta(T_+)$. The features on the right ($T_- \simeq 0.76 T_c$) are due to stimulated Raman scattering of phonons by the $J = 2^+$ mode and occur at a temperature such that $|\omega_1 - \omega_2| = \sqrt{8/5}\,\Delta(T_+)$. The large central feature in Figure 1 (a) is due to the linear coupling of the sound to the $J = 2^+$ mode as a result of particle-hole asymmetry and occurs at a temperature T_o such that $\omega_1 = \sqrt{8/5}\,\Delta(T_o)$. The linear resonance is not shown in Fig. 1 (b) because it occurs at a temperature greater than $0.8 T_c$. Amplification of the low-frequency sound wave is possible at the resonance T_- for the same parameters as Fig. 1(b).[12] An amplification occurs as the high-frequency phonons decay into real squashons and low frequency phonons. In order to reduce heating effects in the experimental cell, it may be desirable to use a smaller sound energy density than the value $U/U_c = 0.2$ used in Fig. 1 (a) and (b). Although a smaller value of U/U_c reduces the size of the nonlinear features they should still be observable since changes in the phase velocity of order one part in 10^6 are presumably detectable.

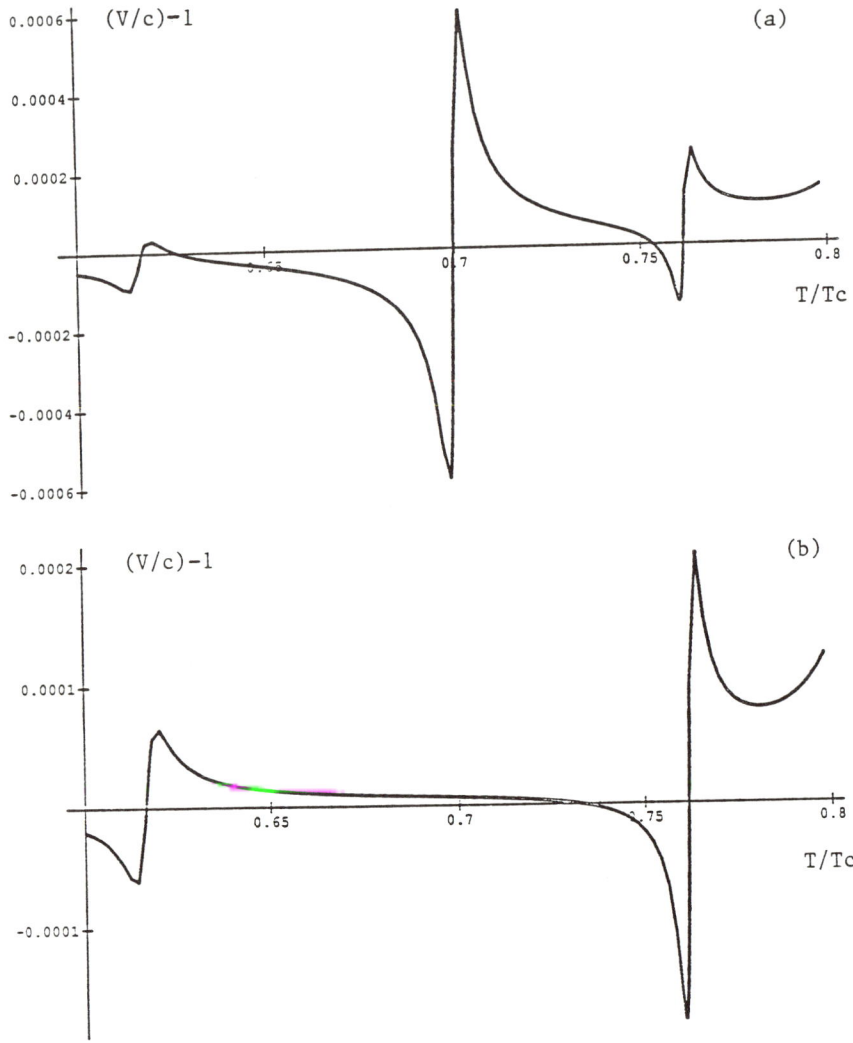

Figure 1. The predicted temperature dependence at zero pressure of $v_{2+}(T)/c_1$, where v_{2+} is the contribution of the $J = 2^+$ mode to the phase velocity of a zero sound wave of frequency ω_1 in the presence of a parallel wave of frequency ω_2 and energy density $0.2U_c$. In (a) $\omega_1=35.4$ MHz and $\omega_2=2.87$ MHz and in (b) $\omega_1=2.87$ MHz and $\omega_2=35.4$ MHz The features at $T/T_c \sim 0.62$ and 0.76 in both graphs are the nonlinear resonances.

SQUEEZING OF ACOUSTIC NOISE

The density oscillation produced by a sound mode of frequency ω can be written in the form

$$\delta n = P_1 \cos \omega t + P_2 \sin \omega t \qquad (11)$$

where for simplicity the spatial dependence is omitted. In general, there are fluctuations in the amplitudes P_1 and P_2 caused by noise of various sources. For example, there is noise in the electrical signal that drives the sound transducer, as well as thermal fluctuations of the density. Generally, the noise will be randomly distributed in phase and the fluctuations ΔP_1 and ΔP_2 in two quadratures will be equal.

We now consider the possibility of producing "squeezed" sound which has unequal noise in the two quadratures. In analogy with nonlinear optics[13] this can be done by four-wave mixing, which makes use of the large nonlinear susceptibility $\chi^{(3)}$ in ^{3}He-B. Two high intensity counter-propagating waves (referred to as pump waves) interact with a second pair of counter propagating waves (signal waves) in a cavity oriented at an oblique angle to the pump waves. The four waves usually have the same frequency. It can be shown that the fluctuations ΔP_1 and ΔP_2 in the quadratures of the signal waves leaving the cavity are related to the initial fluctuations ΔP_o in the signal waves entering the cavity by

$$(\Delta P_1)^2 = (\Delta P_o)^2 \left[e^{-2s} \cos^2 \tfrac{1}{2}\theta + e^{2s} \sin^2 \tfrac{\theta}{2} \right]$$

$$(\Delta P_2)^2 = (\Delta P_o)^2 \left[e^{-2s} \sin^2 \tfrac{1}{2}\theta + e^{2s} \cos^2 \tfrac{\theta}{2} \right] \qquad (1?)$$

where $se^{i\theta} = \chi^{(3)} A^2 L$, A is the amplitude of the pump waves, and L is the interaction length for the cavity. In order to observe significant squeezing of the noise it is desirable to tune $\theta \approx 0$ or π which implies that the $|\text{Im}\chi^{(3)}| \ll |\text{Re } \chi^{(3)}|$. In addition the squeezing parameter s should be of order unity or larger. This requires a large nonlinear, susceptibility, large pump-wave energy densities and a long interaction length. For zero sound in ^{3}He-B with frequency about half the J = 2^+ mode frequency $|\chi^{(3)}|$ is sufficiently large that values of s~1 are possible for interaction lengths of order centimeters and pump wave energy densities U several orders of magnitude smaller than the superfluid condensation energy density U_c. In addition the frequency is far enough from resonance that $|\text{Im}\chi^{(3)}| \ll |\text{Re}\chi^{(3)}|$.

The squeezing of noise can be measured by *homodyne detection*: the signal wave is mixed with another sound wave (known as the local oscillator) of much higher intensity and with a phase ϕ relative to the signal wave. The noise power of this mixed wave is then measured. Suppose that P_A is the noise power in the absence of both signal and pump waves and P_o+P_A is the noise power measured in the absence of the pump wave. Then P_s is given by

$$P_s = P_A + P_o \left[1-\eta+\eta\left(e^{-2s}\sin^2(\phi-\tfrac{\theta}{2})+e^{2s}\cos^2(\phi-\tfrac{\theta}{2})\right)\right] \qquad (13)$$

where η = 1 for classical noise. Thus, as the local oscillator phase ϕ is varied squeezing results in oscillations in the noise power, and so the squeezing of classical acoustic noise should be observable.

Even if the thermal sources of noise mentioned above can be eliminated there will be noise due to quantum fluctuations in the superfluid. It can be shown that the variances ΔP_1 and ΔP_2 must satisfy the uncertainty principle,

$$\Delta P_1 \Delta P_2 \geq 2n \frac{\hbar\omega}{Vmc_o^2} \qquad (14)$$

where n is the equilibrium density of the superfluid, m is the mass of the ^{3}He atom, c_o is the velocity of sound and V the volume of the sound mode. It is an interesting question as to whether it would be possible to reduce the classical noise in superfluid ^{3}He-B to the extent that the equality in (14) is satisfied. In optical systems it is possible to generate *coherent states of* light using stable lasers in which the noise is dominated by quantum fluctuations. Moreover, it has been possible to use four-wave mixing and homodyne detection to produce squeezed quantum states of light.[13] The noise power that is measured is similar to (13) with η equal to the quantum efficiency of the photodetector, i.e. the ratio of the number of incident photons to the number of photons detected. In optical experiments η is usually greater than 0.5 so the oscillations in P_s with ϕ are large enough to be observed. Even if it is possible to produce coherent phonon states, squeezed quantum states of sound will not be observable unless high efficiency phonon detectors are developed. The quartz transducers commonly used have a large acoustic impedance mismatch with superfluid ^{3}He, resulting in a low value for η.

ACKNOWLEDGEMENTS

Some of this work was performed by one of us (R.M.) in partial fulfillment of the requirements for a Ph.D degree at Princeton University. We are particularly grateful to W.P. Halperin and B. Yurke for helpful discussions. This work was supported in part by NSF grant DMR 8518163.

REFERENCES

1. For a review see W.P. Halperin, Physica 109 and 110B 1596 (1982) and for more recent references see those given in B.S. Shivaram, M.W. Meisel, B.K. Sarma, W.P. Halperin, and J.B. Ketterson, J. Low Temp. Phys. 63, 87 (1986).

2. A.J. Leggett, Rev. Mod. Phys. 47, 331 (1975).

3. K. Maki, J. Low Temp. Phys. 24, 775 (1976).

4. O. Avenel, E. Varoquaux, and H. Ebisawa, Phys. Rev. Lett. 45, 1952 (1980).

5. B.S. Shivaram, M.W. Meisel, B.K. Sarma, W.P. Halperin, and J.B. Ketterson, Phys. Rev. Lett. 50, 1070 (1983).

6. N. Schopohl, M. Warnke, and L. Tewordt, Phys. Rev. Lett. 50, 1066 (1983); R.S. Fishman and J.A. Sauls, Phys. Rev. B 33, 6068 (1986).

7. R. Movshovich, E. Varoquaux, N. Kim, and D.M. Lee, Phys. Rev. Lett. 61, 1732 (1988).

8. Y.R. Shen, Principles of Nonlinear Optics (Wiley, New York, 1984).

9. J.W. Serene, in Quantum Fluids and Solids-1983, edited by E.D. Adams and G.G. Ihas, AIP Conference Proceeding No. 103 (American Institute of Physics, New York, 1983), p. 305.

10. R.S. Fishman and J.A. Sauls, Phys. Rev. B 31, 251 (1985).

11. V.E. Koch and P. Wolfle, Phys. Rev. Lett. 46, 486 (1981)

12. Ross H. McKenzie, Ph.D. Thesis, Princeton University, 1988; Ross H. McKenzie and J.A. Sauls, Europhys. Lett., to appear. Ross H. McKenzie and J.A. Sauls, to be published.

13. R. Loudon and P.L. Knight, J. Mod. Opt. 34, 709 (1987).

FERMI PARAMETER ESTIMATES FROM NON-RESONANT MEASUREMENTS OF THE REAL SQUASHING MODE FREQUENCY IN ^{3}He-B

P.N. Fraenkel
Laboratory of Atomic and Solid State Physics, and
The Material Science Center,
Clark Hall, Cornell University
Ithaca, New York 14853-2501

ABSTRACT

We present the first measurements of the real squashing mode (RSQ) frequency to be made with non-resonant ultrasonic transducers. The data are obtained through repeated frequency sweeps that yield the frequency of the RSQ attenuation maximum over a broad range of temperatures, at fixed pressure. Extensive data at low temperatures, where fixed frequency measurements are difficult, facilitate comparison with theory. Our data place restrictions on Fermi parameters, but proper interpretation at high pressures may require non-trivial strong coupling corrections to the mode frequency calculations.

The coupling of zero sound to the collective modes of the ^{3}He superfluids[1] provides a convenient probe of the superfluid order parameter. Experimentally, these modes can be observed as ultrasonic attenuation maxima at eigenfrequencies f_i of the 3×3 ^{3}He order parameter. Generally, $f_i = a_i(T,p)\Delta(T,p)$, where Δ is the temperature dependent energy gap and a_i is a weakly temperature dependent prefactor of order one. The J=2$^+$ or real squashing mode (RSQ) in ^{3}He-B is characterized in the weak coupling limit by[2,3] $a \approx \sqrt{\frac{8}{5}}$, where the equality is exact if the Fermi parameters F_2^a and x_3^{-1} are set equal to zero.

Experimental observations of this mode[4,5] have been difficult to compare with theory because experiments have been restricted to the discrete harmonic frequencies of the quartz transducers traditionally employed. In order to take data over a broad temperature range, using only a few frequencies, experimenters have varied the gap energy by sweeping sample pressure. This allows the resonance condition to be met, but the scarcity of data points at a single given pressure precludes theoretical fits. In the work reported here, we have used non-resonant ultrasonic transducers which allow us to sweep frequency through the RSQ mode at constant temperature and pressure.

Our apparatus has been described elsewhere.[6] We note here only that the thin film transducers are separated by a 1.6mm spacer and that the experimental volume is in close thermal contact with a 100μm pancake geometry torsional oscillator cell, which is used to measure the normal fluid fraction. The determination of ρ_n/ρ is the primary thermometry for the experiment.

Our transducers are not sensitive in a traditional sense of the word. Typically, we drive one transducer with a 20-110MHz pulse, 2.5 V_{pp} in amplitude, 3μsec in duration, at a repetition rate of up to 0.5 Hz. At this drive level, received signals are useable after 8 or fewer averages, and we measure heating to be between 0.2 and 0.5 nJ for each pulse in the series.

After digitization of the envelope of the received pulse, a computer program locates the pulse train in the time record, computes the area beneath the rectified first received pulse, and stores this "amplitude" along with the temperature and frequency at which the measurement was taken. Repeated sweeps, at varying temperatures, allow us to map out $f_{RSQ}(\rho_n/\rho)$ at constant pressure. Measured mode frequencies are considered precise to 0.1 MHz.

In Fig. 1, we present the principal raw result of the experiment: the RSQ frequencies as a function of normal fluid fraction at 7 pressures between 0 and 29 bar. Using resonant transducers, it would be impossible to measure more than a few points on each of these curves. In particular, one would not observe the slow variation and saturation of f_{RSQ} at low temperatures, and a difficult extrapolation to the zero temperature value would be necessary. Our data are limited at high temperatures by the presence of the squashing mode, which reduces signal sufficiently that our technique cannot resolve the real squashing mode.

The solid lines in Fig. 1 are theoretical fits to the coldest 20 percent of the data. In warmer regions, where the variation of frequency becomes steep, temperature inaccuracies become proportionately more important. Theory, as it stands, includes strong coupling corrections to the gap[3], but only Fermi corrections to mode frequency[2] as a function of gap. "Non-trivial" strong coupling corrections, which would go beyond simple substitution of a strong coupling value for the gap into a weak coupling expression for the frequency, have yet to be calculated. In fact, it has been pointed out[3] that the calculations of collective mode frequencies explicitly assume a weak coupling gap, so that this substitution may be mathematically inconsistent.

In the absence of non-trivial strong coupling corrections, the full resonance condition is

$$f = (k_B/h)T_c \times \phi_{rsq}\left[x_3^{-1}, F_2^a, \left\{\frac{\Delta_+}{k_B T_c}\right\}[\Delta C/C, T/T_c]\right] \times \left(1 + \frac{1}{2}\left(\frac{c_{rsq}}{v_F}\right)^2 \left(\frac{v_F}{c_0}\right)^2\right)$$

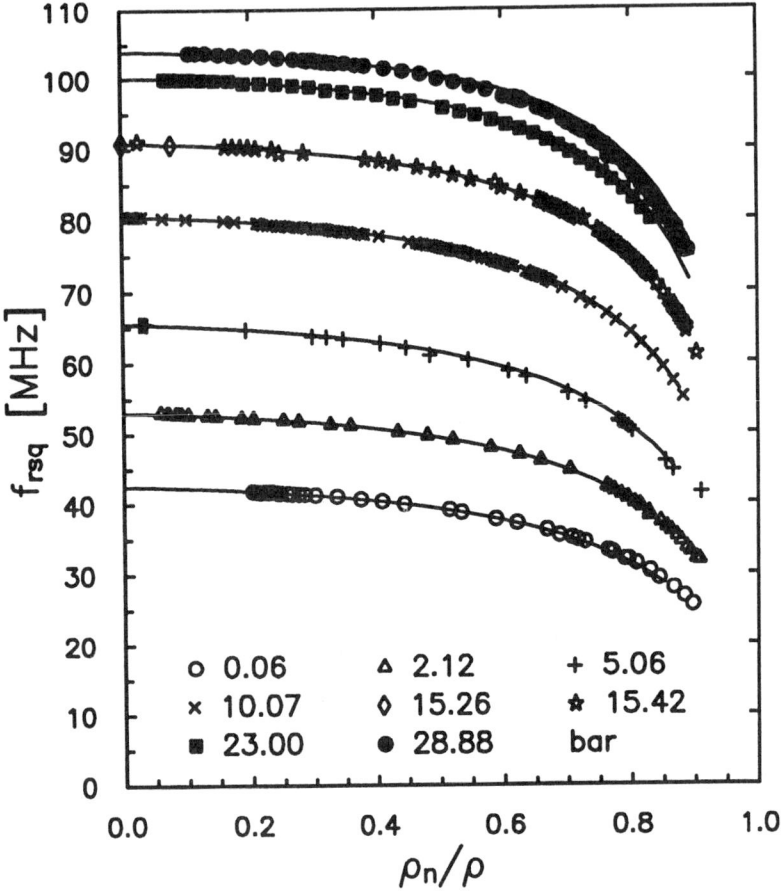

Figure 1. Frequency of RSQ as a function of normal fluid fraction at seven pressures. Lines through symbols are theoretical fits to the coldest 20 percent of the data.

In this expression, the transition temperature T_c appears as an overall scaling factor. The reduced RSQ frequency ϕ_{rsq} depends on $x_3^{-1} = (1/v_1 - 1/v_3)^{-1}$, the cutoff independent f-wave coupling parameter, F_2^a, a second order antisymmetric Fermi parameter, and the gap energy, to which it is loosely proportional. The gap energy Δ_+ is given by the weak-plus theory[3] as a ratio to T_c that depends on reduced temperature T/T_c and on the specific heat jump at the superfluid transition, $\Delta C/C$. The final multiplicative term is a dispersion correction that takes into account the fact that zero sound crosses the RSQ mode at finite wave vector. The ratio of the RSQ mode speed c_{rsq} to the Fermi velocity v_F is, in the weak limit,[7] dependent on magnetic field (zero in our experiment), RSQ quantum number m_J and various Fermi parameters.

This multiplicity of parameters somewhat overspecifies the data, but recent determinations of some parameters make it possible to place constraints on others. We take the transition temperatures to be those measured by Greywall.[8] It is encouraging that his temperature scale, based on measurements of ^{3}He heat capacity, agree closely with that of Fukuyama et al.[9], which derives from susceptibility measurements in platinum. The most elaborate gap calculation to date is that of Serene and Rainer[3]. Their quasi-classical weak-plus gap can be expressed a function of $\Delta C/C$, which in the weak limit approaches 1.43. The weak-plus gap differs from the weak gap by up to 20 percent, and, at a given pressure, the difference is temperature dependent. The most recent confirmation of the weak-plus gap is probably its success in predicting the squashing mode g-factor measured by Movshovich et al..[10]

The dispersion correction is on the order of one percent, which is within the uncertainty in T_c; parameters which enter only through dispersion are relatively unimportant. By contrast, reasonable choices of x_3^{-1} and F_2^a can affect ϕ_{rsq} by up to 10 percent; we vary these parameters to fit our low temperature data. Within the weak-coupling theory, ϕ_{rsq} is identical for a family of parameters, $x_3^{-1} = a + bF_2^a$. The best fitting values of a and b, corresponding to the curves in Fig. 1, are shown in Fig. 2 as a function of pressure.

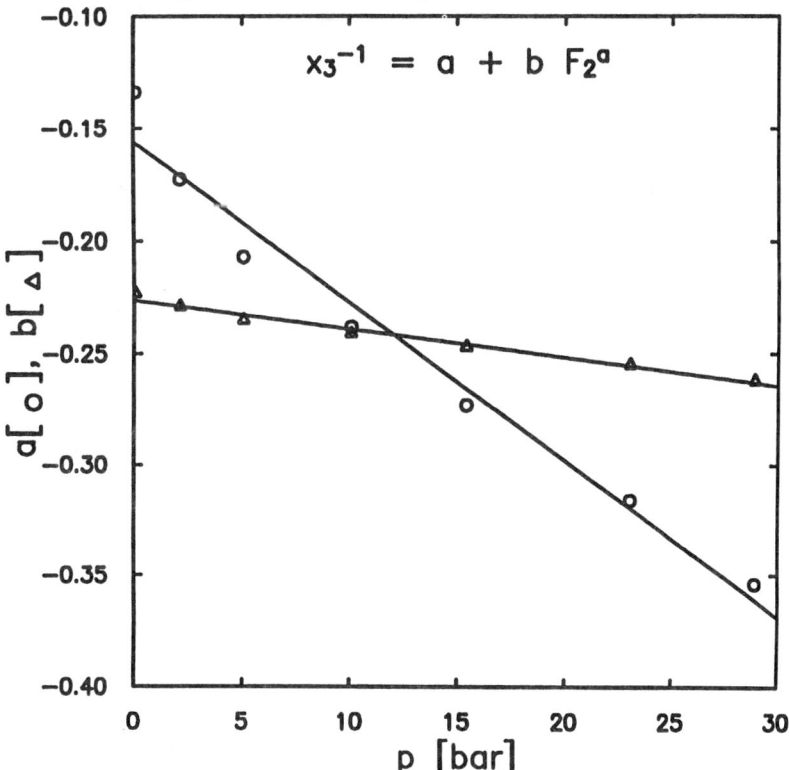

Figure 2. Fits yield a family of parameter values, $x_3^{-1} = a + bF_2^a$.

Fig. 3 shows our data at 15.42 bar, normalized by the theoretical fit. Also shown are predictions using the weak-limit gap and/or without Fermi corrections. The systematic deviation at high temperature is probably due to misdetermination, either theoretical or experimental, of ρ_n/ρ, though this could also be an indication that non-trivial strong coupling corrections are important. We have also plotted, in Fig. 4, the data of Mast *et al.*[11] between 6.5 and 15.5 bar. The data are normalized by theoretical values obtained from the Fermi parameters interpolated from fits to our data. Agreement is good, except, again, at high temperature. Since Mast used more conventional thermometry than we did, the high temperature deviation cannot be easily dismissed.

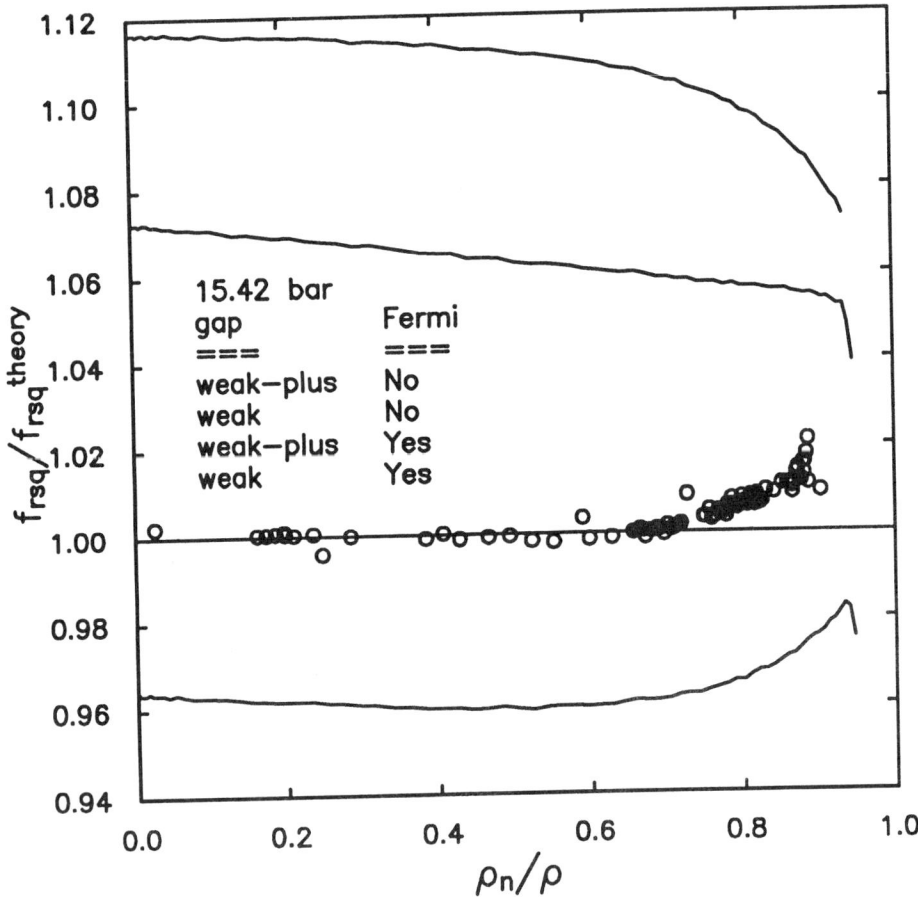

Figure 3. 15.42 bar data normalized by theoretical fit to data below $\rho_n/\rho = 0.35$. Here, we have taken $F_2^a = 0, x_3^{-1} = -0.273$. Solid lines are similarly normalized predictions using the weak BCS gap and/or $F_2^a = x_3^{-1} = 0$.

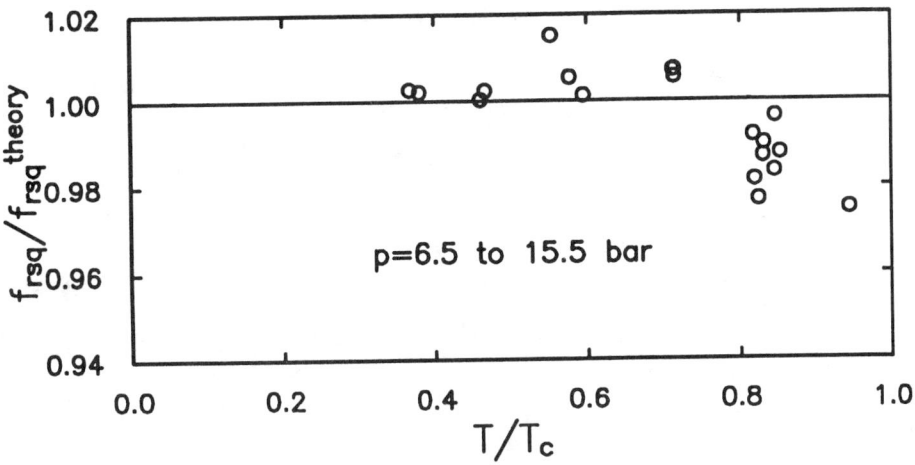

Figure 4. Data of Mast *et al.* between 6.5 and 15.5 bar, normalized by theoretical predictions using parameter values interpolated from our extractions.

Further indication that additional theoretical work is needed is the lack of agreement between Fermi parameters extracted from different experiments.[12] The longitudinal magnetic resonance measurements of Candela *et al.*[13] yield independent estimates of x_3^{-1} and F_2^a that are at variance with our results. At zero pressure, where strong coupling effects should be minimal, the two experiments agree: they find $F_2^a = 0$, $x_3^{-1} = 0 \pm 0.1$; for $F_2^a = 0$, we have $x_3^{-1} = -0.14 \pm 0.05$. At pressures higher than a few bars, the differences do not fall within estimated error limits. At 30 bar, for example the magnetic data yield $x_3^{-1} = -0.35 \pm 0.1$, while we obtain $x_3^{-1} = -1.1 \pm 0.05$.

ACKNOWLEDGEMENTS

The work reported here was carried out at Cornell University in the Laboratory of Atomic and Solid State Physics with support from the National Science Foundation through grant **NSF-DMR-8418605** and with support from the Cornell Material Science Center through grant **NSF-DMR-82-17227A**, MSC Report 6550. I wish to thank Jim Sauls for useful conversations and for providing me with computer programs.

REFERENCES

[1] see, e.g., P. Wölfle, in *Progress in Low Temperature Physics*, edited by D. R. Brewer (North Holland, Amsterdam, 1978), Vol. 7A.

[2] J. A. Sauls and J. W. Serene, Phys. Rev. B. **23**, 4798 (1981).

[3] J. W. Serene and D. Rainer, Phys. Rep. **4**, 221 (1983).

[4] R. W. Gianetta, A. Ahonen, E. Polturak, J. Saunders, E. K. Zeise, R. C. Richardson, and D. M. Lee, Phys. Rev. Lett. **45**, 262 (1980).

[5] D. B. Mast, Bimal K. Sarma, J. R. Owers-Bradley, I. D. Calder, J. B. Ketterson, and W. P. Halperin, Phys. Rev. Lett. **45**, 266 (1980).

[6] P. N. Fraenkel, R. Keolian and J. D. Reppy, Phys. Rev. Lett **62**, 1126 (1989).

[7] R. S. Fishman and J. A. Sauls, Phys. Rev. B. **33**, 6068 (1986).

[9] Hiroshi Fukuyama, Hidehiko Ishimoto, Tetsurou Tazaki and Shinji Ogawa, Phys. Rev. B. **36**, 8921 (1987).

[10] R. Movshovich, E. Varoquaux, N. Kim, and D. M. Lee, Phys. Rev. Let. **61**, 1732 (1988).

[11] W. Halperin, private communication..

[12] See also R. S. Fishman and J. A. Sauls, Phys. Rev. B. **38**, 2526 (1988).

[13] D. Candela, D. O. Edwards, A. Heff, N. Masuhara, Y. Oda, and D. S. Sherrill, Phys. Rev. Lett. **61**, 420 (1988).

COLLECTIVE MODE RESONANCES IN SUPERFLUID HELIUM 3-A

E.R.Dobbs, R.Ling*, J.Saunders and W.Wojtanowski#
Royal Holloway and Bedford New College, Egham, Surrey, TW20 0EX, U.K.

We have previously reported[1] the observation of the clapping mode resonances over a wide frequency range, and both the re-entrant and the high temperature normal flapping mode resonances, in an ultrasonic study of high pressure ^{3}He-A. Here we show that the theoretical interpretation of our data provides evidence for strong coupling corrections to the energy gap far from T_c and to the clapping mode eigenvalue, as well as small $l = 3$ pairing fluctuations.

For the clapping mode we find small corrections to the eigenvalue ω_{cl} near T_c given by:

$$\omega_{cl}/\Delta_0(T) = 1.23 \ [1 - (0.005 - 0.106 x_3^{-1} - 0.052 \ F_2^s)\{\Delta_0(T)/k_BT\}] , \qquad (1)$$

which agrees, with $x_3^{-1}=0$, with that found previously[2]. Here we compare the collective mode resonances that we have measured with theoretical computations of the attenuation maxima associated with the resonances at their reduced temperatures. The parameters in these calculations are $\Delta_0(T)$, F_2^s and x_3^{-1}. Our data was taken at 29.3 bar pressure, where strong coupling corrections apply to the gap. At $T = 0$ the weak coupling gap[3] is $2.03 k_B T_c$, while the strong coupling enhancement is not necessarily the same as that measured[4] near T_c. We have chosen $\Delta_0(0)/k_BT_c = A$ as our unknown parameter to account for the temperature dependent strong coupling effects and write for the gap:

$$\Delta_0(T) = \Delta_0(0) \tanh [\ (\pi/A) \ \{ \ (\Delta C \ / \ C \ ) (1/t - 1) \ \}^{1/2}], \qquad (2)$$

where $\Delta C / C$ is the heat capacity jump at T_c and $t = T/T_c$. We use the heat capacity data and temperature scale of Greywall[4] and obtain F_2^s from Engel and Ihas[5], leaving A and x_3^{-1} to be determined from our data.

The results for the temperature dependence of the clapping mode resonances, normalised to $\Delta_0(T)$, are shown in Fig.1, for two choices of $A = 2.03$ (▲) and 2.64 (●). For both sets of points the absolute value of ω_{cl}/Δ_0 near T_c is significantly smaller than 1.23, in agreement with the data of others[1,6] for $t > 0.98$ analysed with the same parameters. The experimental errors are too small to account for this difference, which could be explained by a non-trivial, strong coupling correction to the mode frequency, reducing it by about 6 %. The parameter A, which controls the strong coupling corrections to the gap away from T_c, now gives the temperature dependence of the experimental values of ω_{cl}/Δ_0, but has a negligible effect on the theoretical curves in Fig.1. The small rise in the upper curve as T falls is due to the F_2^s term, which is cancelled in the lower curve with $x_3^{-1} = -0.4$, in accordance with Eq.(1). Thus agreement between the data and theory requires $\Delta_0(0) = 2.64 k_B T_c$ or 1.3 ± 0.1 times the weak coupling value.

The re-entrant normal flapping resonance is relatively sensitive to x_3, but increasingly insensitive to A at lower frequencies. It has been shown[7,8] that the f-wave pairing interaction induces a distortion of the equilibrium energy gap without changing its symmetry and that[9] it is important to include this gap distortion in calculating ω_{nfl}.

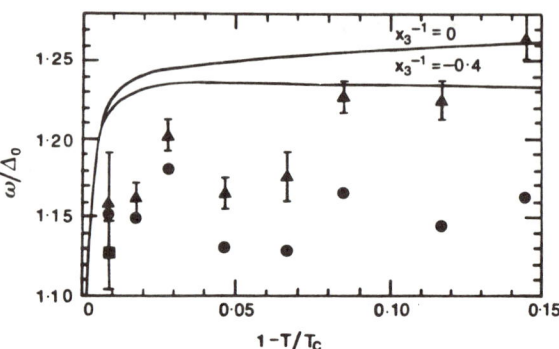

Fig.1. The normalised clapping mode resonances.

We have calculated the reduced temperature of the attenuation maxima due to the re-entrant normal flapping mode, including the effects of x_3 and the gap distortion, as shown in Fig.2. The curves, which are not strongly influenced by F_2^s, show that small f-wave effects must be included to fit the data. With $A = 2.64$ from ω_{cl}, we estimate $x_3^{-1} = -0.1 \pm 0.05$. The high temperature normal flapping peaks have been resolved at 34.2 and 44.2 MHz, but are rather broad with maxima at $t = 0.952$ and 0.892. In these cases the f-wave and strong coupling effects largely cancel and so the data appears to be insensitive to these effects.

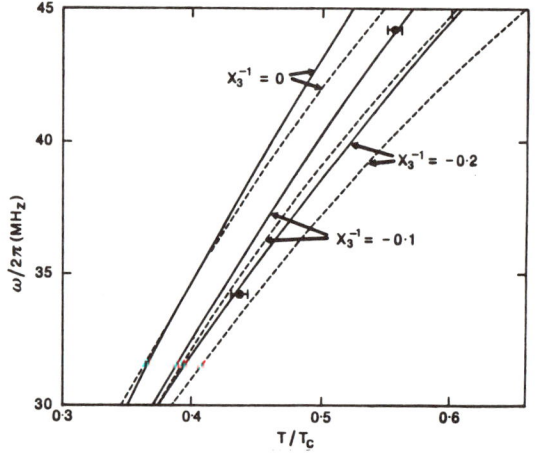

Fig.2. Re-entrant normal flapping mode resonances.

REFERENCES

* Now at Physics Building, University of Sussex, Brighton, Sussex BN1 9QH, U.K.
# On leave from: Institute of Physics, Pedagogical University, Oleska, Opole, Poland.
1. R.Ling, J.Saunders and E.R.Dobbs, Proc.LT18, Jap.J.App.Phys.**26**,Supp.26-3,119 (1987) and references therein.
2. M.Ashida and K.Nagai, Proc.LT17, Ed.U.Eckern, A.Schmid, W.Weber and H.Wuhl.(North Holland,Amsterdam, 1984),p.783.
3. R.Combescot,J.Low Temp.Phys.**18**,537 (1975).
4. D.S.Greywall,Phys.Rev.**B 33**,7520 (1986).
5. B.Engel and G.G.Ihas,Phys.Rev.Lett.**55**,955 (1985).
6. U.E.Israelsson,B.C.Crooker, H.M.Bozler and C.M.Gould,Phys.Rev.Lett.**56**,2383 (1986).
7. W.Wojtanowski and P.Wölfle, Phys.Lett.**A 115**,49 (1986).
8. J.A.Sauls, Phys.Rev.**B34**, 4861 (1986).
9. D.S.Hirashimi, Phys.Rev.Lett.**59**,2386 (1987).

ANISOTROPIC ATTENUATION OF ZERO SOUND IN SUPERFLUID HELIUM 3 - A

J.Saunders, W Wojtanowski[*], R.Ling[#] and E.R.Dobbs.
Royal Holloway and Bedford New College, Egham, Surrey, TW20 0EX, U.K.

The attenuation α of zero sound in ^{3}He-A is anisotropic with respect to the angle β between q, the ultrasonic wave vector, and the **l** vector of the superfluid. It is given by:

$$\alpha(\beta) = \alpha_\parallel \cos^4\beta + 2\alpha_c \cos^2\beta \sin^2\beta + \alpha_\perp \sin^4\beta, \qquad (1)$$

where $\alpha_\parallel, \alpha_c$ and $\alpha_\perp$ are independent functions. The collective modes are related to each of these functions: $\alpha_\perp$ has contributions from the clapping mode, α_c from the flapping mode, while $\alpha_\parallel$ has no contributions from any of the collective modes and so is due solely to pair breaking. Since α is very sensitive to the angle β, it is essential to make measurements on samples with well defined textures. We have propagated zero sound at angular frequency ω through a 250 µm, plane parallel sample of 29.3 bar ^{3}He-A in a sonic cell[1], where an applied magnetic field **B**, that could be oriented at any angle θ to **q**, provided control of the almost uniform texture.

The temperature dependence of the pair-breaking attenuation $\alpha_\parallel$ is shown in Fig.1. For temperatures near T_c (Fig.1a), the attenuation was found to be almost independent of frequency in the range 44-84 MHz. In the collisionless theory[2], $\alpha_\parallel$ is independent of frequency in this range, proportional to $(1-T/T_c)$ and weakly dependent on F_2^s. The inclusion of $F_2^s = 0.8$, from Engel and Ihas[3], is shown in the solid line in Fig.1a to improve the fit to the data at 54.0 MHz, when compared with the broken line where $F_2^s = 0$. Further away from T_c the attenuation reaches a maximum near $\omega = 2\Delta_0(T)$, the maximum energy gap, and then decreases as T falls and the gap expands. As expected, instrumental broadening smears the pair-breaking cusp in the data. In Fig.1b the data at 63.9 MHz is compared with collisionless theory, with and without f-wave interaction; the best fit was obtained with a small f-wave parameter, x_3^{-1}.

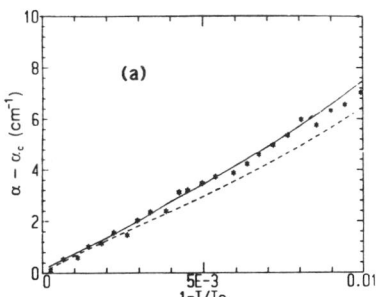
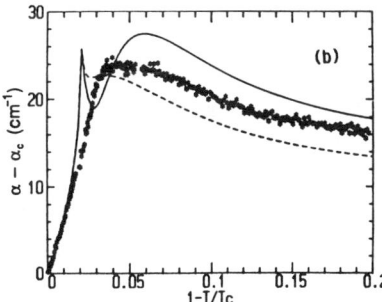

Fig.1. Pair-breaking attenuation, $\alpha_\parallel$: (a) 54.0 MHz, (b) 63.9 MHz, $x_3^{-1} = 0$ (---); - 0.1 (—).

The clapping mode attenuation, $\alpha_\perp$ has been measured at eight frequencies in the range 24 to 94 MHz producing attenuation peaks rising to 100 cm^{-1}. A typical result measured with $\beta = \pi/2$ at 54.0 MHz is shown in Fig.2, where the data is compared with the theory of Wölfle and Koch[4] that includes collisions, but no f-wave interactions. It is seen that the maximum of $\alpha_\perp$ is in good agreement with theory and that the shoulder near T_c is well reproduced. On the other hand the linewidth is rather larger than predicted, as had been that measured previously by Paulson et al.[5].

In order to measure the attenuation due to both of the normal flapping peaks, we stabilized the A phase to low temperatures in magnetic fields in the range 0.35 - 0.45 T. At these low temperatures the damping of the resonances is dominated by pair-breaking processes and so we compare our data at 44.2 MHz with the collisionless theory in Fig.3. It is evident that the general shape of our data is reproduced by the theory, especially when the parameters used to fit the positions of the peaks[6] are : $x_3^{-1} = -0.1$, $\Delta_0(0) = 2.64\, k_B T_c$, as shown by the solid curve, rather than $x_3^{-1} = 0$ and the weak coupling gap, $\Delta_0(0) = 2.03\, k_B T_c$, as shown by the broken curve. However, the total attenuation is significantly smaller than predicted, which we attribute to non-linearity caused by taking this set of data at a power level some 10 dB greater than our other data. Avenel et al.[7] found a similar non-linearity in their study of the normal flapping mode.

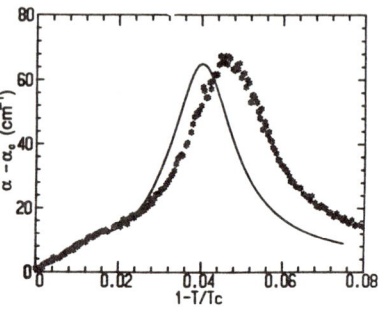

Fig.2. Clapping mode attenuation at 54 MHz.

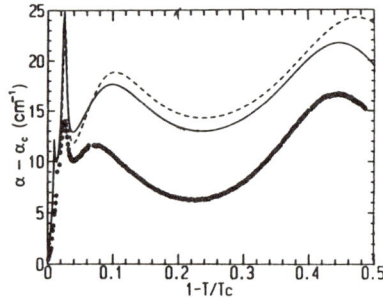

Fig.3. Intermediate angle attenaution, 44.2 MHz.

REFERENCES

* On leave from: Institute of Physics, Pedagogical University, Oleska, Opole, Poland.
\# Now at Physics Building, University of Sussex, Brighton, Sussex BN1 9QH, U.K.
1. J.Saunders, M.E.Daniels, E.R.Dobbs and P.L.Ward, Quantum Fluids and Solids, Eds.E.D.Adams and G.G.Ihas (American Institutre of Physics, New York, 1983),p.314.
2. N.Schopohl, private communication.
3. B.Engel and G.G Ihas, Phys.Rev.Lett. **55**, 955 (1985).
4. P.Wölfle and V.E.Koch, J.Low Temp.Phys. **30**, 61 (1978).
5. D.N.Paulson, M.Krusius and J.C.Wheatley, J.Low Temp.Phys. **26**, 73 (1977).
6. E.R.Dobbs, R.Ling, J.Saunders and W.Wojtanowski, previous paper in this volume.
7. O.Avenel, L.Piche and E.Varoquaux, Physica, **107B**, 689 (1981).

PAIR BREAKING EDGE IN SUPERFLUID ^{3}He

S. Adenwalla, Z. Zhao, and J.B. Ketterson
Northwestern University
Evanston, Illinois 60208

B.K. Sarma
University of Wisconsin-Milwaukee
Milwaukee, Wisconsin 53201

We present the first direct, systematic measurements of the superfluid energy gap in the B-phase of ^{3}He. We compare our measurements to the predictions of the weak-coupling-plus model.[1] The comparison depends on the temperature scale used; however weak-coupling theory does not provide an accurate description for either temperature scale. We find the best agreement occurs if we use a combination of the Greywall temperature[2] scale and the weak-coupling-plus model[1] for the gap.

We have measured the pair-breaking edge, 2Δ, using a c.w., single-ended impedance technique, and alternate presssure and temperature sweeps, described elsewhere.[3] The acoustic cell contains two transducers separated by 190.5 μm and thus the round trip path length is 381 μm. Experiments were performed up to the 13th harmonic of our 12.79 MHz fundamental transducer (167 MHz). We used an LCMN thermometer mounted above the acoustic cell out of the field of the demagnetization magnet.[3] To analyze our data, we used the temperature scale, and the values for $\Delta C/C_N$ as a function of pressure, reported by Greywall.[2]

A typical temperature (pressure) trace is shown in Fig. 1. As we cool into the superfluid, there is a step in the impedance at T_c. Below T_c, the attenuation is high due to damping by the pair breaking mechanism and continues to increase as we cool. At a temperature T_{PB}, where $h\nu = 2\Delta(T_{PB})$, the sound attenuation decreases abruptly and we observe the onset of oscillations due to the presence of standing waves in the cell. (The oscillations are caused by changes in the sound velocity with temperature or pressure and can only appear when the attenuation is low enough that the returning (reflected) wave can cause a measurable shift in the transducer response.) The point at which the oscillations appear (implying the presence of a standing wave pattern in the cell, and hence a sudden decrease in attenuation) is taken as the pair-breaking edge, 2Δ.

In Fig. 2, we have plotted our data for the pair-breaking edge in the pressure-temperature plane. The curves are plotted for the frequencies at which data were taken using both Δ_{BCS} and Δ^+, the weak-coupling-plus model of Rainer and Serene.[1] Our data appear to lie between the two curves, but are much closer to the Δ^+ curves. The scatter in our data is chiefly due to our thermometry.

© 1989 American Institute of Physics

For the pair-breaking edge, the coefficient of the gap, a, (in the expression $h\nu = a\Delta$) must be 2. We have calculated the coefficient for all our data points using both the Helsinki temperature scale[4] and the Greywall temperature scale[2] and find that, if our data points are to cluster around 2, it is necessary to use both the Greywall temperature scale and the weak-coupling-plus gap. These results provide an indirect confirmation of the Greywall temperature scale. The weak-coupling-plus model appears to work well; however, the "center of gravity" of our points appears to be slightly less than 2, which would imply that the weak-coupling-plus gap tends to over-estimate the strong-coupling effects. This can be seen clearly in Fig. 2 (especially for the 141 MHz data) in which our data points lie close to, but not on, the line calculated using the weak-coupling-plus gap.

Acknowledgements: We are grateful to J.A. Sauls for providing the computer program for the weak-coupling-plus gap and to D.W. Hess for useful conversations. This work is supported by NSF grant DMR 86-02857 at Northwestern University and the Office of Naval Research at the University of Wisconsin-Milwaukee.

REFERENCES

1. D. Rainer and J.W. Serene, Phys. Rev. B13. 4745 (1976).
2. D. Greywall, Phys. Rev. B33, 7520 (1986).
3. S. Adenwalla, Z. Zhao, J.B. Ketterson and B.K. Sarma, JLTP, 76 (to be published).
4. T.A. Alvesalo, T. Haavasoja, M.T. Manninen and A.T. Soinne, Phys. Rev. Lett., 44, 1076 (1980).

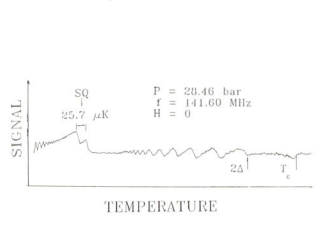

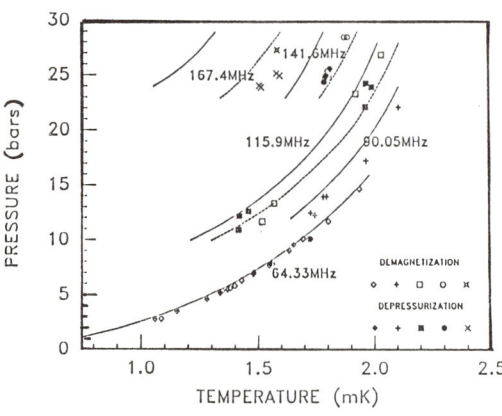

Fig. 1 Typical temperature (pressure) sweep. For details, see text.

Fig. 2 Pair breaking edge in the pressure-temperature plane. The solid curves correspond to $2\Delta_{BCS}$ and the dashed curve to $2\Delta^+$ for each frequency.

OBSERVATION OF A DOUBLET SPLITTING OF THE SQUASHING MODE IN SUPERFLUID ^{3}He–B

Z. Zhao, S. Adenwalla, P.N. Brusov and J.B. Ketterson
Physics and Astronomy Department
Northwestern University
Evanston, Illinois 60208

B.K. Sarma
University of Wisconsin-Milwaukee
Milwaukee, Wisconsin 53201

EXPERIMENT

The squashing mode in ^{3}He - B has been studied using a single ended cw acoustic impedance technique.[1] The sound cell employed two 12.8 MHz x-cut quartz transducers separated by a pair of gold plated tungsten wires resulting in a round trip path length of 381 μm. The spacing was calibrated by measuring the change of the velocity with pressure and was consistent with the measured wire diameter.

The principal thermometer was the susceptibility of LCMN, which was calibrated against the superfluid transition temperature. The pressure was measured in situ with a capacitance gauge that was in turn calibrated (at 1K) against a Paro Scientific Pressure Gauge. The Helsinki temperature scale was adopted. The collective modes were probed in both pressure and temperature sweeps at frequencies of 115.8 MHz and 141.6 MHz. A doublet splitting of the sq-mode in zero field has been observed in both cases as shown in Fig. 1.

DISCUSSION

There are at least three explanations of the observed effect: (1) a dispersion induced splitting of the sq-mode; (2) a texture induced splitting of the $J_z = 0$ branch of the sq-mode and (3) the existence of some other phase near the boundary.

In the first case, a three-fold splitting should be observed as predicted in Ref. 2. However the splitting between the $|J_z| = 1$ and $|J_z| = 2$ branches is only one-quarter of that between the $|J_z| = 0$ and $|J_z| = 2$ branches, which would not be resolved in this experiment. The experimentally observed splitting is about four times larger than that predicted by weak-coupling collective mode theory.[2] A similar situation arose for the rsq-mode[1] and the discrepancy was greatly reduced by incorporating Fermi liquid corrections (FLC).

Two other explanations of the observed two-fold splitting are attractive. A picture similar to the case of the rsq-mode may apply,[1],[3] where a texture induced doublet splitting of the $J_z = 0$ central peak was observed in low magnetic fields. Similar phenomena may occur for the sq-mode in zero field because of textures created by the restricted geometry. (The theory of texture induced splitting of sq-mode needs further development.)

Another possibility is that the additional peak is associated with a collective mode in a boundary induced 2D-phase[4]: one peak would arise from the bulk B-phase sq-mode and the other from the super-flapping (sfl) mode in the 2D-phase. The spectrum of the collective modes in the 2D-phase has been calculated recently.[5] The results show that part of the spectrum in the 2D-phase is the same as in the A-phase (e.g. we have the cl-mode and the pb-mode). It is known that the sfl-mode only appears when strong coupling effects are included.[6] Estimates show that the difference between the sq-mode in ^{3}He - B and the sfl mode in the 2D-phase is of the order of a few tens of μK at T/T_c = 0.7 and is consistent with the observed splitting.

This work was supported by NSF grant DMR-86-02857 and ONR.

REFERENCES

1. B.S. Shivaram, M.W. Meisel, B.K. Sarma, D.B. Mast, W.P. Halperin and J.B. Ketterson, Phys. Rev. Lett. 49, 1646 (1982).
2. Y.A. Vdovin, Proc. of Moscow Engineering Physical Institute (MEPI), Moscow, 94, (1962). P.N. Brusov, V.N. Popov, Sov. Phys. JETP. 51, 1217 (1980).
3. R.S. Fishman, J.A. Sauls, Phys. Rev. Lett. 61, 2871 (1988).
4. T. Fujita, M. Nakahara, T. Ohmi and T. Tsuneto, Progr. Theor. Phys. 64, 396 (1980).
5. P.N. Brusov, M.V. Lomakov, to be published.
6. P.N. Brusov, V.N. Popov, Sov. Phys. JETP, 52, 945 (1980).

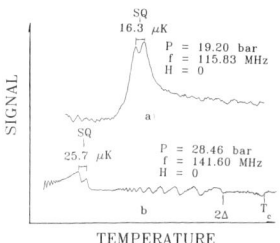

Fig. 1 Doublet splitting of the sq-mode in ^{3}He - B

THE LINEAR ZEEMAN EFFECTS FOR CLAPPING AND PAIRBREAKING MODES IN ^{3}He–A

P.N. Brusov, M.Y. Nasten'ka, T.V. Filatova–Novoselova,
M.V. Lomakov and V.N. Popov
*Department of Physics and Astronomy, Northwestern University,
Evanston, IL 60208-3112*

First theoretical investigations[1] of the collective excitation (CE) spectrum in ^{3}He–A determined the energies of the clapping (cl), flapping (fl), and pair–breaking (pb) modes to be $E = 1.22\Delta_0$, $1.56\Delta_0$, and $2\Delta_0$, respectively (here Δ_0 is the maximum of the energy gap $\Delta = \Delta_0\sin\theta$). These values were obtained without taking any CE damping into account. It is clear that the vanishing of the gap along the orbital anisotropy axis $\vec{l}$ leads to damping of CE, because CE with nonzero energy and small momentum can always decay kinematically into two fermions whose momenta are nearly opposite to each other and close to the axis $\vec{l}$. The whole CE spectrum with this damping was first considered by Brusov and Popov[2], who obtained six cl–modes with energy $E = (1.17 - i0.13)\Delta_0$ and three pb–modes (superclapping modes in other terminology) with $E = (1.96 - i0.31)\Delta_0$. Note that the damping (Im $E \neq 0$) leads to renormalization of Re E via dispersion relations. Another interesting fact first obtained in Ref. 2 is that the number of Goldstone modes (gd) in the weak–coupling approximation is equal to nine rather than five, which takes place in real ^{3}He–A. The existence of four additional quasi–gd spin–orbit modes is a consequence of the latent symmetry of the system as Volovik[3] noted. These four modes are an analogy of the massless W–bosons in the theory of electroweak interaction. In ^{3}He –A the W–bosons have a mass because of the strong coupling corrections[2] (and gd–modes become fl–modes), whereas they acquire a mass in the Weinberg–Salam theory because of the Higgs phenomenon.

Here, by using a path integral methos, we obtain the whole set of equations which describe the CE in ^{3}He–A in a magnetic field for $T_c - T \sim T_c$, and solve them for small fields and for zero CE momenta. The system is described by the hydrodynamic action functional

$$S_h = g^{-1} \sum_{p,i,a} c^{\dagger}_{ia}(p) \, c_{ia}(p) + \frac{1}{2} \ln \det \hat{M}(c,c^+)/M(c^{(0)},c^{(0)+}),$$

where $c_{ia}(p)$ are the Bose fields, and $\hat{M}$ is an operator which depends on quasi–fermion parameters and $c_{ia}(p)$. As a first approximation the CE spectrum is determined by the quadratic part of S_h, which can be obtained by a shift in the Bose fields $c_{ia}(p) \to c^{(0)}_{ia}(p) + c_{ia}(p)$, where $c^{(0)}(p)$ is the condensate wave function.

In the presence of a magnetic field, we must take into account both the additional term in S_h and the distortion of the order parameter[4]. The latter in our case is

$$c_{ia}(p) = c\sqrt{\beta V} \, \delta_{p0} \, (\delta_{a1}\alpha_+ + i\delta_{a2}\alpha_-)(\delta_{i1} + i\delta_{i2}).$$

© 1989 American Institute of Physics

Here $\alpha_\pm = (\Delta_\uparrow \pm \Delta_\downarrow)/2\Delta$, $\Delta^2_{\uparrow\downarrow} = N(0)(\tau \pm \eta h)/2\beta_{245}$ with $\eta = (N(0)/N(0))T_c \ln(1.14\epsilon_0/T_c)$ and $h = \mu_0 H/T_c$, and $\Delta = 2cZ$ is a single fermion spectrum gap determined by the gap equation

$$g^{-1} = -\frac{Z^2}{\beta V} \frac{1}{(\alpha_+^2 + \alpha_-^2)} \sum_p \left[\frac{(\alpha_+ + \alpha_-)^2 \sin^2\theta}{\omega^2 + (\xi - \mu H)^2 + \Delta^2 \sin^2\theta(\alpha_+ + \alpha_-)^2} \right.$$

$$\left. + \frac{(\alpha_+ - \alpha_-)^2 \sin^2\theta}{\omega^2 + (\xi + \mu H)^2 + \Delta^2 \sin^2\theta(\alpha_+ - \alpha_-)^2} \right]$$

The equation $\det Q = 0$, where Q is the matrix of quadratic form for S_h, gives us 18 equations which completely determine 18 collective modes with arbitrary momenta in an arbitrary magnetic field. After solving them for small magnetic field $\vec{H}$ and for $\vec{K} = 0$, we obtain the energies of the cl– and pb– modes

cl: $E_1 = (1.17 - i0.13)\Delta_0$

$E_{2,3} = (1.17 \pm 1.62\gamma H)\Delta_0 - i(0.13 \pm 1.33\gamma H)\Delta_0$

pb: $E_1 = (1.96 - i0.31)\Delta_0$

$E_{2,3} = (1.96 \pm 1.18\gamma H)\Delta_0 - i(0.31 \pm 0.69\gamma H)\Delta_0$

Therefore, for small $\vec{H}$ we have three–fold splitting of the cl– and pb–modes, a linear Zeeman effect. The degeneracy of the pb–modes is completely lifted by the magnetic field, while the cl–modes is split into three doublets.

Note that the magnetic field changes both the real and the imaginary parts of CE enregies, i.e., the CE frequencies and the dampings, even within the linear approximation. The number of gd–modes decreases from nine to six via the appearance of the gap $\sim \mu H$ in their spectrum.

REFERENCES

1. P. Wölfle, Physica **90B**, 96 (1977).
2. P.N. Brusov and V.N. Popov, Sov. Phys. JETP **52**, 943 (1980).
3. G.E. Volovik, JETP Lett. **43**, 693 (1986).
4. M.Y. Nasten'ka and P.N. Brusov, Phys. Lett. A, to be published.

NEW SOUND AND NMR EXPERIMENTS IN ^{3}He–B AND COMMON APPROACH TO THEIR THEORETICAL DESCRIPTION

M.Y. Nasten'ka and P.N. Brusov

*Department of Physics and Astronomy, Northwestern University,
Evanston, IL 60208–3112*

There are two main methods for experimental investigation of the collective properties of the superfluid phases of ^{3}He: sound and NMR. Until recently, the theoretical description of these two kinds of experiments were based on very different schemes: the description of the former was within the theory of collective excitations, while the description of the latter was based on the phenomenological Leggett–Takagi equations.

A few years ago, one of us suggested a new point of view on this problem: a common approach to the description of sound and NMR experiments in superfluid ^{3}He in the language of collective excitations. The complicated order parameter in superfluid ^{3}He leads to (besides the difficulties of the theoretical description of this) the rich spectrum of collective excitations, which consists of 18 branches in each phase. Among these there are nonphonon modes (the oscillations of the self–consistent field) whose frequencies are nonzero at zero excitation momentum and Goldstone modes whose existence is connected with the spontaneous breaking of the symmetry at the phase transition into the superfluid state. In sound experiments the former modes are excited, and in NMR the latter ones (longitudinal and transverse spin waves). If one could create a theory that describes the whole spectrum of collective excitations (nonphonon and Goldstone modes), one could describe both sound and NMR experiments simultaneously within one theoretical scheme. For the description of NMR experiments we must take into account the dipole interaction as well as the gap distortion caused by it. Taking the gap distortion into account is also very important for any external perturbations.

Below we suggest some new experiments following Brusov and Popov[1,3].

1. Dispersion–induced splitting of the squashing mode

In 1980 Brusov and Popov[1] made the first calculation of all 18 collective modes in ^{3}He–B with dispersion corrections. (For 14 modes this has been done by Vdovin and Nagai). These reslults show a dispersion–induced three–fold splitting of the real squashing (rsq) mode and the squashing (sq) mode. The splitting of the rsq–mode has been observed experimentally. The sound absorption by the sq–mode is much larger than by the rsq–mode and it is difficult to observe such a splitting. But is seems possible to observe it using an impedance technique.

2. Experiments in electric fields

a) One could observe splitting of the rsq– and sq–modes at $E \approx 5 \cdot 10^5 - 2.7 \cdot 10^6$ V/cm.

b) Electric fields could change by a detectable amount the separations between the three dispersion–induced peaks for the rsq– and sq–modes in the sound absorption spectrum.

c) One could observe resonant sound absorption at the absorption edge (see below).

d) One could observe a shift of the longitudinal NMR frequency in fields of order $1.5 \cdot 10^5 - 1.5 \cdot 10^6$ V/cm. (The experimental accuracy is of order 10^{-5}).

e) At nonzero momentum $\vec{k}$, it is possible that crossing occurs among the branches of rsq– or sq–modes with different values of $|J_z|$. If one changes the field E, it is possible to observe the disappearance of one peak at some value of E.

3. Experiments in the presence of superflow

a) One could observe a three–fold splitting of the rsq– and sq–modes at $v < v_c$. At velocities close to the critical value $v_c \sim 1$ mm/s such splitting is of the order of $10^{-2}\Delta_0^2$ for the squares of frequencies.

b) One could observe crossing of the branches of rsq–modes with different values of J_z at nonzero momentum $\vec{k}$. The branch with $J_z = 0$ crosses the branches with $|J_z| = 1$ and 2, and the branches with $|J_z| = 1$ cross branches with $|J_z| = 2$.

c) One could observe resonant sound absorption at the absorption edge in the presence of the superflow (see below).

d) It is possible to observe hydrodynamical shifts in NMR experiments.

4. Sound experiments at the absorption edge

In 1983 Daniels et al.[2] experimentally observed the resonant absorption of zero sound at the absorption edge in the presence of magnetic fields. As Brusov has shown, it is possible to observe the same phenomena in the presence of electric field and superflow (for $v_s < \sim 1$ mm/s).

5. Two–dimensional superfluidity in ^{3}He films

A new possibility in the search of two–dimensional (2D) superfluidity exists according to the results obtained by Brusov and Popov[3], who showed by investigation of the Bose–field correlator behavior that 2D superfluidity does exist in the presence of magnetic field.

REFERENCES

1. P.N. Brusov and V.N. Popov, Sov. Phys. JETP **51**, 1217 (1980).
2. M.E. Daniels et al., Phys. Rev. **B27**, 6988 (1983).
3. P.N. Brusov and V.N. Popov, Sov. Phys. JETP **53**, 804 (1981).

GYROSONIC EFFECT IN ROTATING SUPERFLUID ^{3}He-B

M.M. Salomaa and G.E. Volovik
*Low Temperature Laboratory, Helsinki University of Technology
SF-02150 Espoo 15, Finland*

Collective-mode spectroscopy in superfluid ^{3}He-B has emerged as a new tool, serving to probe the ^{3}He-B $\hat{n}$-textures under rotation[1]. We present a phenomenological theory for the propagation of the real squashing modes (rsq) utilizing the minimal number of phenomenological parameters, which allows one to extract information on the textures by analyzing the experimental lineshapes of the rsq modes.

Our approach[2] is based on the phenomenological Lagrangian $\mathcal{L}$ for the symmetric traceless matrix u_{ik} describing the rsq modes with $J = 2$:

$$\mathcal{L} = -\dot{u}_{ik}\dot{u}_{ik} + \omega_0^2 u_{ik}u_{ik} + c_1^2 \vec{\nabla} u_{ik} \cdot \vec{\nabla} u_{ik} + c_2^2 \nabla_i u_{ik} \nabla_j u_{jk}$$
$$+ \epsilon_{ijl} g \tilde{H}_l u_{ik} \dot{u}_{jk} + (\eta_{KW} \nabla_i \nabla_k u_{ik} + \eta_{SS} \omega_i \nabla_k \dot{u}_{ik})\rho, \quad (1)$$

where the field $\tilde{\vec{H}}$ is the "textural" field related with the actual magnification field $\vec{H}$ through the order-parameter matrix $R_{\alpha i}(\vec{r})$ in the texture, i.e., $\tilde{H}_\alpha(\vec{r}) = R_{\alpha i}(\vec{r})H_i$.

The final term in Eq. (1) is the coupling of the rsq mode to density (ρ) fluctuations of the zero sound; this coupling describes the generation of the rsq collective modes by the ultrasound. The term with η_{KW} was first introduced by Koch and Wölfle[3], while the one denoted by η_{SS} describes the effect of the superfluid vs. the normal fluid counterflow $\omega = v_s - v_n$, first introduced by Sauls and Serene[4].

The relative strength of the magnetic-field term, in comparison with the dispersion terms is scaled by the parameter $H_0 = c_2^2 q^2/4g\omega_0$. For $H \ll H_0$, the spectrum is described by the projection m of $J = 2$ on the direction $\hat{q}$ of the wavevector $\vec{q}$, but for large fields $H \gg H_0$ by the projection M of $J = 2$ on the textural field $\hat{h} = \tilde{\vec{H}}/H$, i.e.,

$$E = \frac{\omega^2 - \omega_0^2 - c_1^2 q^2}{2\omega_0 g H_0} = \begin{cases} \frac{HM}{H_0} + \left[\frac{1}{3}(1 + \frac{1}{2}M^2) + (\hat{q}\cdot\hat{h})^2(1 - \frac{1}{2}M^2)\right], & H \gg H_0; \\ \frac{1}{3}(4 - m^2), & H \ll H_0. \end{cases}$$

In the intermediate range of fields, $H \approx H_0$, the spectrum of the rsq modes has been evaluated from Eq. (1) numerically; the intensities of the rsq modes in the ultrasonic absorption spectrum have also been found; for results with $H = 2H_0$, see Fig. 1.

© 1989 American Institute of Physics

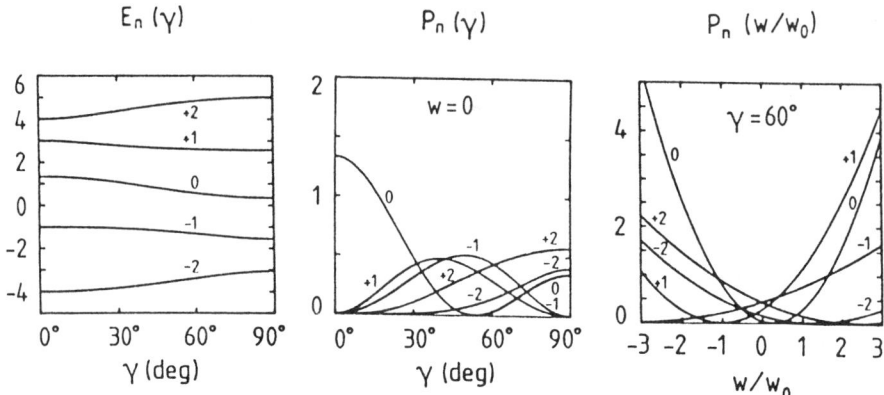

Fig. 1. Spectrum of rsq modes (a) as a function of the angle γ of the texture ($\cos\gamma = \hat{h}\cdot\hat{q}$) and intensity of these lines in the ultrasound absorption (b) in terms of γ, and (c) counterflow w, in the plane perpendicular to $\hat{q}$. Here $\omega_0 = q\eta_{KW}/\omega_0\eta_{SS}$.

Note that the intensities are not even in ω, thus producing a novel kind of *gyrosonic effect*: the mode intensities depend on the direction of rotation even when the ultrasound propagates along the rotation axis. This textural effect is related with the scalar $Q = \omega(\vec{q}\cdot\hat{h})(\vec{\omega}\cdot\hat{h})$, invariant under both time inversion ($\omega \to -\omega, \vec{\omega} \to -\vec{\omega}, \hat{h} \to -\hat{h}$) and space inversion (i.e., $\vec{q} \to -\vec{q}, \vec{\omega} \to -\vec{\omega}$). The mode intensities should not be even in the scalar Q; they therefore change on switching the sign of Q either by reversing the sense of rotation for a given texture, whence $\vec{\omega} \to -\vec{\omega}$, or by transmitting the ultrasound pulse in the opposite direction, with $\vec{q} \to -\vec{q}$, under rotation.

This research has been supported by the Körber-Stiftung (Hamburg, FRG) and by the USSR Academy of Sciences and the Academy of Finland through the project ROTA for the investigation of superfluid ^{3}He under rotation.

REFERENCES

1. R.H. Salmelin, J.P. Pekola, A.J. Manninen, K. Torizuka, M.P. Berglund, J.M. Kyynäräinen, O.V. Lounasmaa, G.K. Tvalashvili, O.V. Magradze, E. Varoquaux, O. Avenel and V.P. Mineev, to be published.

2. M.M. Salomaa and G.E. Volovik, *J. Low Temp. Phys.*, submitted (1989).

3. V.E. Koch and P. Wölfle, *Phys. Rev. Lett.* **46**, 486 (1981).

4. J.A. Sauls and J.W. Serene, *Proc. 17th Int. Conf. Low Temp. Phys.*, ed. U. Eckern, A. Schmid, W. Weber and H. Wühl (North-Holland, Amsterdam, 1984), p. 775.

FINITE AMPLITUDE ACOUSTIC WAVES IN LIQUID ^{3}He AND ^{4}He

M. Chapellier*, J. Joffrin*, M.W. Meisel‡, and A. Schuhl*

*Laboratoire de Physique des Solides, Univ. Paris–Sud, Bat. 510, 91405 Orsay, France; ‡Department of Physics, Univ. Florida, Gainesville, FL 32611, USA.

Liquid ^{3}He and ^{4}He possess strong anharmonic effects when subjected to a coherent plane acoustic sinusoidal wave. Consequently, any finite sinusoidal acoustic wave generates harmonic waves at the expense of the pumping wave.

An acoustic cell with a sample path length of 0.5 mm was constructed to operate at 83, 250 and 425 MHz and down to 20 mK and to register the transmitted signal after various trips through the liquid. Our work is an extension of previously reported investigations.[1][2][3]

If the equation of state of helium is given by

$$\delta p = p - p_0 = \rho_0^2 \left[\left. \frac{\partial^2 u}{\partial \rho^2} \right|_{\rho_0} \delta\rho + \frac{1}{2} \left. \frac{\partial^3 u}{\partial \rho^3} \right|_{\rho_0} \delta\rho^2 + ... \right],$$

then the equation of propagation of a disturbance along the x–axis may be written as

$$\left[-\frac{\partial^2}{\partial t^2} + v_0^2 \frac{\partial^2}{\partial x^2} \right] \frac{\partial a}{\partial x} = \gamma v_0^2 \frac{\partial^2}{\partial x^2} \left[\frac{\partial a}{\partial x} \right]^2 \quad [1]$$

with $\quad \gamma = 1 + \dfrac{\rho_0}{v_0} \left[\left. \dfrac{\partial v}{\partial \rho} \right|_{\rho_0} \right],$

where γ is the effective nonlinear coefficient, a is the amplitude of the disturbance, and the other symbols maintain their usual definitions. The solutions of Eq. 1 may be calculated[4] (Fig. 1), and in this analysis, it is customary to introduce a length $L = v_0^2/(2\gamma\omega\dot{a}_0)$, where $\dot{a}_0$ is the initial velocity of a disturbance of frequency ω. For a propagation distance $x < L$, the wave is distorted; for $x > L$, a shock–wave appears. In this second regime, the amplitude at the pumping frequency is restricted to a limiting value even if the $\dot{a}_0$ is increased to infinity. Under these conditions, the excess energy flows into higher and higher harmonics of the initial frequency (Fig. 2 and 3). When x = L, a crossover of regimes occurs and is labelled by $\sigma = x/L = 1$ (Fig. 1, 2 and 3). The above theoretical discussion is quantitatively valid for a lossless medium (Fig.2) and is qualitatively correct when the sample possesses non–negligible intrinsic attenuation, α (Fig.3).

1. A. Hikata, H. Kwun, and C. Elbaum, Phys. Rev. B <u>21</u>, 3952 (1980).
2. H.A. Kashkooli, P.J. Dolan, Jr., and C.W. Smith, J. Acoust. Soc. Am. <u>82</u>, 2086 (1987).
3. D. Rugar and J.A. Foster, Phys. Rev. B <u>30</u>, 2595 (1984).
4. D. Blackstock, J. Acoust. Soc. Am. <u>39</u>, 1019 (1966).

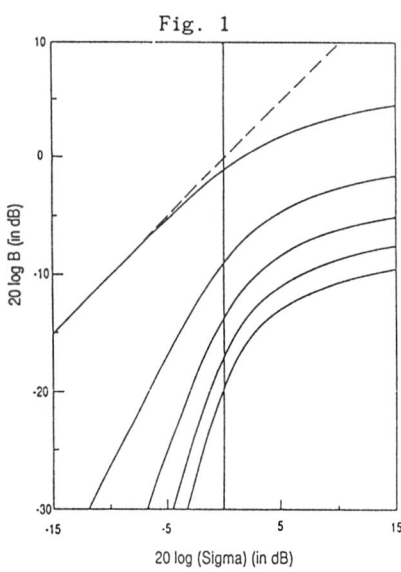

Figure 1. The exciting power (B) is plotted as a function of the resultant signal (Sigma) for several frequencies: f (the exciting frequency), 2f, 3f, 4f, and 5f [Ref. 4]. The vertical line is for $\sigma = 1$ (see the text).

Figure 2. The data for ^{4}He at 83 MHz and 20 mK are plotted on the same axes as defined in Fig. 1. This form of plotting allows the figures to be superposed. This comparison shows excellent agreement between Fig. 1 and 2. The data detected at f, 3f and 5f are shown. The dashed lines refer to an experimentally defined reference power (-20 dBm in helium after careful calibration of the emitter and receiver) and the shaded box shows the error in our ability to reproduce or define this reference point (± 2 dB). The $\sigma = 1$ location is labelled (see text).

Figure 3. The data for ^{3}He at 81 MHz and 250 mK is shown. The axes, dashed lines and shaded box are the same as defined in Fig. 2. The intrinsic attenuation α is labelled. The expected $x = L$ region is labelled by σ', and the measured $\sigma = 1$ is also labelled. Due to α, the 3f data are suppressed compared to the $\alpha = 0$ calculation of Fig. 1.

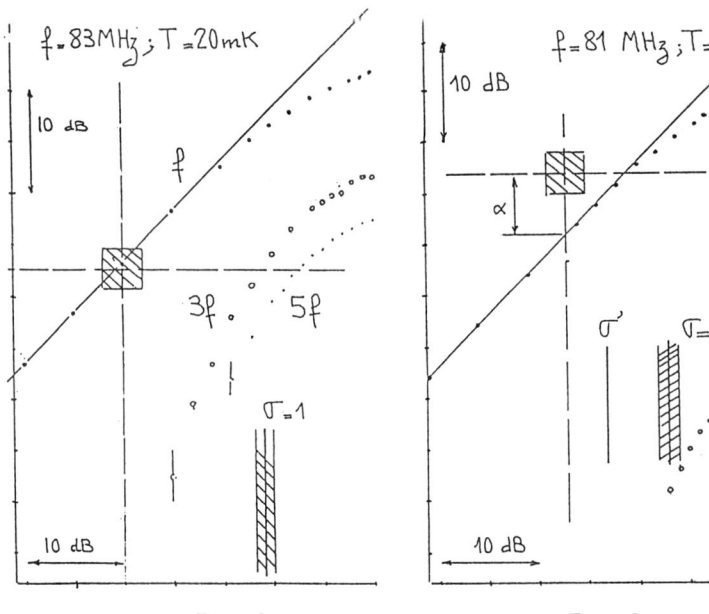

Fig. 2   Fig. 3

INTRINSIC MAGNUS EFFECT IN SUPERFLUID ^{3}He-A

M.M. Salomaa and R.H. Salmelin
*Low Temperature Laboratory, Helsinki University of Technology,
SF-02150 Espoo 15, Finland*

The orbital angular momentum of the coherently aligned Cooper pairs in superfluid ^{3}He-A is transmitted to an object immersed into the condensate. We evaluate the quasiparticle-scattering asymmetry experienced by an electron bubble. This is found to lead to a measurable, purely quantum-mechanical Magnus force deflecting the ion's motion.

The Magnus force is known from classical hydrodynamics: it arises when a fluid flows past a rotating body, and a Bernoulli pressure difference is created across the object for points where the fluid reinforces or counteracts its circulation; the deflecting force is perpendicular to both the directions of motion and the axis of rotation. In ^{3}He-A, the picture is reversed: the condensate itself possesses spontaneous angular momentum, which is experienced by a moving, non-rotating object. Here the circulation is an inherent property of the fluid - thus we suggest[1] to call the ensuing phenomena *intrinsic Magnus effects*.

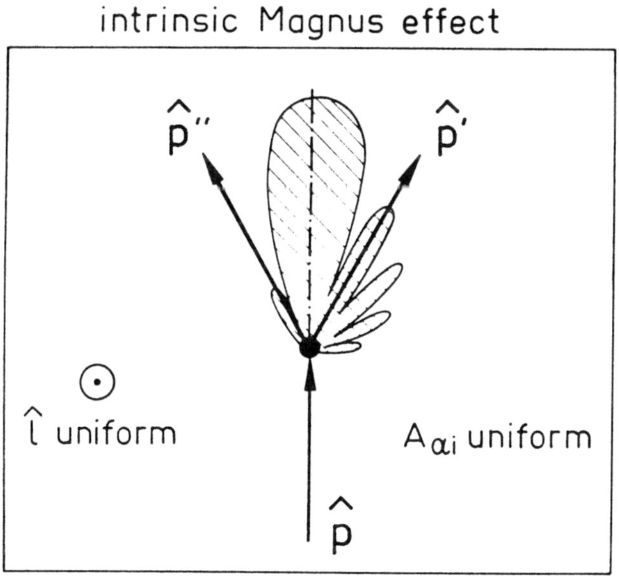

Figure 1. For tiny objects $(R \ll \xi_0)$, there occurs an asymmetry in the superfluid ^{3}He-A quasiparticle scattering, i.e., a quantum-mechanical *intrinsic Magnus effect*.

The calculated ratio of the transverse Magnus force and the component driving the electron bubble along the electric field $\vec{E}$ is seen in Fig. 2. The detectable 5° deflection would be expected already for $T = 0.7T_c$, while for $T = 0.5T_c$, an expression of as much as 26° is predicted. It would be of interest to extend ion-mobility techniques to find this manifestation of the internal Cooper-pair angular momentum in superfluid ^{3}He-A.

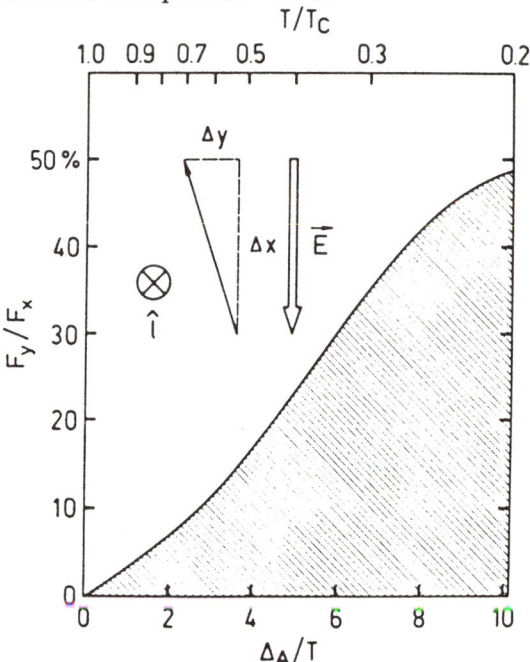

Figure 2. The magnitude of the intrinsic Magnus effect in the plane perpendicular to the $\vec{\ell}$-vector for negative ions in superfluid ^{3}He-A. An ion posses a transverse component Δy, like in the Hall effect, in addition to the motion Δx parallel with the applied electric field. The calculated ratio of forces, F_y/F_x, is shown as a function of Δ_A/T (lower scale) and T/T_c (upper scale, within weak coupling); it is linear in Δ_A/T for $T \to T_c$.

We thank O.V. Lounasmaa, C.J. Pethick, and G.E. Volovik for discussions. This research has been supported by the Körber-Stiftung (Hamburg, FRG) and by the USSR Academy of Sciences and the Academy of Finland through the project ROTA.

REFERENCES

1. R.H. Salmelin, M.M. Salomaa and V.P. Mineev, *Phys. Rev. Lett.*, submitted (1989).

QUANTUM INTERACTIONS

CHAIRMAN

Neil S. Sullivan
Department of Physics
University of Florida

This session was made possible, in part, through
a generous donation provided by

Schmidt, Garden & Erikson, Inc. Architects Engineers Planners
5700 Midnight Pass Road, Sarasota, Florida 34242
(813) 346-0798   FAX (813) 346-0146

SPECIALISTS IN ARCHITECTURE, ENGINEERING, AND PLANNING
FOR LABORATORIES AND TECHNICAL RESEARCH FACILITIES

DESIGNERS OF THE

MICROKELVIN RESEARCH LABORATORY
UNIVERSITY OF FLORIDA
GAINESVILLE, FLORIDA

SPECIFIC HEAT OF LIQUID ^{3}He BUBBLES IN SOLID MATRICES

E. Syskakis, Y. Fujii*, M. Gebhardt, and F. Pobell
Phys. Inst., Universität Bayreuth, D-8580 Bayreuth, FRG

ABSTRACT

We report on specific heat measurements of liquid ^{3}He bubbles in Ag foils as well as in a matrix of solid ^{4}He at millikelvin temperatures. - For the first part we have implanted about 0.1 % ^{3}He in Ag foils. The foils have been annealed between 900 K and 1112 K. This results in high-pressure ^{3}He gas bubbles of diameters between about 40 Å and 110 Å; the ^{3}He liquifies when the metal foils are cooled to low temperatures. After each annealing step we have measured the thermal relaxation time of the foils at 13 mK $\subseteq$ T $\subseteq$ 1.2 K, and have determined in this way the specific heat of the ^{3}He bubbles; it differs from the specific heat of bulk liquid ^{3}He. - In the second set of experiments we have measured the specific heat of a 0.75 % ^{3}He-^{4}He mixture at 25.3 bar (27.1 bar) and 20 mK $\subseteq$ T$\subseteq$ 300 mK; simultaneously we have measured the pressure of this mixture. Our data indicate that at the lower (higher) pressure there are bubbles of a liquid mixture (of liquid ^{3}He) in a solid ^{4}He matrix.

---

* Permanent address: Okayama Univ. of Science, Okayama, Japan

INTRODUCTION

The influence of dimensionality and of finite size on magnetism, superconductivity, superfluidity, or other ordered states is of fundamental importance in physics. In the studies of size effects, helium has played an outstanding role because of its homogeneity and purity, and because of the advanced state of low temperature thermometry. One can study phase transitions of liquid and solid helium with a temperature stability, resolution and homogeneity much better than for any other material. In addition, samples of helium can be produced in almost any size, and their quantum behavior adds additional interest. Therefore, the most detailed studies of the influence of finite size and of dimensionality on phase transitions have been done using in particular liquid helium.[1] The experiments have been performed on thin helium films or on helium confined to the pores of compressed powders or porous glasses. The measured properties have been mostly the specific heat, the superfluid density, or the onset to superflow.

Even though these studies have profoundly influenced, for example, our understanding of the superfluid state as well as the influence of a restricted geometry on phase transitions in general, there are severe shortcomings of many of them. The reason being that the confining geometry of porous systems usually is a tangled, interconnected, often insufficiently known structure with a distribution of sizes (Fig. 1a). These not well-defined geometries can influence the experimental results and sometimes have made their interpretation difficult. In this paper we discuss investigations of size effects on liquid and solid helium at low temperatures in alternative, well defined confining geometries: microscopic, isolated, helium filled bubbles in metal foils or in a matrix of hcp ^{4}He.[2,3]

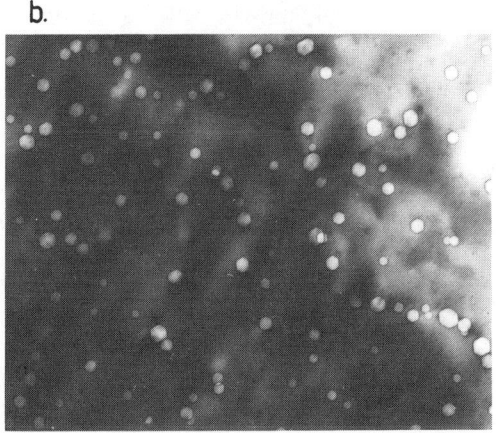

Fig. 1. a) Typical structure of a porous glass, one of the favorite matrices for studies on liquid helium in finite size geometries
b) TEM picture of underfocused helium bubbles in Cu. The bubbles appear as light areas and are clearly faceted (from Ref.2,3).

MICROSCOPIC HELIUM BUBBLES IN METALS

The samples for these studies are produced by shooting ^{4}He or ^{3}He ions with a cyclotron into metal foils where the atoms come to rest.[2] We wobble the ion beam and degrade its energy periodically from 0 to E_{max} (some MeV) to get a homogeneous distribution of about o.1 at % He in typically o.1 mm thick metal foils of 1o mm diameter. The solubility of He in metals is extremely low, typically 10^{-3} at ppm. By annealing the He doped metal, the He precipitates in gas bubbles whose pressure and size can be adjusted by the annealing temperature. The resulting sample can be examined by transmission electron microscopy; we find isolated He bubbles which are not interconnected (see Fig. 1b).

The He pressure P in a spherical bubble of radius r at thermal equilibrium is given by $P = 2 \cdot \sigma_{metal}/r$ (σ_{metal} = surface tension of the metal, 1.17 J/m^2 for Ag). This results in a helium density of order 0.1 g/cm^3 for a typical bubble radius of 50 Å (or about 10^4 atoms). Such a density corresponds to the low temperature density of solid or liquid helium, which means that we obtain bubbles filled with solid or liquid helium when we cool the sample to low temperatures. The advantages of this geometry are that the microscopic systems are isolated from each other and not interconnected, it allows a simple variation of size, and we can study it visually by electron microscopy. Unfortunately, we still have a size distribution (see Fig. 2) and, as a new disadvantage, the pressure or density in the bubbles change when we change their size. - In Fig. 2 we show the mean bubble radius r of ^{4}He in Cu and of ^{3}He in Ag, respectively, as a function of annealing temperature T_a. For the latter combination r seems to be roughly 15 % smaller at the same T_a.

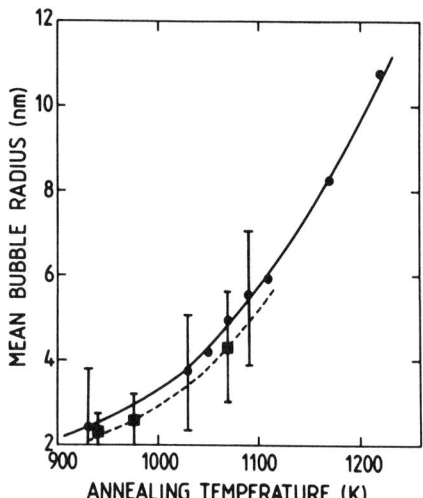

Fig. 2. Average radius of ^{4}He - ● (^{3}He - ■)bubbles in Cu (Ag) for isochronal annealing experiments. The samples were annealed to progressively higher temperatures Ta (ta = 2 h). The bars show typical half-widths of the corresponding size distributions.

In our former studies[2] we had measured the heat capacity of about 1 µg of ^{4}He confined to bubbles in Cu foils (7 mm diam., o.1 mm thickness) in a sensitive microcalorimeter at 1.5 K ≤ T ≤ 7 K after annealing the samples at temperatures between 930 K and 1220 K, resulting in ^{4}He bubbles of 25 Å to 110 Å diameter. By these experiments we could investigate the influence of this new confining geometry on the superfluid transition of ^{4}He. In this paper we present our new results on the heat capacity of ^{3}He filled bubbles in Ag foils at 13 mK ≤ T ≤ 1.2 K.

SPECIFIC HEAT OF MICROSCOPIC LIQUID ^{3}He BUBBLES IN SILVER

The measurements were performed in a relaxation microcalorimeter consisting of a o.5o g Ag sample holder with three different carbon layers (o.8 mg) as thermometers, and about 6 mg further addenda(see Fig. 3). Support of the calorimeter is by four o.1 mm nylon threads and the electrical leads are 10 µm NbTi wires. The thermal link to the dilution refrigerator are two Ag wires (50 µm diam., 12 cm length) with a thermal resistance R = 5.2 · 10^5/T (K/W). Silver with its nuclear spin I = 1/2 was used for the calorimeter and for the samples to avoid possible nuclear quadrupole contributions to the specific heat. The heat capacity C of the samples were obtained by observing the exponential decay of their temperature after applying a heat pulse and by calculating C = τ/R from the measured thermal time constant τ.

Our first samples were Ag foils (9 mm diam., o.25 mm thick) into which about o.2 % ^{3}He were shooted with an energy up to E_{max} = 36 MeV. When we tried to measure their heat capacity, it turned out that the samples showed a "large" heat leak of about 5 nW resulting from radioactivity induced by the ^{3}He implantation. We could not cool the samples below T = 6o mK. We then produced a second set of samples by implanting ^{3}He ions with an energy up to only E_{max} = 12 MeV, which substantially reduced the heat leak. These Ag foils had a 9.5 mm diameter and were 58 µm thick with an implanted depth of 42 μm. The 11 foils used for calorimetry had a Ag mass of m_{Ag} = 0.47 g or 4.4 mmole. The implanted volume (3.2 mmoles Ag) contained about 0.1 % ^{3}He, giving about 3 μmoles or about 9 μg of ^{3}He.

Fig. 3 shows the measured thermal relaxation times τ of our calorimeter, of the Ag foils without ^{3}He, of the Ag foils implanted with ^{3}He, of these foils after annealing them for 2 h at the indicated temperatures, as well as calculated relaxation times.

The data show that we can measure the tiny heat capacity of our calorimeter and foils, and that the data agree reasonably well with the specific heat of Ag at 0.1 K ≤ T ≤ 1 K. We do not have an obvious explanation for the increase of τ ∝ C/T of our calorimeter and bare foils at T ≲ 0.1 K, because there is no nuclear quadrupole interaction in Ag; possibly it results from the addenda or from impurities in the Ag. The data for the foils plus ^{3}He annealed at T_a ≤ 900 K show an extra contribution from the ^{3}He below 0.1 K, which is independent at T_a. But after annealing the foils at T_a ≥ 940 K we see a contribution from the ^{3}He at all T which increases with T_a. Because R ∝ T^{-1} we should have τ = constant if C ∝ T; this is clearly not the case for our ^{3}He data.

In Fig. 4 we have plotted the heat capacity C of the ^{3}He bubbles (calculated from the data in Fig. 3) and compared them to the specific heat of 3 μmoles of bulk ^{3}He at SVP.^{4} For T_a = 1112 K, for example, the data at 0.1 ≤ T ≤ 1 K coincide with the specific heat of bulk liquid ^{3}He if we shift the Fermi temperature of the liquid in the bubbles to T_F = 1.2 K (whereas $T_F^{bulk,SVP}$ = 1.8 K) and reduce the amount of ^{3}He from 3 μmoles to about half this value. Already in our former measurements on ^{4}He in Cu, we had found that the sample showed only about half the heat capacity of the corresponding amount of bulk helium after annealing at 1100 K. But there is clearly an extra contribution at T ≤ 0.1 K for the data at T_a = 1112 K as well as for the data at the other T_a.

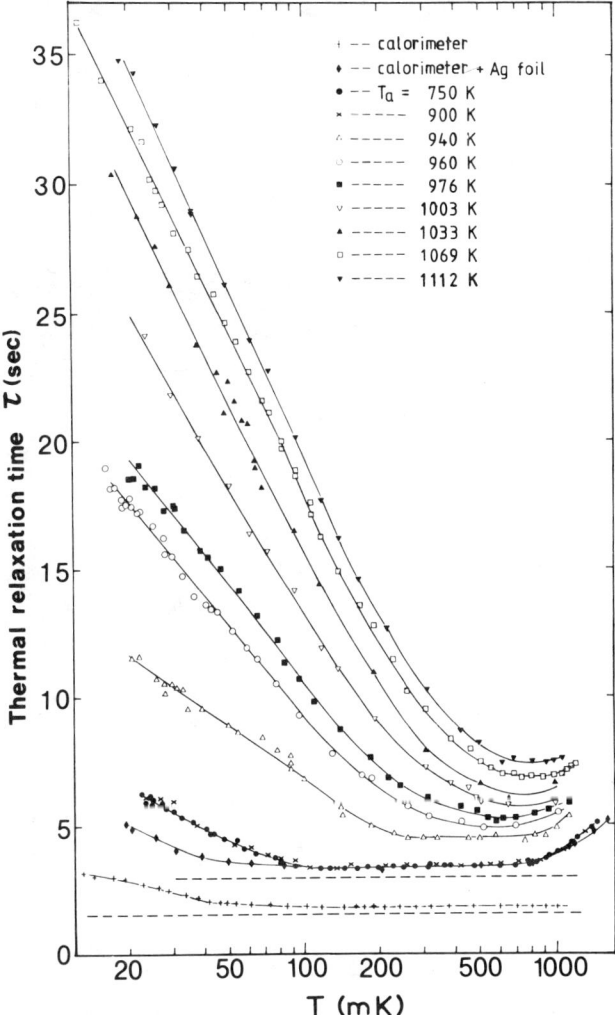

Fig. 3. Thermal relaxation time $\tau = R \cdot C$ as a function of temperature with $R = 5.2 \cdot 10^5/T$ (K/W), the thermal resistance to our calorimeter, and C
 a) heat capacity of the calorimeter
 b) heat capacity of the calorimeter plus 11 Ag foils (without ^{3}He)
 c) specific heat of the calorimeter plus 11 Ag foils implanted with ^{3}He, after annealing the Ag foils for 2 h at the indicated temperatures.
 The dashed lines are calculated values for the calorimeter and calorimeter plus Ag foils, respectively.

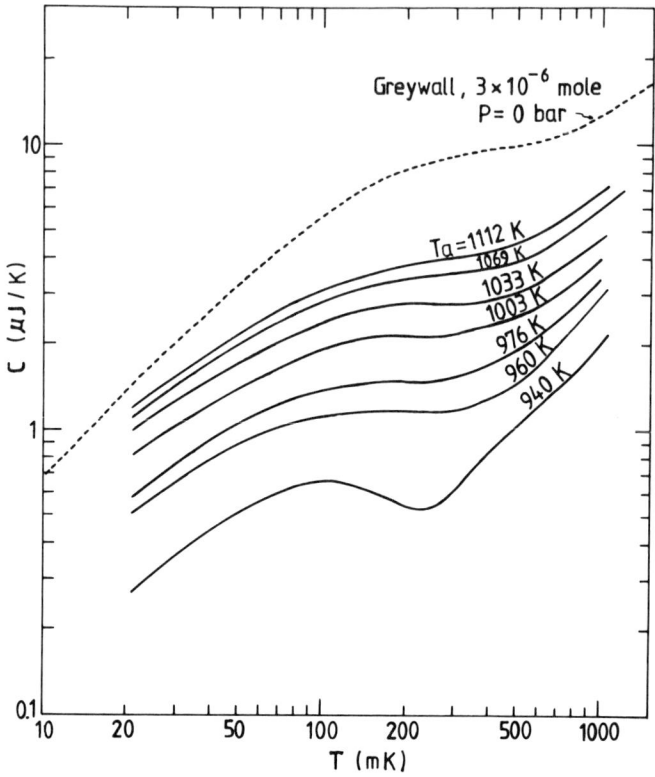

Fig. 4. Heat capacity of ^{3}He bubbles in Ag calculated from the data of Fig. 3 for the given annealing temperatures. The data are compared to the specific heat of 3 μmoles bulk ^{3}He at SVP (from Ref. 4).

For T ≲ 0.1 K we discuss the data in terms of the "layer model" usually applied to analyze data on helium in restricted geometries, particularly helium films.[1] The first layer of helium near a metal wall is a two-dimensional solid at P ~ 400 bar,[1,5] giving negligible lattice and magnetic contributions to the total specific heat of the helium bubbles in the investigated T-range.[5,6] In Ref. 6 it was shown that the second layer of ^{3}He on a Ag substrate shows a constant specific heat $C_2 \simeq 0.2\ N_2 k_B$ at 0.3 mK ≤ T ≤ 7 mK. Higher layers behave like bulk liquid ^{3}He. For our analysis we make the crude approximation that the first layer of d_1 ~ 3.0 Å does not contribute to our data, the second layer of d_2 ~ 3.5 Å gives the result of Ref. 6, C_2 = 0.2 R, also at temperatures of 13 mK ≤ T ≤ 100 mK(!), and that the remaining liquid in the center of the bubbles have the specific heat of bulk liquid ^{3}He at SVP.[4] With these crude assumptions we have calculated the specific heat for 2.2 μmoles ^{3}He in bubbles of 50 Å radius (corresponding to T_a ~ 1100 K) and plotted the result as C/T together with our measured data in Fig. 5; the agreement seems to be remarkably good for this crude model. - The dip near T = 0.2 K in the data at T_a = 940 K may result from melting of solid ^{3}He in small bubbles still present at this low annealing temperature. - More data, in particular from our planned magnetic measurements are necessary for a more detailed discussion.

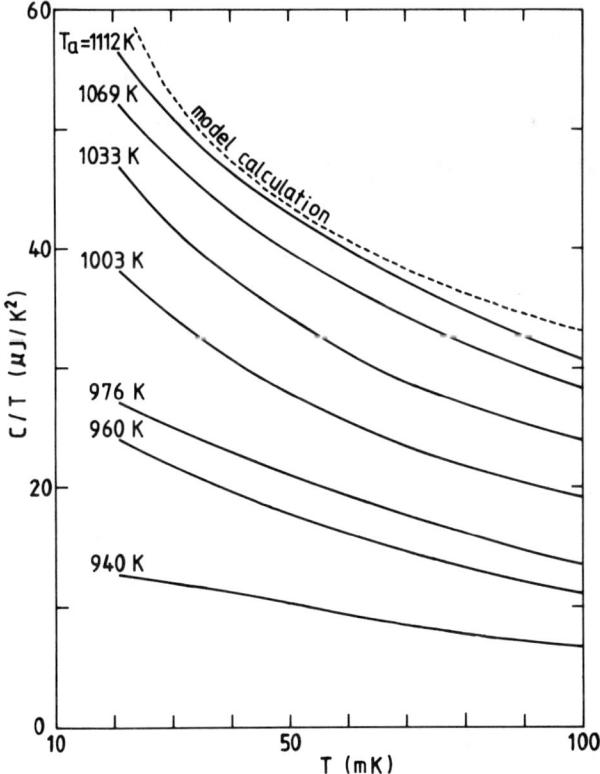

Fig. 5. Heat capacity C divided by temperature T of ^{3}He bubbles in Ag calculated from the data of Fig. 3 for the given annealing temperatures. The broken line are the values for 2.2 μmoles ^{3}He in r = 50 Å bubbles as calculated with the model described in the text.

SPECIFIC HEAT OF LIQUID HELIUM BUBBLES IN SOLID HCP ^{4}He

Liquid as well as solid ^{3}He-^{4}He mixtures phase-separate when cooled to millikelvin temperatures. ^{3}He and ^{4}He have different melting pressures. Therefore there exists a pressure range between about 25 bar and about 28 bar where hcp ^{4}He coexist with liquid ^{3}He or liquid ^{3}He-^{4}He mixtures depending on pressure. Many studies of the complicated phase diagram at these pressures have been published.[7] In Refs. 8 and 9 it was shown that with appropriate experimental conditions, one may succeed in creating liquid helium bubbles in hcp ^{4}He. This was particularly obvious from the specific heat measurement at Grenoble which showed Fermi liquid behavior C = γ T below phase separation, but with a coefficient γ larger than the bulk ^{3}He value. Unfortunately, the pressure was only approximately known for this measurement. We have performed a first set of measurements of the specific heat of a 0.75 % mixture at 20 mK $\angle$ T $\angle$ 3oo mK, and at 18.o bar, 25.3 bar, and 27.1 bar, respectively. The calorimeter contained a liquid volume of 1.9 cm^3, was equipped with carbon thermometers, and linked to a dilution refrigerator by a superconducting heat switch. The measurements are at constant volume because a plug was formed in the fill capillary during cooldown under pressure.

At 18.o bar (not shown) we see a Fermi liquid specific heat linear in T for T $\angle$ 60 mK and temperature independent at 6o mK$\angle$ T $\angle$ 180 mK indicating that our whole mixture is in the liquid state. For P = 27.1 bar we find a linear behavior, C = o.o3 T (J/K), for T $\angle$ 80 mK and a strong signature of phase separation at 80 mK $\angle$ T $\angle$ 22o mK (see Fig. 6a). This result is expected if the mixture separates almost completely into liquid ^{3}He and hcp solid ^{4}He. If the liquid phase would remain pure ^{3}He after we reduce the pressure from 27.1 bar to 25.3 bar, we should observe a decrease of the specific heat by a few percent. Instead we see a dramatic increase to C = o.o7 T (J/K) for T $\angle$ 6o mK, then a slight flattening off, and eventually an onset to phase separation at T $\supset$ 2oo mK (see Fig. 6a). This large linear specific heat indicates that the liquid phase at 25.3 bar is a mixture containing probably about 8 % ^{3}He.

The drastic change of the behavior of the liquid phase in an hcp ^{4}He matrix by changing the pressure by less than two bar is even more obvious from the "excess" pressures at phase separation (see Fig. 6b). Actually, the behavior at the pressure of 25.3 bar may indicate that the sample was in a three-phase region consisting of hcp ^{4}He, a liquid mixture phase, and possibly some pure liquid ^{3}He. - From the behavior of our samples - in particular the fast response to

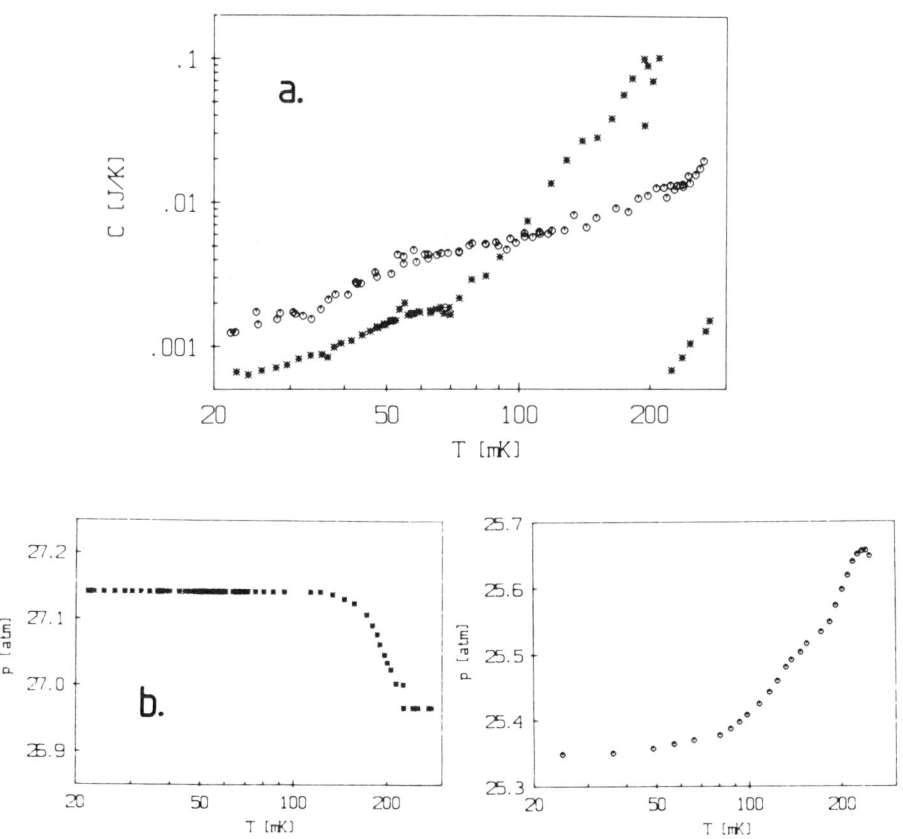

Fig. 6. a) Heat capacity at constant volume of 1.9 cm^3 of a o.75 % ^{3}He-^{4}He mixture at about 25.3 bar (o) and at about 27.1 bar (✶), respectively
b) Pressure in the closed sample cell during the specific heat measurements of which the data are shown in the upper part of the figure (for details see text).

temperature changes – we conclude that liquid bubbles are distributed in the hcp ^{4}He matrix and that the two phases are not totally separated in space. Of course, much more studies are necessary to understand the behavior of this interesting system.

CONCLUDING REMARKS

The discussed data demonstrate the remaining problems but also the possibilities of the two investigated confining geometries to understand the influence of size effects on the Fermi properties of liquid ^{3}He.

There is the unsolved question of the extra contribution to the specific heat of microscopic liquid ^{3}He bubbles in Ag which may result from the second layer, and which we hope to understand better when the results from our planned magnetic measurements on this system are available. But the size distribution of bubbles, the dependence of the helium pressure on the radius of the bubbles, and different contributions from different ^{3}He layers close to the substrate may make quantitative interpretation of experimental results difficult. – There is also the question of the possible existence of superfluidity of ^{3}He in this restricted geometry, possibly of another symmetry than for the bulk superfluid states. We have to remember that the typical ^{3}He bubble radii in a metal matrix are smaller than the coherence length of the bulk superfluid state of ^{3}He.

The measurements on liquid helium bubbles in hcp ^{4}He will first be extended to more pressures to understand the phase diagram and eventually we are interested in the bubble size and bubble formation, and the dynamics of phase separation. The advantage of this system is the fact that the surface tension of hcp ^{4}He is much smaller than the surface tension of metals. Hence in principle one has access to substantially smaller liquid bubbles. Unfortunately, very little is known about the dynamics, size, and isotopic concentration of the bubbles in this system.

ACKNOWLEDGEMENT

We gratefully acknowledge performance of the ^{3}He implantation as well as the TEM measurements by Dr. P. Jung, Dr. H. Schröder and Prof. H. Ullmair (KFA Jülich). - This work was partly supported by the Deutsche Forschungsgemeinschaft.

REFERENCES

1. D.F. Brewer, J. Low Temp. Physics 3, 2o5 (1970); and in "The Physics of Liquid and Solid Helium", Part II, p. 573, ed. K.H. Bennemann and J.B. Ketterson; J. Wiley and Sons, New York (1978).
2. E.G. Syskakis, F. Pobell, and H. Ullmaier, Phys. Rev. Lett. 55, 2964 (1985); E.G. Syskakis, Ph.D. Thesis, KFA Jülich, Report JÜL - 2012 (1985).
3. A preliminary report of this work is given in E. Syskakis, M. Gebhard, and F. Pobell, Proc. Int. Conf. on Polarized Quantum systems, Torino, June 1988.
4. D.S. Greywall, Phys. Rev. B 27, 2747 (1983).
5. D.F. Brewer, A. Evenson, and A.L. Thomson, J. of Low Temp. Phys. 3, 603 (1970).
6. D.S. Greywall and P.A. Busch, Phys. Rev. Lett. 60, 1860 (1988).
7. P.M. Tedrow and D.M. Lee, Phys. Rev. 181, 399 (1969); V.L. Vvedenskii, JETP Lett. 24, 132 (1976); B. v.d. Brandt, W. Griffioen, G. Frossati, H.V. Beelen and R. de Bruyn Ouboter, Physica 114B, 295 (1982); V.N. Lopatnik, Sov. Phys. JETP 59, 284 (1984); D.O. Edwards and S. Balibar, Phys.Rev.B39, 4083 (1989)
8. A.S. Greenberg, W.C. Thomlinson, and R.C. Richardson, J. Low Temp. Phys 8, 3 (1972).
9. B. Hebral, A.S. Greenberg, M.T. Beal-Monod, M. Papoular, G. Frossati, H. Godfrin, and D. Thoulouze, Phys. Rev. Lett. 46, 42 (1981).

FRACTIONAL STATISTICS AND ANALOGS OF QUANTUM HALL EFFECT IN SUPERFLUID ^{3}He FILMS

G. E. Volovik

*L.D. Landau Institute of Theoretical Physics,
117334 Moscow, USSR*

ABSTRACT

The superfluid ^{3}He phases provide us with systems which share many of the properties of the physical vacuum in particle physics including bosons, fermions, gauge fields, gravity, chiral anomaly, zero-charge effect, etc. The superfluid phases which may exist in thin ^{3}He films have additional properties of a two-dimensional origin, such as the fractionally quantized Hall conductivity. This quantum Hall effect (QHE) is representative of a whole class of phenomena caused by the quantization of physical parameters. They include fractional charge and spin of the particle-like solitons in superfluid ^{3}He films and QHE for spin current. All quantized physical parameters (charge, spin, Hall conductivity) have step-like dependence on the film thickness. Also the quantum statistics of solitons changes with film thickness, alternating between Fermi-, Bose- and possibly parastatistics.

INTRODUCTION

Three dimensional superfluid phases of liquid ^{3}He are systems with the maximal known broken symmetries in condensed matter physics. A large number of different elementary excitations ("elementary particles"), fermions and bosons, can exist in these liquids, as well as stable nonuniform structures for the order-parameter field – *nonuniform vacua* – such as vortices, solitons, boojums, etc. These phases share many properties which were solely the privilege of particle physics, such as chiral anomaly, zero-charge effect, gauge fields, gravity, etc.[1]

New concepts appear if one proceeds to lower dimensionality. In the relativistic quantum field theory (QFT) the two-dimensional system is not realistic and may serve only as a model, while in condensed matter physics this is a reality with a number of new phenomena, such as integer and fractional quantum Hall effect (QHE), and charge and spin fractionalization. In two-dimensional systems of superfluid ^{3}He with its unprecedently rich order parameter one may expect a wealth of unique phenomena which are impossible in other systems.

For example the particle-like solitons in superfluid ^{3}He-A film should be fermions in some ranges of film thickness[2] while in conventional antiferromagnets the same solitons (skyrmions) always obey Bose statistics[3]. This is because of the combination of the antiferromagnetic properties of the spin subsystem of ^{3}He-A with the ferromagnetic properties of the orbital subsystem of this quasi two-dimensional condensed matter. As a result a proposed pairing mechanism of the high-T_c superconductivity based on the Ferm statistics of solitons[4] in magnets can be realized only in such magnetic materials which share the magnetic properties of ^{3}He-A.

In ^{3}He-A the quantum statistics of solitons depends on the film thickness: the spin of the soliton increases with the thickness in a step-like manner. It is important that manipulation of the film thickness of the ^{3}He liquid allows us to investigate both theoretically and experimentally the intricate processes of transformation of specific two-dimensional properties to essentially different properties of the bulk liquid. Here we review some of the properties of quasi two-dimensional superfluid ^{3}He films, which seem to be important for relativistic QFT as well as for two-dimensional systems in condensed matter.

FRACTIONAL CHARGE, SPIN AND STATISTICS IN SUPERFLUID ^{3}He FILMS

Here we consider a very thin film of superfluid ^{3}He on a substrate. In order to observe the variety of effects of parameter quantization[2] resulting from the quasi two dimensional nature of the film, its thickness should be at least less than 20-30 interatomic spaces or 100 Å, which is less than the coherence length for superfluidity. To preserve the p-wave superfluidity in such conditions the requirement of perfect smoothness of the substrate surface should be fulfilled.

In the ^{3}He film two superfluid phases are possible in equilibrium[5]: ^{3}He-A and a planar state. Also the A_1 state should exist in a large applied magnetic field or possibly for a film on a magnetic substrate.

The phenomena discussed here are defined by the topological properties of the quasiparticle spectrum. In a film the transverse motion – along the normal to the film – is quantized. As a result the Bogoliubov matrix $\hat{H}$ for the quasiparticles and their Green's function $\hat{G}$ acquire the indices $mm\prime$ of the transverse levels

$$\hat{H}_{mm\prime}(\vec{k}) = \begin{pmatrix} \varepsilon_{mm\prime}(\vec{k}) & \Delta_{mm\prime}(\vec{k}) \\ \Delta^{\dagger}_{mm\prime}(\vec{k}) & \varepsilon_{mm\prime}(\vec{k}) \end{pmatrix} \quad , \hat{G}^{-1}_{mm\prime}(\vec{k},\omega) = i\omega\delta_{mm\prime} - \hat{H}_{mm\prime} \quad . \quad (1)$$

Here $\vec{k}$ is the two-dimensional vector in the plane of the film and $\varepsilon_{mm\prime}(\vec{k})$ is the Hamiltonian for the excitations in the normal Fermi liquid. In the simplest case of noninteracting levels this is a diagonal matrix:

$$\varepsilon_{mm\prime}(\vec{k}) = (\varepsilon_m(\vec{k}) - \mu)\delta_{mm\prime}$$

where in the Fermi gas approximation $\varepsilon_m(0) = \pi^2 m^2 \hbar^2 / 2m_3 a^2$ is the m-th energy level of transverse motion and a is the film thickness. In this approximation of noninteracting levels there are n independent two-dimensional Fermi systems (n families of fermions in analogy with particle physics) where $n \approx k_F a/\pi\hbar$ corresponds to the highest level below the Fermi level μ.

In each Fermi system an A-phase superfluid is formed with a different gap Δ_m:

$$\Delta_{mm\prime}(\vec{k}) = \delta_{mm\prime} \frac{\Delta_m}{k_F} i\sigma_2 \vec{\sigma} \cdot \vec{d}(k_x \pm ik_y) \quad , \quad (2)$$

but with a common orbital vector $\vec{l}$ of the orbital ferromagnetism fixed along the normal to the film, $\vec{l} = \pm \hat{z}$, and with a common $\vec{d}$ vector of spin antiferromagnetism which is free to rotate.

The most important difference between the bulk A phase and the A phase film is that the quasiparticle spectrum in the film

$$E_m(\vec{k}) = ((\frac{k^2}{2m_3} + \varepsilon_m(0) - \mu)^2 + \Delta_m^2 \frac{k^2}{k_F^2})^{\frac{1}{2}} \qquad (3)$$

has no gap nodes due to the quantization of motion along $\vec{l}$. Therefore the dynamics of the A phase in the film has no singularities which are of utmost importance in the bulk A phase due to topologically stable nodes of the gap in momentum space. Nevertheless the A phase film also has a nontrivial topology in momentum space which results in many exotic physical properties. The topological invariant which is behind these properties is

$$N = \frac{1}{24\pi^2} e^{\mu\nu\lambda} \int dk_x\, dk_y\, d\omega \text{ tr } G\partial_{k_\mu}G^{-1}G\partial_{k_\nu}G^{-1}G\partial_{k_\lambda}G^{-1} . \qquad (4)$$

Here $\mu = 0, 1, 2$ and $k_0 = \omega$, $k_1 = k_x$ and $k_2 = k_y$.

This integer invariant describes the nontrivial mapping of the three dimensional momentum space (k_x, k_y, ω) into the space of the nondegenerate matrices $\hat{G}_{mm'}(\vec{k}, \omega)$. In the approximation of noninteracting levels with an A phase at each level this invariant proves to be expressed in terms of the number of families of fermions n:

$$N = 2n(\vec{l} \cdot \hat{z}) , \qquad (5)$$

where $(\vec{l} \cdot \hat{z}) = \pm 1$. In the general case of real Fermi liquid with the A-phase superfluidity this is the number of the transverse levels below the Fermi energy for the Fermi gas state from which the Fermi liquid state is obtained when the interaction is adiabatically switched on. The adiabatical switching on of the interaction means here that in this process the quasiparticle energy E in Eq.(3) never turns to zero. As an example of nonadiabatic process one can consider a case in which one of the levels of the transverse motion $\varepsilon_m(0)$ crosses the Fermi level μ. At the moment of crossing the quasiparticle energy spectrum E touches zero at $\vec{k} = 0$ and at the same time the topological invariant abruptly changes from $2n$ to $2(n \pm 1)$.

The same adiabatic process takes place if one increases the film thickness a: the topological invariant N is conserved when there is a gap in the quasiclassical spectrum E and abruptly changes at some critical value of $a = a_m$ when the m-th branch of the quasiparticle spectrum crosses the Fermi level. In the Fermi gas approximation this critical value $a_m = \hbar\pi m/\sqrt{2m_3\mu}$. This is an example of plateau-like behavior of the physical parameter with a step-like change at critical points.

Here we encounter an important concept of the momentum space topology. It is important that each superfluid phase of ^{3}He has its own nontrivial topology in momentum space both in two- and three-dimensional systems, which gives rise to many exotic phenomena such as the chiral anomaly[1] in the A phase. In the films there are two groups of phenomena resulting from the existence of the integer topological invariant N in momentum space. The first group is related with the properties of the topological particle-like solitons in the field of the $\vec{d}$ vector. These solitons are characterized by an integer-valued topological invariant Q in real space x, y of the film

$$Q = \frac{1}{4\pi} \int dxdy (\vec{d} \cdot \partial_x \vec{d} \times \partial_y \vec{d}) . \tag{6}$$

The simplest realization of the soliton, also known as skyrmion, with the topological charge $Q = 1$ is the cylindrically symmetric fountain-like $\vec{d}$ texture

$$\vec{d}(x,y) = \hat{z} \cos\beta(r) + \hat{r} \sin\beta(r) , \tag{7}$$

with $\beta(0) = 0$ and $\beta(\infty) = \pi$.

From an investigation of the effective hydrodynamical action for the $\vec{d}$ field it follows[2] that permutation of two identical solitons with topological charges Q changes the wave function of the soliton by the factor $(-1)^{NQ^2/2}$ with $N = \pm 2n$. This means that at odd n the solitons with odd Q must behave as fermions under permutation. Also under 2π spin rotation of the $\vec{d}$ field of the soliton around the z-axis the wave function is multiplied by the factor $(-1)^{NQ/2}$, which means that the spin of the soliton is $s = \hbar n Q/2$. This value of spin is in correspondence with the conventional relation between the spin and the statistics of the topological solitons[6]. Thus the quantum statistics of elementary solitons in Eq. 7 depends on the film thickness and changes when one of the transverse levels crosses the Fermi energy.

The term in the hydrodynamical action which gives rise to the unusual quantum-statistical properties of the solitons is the so-called θ-term[6], which is expressed in terms of $\vec{d}$ field through an auxiliary "gauge" field $\tilde{A}_\mu$ ($\mu = 0, 1, 2$):

$$S_\theta = \theta H^{Hopf} , \quad \theta = \frac{\hbar N}{2} , \quad H^{Hopf} = \frac{1}{32\pi^2} \int d^2x \, dt \, e^{\mu\nu\lambda} \tilde{A}_\mu \tilde{F}_{\nu\lambda} , \tag{8}$$

$$\tilde{F}_{\nu\lambda} = \partial_\nu \tilde{A}_\lambda - \partial_\lambda \tilde{A}_\nu = \vec{d} \cdot \partial_\nu \vec{d} \times \partial_\lambda \vec{d} . \tag{9}$$

Here H^{Hopf} is the Hopf invariant describing the mapping of the three dimensional space-time (x, y, t) onto the sphere S^2 of the unit vector $\vec{d}$. The permutation of two identical solitons with topological charges Q changes their linking number and therefore the value of the Hopf invariant by Q^2. As a result the wave function $\exp(iS/\hbar)$ changes by the factor $(-1)^{nQ^2}$, thus defining the statistics of the soliton.

The existence of such topological θ-term in the action reflects the unique symmetry of the A-phase film. The integrand in Eq. 8 is odd both under a time inversion $(t \to -t)$ and under the orbital rotation by angle π around the x axis $(x \to x, y \to -y, z \to -z)$; therefore the θ-term does not exist in conventional antiferromagnets[3]. In the ^{3}He-A film this term is allowed by symmetry since the spin antiferromagnetism is accompanied by the orbital ferromagnetism. The change of the sign of the integrand under the transformations is compensated by the change of the sign of N in Eq. 5 due to the change of the ferromagnetic

$\vec{l}$-vector direction into the opposite $(\vec{l} \to -\vec{l})$. Therefore S_θ is invariant under the symmetry operations.

Another important property of the θ-term is that the integrand in Eq. 8 is not invariant under the "gauge" transformation

$$\tilde{A}_\mu \to \tilde{A}_\mu + \partial_\mu \alpha , \qquad (10)$$

while the total integral, the Hopf invariant, is invariant under this transformation. The gauge corresponds to spin rotation by an arbitrary angle $\alpha(x,y,t)$ about the axis $\vec{d}$ which does not change the physical state of the system and therefore should not change the hydrodynamical action.

The gauge property of the integrand results in the following important consequence. The parameter θ can not depend on the space and time coordinates, otherwise the whole integral depends on the gauge. Therefore this parameter in the hydrodynamical action should be fundamental for the ^{3}He-A film. The parameter θ in Eq. 8 meets this requirement: it is expressed through the fundamental constant h and therefore does not change in certain regions of external parameters, such as film thickness a, and abruptly changes to another fundamental value at some critical a. Such topological quantization of the hydrodynamical parameter (parameter in the hydrodynamical action) reminds us of the behavior of the Hall conductivity σ_{xy} in the QHE. The analogs of QHE represent the second group of the phenomena related to the momentum space invariant N.

PARTICLE CURRENT AND SPIN CURRENT QHE IN SUPERFLUID ^{3}He FILMS

Due to the momentum space topological invariant N in the ^{3}He-A film there exist two types of QHE. The first one is the analog of the conventional QHE in a two dimensional system of electrons: in an applied gradient of the chemical potential $\partial \mu$, which for the electrically neutral liquid plays the role of the electric field, the particle current appears in the perpendicular direction with the quantized Hall conductivity[7] σ_{xy}:

$$j^x = \sigma_{xy} \partial_y \mu , \quad \sigma_{xy} = \frac{N}{8\pi\hbar} . \qquad (11)$$

As distinct from the conventional QHE in two dimensional system of electrons this is an anomalous QHE in the sense that it takes place without any magnetic field and is produced by the orbital ferromagnetism of ^{3}He-A, i.e., instead of the magnetic field direction, the direction of the orbital momentum $\vec{l}$ defines the sign of the Hall current: according to Eq. 5 σ_{xy} is odd under time reversal, since it depends on the direction of $\vec{l}$.

For the electrically charged A-phase film (p-wave superconducting film in the A state) this is the real Hall effect. In this case the Hall current may be found from the following term[7] in the hydrodynamic action for the film in the presence of an external electromagnetic field A_μ which reminds us of the θ-term in Eq. 8:

$$S_{em} = \frac{e^2 N}{8h} \int d^2x \, dt \, e^{ij} (A_0 \partial_i A_j + A_i \partial_j A_0) , \qquad (12)$$

The variation of S_{em} over A_x produces an electric Hall current $j^x = \frac{e^2 N}{4h} E_y$ (e is the charge of the electron and $\vec{E}$ is the electric field) in accordance with Eq. 11. Note that for $N = 2$ this corresponds to the half-integer Hall effect, while for the A_1 phase where N may be an odd integer one has the fractional QHE with denominator 4 (Ref. 7).

With this term in Eq. 12 one may also find the electric charge of the quantized Abrikosov vortex in the superconducting film with an A phase structure. After variation of S_{em} over A_0 which gives the electric charge density and integrating this density over the singly quantized Abrikosov vortex, one obtains the following electric charge of the vortex

$$q = \frac{e^2 N}{8h} \int d^2x \, e^{ij} \partial_i A_j = e\frac{N}{16}. \tag{13}$$

Another type of QHE is related to the spin current. This may be obtained from the θ-term in Eq. 8 by variation over the gradient of the spin rotation angle in the same manner as the particle density and particle current are obtained from the electromagnetic term in Eq. 12. However, it is more instructive to relate the spin current QHE with the particle current QHE in Eq. 11. Let us use the representation of ^{3}He-A in terms of two superfluid components with spin up and spin down[8]. In this representation the spins of the particles are perpendicular to the $\vec{d}$ vector. The advantage of this picture of ^{3}He-A is that the spin current is just the counterflow of two spin components, $\frac{1}{2}\hbar(j_\uparrow - j_\downarrow)$.

According to Eq. 11 for each component there exists the QHE

$$j^x_\uparrow = \frac{1}{2}\sigma_{xy}\partial_y \mu_\uparrow \, , \quad j^x_\downarrow = \frac{1}{2}\sigma_{xy}\partial_y \mu_\downarrow \, . \tag{14}$$

In an applied magnetic field, which in neutral ^{3}He liquid interacts with the ^{3}He nuclear spins only, the chemical potentials for the components split. If a magnetic field is directed along the spin quantization axis then

$$\mu_\uparrow = \mu - \frac{1}{2}\hbar\gamma H \, , \quad \mu_\downarrow = \mu + \frac{1}{2}\hbar\gamma H \, . \tag{15}$$

As a result a current of the spin projection on the magnetic field arises

$$\frac{1}{2}\hbar(j^x_\uparrow - j^x_\downarrow) = -\frac{N\hbar}{32\pi}\gamma\partial_y H \, . \tag{16}$$

To write down the general vector form of the spin current, one must take into account that the spin quantization axis in the two component representation of ^{3}He-A is perpendicular to the $\vec{d}$ vector. Therefore in the general expression for the spin current one must change $\vec{H} \to \vec{H}_\perp = \vec{H} - \vec{d}(\vec{d} \cdot \vec{H})$:

$$\vec{j}^x_{spin} = -\frac{N\hbar}{32\pi}\gamma\partial_y(\vec{H} - \vec{d}(\vec{H} \cdot \vec{d})) \, . \tag{17}$$

Thus the factor in the response of the spin current to the gradient of magnetic field is also quantized in terms of the momentum space topological invariant N.

The two spin component representation is also useful for the investigation of the parameters quantization in another possible superfluid phase in the ^{3}He film. In this representation the planar state has two superfluid components with the opposite spin projections, $\vec{s}_\uparrow = -\vec{s}_\downarrow = \hat{s}$, and also with the opposite orbital momenta: $\vec{l}_\uparrow = -\vec{l}_\downarrow = \vec{l}$. Therefore the time inversion symmetry is not broken in this state; however the symmetry of this phase allows for the response of particle current on the gradient of the magnetic field.

As distinct from the A phase where the momentum space topological invariants for both spin components are equal, $N_\uparrow = N_\downarrow = n$ with $N = N_\uparrow + N_\downarrow$, in the planar phase the $\vec{l}$ vectors of the components are opposite to each other and therefore the momentum space topological invariants have different signs, $N_\uparrow = -N_\downarrow = n$. This leads to an interplay between spin and gauge variables in the QHE : the particle current has quantized response on the gradient of the magnetic field:

$$j^i = (j^i_\uparrow + j^i_\downarrow) = -\frac{n}{8\pi} e^{ijk} \vec{l}_j \gamma \partial_k (\vec{H} \cdot \hat{s}) \; , \tag{18}$$

while the quantization of the spin current response on the gradient of the chemical potential takes place

$$\vec{j}^i = \frac{\hbar}{2}(j^i_\uparrow - j^i_\downarrow) = \frac{n}{8\pi} \hat{s} e^{ijk} \vec{l}_j \partial_k \mu \; . \tag{19}$$

In the planar state there is also a close relation between the quantization of the hydrodynamical parameters and the quantization of the characteristics of the solitons, which have the same origin – the existence of the momentum space topological invariant,

$$\tilde{N} = N_\uparrow - N_\downarrow = \frac{1}{24\pi^2} \int \text{Tr} \; \tau_3 \sigma^3 \; \mathbf{G}\partial \mathbf{G}^{-1} \wedge \mathbf{G}\partial \mathbf{G}^{-1} \wedge \mathbf{G}\partial \mathbf{G}^{-1} \; . \tag{20}$$

For example the 4π nonsingular disclination has fractional fermionic charge $n/2$.

The A_1 phase contains only one spin component with the Cooper pair spin $\hat{s}$ and orbital momentum $\vec{l}$. This phase is a combination of spin and orbital ferromagnets. The particle current and the spin current coincide in this phase. The momentum space topological invariants for this state are $N_\uparrow = n$ and $N_\downarrow = 0$ which give $\tilde{N} = N = n$. As a result the nonsingular disclination, which is simultaneously a 4π nonsingular vortex, has both fractional spin $n\hbar/4$ and fractional charge $n/4$, which are half the values for the A phase and the planar phase, respectively.

DISCUSSION

It is generally accepted that the properties of any degenerate condensed matter are defined mainly by the symmetry of its order parameter and therefore the symmetry classification is one of the most important tools in condensed matter physics. However from the experience with the integer and fractional

QHE one can see that this is not the whole story: there are more subtle characteristics which may discriminate between the systems within the same symmetry class and they manifest themselves mostly in two-dimensional systems. These are topological features of the configuration manifold of the many body wave function: Laughlin function[9] for the fractional QHE (on the topological invariant for QHE see Ref.10), the ground state of nondegenerate spin liquid (RVB state[11]) or the ground state wave function for unconventional superconductors. One of such characteristics is what we have considered here – the momentum space topology of the one- particle Green's function of the two-dimensional system.

The topology of the one-particle Green's function has been already exploited by Luttinger and Ward[12] in the normal Fermi liquid who found the relation between the volume of the momentum space inside the Fermi surface and the particle density. From the topological point of view the Fermi surface is the topologically stable surface of the singularities in the Green's function, described by Π_1 homotopy group. Many different phase transitions are related with the change of the Fermi surface topology including the Lifshitz transition and the metal-insulator transition. In the degenerate systems of ordered magnets and pair-correlated superconducting states, the singular surfaces in momentum space disappear but the gap nodes, singular lines and points of the Green's function, described correspondingly by higher homotopy groups Π_2 and Π_3, may appear producing a variety of phenomena.

In the superfluid films the momentum space topology with higher homotopy groups plays a decisive role in the properties of these degenerate two-dimensional systems. In particular it is found that the quantum statistics of the solitons in ^{3}He-A films essentially depends on the film thickness, abruptly changing at some critical values of the thickness where the integer momentum space topological invariant changes its value. At the moment of transition between Fermi and Bose statistics the quasiparticle energy gap disappears, i.e. the system passes through a dissipative state, in a complete analogy with the QHE. The nonzero topological invariant in momentum space gives rise to the nonzero topological θ-term in the action, which is responsible for quantum statistics and also leads to the quantum Hall effect for spin current. This also gives a new meaning for the notion of the adiabatic process in the many-body system: this is the process without a change in the internal topology.

The other superfluid phases in ^{3}He films may also exhibit the quantization of parameters, with fractional spin and fermionic charge, since they also have the momentum space invariants which may be quite different from that in ^{3}He-A. The symmetry is also important here since it may completely prohibit the existence of invariants in momentum space. In this sense all the possible superfluid phases in the film have proper symmetry. All these phases combine the properties of spin and orbital magnets in different ways resulting in different momentum space invariants and therefore in different properties of topological objects and different types of QHE. For example 1) the spin disclination in ^{3}He-A, which is a spin antiferromagnet and an orbital ferromagnet, has a fractional spin but no fermionic charge; 2) in the planar phase, which is spin and orbital antiferromagnet, the corresponding spin disclination has a fractional fermionic charge without a spin; 3) in ^{3}He-A_1 phase, which is a combination of spin and orbital ferromagnets, this spin disclination has both a fractional spin and a fractional charge.

In the relation of the fractional charge and spin of the solitons the ^{3}He film should not be unique. Among the magnets there should exist quasi two-dimensional electron systems where the spin ferro- or antiferromagnetism is combined with the orbital ferro- or antiferromagnetism in such a way that the symmetry allows for either the θ-term in the action and therefore fractional statistics of solitons or the fractional charge for topological objects. The possibility of existence of neutral objects obeying fractional statistics and charge-e bosons was proposed in the resonating-valence-bond state[14]. The analogy between the ground states of the nondegenerate magnets and the fractional quantum Hall states[9] with respect to fractional statistics was discussed in Refs. 12 and 15.

In our case of He-3 films the fractional spin and statistics appear, however, in the degenerate system with the long-range order. We obtained the QHE in the ^{3}He-A film with the denominator 2 of the fractional factor $\nu = n/2$ in the Hall conductivity $\sigma_{xy} = \nu/h$, where n is the number of "families" of fermions. From this one may conclude that in the pair-correlated systems the quantized Hall conductivity is defined by the product of the number of families and the z projection of the orbital angular momentum per particle, which is $m = 1/2$ for the A phase where the orbital momentum projection of the Cooper pair is unity, i.e. $m = 1/2$ per each particle. In the general case of Cooper pairing with an arbitrary momentum projection the Hall conductivity should be given by $\nu = nm$.

This equation shows how to obtain the Hall coductivity in a correlated system with an arbitrary denominator. For example the denominator 3 is obtained if one has a six-particle-correlated superfluidity, i.e. the condensate of bosons with each boson made of six fermions. If the boson has the orbital projection say $L_z = 2$ then the orbital momentum per particle is $m = 1/3$ leading to the fractional QHE with the denominator 3. And in the general case of q-particles-correlated states with momentum L_z per one boson one has $\nu = nL_z/q$. Note that two states with very different symmetries, such as many-particle correlated ^{3}He film and the Laughlin state, may have the same Hall conductivity. This again shows that the quantization of the physical parameters are defined rather by the internal topology of the state than by its symmetry.

Many-particle correlated states should occur in a ^{3}He film on a rough substrate. The roughness of the solid surface plays the role of impurities which tend to destroy the anisotropic order paramater $A_{\alpha i}$. At decreasing film thickness the impurities become more effective in destroying the Cooper pairing and at some critical thickness the spatial average of the order parameter would disappear, $<A_{\alpha i}> = 0$. However the correlators of the order parameter which are more symmetric than the order parameter itself still may survive, such as

$$< A_{\alpha i} A_{\alpha i} > , \tag{21}$$

$$e_{\alpha\beta\gamma} < A_{\alpha i} A_{\beta i} > , \ e_{ijk} < A_{\alpha i} A_{\alpha j} > , \tag{22}$$

$$e_{\alpha\beta\gamma} < A^*_{\alpha i} A_{\beta i} > , \ e_{ijk} < A_{\alpha i} A^*_{\alpha j} > , \ e_{ijk} e_{\alpha\beta\gamma} < A^*_{\alpha i} A_{\beta j} > . \tag{23}$$

They also correspond to some long-range order: e.g., Eq. 21 describes the isotropic superfluidity of the four-particle-correlated state, while Eqs. 22 describe the superfluid four-particle-correlated state with corresponding spin and orbital anisotropies. The correlators in Eqs. 23 correspond to the nonsuperfluid states with the magnetic or/and orbital properties. So with decreasing ^{3}He film thickness one may expect a sequence of transitions into the exotic states of many-particle correlated Bose condensate.

In addition to the ^{3}He films there are several other examples of quasi-two-dimensional systems in superfluid ^{3}He which have many features in common. These are:

1) The domain walls between two different bulk vacua of ^{3}He, such as A-B phase boundary, B-B interface and A-A interface[16]. For the Fermi excitations which are bound to the wall the dynamics is pure two-dimensional and one may expect for them the anomalous behavior, which is characteristic for (2 + 1) - quantum field theory.

2) The superfluid states on the surface of bulk superfluids. The more interesting situation here is when the surface state is more degenerate than the bulk liquid. This may be either the A phase or the axiplanar state on the surface of the B phase. In this case one has specific surface topological defects: point vortices (boojums, see Ref.17) or linear defects. In the presence of these topological defects the QFT of the system becomes more complicated.

3) The superfluid surface state on the surface of the bulk normal liquid. They have the maximal degrees of freedom for the order parameter as compared with the other two-dimensional systems of ^{3}He since there is no coupling with the bulk order parameter. As in the case of the films the QFT in all possible p-wave states here, A phase, planar state and A_1 phase, have peculiar properties with fractional charge and/or fractional spin and statistics of the topological objects and with quantization of parameters leading to variety of different types of QHE.

REFERENCES

1. G. E. Volovik *J. Low Temp. Phys.* **67**, 331 (1987)
2. G. E. Volovik and V. M. Yakovenko 1989, submitted to *J. Phys. C*
3. F. D. M. Haldane *Phys. Rev. Lett.* **61** 1029 (1988)
4. I. Dzyaloshinskii, A. Polyakov and P. Wiegmann *Phys. Lett.* **A127** 112 (1988)
5. P. N. Brusov and V. N. Popov *ZhETF* **80** 1564 (1981); [*Sov. Phys. JETP* **53** 804].
6. F. Wilczek and A. Zee *Phys. Rev. Lett.* **51** 2250 (1983)
7. G. E. Volovik *ZhETF* **94** 123 (1988)
8. A. J. Leggett *Rev. Mod. Phys.* **47** 331 (1975)
9. R. B. Laughlin *Phys. Rev. Lett.* **50** 1395 (1983)
10. J. E. Avron and R. Seiler *Phys. Rev. Lett.* **54** 259 (1985); Q. Niu, D. J. Thouless and Y. S. Wu 1985 *Phys. Rev. B* **31** 3372; R. Tao and F. D. M. Haldane 1986 *Phys. Rev. B* **33** 3844
11. P. W. Anderson *Mater. Res. Bull.* **8** 153 (1973)
12. V. Kalmeyer and R. B. Laughlin *Phys. Rev. Lett.* **59** 2095 (1987)
13. J. M. Luttinger and J. C. Ward *Phys. Rev.* **118** 1417 (1960)
14. Z. Zou and P. W. Anderson, *Phys. Rev. B* **37** 627 (1988)
15. D. P. Arovas, A. Auerbach and F. D. M. Haldane *Phys. Rev. Lett.* **60** 531 (1988)

16. M. M. Salomaa and G. E. Volovik *Phys. Rev.* B **37** 9298 (1988); *J. Low Temp. Phys.* **74** 319 (1989)
17. M. M. Salomaa and G. E. Volovik G. E. submitted to *J. Low Temp. Phys.*

INTERACTION OF VORTICES WITH THE HOMOGENEOUSLY PRECESSING RESONANCE MODE IN SUPERFLUID ³He–B

J.S. Korhonen, Z. Janú, Y. Kondo, and M. Krusius
Helsinki University of Technology, 02150 Espoo, Finland

Yu.M. Bunkov, V. Dmitriev, and Yu. M. Mukharskiy
Institute for Physical Problems, 117334 Moscow, USSR

In ³He-B the scale of the order parameter texture is set by the magnetic coherence lenght which is generally larger than the distance between quantized vortex lines in the rotating state. In conventional cw NMR the presence of vortices is therefore observed only indirectly via their influence on the texture.[1] In the B-phase also a second uniquely different resonance mode can be sustained continuously in which the total magnetization precesses uniformly with an intense applied rf excitation field within a coherent domain, called the homogeneously precessing domain (HPD).[2] We have studied resonance absorption in the HPD in a rotating cryostat and find that it increases directly proportional to the number of vortices.

The HPD is established at fixed rf frequency ω_{rf} and amplitude by maintaining a constant linear field gradient ∇H over a closed resonance cell of length L, superimposed to the homogeneous field H. When H is swept downward the HPD domain boundary first forms at the down-field cell wall, once the resonance condition $\omega_{rf} = \gamma(H+\nabla H \cdot L/2)$ is fullfilled. If one continues to reduce H the HPD boundary moves through the cell, the resonance condition is met at the domain wall and the reduction in Zeeman energy within the HPD is compensated by the dipolar interaction. By recording the out-of-phase rf absorbtion vs the in-phase dispersion signal during the downward H sweep the HPD characteristics shown in Fig. 1 are traced. At the minimum point to the right the HPD fills the cell. During rotation at equilibrium vortex density we can record similar absorption - dispersion curves where the increase in absorption is linear with HPD length and rotation velocity Ω.

In the measurements the bias value for H is chosen to be the minimum point shown in Fig. 1. An equilibrium vortex density is secured by rapidly accelerating at $\Omega = 0.1$ rad/s² to our maximum Ω of 3 rad/s and then decelerating to 1-2 rad/s. The linear dependence of the additional absorption in rotation on Ω is next established by monitoring the absorption level during deceleration to zero, as shown in Fig. 2. In Fig. 3 the vortex absorption is plotted vs T/T_c. The HPD is a stable resonance mode only above 0.4 T_c.[2] Vortices cannot be created

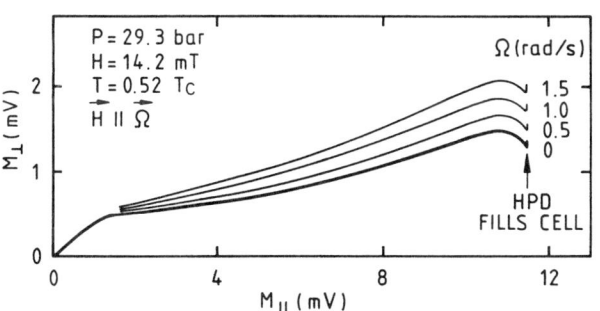

Fig. 1. Measured HPD absorption vs dispersion while sweeping downward the magnetic field.

in the present experiment above 0.55 T_c - 0.70 T_c with isothermal rotation procedures. The results show a drop by 60% on warming through T_v = 0.60 T_c at 29.3 bar. Here the singular vortex core is known to undergo a 1st order phase transition in which the nonaxisymmetric core below T_v transforms to an axisymmetric structure above T_v.[1] The vortex absorption appears to be concentrated in the singular vortex core of some superfluid coherence lengths in diameter. The effect is larger for the less symmetric core and increases towards higher pressures, inspite of decreasing core size. The dominant two relaxation processes at $\Omega = 0$ are spin diffusion through the domain wall and Leggett - Takagi relaxation in the bulk HPD. Both mechanisms are also expected to contribute to absorption within the inhomogeneous order parameter distribution with magnetic anisotropy in the vortex core. Structural information about the superfluid vortex cores in the B-phase has previously been obtained from similar discontinuities at T_v in the spontaneous and induced core magnetizations, which can be extracted from conventional NMR measurements and texture analysis.[1]

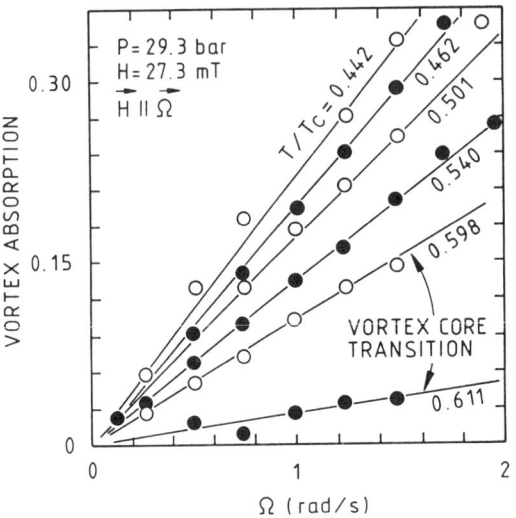

Fig. 2. Vortex absorption in HPD vs Ω at different temperatures.

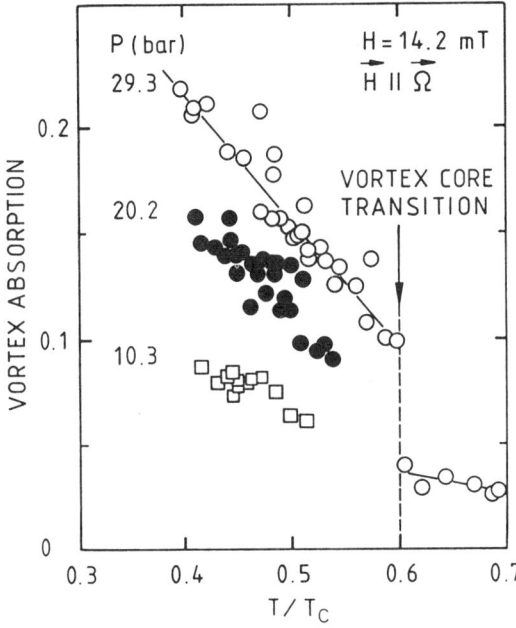

Fig. 3. Vortex absorption at Ω = 1 rad/s vs temperature at different pressures.

Vortex absorption in the HPD depends on the angle between the field and the core axis. In the case of an ordered array of vortices it can be used to monitor quantitatively the amount of core material in the sample, eg. for studying vortex formation and decay.

1. P.J. Hakonen et al., J. Low Temp. Phys. 76, N 3/4 (1989).
2. A.S. Borovik-Romanov et al., these proceedings.

VORTEX NUCLEATION IN SUPERFLUID ^{4}He

R.M. Bowley

Department of Physics, University of Nottingham, Nottingham NG7 2RD, UK

ABSTRACT

Muirhead, Vinen and Donnelly have argued that vortex nucleation by negative ions in superfluid ^{4}He involves small vortex loops emerging in the equatorial plane of the ion. These loops have to overcome an energy barrier of about 3°K. They do this either by thermal activation (involving a phonon or roton) or by tunnelling. Recent experiments have observed this thermally activated process and found an energy barrier of order 3°K.

Another feature of their model is that vortex lines undergo a sort of cyclotron motion at a frequency $\omega_c = \hbar/(m\pi a^2)$ where a is the vortex core radius. When quantised these cyclotron levels are about 50°K apart. The presence of these levels can be inferred from measurements of the vortex nucleation rate at low temperatures as a function of electric field.

INTRODUCTION

Superfluidity of ^{4}He breaks down when the liquid flows through a tube at a sufficiently high velocity. Either rotons are produced, for velocities above the Landau velocity, or vortices are produced or expanded. Superfluid ^{4}He invariably contains a self sustaining tangle of vorticity[1], so it is seldom clear whether a new vortex is created, or a pre-existing vortex is enlarged. If vortices are created in geometries where the relevant length scale is much less than the distance between pre existing vortices, then it is reasonable to suppose new vortices are nucleated. Such is the case for a tiny orifice separating two baths of superfluid ^{4}He, as in the beautiful experiment of Varoquaux et al[2], or for nucleation of vortices by negative ions.

Flow through a tiny orifice has many appealing features. In particular individual phase slip, events, can be detected as each vortex is nucleated and move across the orifice. The main disadvantage is that the experiment does not allow one to measure the nucleation rate. All one can get is the flow velocity through the orifice when the nucleation rate equals the time of oscillation of the membrane used to drive the flow.

In contrast the rate of vortex nucleation by ions can actually be measured because ions carry charge. An ion which forms a vortex is trapped by it and slows down dramatically to a speed of < 1 ms^{-1}, whereas bare ions travel at speeds of ~ 50 ms^{-1} limited by the Landau velocity for roton emission. The current is then

$$I = -e\,(n_b(t)\,\bar{v}_b + n_v(t)\,\bar{v}_v) \simeq -e\,n_b(0)\,\bar{v}_b\,e^{-\nu t} \qquad (1)$$

© 1989 American Institute of Physics

where $n_b(t)$ is the number of bare ions at time t, $\bar{v}_b$ their average velocity, with n_v and $\bar{v}_v$ being the corresponding quantities for ions trapped on vortices. Since $n_b(t)$ decays exponentially away, we can measure ν by measuring the electrical current. This is the basis of the technique used by McClintock and his group[3].

The information contained in the variation of ν with electric field, E, and temperature T, enables us to study in great detail different nucleation mechanisms.

THE THEORY OF MUIRHEAD, VINEN AND DONNELLY

Significant advances in the theory of vortex nucleation by ions were made by Muirhead Vinen and Donnelly[4] (MVD). They showed that the favoured configuration was a vortex loop, that is a segment of a circular vortex ring, emerging from the equator of the ion (the north pole is in the direction of motion). Because of translational invariance of the system, the momentum, or more precisely the impulse, is conserved. MVD calculated the energy of the system for circular loops of different radii for constant impulse. An ion with a loop of zero radius is of course a bare ion, so a convenient way of specifying the impulse is to quote the speed of the bare ion.

MVD's results for a pressure of 17.5 bar is shown if figure 1 for a different bare ion speeds. It can be seen that there is an instability in the systems when v exceeds 57 ms^{-1} when a loop of radius 5 A is formed. However to go from a bare ion to an ion attached to a vortex loop the system has to cross a barrier of about 3.1°K. It can do this in two ways: either the vortex tunnels through the barrier, or it is thermally excited over the barrier with the help of a thermal phonon or roton.

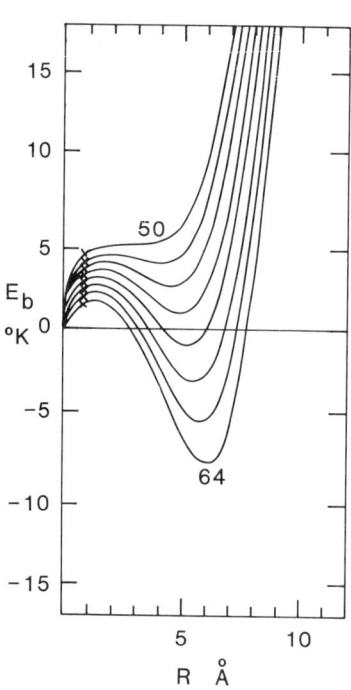

Fig 1. The variations of the energy as a function of the radius of the vortex loop. The upper curve is for a bare ion velocity of 50 ms^{-1} and successive curves are for velocities larger by 2 ms^{-1}.

Another idea contained in MVD is that the vortex moves as if it were a charged particle in crossed electric and magnetic fields.

This is easiest to understand for a vortex in a 2 dimensional film of helium. The vortex is taken to have a hollow core of radius a, and an effective mass per unit length of $m=\pi a^2 \rho_s$. The vortex is acted on by the Magnus force, so its equation of motion is

$$m\ddot{\underline{r}} = \rho_s (\underline{v}_s - \dot{\underline{r}}) \times \underline{K} \qquad (2)$$

where $\underline{r}$ is the position of the vortex and $\underline{K}$ is the circulation vector whose direction lies parallel to the vortex core.

This equation is analogous to that the a charged particle in crossed electric and magnetic fields. The charge is q=K, the "electric field" $\underline{E} = \rho_s \underline{v}_s \times \hat{\underline{K}}$ and "the magnetic field" $\underline{B} = -\rho_s \hat{\underline{K}}$ ($\hat{\underline{K}}$ is a unit vector in the direction of $\underline{K}$). Now charged particles rotate about the $\underline{B}$ field at the cyclotron frequency $\omega_c = \rho_s K/m = K/\pi a^2$. Consequently if we can measure ω_c we can get an estimate of the hollow core radius a. It is expected to be about 1 A.

The same cyclotron motion occurs in the case of a vortex loop created on the equator of the ion. The only difference is that the circulation $\underline{K}$ is not a constant vector but bends since it lies parallel to the vortex core. The cyclotron motion can be described as a circular motion of the vortex core around its average position. Suppose the vortex loop starts displaced to the south pole. Its radius then shrinks as it moves back to the equator and the ion is accelerated. The vortex loop then expands as it is displaced towards the north pole, before returning to its starting point with an enlarged radius.

As is well known the circulation of a vortex is quantised because the superfluid order parameter has to be a single valued function. Its phase changes by 2π on passing around the core of a vortex. This quantisation of circulation is analogous to the quantisation of electric charge. MVD treat the vortex loop <u>as if it were a quantum object itself</u> just as one would treat an electron. As such a vortex loop can tunnel and it can exist in various quantum mechanical states. In particular a vortex can exist in quantised cyclotron levels of energy $\hbar\omega_c(n+\tfrac{1}{2})$.

Of course it may be questioned whether it is valid to apply quantum mechanics to a "classical" object like a vortex loop, an object which itself arises from characteristic quantum effects. Such a question could be applied equally to a flux quantum in a SQUID, and since 1980 there has been an industry showing that these "classical" flux quanta themselves show quantum effects[5].

At low temperatures all nucleation processes must involve tunnelling; in practice these tunnelling processes dominate below 150mK. There will be different critical velocities for tunnelling into these different cyclotron levels. Suppose the critical velocity for the n=0 level is v_{co} and that for the n=1 level is v_{c1}. Then energy conservation requires

$$\tfrac{1}{2} M_{ion}(v_{c1}^2 - v_{co}^2) = \hbar\omega_c = \hbar K/\pi a^2 \qquad (3)$$

If we can obtain values of v_{c1} and v_{c0} we have a way of measuring ω_c and of obtaining a value for the vortex core radius.

We do this in the following way. First we solve the Boltzmann equation for the distribution function of ionic velocities[6] $f(v,E)$. This Boltzmann equation involves the rate of roton emission, and this rate can be determined from measurements of the drift velocity $\bar{v}_b$ as a function of the applied electric field E. Let us suppose that vortex nucleation into the ground state occurs at a rate R_0, and into the first excited state at a rate R_1. If all ion-vortex complexes which are formed are stable then the rate of nucleation is [3].

$$\nu = \int dv \; f(v,E) \; (R_0 \theta(v-v_{c0}) + R_1 \theta(v-v_{c1}) + \ldots) \qquad (4)$$

The rate of R_1 is expected to be much larger than R_0 because tunnelling is so much easier into the first excited state, partly because the energy barrier is much lower and partly because a smaller loop is formed. If we compute ν using equation (4) then we can adjust v_{c0} and v_{c1} to get the best fit to the data.

Before turning to the experimental results let's summarise the predictions of MVD. They estimate the critical velocity v_{c0} to be 57 ms^{-1} at 17.5 bar, and the energy barrier to be 3.1°K. They propose that both phonons and rotons can assist the nucleation of a vortex by the ion. Finally they treat the vortex as a quantum object and show that it undergoes cyclotron motion at frequency ω_c.

EXPERIMENTAL RESULTS

In figure 2 we show a typical set nucleation rates, here taken at a pressure of 15 bar[7]. At low temperatures ν is a constant wheres for high temperatures the nucleation rate rises exponentially fast. The temperature range shown is below 0.5°K so there are few rotons present. Consequently we expect the nucleation rate to be

$$\nu = \nu_0 + AT \exp(-E_b/kT) \qquad (5)$$

where E_b is the barrier height. Here we have assumed that all phonons of energy larger than E_b will, on colliding with the ion, have the same cross section for producing a vortex. The temperature independent term ν_0 is interpreted as arising from tunnelling. In fact all the data in the range 15-19 bar shows a small dip in ν for temperatures around 0.25K-0.3K. This dip can be adequately described by multiplying ν_0 equation (5) above by a factor $\exp(-\gamma T^3/E)$. When we do this the best estimate of E_b for 15 bar is 2.1°K. If we modified equation (5) by multiplying instead by $\exp(-\gamma T^4/E)$ then E_b becomes 2.4°K. Both values are in fairly good agreement with MVD's estimate of 3.1°K at 17.5 bar. However there is a puzzle: the pressure dependence of E_b is quite large. At 13 bar, using the factor $\exp(-\gamma T^3/E)$ one finds $E_b=1.9$°K, at 17 bar $E_b = 3.4$°K and at 19 bar $E_b = 5.1$°K. Eventually at higher pressures E_b is indistinguishable from the roton energy, and phonon assisted tunnelling disappears.

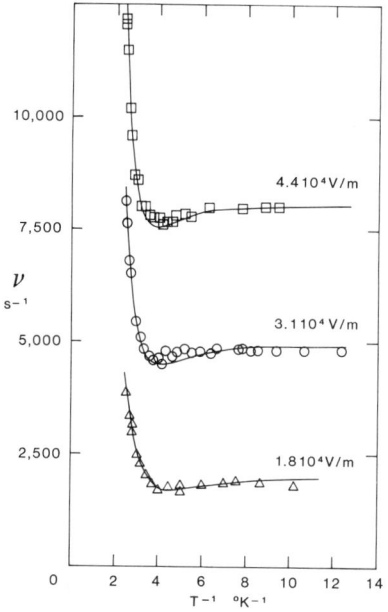
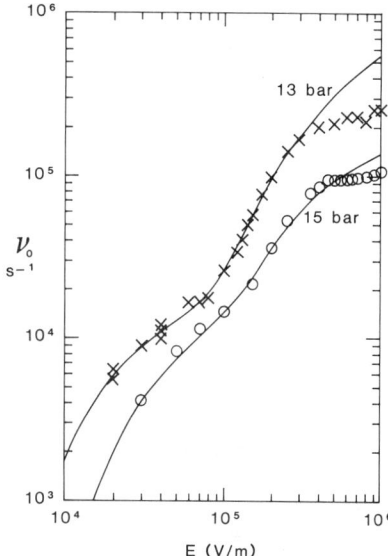

Fig 2. The observed nucleation rate at a pressure of 15 bar for three different electric fields. The solid lines show the best theoretical fit to the data.

Fig.3 The nucleation rate below 150 mK as a function of electric field for two different pressures.

Nevertheless the very fact that the order of magnitude of E_b is that predicted by MVD is confirmation that vortex loops are produced and not large encircling rings. It also makes more plausible the idea that a vortex loop could tunnel out of the ion, since the energy barrier is not prohibitively large.

Now let's return to the low temperature region where the nucleation rate is independent of temperature. By increasing the field E, one increases the drift velocity of the ions and hence the nucleation rate. Figure 3 shows some recent data of Hendry et al[7] for pressures of 15 and 13 bar (other data at higher pressures is presented in reference 3). The most striking factor of this data is the structure which appears for fields around 10^5 V/m. We can fit equation (4) to the data, adjusting R_o and v_{co} to fit the low field data, and R_1 and v_{c1} to fit the high field data. The fit is quite good, bearing in mind the crudeness of the model we are using. Deviations occur at very high fields: it seems that above $E \simeq 10^6$ V/m the ion-vortex complex is unstable and the vortex is shed by the ion. Consequently there is a dramatic drop in $\nu(E)$ for these high

fields.
Values of v_{c1} and v_{co} are given in the table.

Table 1 Values of the critical velocity derived from the data.

Pressure	v_{co} ms^{-1}	v_{c1} ms^{-1}	$\hbar\omega_c/k_B$ °K	a A
15 bar	59	74	50	0.7
13 bar	58	84	90	0.5

Notice that v_{co} is very close the value predicted by MVD for a pressure of 17.5 bar, and that the value of a, the radius of the hollow core, is very plausible. We see then that the predictions of MVD are in very good agreement with the experimental results.

If we try to model the dramatic drop in $\nu(E)$ for high fields, then this reduces the values of v_{c1} and consequently increases the value of a. The values quoted in the table above are only preliminary estimates of the best fit.

There is one problem, which we have not yet fully discussed. Why is there a small dip in $\nu(T)$ for temperatures around 0.25K? At first it was thought that phonon scattering off the ion caused it, but detailed calculation[8] showed this to be too small an effect. Another possibility is that phonon damping, reduces the tunnelling rate[9], as argues by MVD. The tunnelling rate is expected to decrease by a factor[9]

$$\exp(-\phi\eta q^2_o)$$

where ϕ is a dimensionless number, η is the dissipation coefficient and q_o is the distance to be traversed under the barrier. We expect the damping of the vortex is caused by phonon scattering, as so η should vary as T^3. The quantity q_o will decrease as the velocity of the ion increases and so as the electric field gets larger one expects q_o^2 to decrease. This is qualitatively in agreement with the factor $\exp(-\gamma T^3/E)$ which we have introduced in an ad hoc manner in order to fit the data.

Another possible explanation is that the ion vortex complex, when it is just formed, has a finite probability $P=\exp(-\gamma T^3/E)$ of retunnelling back to a bare ion state. It can do this if its energy for a given impulse of the system is equal to the bare ion energy. Phonon scattering off the ion will tend to reduce the impulse and thereby increase the probability of retunnelling. In contrast the electric field increases the momentum of the ion-vortex complex and so it makes it more stable. This physics would give rise to the factor $\exp(-\gamma T^3/E)$ which we have introduced.

Can MVD's approach, which is so successful in explaining the nucleation rates for ions, tell us anything about nucleation in orifices? The first point to be made is that there is no translational invariance and so impulse is not conserved. The criterion, for stability against the formation of a vortex loop, just concerns the energy of the system. When an instability appears, vortices are produced at a rate $\nu=\nu_o(v)+A\exp(-E_b(v)/kT)$ that is we expect both tunnelling and thermally activated processes to occur. If the energy barrier is too large then the nucleation rate is unobservably small.

The experimental criterion for a critical velocity v_c is that ν is equal to roughly the rate of oscillation of the membrane used to drive the superfluid, ν_m. Consequently v_c is given by the solution of

$$\nu_m = \nu_o(v_c) + A\exp(-E_b(v_c)/kT) \qquad (6)$$

Varoquaux et al[2] ignore the tunnelling process and then argue that $E_b(v)$ must vary linearly with v in order to explain the temperature dependence of v_c. This neglect of $\nu_o(v)$ is likely to be seriously in error at low temperatures where nucleation should take place by tunnelling. Further work is needed to clarify this problem.

ACKNOWLEDGEMENTS

This work was done in close collaboration with Dr P V E McClintock and I am grateful to him for his persistent questioning. This work was supported by grants from SERC.

REFERENCES

1. D Awschalom and K W Schwarz, Phys Rev Lett 52 49 (1984)
2. E Varoquaux, M W Meisel and O Avenal, Phys Rev Lett 57 , 2291 (1986)
3. R M Bowley, P V McClintock, F E Moss, G G Nancolas and P C E Stamp
4. C M Muirhead, W F Vinen and R J Donnelly, Phil Trans Roy Soc London A311, 433 (1984)
5. A J Leggett., Proc 18th Int. Conf. on Low Temp Physics, Kyoto, 1986 (1987)
6. R M Bowley, F W Sheard, Phys Rev B16, 244 (1977)
7. P C Hendry, N S Lawson, P V E McClintock, C D H Williams and R M Bowley (in preparation)
8. P C Hendry, N S Lawson, C D H Williams, P V E McClintock and R M Bowley, in "Elementary excitations in Quantum fluids" edited by K Obayaski and M Watanabe (Springer Verlag, Berlin to be published)
9. A O Caldeira and A J Leggett., Phys Rev Lett 46 211 (1981) and Ann Phys (NY) 149 374 (1983)

ANOMALOUS ^{3}He SOLUBILITY IN SUPERFLUID ^{4}He AT LOW TEMPERATURES

Hiroshi Fukuyama, Ikuya Kurikawa,[*] Hidehiko Ishimoto, and Shinji Ogawa
Institute for Solid State Physics, University of Tokyo,
Roppongi, Tokyo 106, Japan

The phase separation curve of ^{3}He-^{4}He mixtures on the ^{4}He-rich side at the saturated vapor pressure gives a useful information on the ^{3}He quasiparticle interactions in the ^{3}He dilute phase.[1,2] Along this curve, the ^{3}He chemical potential in the dilute phase must be equal to that in the ^{3}He concentrated phase. The latter can be evaluated from tha specific heat of pure ^{3}He which is nearly identical to the concentrated phase below about 60 mK. Presently, we have accurate specific data[3] of pure ^{3}He up to the melting pressure. The phase separation curves at elevated pressures will, therefore, give us a useful knowledge on the interactions in the dilute phase under pressure. We have made the first quantitative measurements[4] of the phase separation curve at non-zero pressures up to 22 bars in a temperature range between 250 and 7 mK, extending the lowest temperature four times colder than in any previous work.

Mixture samples were cooled in a cylindrical copper sample cell (illustrated in Fig.1) by a dilution refrigerator. An admixture of fine platinum powder (200 A diameter) and silver powder (4000 A diameter) served as a heat exchanger with a surface area of 115 m^2. A capacitive concentration gage, which was made of two copper electrodes separated by a gap of 12 μm, was located on the bottom inside the cell. A special attention was paid to reduce the background temperature variation of the empty gage by making a reference capacitor identical with the gage.

When the gap is filled with a phase separated ^{3}He dilute mixture, capacitance measurement of the gage

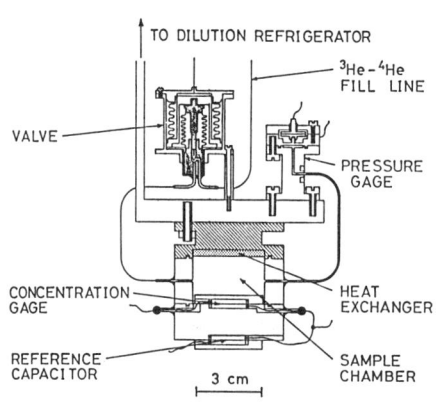

Fig. 1. Sample cell.

[*]Present address: Goldman Sachs (Japan) Corp., Tokyo, Japan.

allows us to know the molar volume v of the dilute phase through the Clausius-Mossotti equation and then the ^{3}He concentration X from
$$v(T,P,X) = v_4(P)[1 + X\alpha(T,P,X)]. \tag{1}$$
We used here the pressure dependence of the BBP parameter[1] α and calculated its slight temperature variation according to Watson et al.[5] The obtained temperature dependence of the concentration X can be fairly expressed by
$$X(T,P) = X_0(P)[1 + \beta(P)T^2] \tag{2}$$
above about 30 mK for all the pressures. This T^2 dependence is expected from the Fermi liquid theory. For P = 0 bar we obtain X_0 = 6.57 % and β = 9.61 K^{-2} which are in excellent agreement with previous measurements[5-7] carried out down to 25 mK. Also $X_0(P)$ has a maximum around 10 bars, again in good agreement with previous works.[5,8]

The most remarkable feature in our data under pressure is an existence of a solubility minimum around 20 mK, below which X increases as the temperature decreases. The deviation from Eq.(2), typically 0.1 at% of ^{3}He, seems to increase with pressure. At present, we have no microscopic explanation for this anomaly. A possible origin might be size effects of the dilute phase in the capacitor gap or some unknown mechanism working at the phase separation interface between the dilute and the concentrated phases.

REFERENCES

1. J. Bardeen, G. Baym, and D. Pines, Phys. Rev. 156, 207 (1967).
2. C. Ebner, Phys. Rev. 156, 222 (1967).
3. D. S. Greywall, Phys. Rev. B 27, 2747 (1983).
4. Above 70 mK semiquantitative measurements were made by B. M. Abraham, O. G. Brandt, and Y. Eckstein, in Proceedings of the 12th International Conference on Low Temperature Physics, Kyoto, 1970, edited by E. Kanda, p.161.
5. G. E. Watson, J. D. Reppy, and R. C. Richardson, Phys. Rev. 188, 384 (1969).
6. D. O. Edwards, E. M. Ifft, and R. E. Sarwinski, Phys. Rev. 177, 380 (1969).
7. B. M. Abraham, O. G. Brandt, Y. Eckstein, J. Munarin, and G. Baym, Phys. Rev. 188, 309 (1969).
8. J. Landau, J. T. Tough, N. R. Brubaker, and D. O. Edwards, Phys. Rev. Lett. 23, 283 (1969).

LOW DIMENSIONAL HELIUM

CHAIRMAN

Gary A. Williams
Department of Physics
University of California, Los Angeles

This session was made possible, in part, through
a generous donation provided by

PIONEERS AND LEADERS IN HELIUM TECHNOLOGY

Our committment to reliable supply and service
of Liquid Helium requirements big or small worldwide

Headquarters
UNION CARBIDE INDUSTRIAL GASES
39 Old Ridgebury Road
Danbury, Connecticut 06817
(203) 794-2000   FAX (203) 794-4633

In Florida
LINDE GASES OF FLORIDA
1201 No. 22nd Street
Tampa, Florida 33605
(813) 248-4931   FAX (813) 247-5255

THE SUPERFLUID TRANSITION IN POROUS MEDIA

J. Machta*
Center for Theoretical Physics and Institute for Physical Science and Technology, University of Maryland, College Park, MD 20817

R. A. Guyer
Laboratory for Low Temperature Physics, Department of Physics and Astronomy, University of Massachusetts, Amherst, MA 01003

ABSTRACT

The superfluid transition for ^4He films adsorbed in porous materials is discussed using a theory which emphasizes topological excitations of the superfluid associated with the multiple connectivity of the porous material. The coupling constant for these excitations is renormalized by the polarization of quantized vortices and is obtained from a generalization of the Kosterlitz-Thouless theory. Within this picture, the crossover from two-dimensional, Kosterlitz-Thouless behavior, to three-dimensional critical behavior may be understood .

INTRODUCTION

In this contribution we discuss a theory of the superfluid transition for ^4He films adsorbed in porous media. In recent years experiments have been performed on films in several materials such as Vycor, silica gels, and packed powders. The experimental picture[1,2] reveals a continuous phase transition characterized by a vanishing superfluid density and a sharp feature in the specific heat. The transition temperature, T_c, depends on the film thickness--for thick films T_c is above 1K while T_c extrapolates to zero for films of approximately two atomic layers[3].

In the experimental situations of interest, the film thickness is considerably less than the characteristic radius of curvature of the porous material, thus, on short length scales the film is two-dimensional. On longer scales (100Å - 1000Å) the materials used in the experiments are three dimensionally connected. Early discussions of these systems have emphasized either their two or three dimensional aspect. Kotsubo and Williams[4] developed a theory based on a finite size rounded Kosterlitz-Thouless (KT) phase transition which gives reasonably good fits to the data. This theory explains the fact that the onset of the transition generally falls along the Kosterlitz-Nelson universal jump line defined by K = 2/π where

$$K = \sigma(\hbar/m)^2/k_B T \ . \qquad (1)$$

Here σ is the areal superfluid density and m the mass of a ^{4}He atom. The finite size KT theory is also in agreement with recent ultrasonic attenuation measurements of Beamish[1,2]. On the other hand, Reppy and collaborators[5], emphasized the three-dimensional nature of the transition and showed that, in Vycor, the superfluid density vanishes as a power law in the reduced temperature with an exponent very near the bulk value. A more comprehensive theory of superfluidity for ^{4}He films in porous media must take into account both the two and three dimensional aspects of the system. Within the past two years such a picture of the transition has been developed by several groups[6-8].

STATISTICAL MECHANICS OF SUPERFLUIDS IN POROUS MEDIA

In the present analysis[7] we focus on the role of the multiple connectivity of the superfluid film and, for the moment, ignore the effects of randomness. As a simple model of a multiply connected surface we consider a "jungle gym" or regular cubic lattice of cylinders as shown in Fig. 1 with cylinder radius, r_0, and lattice constant, c. The ^{4}He is taken to reside on the

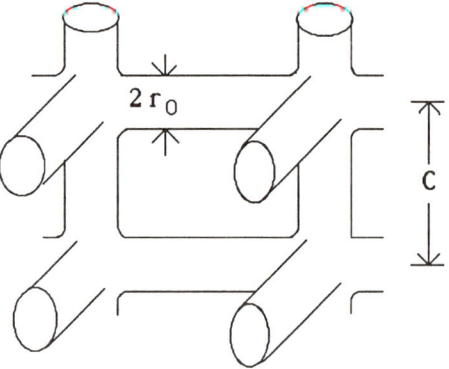

Figure 1. The "jungle gym" model for porous media

surface of the cylinders and to be described by local superfluid velocity field, v(r), which is a well-defined function of r, the two-dimensional coordinate on the manifold given by the surface of the jungle gym. Under the same set of approximations as in the

KT treatment of planar ^{4}He films[9], we take the Hamiltonian, H, to be the kinetic energy of the flow plus an energy associated with vortex cores,

$$H = \frac{\sigma_0}{2} \int d^2r \ v(r)^2 + \text{core energy} \qquad (2)$$

The velocity field is irrotational except at a discrete set of vortex cores and the circulation around all closed loops is quantized in units of $2\pi\hbar/m$. The bare superfluid density, σ_0, differs from the total areal density of Helium due to the presence of an immobile layer and because of short wavelength thermal excitations contributing to the normal fluid. σ_0 is taken to be a slowly varying function of temperature in the region of T_c.

Our object is to understand the statistical mechanics of this system. As a first step, consider the possible excitations of the velocity field. For planar film these consist firstly of "core" vortices which are incompressible flows with quanta of circulation around paths which can be continuously shrunk down to a vortex core ($\nabla \cdot v = 0$ but $\nabla \times v \neq 0$ in the core region). A second class of excitations are "spin waves" or, roughly speaking, third sound modes. These are compressible, irrotational flows ($\nabla \times v = 0$ and $\nabla \cdot v \neq 0$) which lead to height fluctuations in the film. Since no energy is associated with height fluctuations in (2), the third sound velocity is zero in this model.

For a multiply connected film a third class of excitations are possible with quanta of circulation around closed loops in the film that cannot be deformed to a point. These flows are both incompressible and irrotational ($\nabla \times v = 0$ and $\nabla \cdot v = 0$) and we refer to them as "pore" vortices. The simplest multiply connected surface, the torus, has two independent pore vortices illustrated in Fig. 2. We refer to these modes as azimuthal and axial pore vortices. More

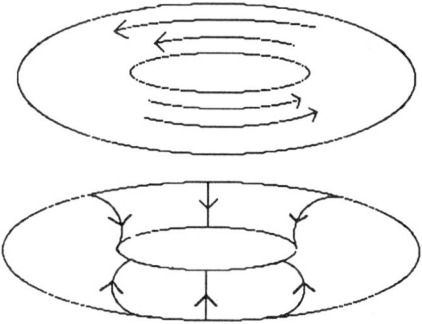

Figure 2. Axial (top) and azimuthal (bottom) pore vortex modes on a torus.

generally, for a surface of genus p there are 2p independent pore vortex modes which, for the jungle gym, can be divided into azimuthal and axial modes. Any allowed configuration of the velocity field can be uniquely decomposed into a spin wave, core vortex and pore vortex part.

A mathematically correct description of the velocity field on a multiply connected surface and its decomposition into the three classes of flows can be formulated within the theory of differential forms on Riemann surfaces[10]. The velocity field is represented by a first order differential; spin waves by exact differentials, core vortices by co-exact differentials and pore vortices by harmonic differentials.

What is the role played by the three types of excitations? Because the Hamiltonian is quadratic, the spin waves are non-interacting and do not renormalize the superfluid density or contribute to the singular part of the specific heat. For planar films, the unbinding of core vortices leads to the KT phase transition and the polarization of vortex pairs renormalizes the superfluid density below T_c. The unbinding process leads to a weak essential singularity in the specific heat.

For the jungle gym or any other multiply connected surface with a characteristic pore size, there can be no core vortex unbinding transition. The reason for this is illustrated in Fig. 3 which shows a pair of oppositely charged core vortices on a piece of a jungle gym. These vortices are connected by a path in the film (the dotted line) which is of length L. The crucial point is that there must be a connecting path such that any loop which cuts this path carries one quantum of circulation. If the length of the path is L, the kinetic energy in the "string" of circulation connecting the two

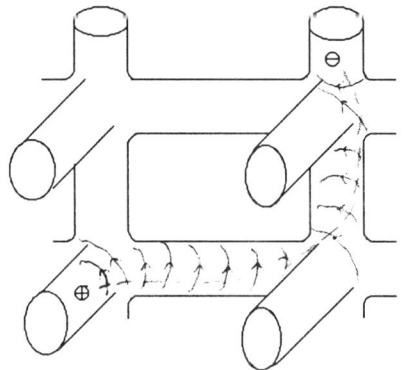

Figure 3. A pair of core vortices and the "string" connecting them.

vortices must scale as KL/r_0. Since the interaction potential for vortex pairs is asymptotically linear in L whereas the entropy is logarithmic in L vortex unbinding is not possible.

To understand the phase transition then, we must look to the pore vortex excitations. The system of pore vortices and long wavelength spin waves can be represented by a spin model on the cubic lattice underlying the jungle gym. To understand this, first suppose that we have suppressed the core vortex and short wavelength spin wave degrees of freedom. Then the Hamiltonian, (2) can be approximated by

$$H = K_0 (\pi r_0/c) \sum_{<ij>} (\phi_i - \phi_j - 2\pi n_{ij})^2 \qquad (3)$$

In this equation, K_0 is obtained from K, defined in (1) by replacing σ by the bare superfluid density, σ_0. The phases, ϕ_i, are associated with the sites of the lattice and lie in the range $-\pi \le \phi_i < \pi$, n_{ij} are integer variables associated with the bonds of the jungle gym lattice and <i j> indicates a sum over all pairs of nearest neighbor sites.

Equation (3) defines an effective Hamiltonian for the axial pore vortices and is the kinetic energy of flows with velocity, v_{ij}, down cylinder ij,

$$v_{ij} = (\hbar/mc)(\phi_i - \phi_j - 2\pi n_{ij}) \ . \qquad (4)$$

This form for v_{ij} insures that circulation is quantized about all loops on the cubic lattice underlying the jungle gym. The effective Hamiltonian does not include azimuthal pore vortices. For cylinder aspect ratios larger than one the excitation energy of these modes is larger than that of the axial pore vortices and it should be a good approximation to ignore these modes. When the pore lengths and diameters become comparable the full set of pore modes must be included leading to a more complicated Hamiltonian however the qualitative conclusions should be unchanged.

Thus far, in writing down the effective Hamiltonian we have ignored the core vortices. The polarization of core vortices by the flow field associated with the pore vortices leads to a renormalization of the superfluid density appearing in (3). A quantitative treatment of the superfluid density for a film on a cylindrical surface can be carried out by generalizing the KT theory and is described in detail in ref. 11. For our purposes the main conclusion of this theory is that K_0 in (3) should be replaced

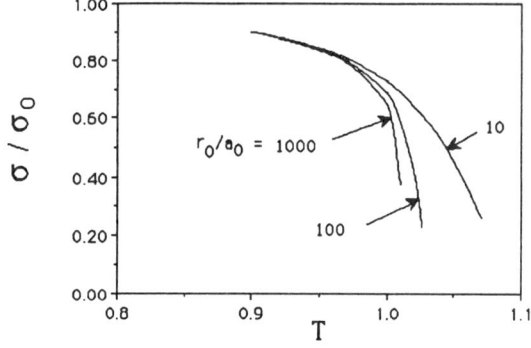

Figure 4. The superfluid density vs. temperature on a cylindrical surface for three ratios of the pore radius, r_0, to the vortex core radius, a_0.

by, K, the renormalized coupling constant obtained from the planar KT recursion relations carried only as far as the length scale corresponding to the diameter of the cylinder. This leads to a finite size rounded KT phase transition with the rounding increasing as the ratio of the core radius to the pore radius increases. The onset of the transition occurs along the Kosterlitz-Nelson universal jump line, $K=2/\pi$. A graph of σ vs T, obtained from the cylinder KT theory of ref. 11 is shown in Fig. 4 for a cylinder of radius ten times the core radius

The effective Hamiltonian is a Villain model[12] with a coupling constant, $K\pi r_0/c$, that is a function of the temperature, the film thickness and the geometry of the pore space. The Villain model is in the XY universality class so that the critical exponents are the same as for the bulk superfluid transition in agreement with experiments on Vycor. The critical coupling of the Villain model[12] is about 1/6 thus, the transition occurs at T_c such that,

$$K(T_c) \approx c/6\pi r_0 \qquad (5)$$

What are the consequences of (3) with K_0 replaced by K? For the systems of interest the typical aspect ratio, c/r_0, is less than 12, thus the three dimensional transition does not happen until the core vortices have significantly renormalized the coupling constant and the transition occurs at a temperature somewhat beyond the point where $K=2/\pi$ on the steep part of the curves in Fig. 4. Furthermore, the 3-d critical region is compressed into a small temperature range because of the steep decline of K vs T above $K=2/\pi$. For temperatures outside of the 3-d critical region or for high frequencies the system behaves like a finite size planar system with the length scale set by the pore radius, r_0. The temperature width of the critical region shrinks as r_0 increases and the finite size KT transition becomes sharper.

A quantitative reflection of the compression of the critical region is seen in the critical amplitudes of the superfluid density (mass/ volume), ρ, and the specific heat per unit volume, C. Starting from the effective Hamiltonian it is straightforward to obtain an expression for leading singular behavior of the heat capacity in terms of the behavior of a reference Villain model with a constant coupling energy. The result is

$$C = c_0^{-3} k_B (a/\alpha) \gamma^{3\nu} t^{-\alpha} \qquad (6)$$

with the compression factor, γ given by

$$\gamma = d \log K / d \log T \,|\, _{T=T_c} \qquad (7)$$

here $t = |1-(T/T_c)|$ is the reduced temperature and a is a constant of order one. Note that the singular part of the heat capacity is proportional to the density of pore vortex degrees of freedom, c_0^{-3}. The singular behavior of the superfluid density, ρ can be obtained from two scale factor universality[13],

$$\rho = \frac{k_B T}{c_0} \left(\frac{m}{\hbar}\right)^2 \left(\frac{a^{1/3}}{R}\right) \gamma^\nu t^\nu \qquad (8)$$

where R is a universal amplitude ratio which is roughly unity. For typical experimental parameters, the compression factor is in the range 1 to 10. These quantitative prediction remain to be tested experimentally.

DISCUSSION

By considering the full set of topological excitations for a multiply connected superfluid film we have achieved an understanding of how to generalize the KT theory to apply to superfluid films in porous media. There is no core vortex unbinding transition but rather, the role of the core vortices is to determine the coupling constant in an effective Hamiltonian for the pore vortices. Pore vortices are flows similar to the vortex excitations of bulk ^{4}He and the effective Hamiltonian for these excitations is a Villain model which is in the XY universality class. Thus we expect to observe the usual three dimensional critical exponents however the range in temperature over which these may be observed can be considerably compressed because of the influence of the core vortices.

The picture presented here is in reasonable agreement with the results for Helium in Vycor however, for the more open materials such as xerogel and aerogel, experiments yield exponents which differ from the bulk exponents. One can speculate that this is the result of a fractal pore structure. Such a situation could be treated using the present ideas by replacing the cubic lattice jungle gym by a fractal jungle gym

The theory presented here may also be extended to treat full pores and to discuss dissipation process. For the case of full pores we also expect to obtain an effective Villain model however the coupling constant in this model is now related to the superfluid density for Helium filling a cylindrical cavity. Critical velocities can be treated following the Langer-Fisher[14] theory except that the bulk vortex loops of that theory are replaced by pore vortex loops. Ultrasonic attenuation can be treated by generalizing the Ambegaokar et. al.[15] theory to a cylindrical geometry[11].

REFERENCES

*Permanent address: Department of Physics and Astronomy, University of Massachusetts, Amherst, MA 01003. Supported in part by NSF Grant DMR 8702705.
1. M. H. W. Chan, K. I. Blum, S. Q. Murphy, G. K. S. Wong and J. D. Reppy, Phys. Rev. Lett. 61, 1950 (1988); D. Finotello, K. A. Gillis, A. Wong and M. H. W. Chan, *ibid.* 61, 1954 (1988); N. Mulders and J. R. Beamish, *ibid.* 62, 438 (1989).
2. See accompanying articles by M. H. W. Chan; J. R. Beamish and K. Shirahama et. al.
3. The existence of a minimum film thickness for superfluid behavior suggests that there is a zero temperature quantum phase transition which is discussed in the accompanying article by M. P. A. Fisher. In the present paper we restrict our attention to films significantly thicker than two atomic layers where the influence of the zero temperature phase transition should be unimportant.

4. V. Kotsubo and G. A. Williams, Phys. Rev. Lett. 53, 691 (1984); Phys. Rev. B33, 6106 (1986).
5. J. E. Berthold, D. J. Bishop and J. D. Reppy, Phys. Rev. Lett. 39, 348 (1977); D. J. Bishop, J. E. Berthold, J. M. Parpia and J. D. Reppy, Phys. Rev. B24, 5047 (1981).
6. T. Minoguchi and Y. Nagaoka, Jpn. J. Appl. Phys. 26 (1987), Suppl. 26-3, 293 (Proceedings of the 18th International Conference on Low Temperature Physics LT-18); Prog. Theor. Phys. 80, 397 (1988).
7. J. Machta and R. A. Guyer, Phys. Rev. Lett. 60, 2054 (1988).
8. F. Gallet and G. A. Williams, Phys. Rev. B39, 4673 (1989).
9. for a review see D. R. Nelson in *Phase Transitions and Critical Phenomena*, edited by C. Domb and J. Liebowitz (Academic, New York, 1983), Vol. 7.
10. G. Springer, *Introduction to Riemann Surfaces* (Chelsea, New York, 1981).
11. J. Machta and R. A. Guyer, J. Low Temp. Phys. 74, 231 (1989).
12. J. Villain, J. Phys. (Paris) 36, 581 (1975); C. Dasgupta and B. I. Halperin, Phys. Rev. Lett. 47, 1556 (1981).
13. P. C. Hohenberg, A. Aharony, B. I. Halperin, and E. D. Siggia, Phys. Rev. B13, 2986 (1976).
14. J. S. Langer and M. E. Fisher, Phys. Rev. Lett. 19, 560 (1967).
15. V. Ambegaokar, B. I. Halperin, D. R. Nelson and E. D. Siggia, Phys. Rev. B21, 1806 (1980).

SUPERFLUID TRANSITION OF ^{4}He CONFINED IN POROUS GLASSES

M. H. W. Chan
Department of Physics
The Pennsylvania State University
University Park, PA 16802 U.S.A.

ABSTRACT

The result of recent superfluid density and heat capacity measurements of ^{4}He confined in three different porous glasses, Vycor, xerogel and aerogel, are reviewed in this paper. These measurements appear to support the suggestion that the superfluid transition of ^{4}He confined in Vycor belongs to the same universality class as the lambda transition of bulk ^{4}He. In contrast, the critical exponent describing the superfluid density of ^{4}He confined in the two other glasses near T_c is quite different from the bulk value.

INTRODUCTION

The superfluid transition of bulk ^{4}He is a well studied model system of critical phenomena; the major virtue being the extreme purity of the fluid. Indeed, the critical exponent obtained for this system is of extremely high quality, e.g., ζ, the exponent describes the temperature dependence of the superfluid fraction ρ_s near the transition temperature T_c and α the exponent describing the divergence of the specific heat are determined to be 0.6717 ± 0.0004[ref. 1] and $-0.0127 \pm .0026$[ref. 2], respectively. These values are found to be consistent with the hyperscaling relation $2 - \alpha = d\zeta$[ref. 3] and are in excellent agreement with theoretical predictions.[4] By varying the pressures of the coexisting vapor (along the λ line), a family of superfluid transitions with slightly different T_c are found. This is convenient in checking the prediction of amplitude ratios for systems in the same universality class. In particular, measurements along the λ line confirms the prediction of hyperuniversality or two-scale-factor universality.[ref. 5] Hyperuniversality states that if two systems belong to the same universality class, then the free energy per correlation volume ξ^3 should be the same. The result of this idea is that for two (primed and unprimed) systems in the same universality class,

$$\frac{A'_\alpha}{A_\alpha} = \left(\frac{\rho'_{so} T_c}{\rho_{so} T'_c}\right)^3 \tag{1}$$

A_α and ρ_s, expressed in units of [erg/k-cm^3] and [gm/cm^3], are, respectively, the amplitudes of the specific heat divergence and the superfluid fraction coarse grain averaged over the entire experimental cell.

The relevant power law expressions as T→T$_c$ are,

$$C = \frac{A_\alpha}{\alpha}(t^{-\alpha} - 1) + B \qquad (2)$$

$$\rho_s = \rho_{so} t^\zeta \qquad (3)$$

In 1975, Hall, Kiewiet and Reppy[6] examine carefully the temperature dependence of the superfluid density of ^{4}He confined in Vycor glass with pores completely filled. They found ρ_s is describable by an exponent of ζ = 0.65 ± .03 into a reduced temperature t = (T$_c$ - T)/T$_c$ of 5×10^{-3}. Since the exponent is rather close to the bulk value, this result suggests the possibility that the transition of ^{4}He in Vycor belongs to the same universality class as bulk ^{4}He. Subsequently, torsional oscillator measurement was made to study the superfluid transition of low coverage ^{4}He coating the surface of the internal Vycor pores. By varying the coverage the superfluid transition was studied over a wide range of transition temperatures. These studies also found a ζ that is very close to the bulk value (0.67 ± 0.03) for all films with T$_c$ ≥ 76 mK.[7] There were concerns, however, about the 3D transition interpretation. It was argued[8] that the superfluid transition of ^{4}He film in Vycor is basically 2D and describable by the Kosterlitz-Thouless theory. The observed power law in ρ_s is a smearing of the characteristic Kosterlitz-Thouless sharp drop in ρ_s at T$_c$ by the finite curvature of the pores.

Experimental and theoretical development in the last two years brought some substantial understanding and also raise new questions concerning the effect of disorder due to the host porous media on the superfluid transition. As in the case of superfluid transition of bulk ^{4}He, heat capacity experiment provide a stringent test on the nature of the transition. If the transition of ^{4}He in Vycor is a 3D critical phase transition, there should be a sharp heat capacity peak at T$_c$. If the transition is 2D and describable by Kosterlitz-Thouless, there should only be a broad anomaly, related to the unbinding of the vortices well above T$_c$. Two attempts, one at Sussex by Brewer and coworkers,[9] and the other at SUNY, Buffalo, by Joseph and Gasparini[10] were made to look for the sharp heat capacity peak at T$_c$ (= 1.955K) for ^{4}He filled Vycor glass. These attempts have been unsuccessful, instead they found a broad anomaly centering near 2.1K. Recently, a third attempt were made at Penn State, despite a precision of 0.1%, no evidence of a sharp peak at T$_c$ is found.[11]

EXPERIMENTAL RESULTS

Fig. 1 shows the broad anomaly is exceedingly similar to that found by the two earlier experiments. The heat capacity is expressed per unit volume of the Vycor glass sample, containing both SiO$_2$ and ^{4}He. (For the lambda transition in bulk ^{4}He, the heat capacity values in corresponding units are 1.96 J/K-cm^3 and 2.9 J/K-cm^3 at 2 mK and 20 μK below Tc, respectively.) We have chosen to express the heat capacity values in this unit to facilitate comparison of the superfluid transition in the various porous systems. It is also interesting to note that the broad anomalies shown in Fig. 1 for ^{4}He in Vycor and xerogel are quite similar to that found in isolated ^{4}He bubbles of sizes ranging from 40 to 100 Å imbedded in Cu.[12] The results of ^{4}He in Cu are interpreted in terms of finite size broadening of the λ transition.

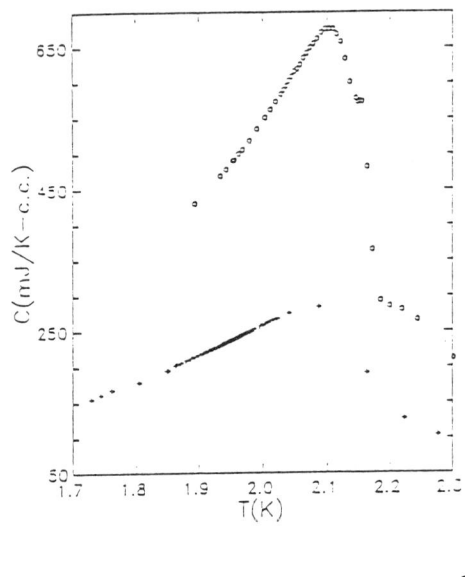

Fig. 1. Heat capacity per unit volumes of sample cells of ^{4}He in Vycor (crosses) and xerogel (circles), where the pores are completely filled. There is no evidence of sharp peak at the superfluid transition temperatures (1.952K for Vycor and 2.088K for xerogel). Small kink near 2.17K of the xerogel plot is due to small quantity of bulk.

Fig. 2. Heat capacity of per unit volume vs. temperatures for three coverages of ^{4}He (in μmole per m^2 of surface area) on Vycor. In addition to the sharp signature at T_c, broad maxima for the two higher coverages are clearly visible. The lowest coverage (Δ) is shifted upward by 1.30 mJ/K-cm^3 and the highest coverage by 2.60 mJ/K-cm^3.

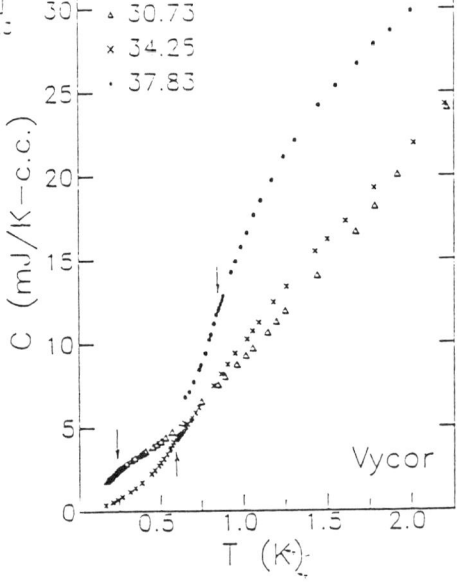

When our Vycor heat capacity experiments were repeated at lower film coverages, a sharp heat capacity peak, in addition to the broad anomaly, is found. The broad anomaly decreases rapidly with coverage and is not visible for films with ^{4}He coverage less than 33.54 μmole of ^{4}He per m^2 of Vycor surface (T_c = 0.495K). Figure 2 shows heat capacity versus temperature for a number of low ^{4}He coverages. Figure 3 shows more clearly the sharp heat capacity peaks. When normalized at a particular coverage, the temperatures of the sharp heat capacity peak for various coverages are found to coincide with T_c as defined by ρ_s measurements.[11] More recently, the simultaneity

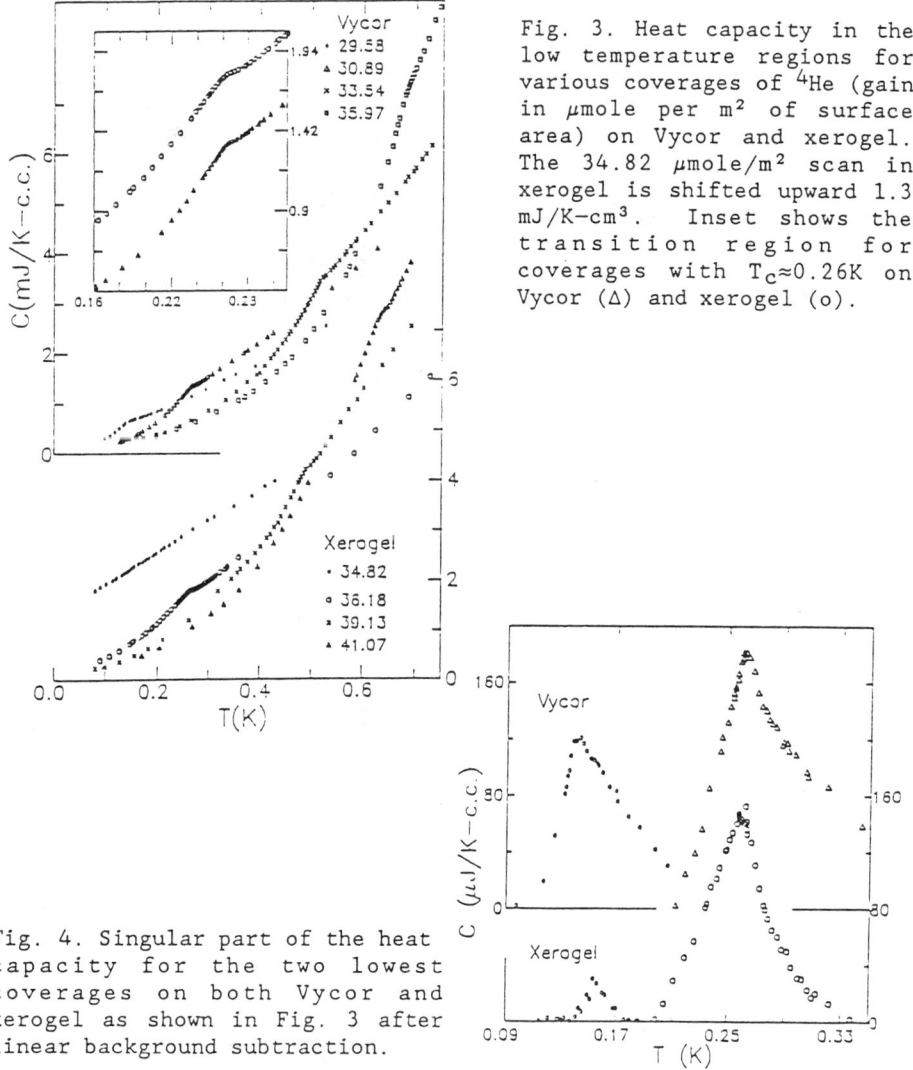

Fig. 3. Heat capacity in the low temperature regions for various coverages of ^{4}He (gain in μmole per m^2 of surface area) on Vycor and xerogel. The 34.82 μmole/m^2 scan in xerogel is shifted upward 1.3 mJ/K-cm^3. Inset shows the transition region for coverages with $T_c \approx 0.26$K on Vycor (△) and xerogel (o).

Fig. 4. Singular part of the heat capacity for the two lowest coverages on both Vycor and xerogel as shown in Fig. 3 after linear background subtraction.

of these signatures is confirmed by performing heat capacity and ρ_S studies on the same piece of Vycor during the same experimental run.[13]

In Figure 4, the singular heat capacity near T_c for two lowest coverages are highlighted by subtracting away linear background terms from that shown in Fig. 3. We have attempted analyses of these sharp signature without arriving at an quantitative conclusion on the value of α. The major difficulty is the substantial rounding of the heat capacity data near T_c that limit the temperature range for reliable power law analysis. The sinusoidal heating power of our ac technique is responsible for rounding on the order of 1mK for measurement with T_c near 145 mK. Rounding on the order of 5×10^{-3} in reduced temperature, possibly related to long length scale inhomogeneity (as opposed to pore size distribution) is also seen in ρ_s measurements.

The magnitude of the singular heat capacity peak of the low coverage Vycor peak is found to be in reasonable (within a factor of 3) agreement with that predicted via the two-scale-factor universality. The comparison is made against the lambda transition of bulk ^{4}He. The values of ρ'_{so} (of ^{4}He in Vycor), and A_α and ρ_{so} (of bulk ^{4}He) in equation (1) are taken from existing experiments.[14] Since there is substantial rounding in our heat capacity peak, $A\alpha$ is evaluated sufficiently far from T_c ($t > 1\times10^{-3}$) to find an estimated value for A'_α. Reasonable agreement, (i.e., in the same order of magnitude), between the estimated and measured amplitude of the singular heat capacity peak is found for all films in Vycor with T_c ranges from 0.145K to 1K. This agreement supports the claim that ^{4}He in Vycor belongs to the same universality class as the lambda transition. This is remarkable since the density of mobile ^{4}He (above the localized layer) for a film with T_c = 0.145K is close to be 1000 times less dense than bulk ^{4}He at the λ point! The reason for the apparent absence of the peak in the filled pore case is due to the fact that the expected amplitude is too small to be resolved when compared with the large broad anomaly. Figure 2 and 3 show that the broad anomaly grows rapidly with coverage and swamps the sharp peak at T_c.

It is noteworthy that in both Vycor and xerogel the heat capacity on the superfluid side decreases with increasing coverage, in contrast to the behavior above T_c. The heat capacity eventually drops below that of the localized layer (the coverage for which T_c=0) making the separation of the mobile-film and localized layer heat capacity contribution inappropriate and impossible. Apparently when the coverage is increased the excitation spectrum in the film is changed in such a way that the entropy of the entire film is lowered. Hydrodynamically, however, the superfluid and localized layer density can be distinguished by imposing a relative velocity between them, e.g. with a torsional oscillator.

In the last two years, there emerges a number of models of superfluid transition of films in porous media focusing on the role of vortices on a lattice of interconnected cylinders.[15-17] The common feature of these models is that whereas there are "2D" characteristics at low temperatures and short length scales, at a temperature sufficiently close to T_c, the transition becomes a genuine 3D critical phase transition so that power law behavior of ρ_S

and sharp heat capacity peak are expected. Recent ultrasonic measurements[18] of ^{4}He in Vycor found similar (i.e., bulk-like) value for ζ as in torsional oscillator experiments. They found very little dissipation when the measurements are made at 5MHz and below. Substantial dissipation is found above 10MHz. These findings are consistent with the idea that the film displays 3D critical characteristics at long length scales. For a Kosterlitz-Thouless transition dissipation is expected and found down to frequencies of the torsional oscillators, typically 1KHz and below. No evidence of dissipation is found near T_c in the torsional oscillator measurement of ^{4}He confined in Vycor. The lattice model[16] can also be used to estimate the area under the heat capacity peak. The value they predicted is on the order of Nk_BT_c where N is the number of cylindrical intersections. (N can be estimated from the knowledge or pore size, surface area, open volume and the assumption of a square lattice.) The predicted area is about an order of magnitude smaller than the experimental value.

The obvious limitation of these models is, of course, there is no disorder in the cylindrical lattice. Indeed recent torsional oscillation measurement of ^{4}He confined in two other glasses found completely unexpected results revealing the crucial role of the nature of the disorder.[14] In contrast to the Vycor glass case, the exponent ζ in xerogel and aerogel glasses were found to be very different from the bulk value, namely 0.89 ± 0.02 and 0.813 ± 0.009, respectively. The transitions of these two glasses are even "sharper" than that in Vycor in that simple power law behavior of ρ_s is found as close as 1×10^{-3} for xerogel and 8×10^{-5} for aerogel in reduced temperature. In Vycor, rounding set in near 5×10^{-3} in reduced temperature. The heat capacity result of ^{4}He in xerogel is similar to that found in Vycor, i.e., no evidence of a sharp heat capacity for the case when the pores are completely filled and clear evidence of sharp signature for low coverage films with onset temperatures below 1K (See Figures 1, 3 and 4). The data again is not of sufficient quality to yield a definitive exponent. In contrast to Vycor and xerogel, a prominent and sharp heat capacity peak is found recently at T_c for ^{4}He confined in aerogel.[19] Preliminary analysis found a value −0.69 ± 0.15 for α.

It is not clear why the result obtained with the xerogel and aerogel glass are completely different from that of Vycor. The manufacturing process of xerogel and aerogel is different from that of Vycor glass. Formation of Vycor glass starts with a molten mixture of SiO_2 and B_2O_3. Upon cooling, spinoidal phase separation occurs leaving bicontinuous interconnected B_2O_3 and SiO_2 regimes. The porous glass is obtained by leaching out the B_2O_3 phase after quenching from a specific temperature. The resulting pores typically have dimensions of 70Å, By contrast, aerogel and xerogel (also called silica-gel) are formed by the sol-gel process. The gel is formed through diffusive aggregation of clusters of molecules in a solvent. The formation of a gel network that involves a macroscopic fraction is called the sol-gel transition. If the withdrawal of the solvent is accomplished by simple drying in air, the resultant product is called xerogel or dry gel. In this process the delicate aspect of the structure collapses due to the surface tension of the liquid-vapor interface. In the process, porous glass with rather uniform pore and large porosity can be made. The xerogel samples used in the heat capacity and torsional oscillator experiments has a porosity of 60% and typical pore diameter of 100Å.[20] In order to preserve the delicate structure, hypercritical drying, avoiding liquid-vapor surface tensions is used. Aerogel is formed by hpercritical drying. The aerogel sample used in the ρ_s experiment has a porosity of 95%![21] In contrast to Vycor and xerogel it is not meaningful to characterize aerogel via pore diameter since the separation of the SiO_2 strands in aerogel range from 200Å to 2000Å.

Examined by itself, the findings in Vycor, namely a ζ equal to that of the bulk system, and the agreement between the observed and predicted (via hyperuniversality) amplitude of the singular heat capacity peak of thin film, can be considered to be consistent with the Harris criterion.[22] This condition states that disorder does not have any effect on the critical phase transition — on the values of the critical exponent, for example — if the pure system (before disorder is introduced) has a negative value for the specific heat exponent. For bulk ^{4}He, α is indeed negative ($\alpha = -0.0127 \pm .0026$). However, the findings in xerogel and aerogel, where ζ is clearly not the bulk value, a different explanation must be advanced. In most studies of disordered systems, including the one by Harris, the disorder is regarded as truly random. In this case, the probability that there is an impurity or defect at a given point is regarded as independent of the locations of other impurities and defects. Disorder in a system can also be correlated.[23] Such correlation can be probed by small angle neutron and x-ray studies. These studies have been performed on the three glasses of interest. For Vycor, in a plot of scattering intensity vs wave vector, a prominent peak is found near $q = 2 \times 10^{-2}$Å^{-1} corresponding to a dominant length scale near 250Å.[24] The intensity rises monotonically with decreasing wave vector for $q > 2 \times 10^{-2}$Å^{-1} and decreases for lower q's. In contrast, the scattering result on a xerogel sample made in the similar fashion as the sample used in the ρ_s and heat capacity experiments show no dominant length scale.[25] The intensity rise monotonically with decreasing wave-vector and the intensity is found to level off near $q = 5 \times 10^{-3}$Å^{-1}, corresponding to a length of 1000Å. Similar behavior is found for aerogel.[26] The leveling of the intensity indicates the absence of correlation for length larger than 1000Å. It is tempting but not easy to correlate the difference in ζ in Vycor and the silica-gel substrate with the observed difference in the small angle data.

It is unclear at this point, considering the difference in the observed value of ζ in xerogel and aerogel (0.89 vs 0.81), whether they belong to the same universality class or there is a host of different universality classes dependent on the detailed differences in the preparation of these glasses. It is our hope that future theoretical studies and experiments will be helpful in developing an adequate understanding of the effect of disorder on the superfluid transition.

The work described above was done in collaboration with Ken Blum, Sheena Murphy, Gane Wong and John Reppy (Cornell), Keith Gillis, Dan Finotello, Apollo Wong, and Wen Ma (Penn State). It has been my pleasure to work with them. I also wish to acknowledge an instructive conversation with Jonathan Machta. The work done at Penn State is supported by the National Science Foundation through Grant No. DMR-8701651 (low temperature physics).

REFERENCES

1. A. Singsaas and G. Ahlers, Phys. Rev. B30, 5103 (1984); D. S. Greywall and G. Ahlers, Phys. Rev. A7, 2145 (1973).
2. J. A. Lipa and T. C. P. Chui, Phys. Rev. Lett. 51, 2291 (1983).
3. B. D. Josephson, Phys. Lett. 21, 608 (1966); M. E. Fisher, M. N. Barber and D. Jasnow, Phys. Rev. A8, 1111 (1978).
4. P. C. Hohenberg, Physica 109 and 110 B, 1436 (1982).
5. P. C. Hohenberg, A. Aharony, B. I. Halperin and E. D. Siggia, Phys. Rev. B13, 2986 (1976); D. Stauffer, M. Ferer and M. Wortis, Phys. Rev. Lett. 29, 345 (1972).
6. C. W. Kiewiet, H. E. Hall and J. D. Reppy, Phys. Rev. Lett. 35, 1286 (1975).
7. D. J. Bishop, J. E. Berthold, J. M. Parpia, J. D. Reppy, Phys. Rev. B24, 5047 (1981); B. C. Crooker, B. Hebral, E. N. Smith, Y. Takano, J. D. Reppy, Phys. Rev. Lett. 51, 666 (1983).
8. V. Kotsubo and G. A. Williams, Phys. Rev. Lett. 53, 691 (1984), and Phys. Rev. B33, 6106 (1986); C. Wang and Lu Yu, Phys. Rev. B33, 599 (1986).
9. D. F. Brewer, J. Low Temp. Physics 3, 205 (1970).
10. R. A. Joseph and F. M. Gasparini;, J. Phys. (Paris), Colloq. 39, C6-310 (1978).
11. D. Finotello, K. A. Gillis, A. Wong and M. H. W. Chan, Phys. Rev. Lett. 61, 1954 (1988).
12. E. C. Syskakis, F. Pobell and H. Ullmaier, Phys. Rev. Lett. 55, 2964 (1985).
13. S. Q. Murphy and J. D. Reppy, to be published.
14. M. H. W. Chan, K. I. Blum, S. Q. Murphy, G. K. S. Wong and J. D. Reppy, Phys. Rev. Lett. 61, 1950 (1988).
15. T. Minoguchi and Y. Nagaoka, Jpn. J. App. Phys. 26, Suppl. 26-3, 327 (1987); Prog. Theor. Phys. 80, 397 (1988).
16. J. Machta and R. A. Guyer, Phys. Rev. Lett. 60, 2054 (1988). See also paper in this conference.
17. F. Gallet and G. A. Williams, Phys. Rev. B39, 4673 (1989); H. Cho, F. Gallet and G. A. Williams, this conference.
18. N. Mulders and J. Beamish, Phys. Rev. Lett. 62, 438 (1989).
19. G. K. S. Wong and J. D. Reppy, Cornell University Materials Science Center Report No. 6619 (1989), and to be published.
20. Sample provided by Merrill Shafer and David Awswchalom of IBM, Yorktown Heights. M. W. Shafer, D. D. Awschalom, J. Warnock and G. Ruben, J. Appl. Phys. 61, 5438 (1987).
21. Sample provided by A. Poelz of DESY. G. Poelz and R. Riethmüller, Nucl. Inst. Methods 195, 491 (1982).
22. A. B. Harris, J. Phys. C7, 1671 (1974).
23. A. Weinrib and B. I. Halperin, Phys. Rev. B27, 413 (1983).
24. D. W. Shaefer, B. C. Bunker and J. P. Wilcoxon, Phys. Rev. Lett. 57, 284 (1987); P. Wiltzius, F. S. Bates, S. B. Dierker and G. D. Wignall, Phys. Rev. A36, 2991 (1987).

25. P. Schmidt, Private communication.
26. D. W. Shaefer, K. D. Keefer, Phys. Rev. Lett. $\underline{56}$, 2199 (1986); R. Vacher, T. Woigner, J. Pelous and E. Courtens, Phys. Rev. B$\underline{37}$, 6500 (1988).

BOSON LOCALIZATION AND THE SUPERFLUID-INSULATOR TRANSITION

Matthew P.A. Fisher
IBM Research Div., T.J.Watson Research Center, Yorktown
Hts.,NY 10598

ABSTRACT

Summary of oral presentation.

SUMMARY

Historically, the λ transition in bulk ^{4}He and the corresponding transition in thin films[1] have served as a cornerstone in testing numerous theoretical concepts in static and dynamic critical phenomena[2]. More recent experiments on ^{4}He absorbed in various porous media[3-5] should be equally fruitful in clarifying the effects of disorder on such continuous phase transitions. In these more recent experiments it is found that a critical density of ^{4}He, n_c, is needed for the system to become superfluid at any positive temperature. The implied superfluid transition at $T=0$, as the density of ^{4}He is increased, is a transition controlled entirely by quantum fluctuations. Since, for $n < n_c$, the system is presumably not conducting at $T=0$, we refer to this $T=0$ transition as the superfluid-insulator transition. Below, I briefly highlight some recent theoretical progress[6,7] made in understanding both the critical properties[8-10] of this transition and properties of the insulating phase[11].

Repulsively interacting bosons moving in a random potential are believed to exhibit two possible phases at $T=0$. At low densities an insulating "Bose glass" phase[6,7] is expected. This phase is insulating because of the localization effects of the randomness, and analogous to the Fermi glass phase of interacting fermions in a strongly disordered potential[12]. The Bose glass phase is characterized by a finite compressibility, gapless low-lying quasiparticle excitations, but an infinite superfluid susceptibility[6,7,11]. At higher densities a superfluid phase is expected.

The $T=0$ superfluid-insulator transition from the Bose glass to superfluid phase is characterized by three exponents[6,7]; a dynamic critical exponent, z, predicted to be exactly equal to the spatial dimensionality d, a correlation length exponent ν which satisfies the bound

$v \geq 2/d$, and an order-parameter correlation exponent η bounded above by $2-d$. These exponent relations can be placed in the framework of a scaling theory[6-8], enabling explicit and verifiable predictions for various measurable properties at low temperatures and densities near n_c.

Experimentally, the most directly accessible quantity is the superfluid density. Near the superfluid-insulator transition this satisfies a scaling law[6-8],

$$\rho_s(T,n) \sim \delta^{(d+z-2)\nu} P(T/\delta^{z\nu}), \qquad (1)$$

where $\delta \equiv n - n_c$ and $P(0) = 1$. For $d \geq 2$, the crossover scaling function P has a singularity at some finite value of its argument, corresponding to the critical line (in the n – T plane) of λ transitions at $T > 0$, which terminate at the superfluid-insulator transition. The critical temperature T_λ thus vanishes as $T_\lambda \sim \delta^{z\nu}$ for $n \to n_c$.

At $T = 0$, Eqn. (1) predicts $\rho_s(0) \sim \delta^\zeta$ with $\zeta = (d + z - 2)\nu$, which is a generalization of the familiar Josephson relation at the λ transition[13]. Using the above exponent relations[6,7] gives $\zeta = 2(d-1)\nu \geq 8/3$ in 3d. Another measurable exponent, x, can be defined as $T_\lambda \sim \rho_s(0)^x$, with $x = z/(d+z-2)$. Since this depends only on z, the exponent relation $z = d$ implies $x = 3/4$ exactly in 3d. Scaling relations for numerous other observables can likewise be obtained[6,7].

The experience of studying the λ transition in ^{4}He suggests that the $T = 0$ superfluid-insualtor transition may lead to insights into more complicated $T = 0$ transitions, such as the superconductor-insulator transition in granular or amorphous superconductors[14] and the metal-insulator transition in Fermi systems[12].

The work summarized above was done in collaboration with P. B. Weichman, G. Grinstein and D. S. Fisher. I am grateful to them all.

REFERENCES

1. D.J. Bishop and J.D. Reppy, Phys. Rev.B 22, 5171 (1980).
2. J.M. Kosterlitz and D.J. Thouless, J. Phys. C 6, 1181 (1973).
3. J. D. Reppy, Physica 126B, 335 (1984).
4. D. Finotello, K. A. Gillis, A. Wong and M. H. W. Chan, Phys. Rev. Lett. 61, 1954 (1988).
5. M. H. W. Chan, K. I. Blum, S. Q. Murphy, G. K. S. Wong and J. D. Reppy, Phys. Rev. Lett. 61, 1950 (1988).
6. D. S. Fisher and M. P. A. Fisher, Phys. Rev. Lett. 61, 1847 (1988).

7. M. P. A. Fisher, P.B. Weichman, G. Grinstein and D. S. Fisher, to appear in Phys. Rev. B (1989).
8. M. Ma, B. I. Halperin and P. A. Lee, Phys. Rev. B34, 3136 (1986). This work derives scaling laws for various properties near the superfluid-insulator transition and undertakes an ε-expansion calculation for the critical properties. As pointed out in Ref. 7, a crucial term was dropped in the effective field theory, invalidating the ε- expansion results.
9. T. Giamarchi and H. J. Schulz, Europhys. Lett. 3, 1287 (1987) and Phys. Rev. B37, 325 (1988). These authors have studied the critical properties of the superfluid-insualtor transition in one dimension, applying a renormalization group perturbative in the strength of the disorder.
10. A. Gold, Z. Phys. B 52, 1 (1983).
11. L. Zhang and M. Ma, Phys Rev. A 37, 960 (1988).
12. See P. A. Lee and T. V. Ramakrishnan, Rev. Mod. Phys. 57, 287 (1985), and references therein.
13. See for example, M. E. Fisher, M. N. Barber and D. Jasnow, Phys. Rev. A8, 1111 (1973); Sec. VI.
14. S. Doniach, Phys. Rev. B24, 5063 (1981).
15. M. P. A. Fisher and G. Grinstein, Phys. Rev. Lett. 60, 208 (1988), and references therein.

ULTRASONIC STUDIES OF HELIUM IN POROUS GLASSES

J.R. Beamish and N. Mulders
Department of Physics and Astronomy
University of Delaware
Newark, DE 19716, USA

ABSTRACT

We have used an ultrasonic technique to study the critical behavior of superfluid ^{4}He in porous Vycor glass and in a silica aerogel. We determine the superfluid density from the increase in the transverse sound velocity when the superfluid decouples and the damping from the corresponding attenuation. At the lowest frequencies (around 3 MHz), we observe transitions (considerably sharper in aerogels than in Vycor) with critical exponents similar to those found in torsional oscillator measurements. At higher frequencies the transitions become noticeably rounded (by as much as 0.1 K at 196 MHz in Vycor). In addition, there are clear attenuation peaks at the transitions and these also broaden and shift to higher temperatures as the sound frequency increases. We compare our results for full pores and for films to other experiments and discuss recent theories of the superfluid transition in porous media.

INTRODUCTION

Superfluidity of helium in restricted geometries has been the object of much theoretical and experimental interest in recent years. Helium has long provided a testing ground for theories of phase transitions. For example, bulk helium exhibits well understood three-dimensional (3D) critical behavior near the lambda transition, while helium films on flat substrates behave two-dimensionally (2D), with a vortex-unbinding transition of the Kosterlitz-Thouless (KT) type[1,2]. When helium is adsorbed in a porous medium, either as a film or filling the pores, its behavior may be changed in a number of ways. Finite size effects can shift or even entirely smear out the phase transition, the multiply-connected substrate geometry may change the effective dimensionality, or the disorder introduced by the porous material may change the nature of the transition. In the work reported here, we have used an ultrasonic technique to study the frequency dependence of the superfluid transition for helium in the pores of Vycor glass and a silica aerogel.

A number of recent experiments have studied ^{4}He on porous substrates. The results for both alumina[3] and platinum[4] powders were similar to the flat-substrate case, namely a sudden decrease in the superfluid density at the superfluid transition accompanied by substantial dissipation. These features were interpreted in terms of the Kosterlitz-Thouless theory, with a broadening of the transition due to finite size effects in the small powders. On the other hand, torsional oscillator experiments[5] using a porous Vycor glass substrate found a relatively sharp superfluid transition, near which the superfluid density disappeared with a critical exponent close to that of bulk helium, apparently indicating 3D critical behavior, even for thin films. In these experiments, the dissipation peak characteristic of 2D vortices on flat substrates was not observed. Further experiments on other porous glasses (silica xerogels and aerogels) have complicated the picture by showing even sharper transitions, but the critical exponent was different for each of the substrates[6]. In addition, specific heat measurements on thin films adsorbed on Vycor and xerogel found sharp cusps at the superfluid onset temperatures[7], in marked contrast to the predictions of the KT theory for 2D films.

There have been a number of different theoretical approaches to the problem. Several authors have extended the KT theory to finite geometries. Initial work[3,8] considered vortices on the surface of isolated spheres and predicted a broad transition with negligible dissipation in pores as small as those of Vycor. Recently, the theory has been generalized to the case of an interconnected network of cylinders[9,10,11]. This model combines the 2D nature of the film at small length scales (where the substrate is locally flat) with the multiply connected geometry which at large length scales can lead to 3D critical behavior. A quite different approach is to consider the substrate as a random potential acting on the helium[12,13] and it has been suggested[6,14] that it is the long-range structure of this "quenched disorder" which is responsible for the different critical exponents for different substrates.

The frequency dependence of the superfluid transition can provide valuable information about the possible role of vortices since their response depends on the measurement frequency as well as the pore size. In 2D films, most of the dissipation near the transition is due to vortex pairs with separations equal to the diffusion length $r_D = (2D/\omega)^{1/2}$ where D is the vortex diffusivity and ω is the drive frequency. This length is of the order of microns at torsional oscillator or third sound frequencies. In porous media, the restriction of vortex motion by the pore geometry is expected to broaden the transition and reduce the associated dissipation. These finite size effects become less important than finite frequency effects when the diffusion length becomes smaller than the pore size. In Vycor, this would occur at frequencies above about 12 MHz (estimating the diffusivity[15] D as $\hbar/m$ and the characteristic length scale of Vycor[16] as 200 Angstroms). In aerogels, the surface has comparable radii of curvature but there is also structure with much longer length scales[17] which may be important. In order to see whether such a crossover occurs for helium in Vycor and aerogels, we have used an ultrasonic technique to make measurements of the superfluid density and dissipation at frequencies from 3 to 200 MHz.

EXPERIMENTAL TECHNIQUE

The superfluid density and dissipation in the helium were determined from measurements of the velocity v and attenuation α of transverse ultrasonic waves propagating in the porous substrate. When the helium in small pores becomes superfluid, only the normal fluid fraction remains viscously locked to the substrate and so the sound velocity increases by an amount to the decoupled superfluid fraction. For films, this increase is given by

$$\Delta v/v = \frac{1}{2}(1 - \chi)\sigma_s S \qquad (1)$$

where σ_s is the areal density of the superfluid component and S is the sample's specific surface area. The fraction χ of the superfluid which remains inertially locked to the substrate at zero temperature is determined by the pore tortuousity. When the pores are filled, the quantity $\sigma_s S$ is replaced by $\phi \rho_s/\rho_p$, where ϕ is the porosity and ρ_s and ρ_p are the superfluid and porous substrate densities. The dissipation in the helium can also be determined since the sound attenuation is related to the damping Q^{-1} by

$$Q^{-1} = 2v\alpha/\omega \quad . \qquad (2)$$

Thus, ultrasonic measurements provide essentially the same information as does a torsional oscillator only at much higher measurement frequencies.

The ultrasonic measurements were made using a pulse technique with two LiNbO$_3$ shear transducers bonded to opposite ends of the samples for sending and receiving. We used a heterodyne phase sensitive system[18] which made simultaneous measurements of the sound velocity and attenuation at a number of different frequencies (in the range 2 to 200 MHz) at each temperature. The velocity resolution of the system was $\Delta v/v \approx 10^{-6}$ and it could detect attenuation changes of 0.005 dB. This meant that in a 1 cm Vycor sample we could detect the decoupling of 4 x 10^{-4} monolayers of superfluid or damping changes of ΔQ^{-1} of 4 x 10^{-7} (at 100 MHz). All measurements were made at low signal amplitudes (peak power < 10^{-6} W, duty cycle < 0.1%) where there was no sample heating or amplitude dependence. About 15 minutes were required to take data at all frequencies, during which time the temperature was controlled to within about 0.1 mK. We estimate our temperatures to be accurate to within about 5 mK.

The Vycor samples (Corning 7930 "Thirsty Glass") were rods about 0.3 cm in diameter and 1 cm long. They were cleaned in a hydrogen peroxide solution, air dried and then evacuated at room temperature for several hours before mounting in the sample cell. A previous characterization of an identical sample[19] gave a BET surface area S = 105 m^2/g and a porosity ϕ = 29%. The silica aerogel[20] had a specific surface area of 780 m^2/g and a porosity of about 84% (density 0.35 g/cm^3). Samples about 1.2 cm long were carefully cut and polished, mounted and then evacuated before cooling.

RESULTS

We present here some results for the Vycor and aerogel samples described above. Some of the Vycor results have recently been published[21].

Figure 1 presents some raw data (the transverse sound velocity and attenuation at a frequency of 30 MHz) for a helium film on Vycor. Below the superfluid transition temperature (T$_c$ = 0.84 K), we see the expected increase in the velocity due to decoupling of the superfluid component. In contrast to torsional-oscillator measurements we also see an attenuation peak centered at the transition. These changes due to superfluidity are superimposed on smooth background variations of the velocity and attenuation which are due to two level systems (TLS) in the glass. Since the presence of helium in the pores is known[22] to increase the relaxation rates of the TLS in Vycor we cannot simply subtract the empty-Vycor data. However, in measurements with different film thicknesses we found that these background variations do not change at the onset of superfluidity. The solid lines in Fig. 1 are the data for a non-superfluid helium film on Vycor (slightly thinner than the critical "inert layer" coverage). Since the background varies slowly and smoothly, we simply fit the data well away from the transition and subtract. For the sound velocity, we extrapolate the linear temperature dependence observed above the transition, and for the attenuation, we fit the data on both sides of the peak to a low order polynomial. This procedure is reliable close to the transition but leads to some uncertainties if extrapolated to too low a temperature.

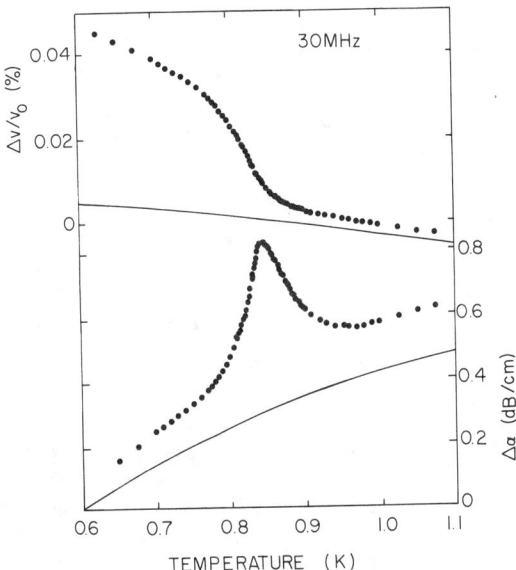

Figure 1 Sound velocity and attenuation of 30 MHz shear waves in Vycor glass with a superfluid ^{4}He film.

Figure 2 shows the results for this film on Vycor (after subtracting the fitted background) at frequencies from 5 to 196 MHz. The sound velocity (i.e. the superfluid density) in Fig. 2a shows quite a sharp transition at the lowest frequencies which becomes progressively more rounded at higher frequencies. At 5 MHz there is a tail region extending about 5 mK above the transition. This is comparable to the rounding seen[5] in torsional-oscillator experiments at 1570 Hz, so that there can be essentially no frequency dependence up to 5 MHz. From the low frequency data close to the transition, we can find a value of the critical exponent for the superfluid density. For all films, is close to the bulklike exponent of 2/3 observed in torsional-oscillator experiments, but the rounding near the transition prevents us from obtaining the exponent very accurately.

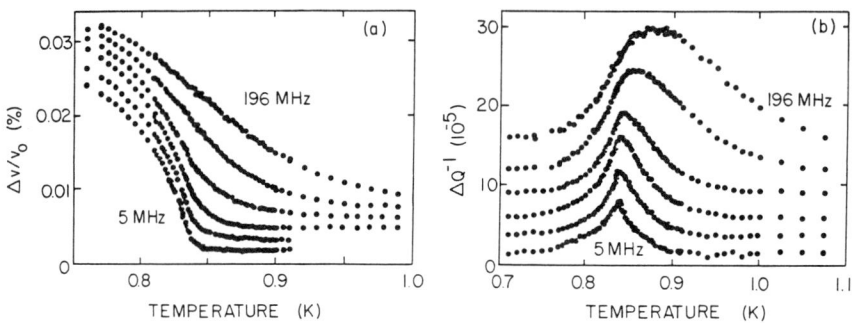

Figure 2 (a) Velocity and (b) damping of ultrasonic shear waves in Vycor. Curves for different frequencies have been vertically displaced for clarity. From bottom to top, the sound frequencies are 5, 8, 16, 83, and 196 MHz.

Figure 2b shows the corresponding attenuation changes (converted to damping ΔQ^{-1} by using Eq. (2)). The peak is a sharp cusp at the lowest frequencies and its maximum occurs at the same temperature as the transition determined from the velocity data. As the frequency increases, the peaks grow and become rounded. They become much wider and the maximum shifts to higher temperature. The broadening is asymmetric, occurring almost totally above the transition temperature.

We made measurements for a number of helium coverages. Figure 3 shows the resulting damping peaks at 30 MHz. Transition temperatures ranged from 0.15 K for the thinnest films to 1.94 K for full pores. In the thinner films (those with T_c below 1 K), the peak heights scale with the transition temperature and hence with the amount of superfluid. The shape and frequency dependence of these peaks are similar in all of these films. Above 100 MHz the dissipation peak heights are roughly constant but at lower frequencies they decrease. This is consistent with the fact that torsional-oscillator experiments in Vycor (at kHz frequencies) did not see similar peaks.

For the thickest film (T_c = 1.67 K), the large dead

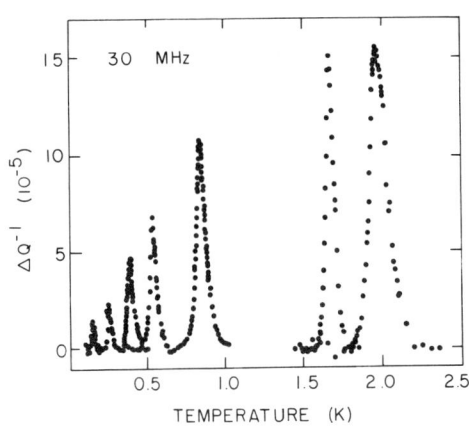

Figure 3 Damping peaks at 30 MHz for different helium coverages in Vycor.

volume of our cell allowed a significant amount of the helium to desorb at higher temperatures, artificially narrowing the peak. In this film and for full pores, the peak height is not proportional to the amount of superfluid in the pores (about 5 times as much in full pores as in the film with $T_c = 0.84$ K). The full pore peaks are somewhat more symmetrical and broaden less with frequency than those for thin films, indicating a change in the dissipation mechanism when the pores are filled.

Although the peaks have some resemblance to the critical sound attenuation seen in liquid helium at the lambda transition, the scaling with frequency is quite different. In bulk helium, the attenuation peak is much narrower and occurs slightly below the transition temperature. As the frequency increases this peak broadens on both sides of the transition and its maximum shifts to lower temperature, in contrast to the behavior shown in Fig. 2.

The behavior of the films on Vycor can be compared to that expected if 2D vortices are important. Torsional oscillator measurements have shown that in films on flat substrates the effect of a finite measurement frequency is to shift the transition to temperatures slightly above its zero frequency position and give the dissipation peak a finite width. In porous media where the geometry restricts vortex motion, such finite frequency effects will be important only at frequencies where the vortex diffusion length is smaller than a characteristic pore size. The characteristic frequency for films on Vycor was estimated above to be about 12 MHz. Our data (Fig. 2) show just the expected broadening and shift to higher temperatures at frequencies above 16 MHz, so that our results are consistent with the predictions of the Kosterlitz-Thouless theory in finite geometries. The similar frequency dependence in the different films can be taken to indicate that the vortex diffusivity D does not depend strongly on film thickness. The motion of "pore vortices" which is considered in recent theories[9,10,11] occurs on longer length scales and so will be important at much lower frequencies.

The behavior of helium in Vycor can be compared to that in our aerogel sample. Figure 4 shows the sound velocity at 3.1 MHz when the aerogel pores are filled with helium. The large porosity and small elastic moduli of the aerogel result in rather large changes in the sound speeds when the superfluid decouples. The transition (at 2.145 K) is much sharper than in Vycor, with a rounding no more than a mK. This allows us to determine the critical exponent quite accurately. Figure 5 is a log-log plot of the normalized superfluid density (determined from the sound velocity of Fig. 4) versus the reduced temperature $t = 1 - T/T_c$. For reduced temperatures less than about 0.05 the data follow a well-defined power law of the form $\rho_s = \rho_{s0} t^\zeta$ with a critical exponent $\zeta = 0.95 \pm 0.02$. Essentially the same behavior was observed in torsional oscillator measurements on a different aerogel sample[6], but with an exponent $\zeta = 0.813 \pm 0.009$. We thus have one more example of a porous glass with a sharp superfluid transition but with yet another critical exponent. To clarify the role of porosity, pore size and micro-structure in the critical behavior of helium, we plan to carry out further ultrasonic studies on a range of aerogel samples, with both helium films and full pores.

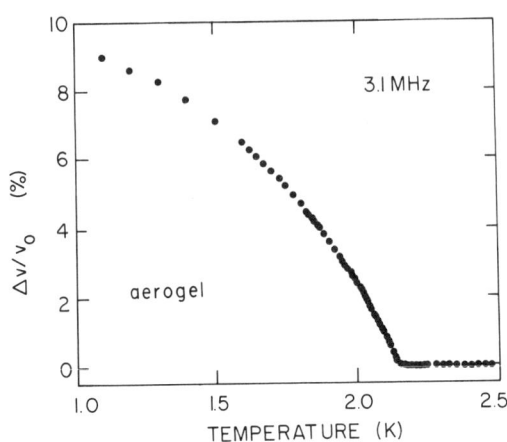

Figure 4 Transverse sound velocity in a helium filled aerogel sample.

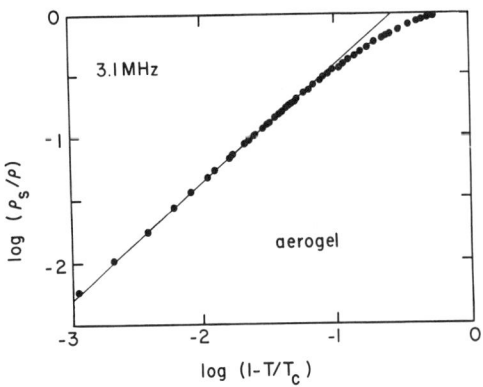

Figure 5 The superfluid-density data for aerogel as a function of reduced temperature. The straight line represents a power-law fit to the data.

We have not yet studied the aerogel samples in the same detail as we did Vycor. Figure 6 shows some preliminary attenuation measurements over a narrow frequency range for the sample of Fig. 5. There are several obvious features. First, the attenuation above the superfluid transition is very large and increases rapidly with frequency but below the transition it drops quickly. We have observed qualitatively similar features in a porous ceramic sample[23] where we attributed them to viscous losses from the motion of the normal fluid component in the relatively large pores (hydraulic radius about 1500 Angstroms). In the aerogel, measurements with a thick superfluid film (T_c = 1.53 K), rather than full pores, showed only a critical peak superimposed on a smooth background, as in the Vycor experiments. This background, presumably also due to the TLS in the glass, is responsible for the variation of the attenuation above the transition and at the lowest temperatures in Fig. 6. In addition to these features, at each frequency there is a narrow critical peak centered at the transition temperature. These peaks are much narrower than in Vycor, with a width (FWHM) of less than 10 mK, supporting the conclusion from the velocity measurements that the transition in aerogels is very sharp.

The other intriguing feature in Fig. 6 is the presence of broad peaks in the attenuation at temperatures well below the superfluid transition. These peaks are reminiscent of those found in the

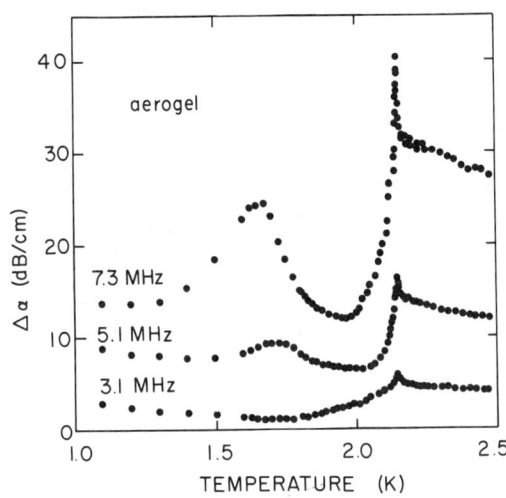

Figure 6 Attenuation of transverse sound in helium-filled aerogel at frequencies of 3.1, 5.1, and 7.3 MHz.

attenuation of longitudinal sound in bulk superfluid 4He which arise[24] from the relaxation of elementary excitations (phonons and rotons). However, the peaks in Fig. 6 appear to shift to higher temperatures as the frequency decreases (from about 1.6 K at 7.3 MHz to 1.7 K at 5.1 MHz), opposite to the behavior expected for a relaxation process. Also, compared to the relaxation peak in bulk helium, this peak shrinks very rapidly with frequency, essentially disappearing at 3.1 MHz. However, the measurements shown in Fig. 6 differ from the sound propagation in bulk helium in a number of respects. First, the sound waves are transverse in the aerogel, and propagate in the glass matrix, where most of the acoustic energy is stored. Secondly, there are two length scales in the aerogel case which are not present in the bulk liquid, namely the pore size and the viscous penetration depth. A hydrodynamic description of the dissipation in the helium breaks down whenever the mean free path of elementary excitations exceeds either of these lengths. This mean free path becomes larger than the viscous penetration depth (about 500 Angstroms at 7.3 MHz) around 1.6 K. Associating the peaks with this crossover is consistent with the fact that there is no corresponding peak in the attenuation when there is only a superfluid film on the aerogel surface. More detailed ultrasonic measurements are planned to clarify the origin of this peak, as well as to study the behavior in the critical region.

We have demonstrated that ultrasonic measurements are a useful probe of the dynamics of superfluid helium adsorbed on porous substrates. At the lowest frequencies (a few MHz), the transitions in both Vycor and silica aerogels remain sufficiently sharp that critical exponents for the superfluid density can be determined from the sound velocities. For helium films on Vycor, we observed a shift and broadening of the transition at frequencies above about 16 MHz, consistent with the expected behavior 2D vortices in porous media.

ACKNOWLEDGEMENTS

This work was supported in part by a grant from the Research Corporation.

REFERENCES

1. J.M. Kosterlitz and D.J. Thouless, J. Phys. C 6, 1181 (1973); V. Ambegaokar, B.I. Halperin, D. Nelson, and E.D. Siggia, Phys. Rev. B 21, 1806 (1980).
2. D.J. Bishop and J.D. Reppy, Phys. Rev. B 22, 5171 (1980).
3. V. Kotsubo and G.A. Williams, Phys. Rev. B 33, 6106 (1986).
4. K. Shirihama, N. Wada, Y. Takano, T. Ito, and T. Watanabe, Jpn. J. Appl. Phys. Suppl. 26-3, 293 (1987).
5. D.J. Bishop, J.E. Berthold, J.M. Parpia, and J.D. Reppy, Phys. Rev. B 24, 5047 (1981).
6. M.W.H. Chan, K.I. Blum, S.Q. Murphy, G.K.S. Wong, and J.D. Reppy, Phys. Rev. Lett. 61, 1950 (1988).
7. D. Finotello, K.A. Gillis, A. Wong, and M.H.W. Chan, Phys. Rev. Lett. 61, 1954 (1988).
8. C.-T. Wang and L. Yu, Phys. Rev. B 33, 599 (1986).
9. T. Minoguchi and Y. Nagaoka, Prog. Theor. Phys. 80, 397 (1988).
10. J. Machta and R.A. Guyer, Phys. Rev. Lett. 60, 2054 (1988); J. Low Temp. Phys. 74, 231 (1989).
11. F. Gallet and G.A. Williams, Phys. Rev. B 39, 4673 (1989).
12. M. Ma, B.I. Halperin, and P.A. Lee, Phys. Rev. B34, 3136 (1986).
13. D.S. Fisher and M.P.A. Fisher, Phys. Rev. Lett. 61, 1847 (1988).
14. P. Weichmann and M.E. Fisher, Phys. Rev. B 34, 7652 (1986).
15. P.W. Adams and W.I. Glaberson, Phys. Rev. B 35, 4633 (1987).
16. P. Wiltzius, F.S. Bates, S.B. Dierker, and G.D. Wignall, Phys. Rev. A 36, 2991 (1987).
17. R. Vacher, T. Woignier, J. Pelous, and E. Courtens, Phys. Rev. B 37, 6500 (1988).
18. N.M. Mulders and J.R. Beamish, to be published in these proceedings.
19. K.L. Warner and J.R. Beamish, J. Appl Phys. 63, 4372 (1988).
20. Made by C.A.M. Mulder and J.G. van Lierop of Philips Research Laboratories.
21. N. Mulders and J.R. Beamish, Phys. Rev. Lett. 62, 438 (1989).
22. J.R. Beamish, A. Hikata, and C. Elbaum, Phys. Rev. Lett. 52, 1790 (1984).
23. K.L. Warner and J.R. Beamish, Phys. Rev. B 36, 5698 (1987); see also the paper by K.L. Warner, N. Mulders, and J.R. Beamish, to be published in these proceedings.
24. I.M. Khalatnikov, Zh. exsp. teor. Fiz. 20, 243 (1950).

ATTENUATION MECHANISMS IN POROUS MEDIA CONTAINING HELIUM

K. Warner, N. Mulders and J.R. Beamish
Department of Physics and Astronomy
University of Delaware
Newark, DE 19716, USA

Acoustic techniques are widely used to probe both the structure of materials and the properties of fluids, for example superfluid helium in porous media. Biot[1] developed a phenomenological theory of sound propagation in a fluid-filled porous medium to describe the effects of relative motion between the fluid and the porous matrix (sloshing), but other loss mechanisms in either the porous solid or the fluid may contribute to the sound attenuation. We have measured the attenuation of transverse ultrasonic waves in three different porous materials containing liquid helium. Different dissipation mechanisms are important in each sample depending on the pore size and whether the helium fills the pores or is adsorbed as a film.

Figure 1 shows the attenuation in a porous alumina ceramic filled with liquid helium[2]. This ceramic has a porosity of 36% with pore sizes up to a micron. This is large enough that the fluid can move with respect to the solid with a resulting attenuation of sound, as described by Biot. Below the superfluid transition, only the motion of the normal (viscous) component of the fluid contributes to the attenuation. As the temperature is reduced, the normal fluid fraction becomes smaller and the attenuation decreases.

Figure 2 shows the attenuation in helium-filled Vycor, a silica glass with 29% porosity and an average pore diameter of about 40 Angstroms. Since the viscous penetration depth at these frequencies is much larger than the pore size, the normal fluid is locked to the substrate and the viscous attenuation of Fig. 1 disappears. The remaining attenuation consists of a critical peak at the superfluid transition (which occurs at a temperature lower than in bulk helium) and a frequency dependent background attenuation due to interaction of the sound wave with two-level systems in the glass itself.

Figure 3 shows the attenuation in an aerogel sample (porosity 84%). Aerogels have very open structures with a wide range of pore sizes. Two of the attenuation mechanisms discussed above are significant. There is a narrow critical attenuation peak at the superfluid transition both when the pores are filled and when there is only a helium film on the pore surface. In the full pore case we also observe, as in the large-pore ceramic, additional attenuation which decreases rapidly below the superfluid transition as the normal fluid component disappears. Apparently, a significant fraction of the open volume has dimensions exceeding the viscous penetration depth (about 500 Angstroms at the transition).

1. M.A. Biot, J. Acoust. Soc. Am. **28**, 168 (1956); **28**, 179 (1956).

2. K.L. Warner and J.R. Beamish, Phys. Rev. **B36**, 5698 (1987).

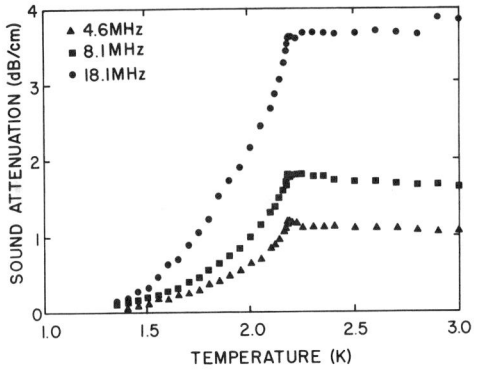

Figure 1 Attenuation of transverse ultrasonic waves in a porous ceramic filled with liquid ^{4}He.

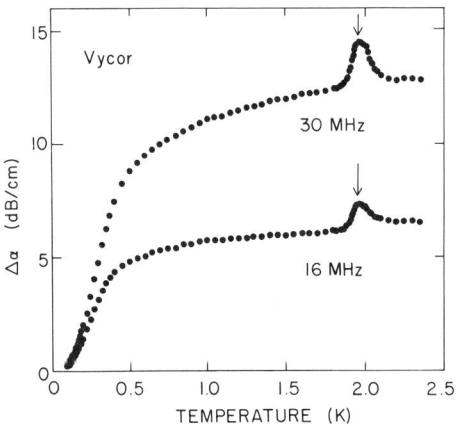

Figure 2 Attenuation of transverse ultrasonic waves in ^{4}He filled porous Vycor glass. The arrows mark the temperature of the superfluid transition (1.94 K).

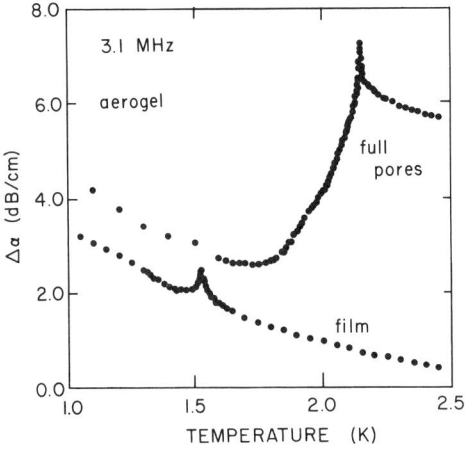

Figure 3 Attenuation of 3.1 MHz transverse ultrasonic waves in a silica aerogel containing ^{4}He. Upper curve - full pores with T_c = 2.145 K. Lower curve - a helium film with T_c = 1.53 K.

VORTEX STRING DISSIPATION IN ⁴He FILMS ADSORBED ON POROUS MATERIALS

H. Cho, F. Gallet, and G.A. Williams
Physics Dept., UCLA, Los Angeles, CA 90024

We present the results of an initial calculation of finite-frequency dissipation near the superfluid transition of ⁴He films adsorbed on porous substrates. A recent model[1] of the transition utilizing vortex string excitations is employed, and we refer the reader to that paper for background material and for the notation used below. The dissipation from an a.c. flow is found by calculating the frequency-dependent dielectric constant $\epsilon(\omega)$,

$$\epsilon(\omega) = 1 + \int_{a_o}^{2a} \frac{\partial \epsilon_2}{\partial r} g_2(r,\omega) dr + \int_{2a}^{\infty} \frac{\partial \epsilon_3}{\partial r} g_3(r,\omega) dr$$

Here 2a is the grain size/lattice constant, ϵ_2 is the 2D static dielectric constant calculated from the Kosterlitz-Thouless recursion relations, and g_2 is the vortex pair response function that is a solution of the Fokker-Planck equation[2].

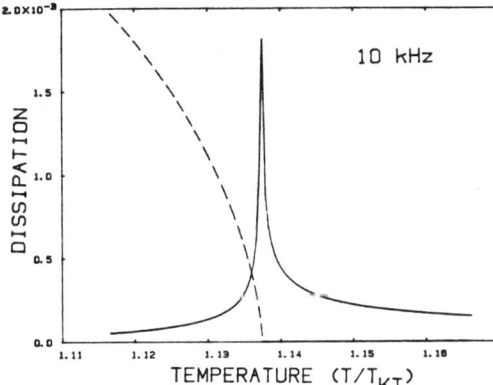

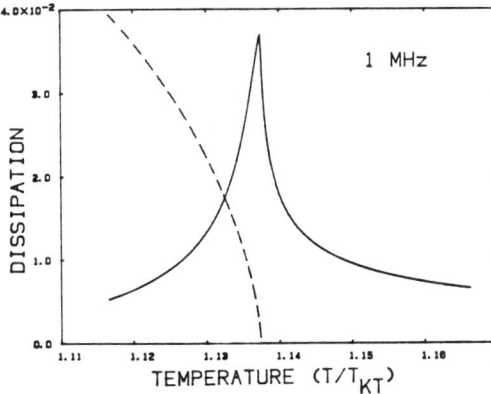

Fig.1: String dissipation (solid curve) at 10 kHz and 1MHz. The dashed curve shows the superfluid fraction.

ϵ_3 is the 3D dielectric constant obtained from the vortex string recursion relations of Ref.1, and g_3 is the string response function obtained by neglecting derivative terms in the Fokker-Plank equation for the strings, $g_3 = \pi K_r/(iar/r_D^2 - \pi K_r)$, where K_r is the normalized superfluid density[1] and $r_D = \sqrt{2D/\omega}$ is the diffusion length.

Fig. 1 shows the results of the calculation for two frequencies where the diffusion length is larger than the grain size and all of the dissipation is from the 3D strings.

The quantity plotted is $\text{Im}(K_o \epsilon(\omega))^{-1}$, with K_o the "bare" normalized superfluid density. The parameters used are the same as in Fig.3 of Ref.1, with also 2a = 150Å, a_o = 10Å, and D = 0.8 ℏ/m. The dissipation arises when the mean string length becomes equal to the diffusion length, and at 10 kHz (where r_D ~ 6000Å) this is only a very narrow temperature range about T_c. At 1MHz r_D is 10 times smaller, and as a result the cusp-like peak is wider and a factor of 20 larger in magnitude.

When the frequency is further increased such that r_D becomes smaller than the grain size 2a, there is then a crossover from the string dissipation to the more usual 2D pair dissipation, as shown in Fig.2. There is also a strong broadening of the superfluid density due to the finite-size cutoff of the KT recursion relations at r_D. The theory is compared with the experimental results of Ref.3 that use ^{4}He films in porous Vycor glass (~ 150Å grain diameter), and it is evident that the main features of the data are well explained by the calculation. The only parameter that was really adjusted to give these curves was the diffusion constant D, which was changed from the nominal value of $\hbar/m$ to 0.8 $\hbar/m$ to better match the experimental curves. There is a "glitch" in the theoretical curves at the static T_c, which we believe arises from our neglect of the derivative terms in the Fokker-Planck equation. At that point the length-dependent superfluid density begins to vary rapidly in space, and the derivatives of g_3 become important.

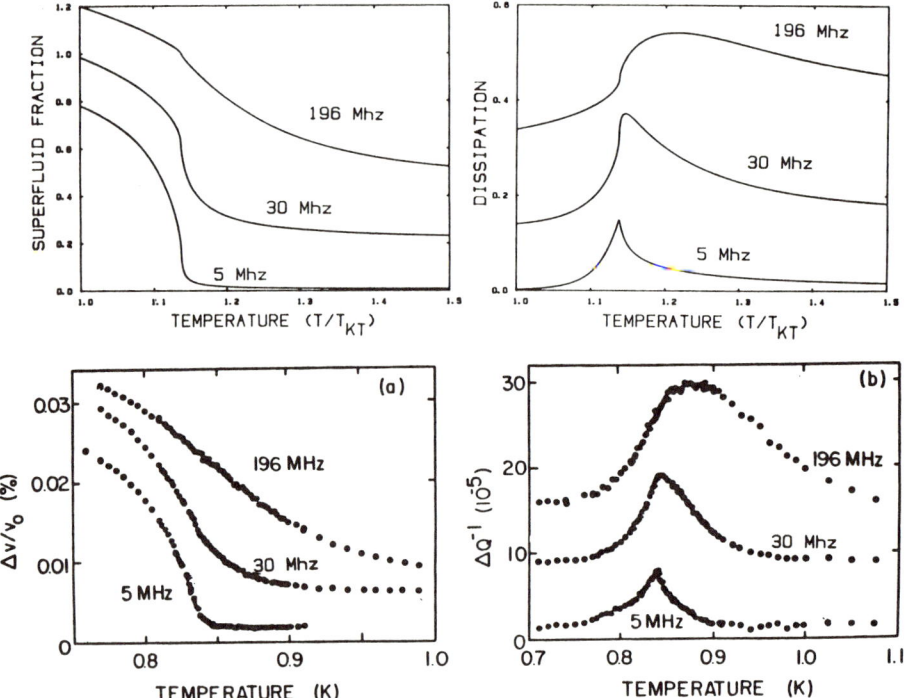

Fig. 2 Dissipation and superfluid fraction between 5 MHz and 196 MHz, compared to data from Ref.3.

1. F. Gallet and G.A. Williams, Phys. Rev. B **39**, 4673 (1989).
2. V. Ambegaokar and S. Teitel, Phys. Rev. B **19**, 1667 (1979).
3. N. Mulders and J. Beamish, Phys. Rev. Lett. **62**, 438 (1989).

PORE-SIZE DEPENDENCE OF THE SUPERFLUID TRANSITION IN ^{4}He FILMS ADSORBED ON POROUS GLASSES

K. Shirahama, N. Wada$^+$, M. Kubota, S. Ogawa,
T. Watanabe and K. Eguchi*

ISSP, Univ.of Tokyo, Roppongi, Minato-ku, Tokyo 106, Japan
+):Dept.of Phys., Fac. of Sci., Hokkaido Univ., Sapporo 060,Japan
*): Government Industrial Res. Inst. Osaka, Ikeda, Japan

The question of the dimensionality of the superfluid transition in ^{4}He films adsorbed on porous materials is of considerable current interest. Reppy and coworkers have interpreted that the transition of the ^{4}He films in porous Vycor glass (pore diameter~60A) as a three dimensional (3D) transition due to a multiple connectivity of the films[1]. However, more delicate explanations have been proposed which have considered unbinding of vortex pairs and a 3D connectivity of the films[2,3]. Studies of the pore-size dependence of such transitions are very important because the pore size should play a crucial role on the dimensionality of the transitions. Indeed, in our previous torsional-oscillator experiment using Pt packed powders as porous substrates, the transition region broadens as the powder size decreases[4].

In order to study the pore-size dependence quantitatively, two conditions must be satisfied:(1) the film thickness should be kept constant for several different pore-size materials, because the properties of the films (especially the transition temperature T_c) depend on the film thickness as well as the pore size. (2) the pore-size distributions should be kept as sharp as possible. Previous measurements employing packed powders did not satisfy both of the conditions so well. The film-thickness estimation by a vapor pressure measurement[5] has an unavoidable uncertainty because the pressure-thickness relation includes the powder size , which span a wide range, as a parameter. The accuracy of our previous coverage estimation[4] depended on the precision of the surface area determined by N_2 adsorption isotherm. In order to distinguish the pore-size dependence from the film-thickness dependence, we use a quite simple method. That is two torsional oscillators containing different pore-size materials with a common ^{4}He gas inlet. This setup realizes an equivalent ^{4}He film thickness on both materials.

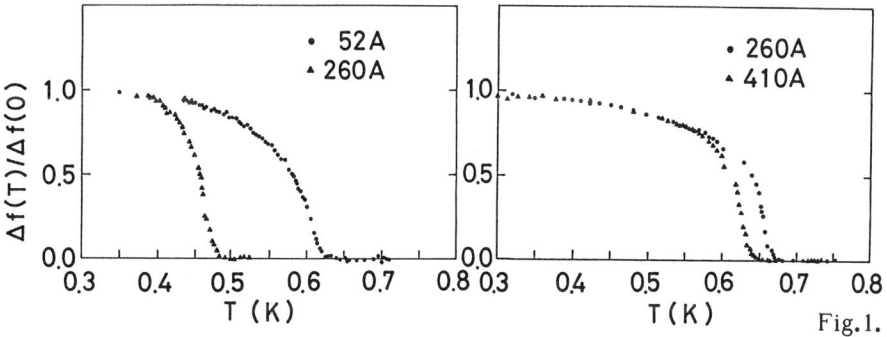

Fig.1.

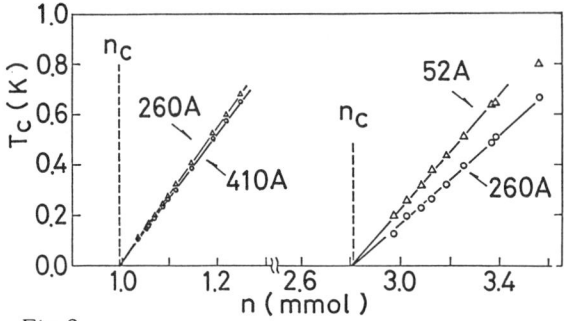
Fig.2.

We employ porous glasses which have much sharper pore-size distributions than the packed powders.

We report here the first quantitative determination of the pore-size dependence of T_c in ^{4}He films.

The porous glass samples used are 52, 260 and 410A in pore diameter. Each of the torsional oscillators has different sample of different pore diameter mounted on a Cu platform. The ^{4}He gas inlet line is connected to each sample through hollow Be-Cu torsion rods. We measure the resonant frequencies as a function of temperature, for combinations of 52 and 260A, and 260 and 410A-samples.

In Fig.1, we show the temperature dependence of the frequency shifts (i.e. the superfluid densities) for a typical coverage. We make such measurements for many different ^{4}He coverages n. In Fig. 2, we show the n-dependence of T_c. For all porous glasses, T_c rises linearly as the n increases in the range of 0.1K<T_c<0.7K The most striking feature of these measurements is the pore-size dependence of T_c. T_c becomes progressively higher as the pore diameter decreases. Nonzero intercepts of T_c-n lines in Fig.2 were found to coincide for both porous glasses within 1%. This coincidence shows that the thicknesses of the 'nonsuperfluid' and 'superfluid' layers are the same for all of the porous glasses. The pore-size dependence of T_c is therefore the intrinsic properties of ^{4}He films in porous glasses.

The observed pore-size dependence of T_c is consistent with the consequence of the vortex unbinding theory in a multiply-connected network of cylinders[2,3]. Our experimental method makes it possible to compare further quantitatively the results with the theoretical prediction. So far, we have made measurements for only limited number of different kinds of porous glasses. Measurements using the larger pore-size samples are being continued.

We have reported systematic measurements of ^{4}He films adsorbed on porous glasses using two torsional oscillators. The observed pore-size dependence of T_c suggests that the vortex-pair unbinding occurs at the transition. Experiments using porous glasses of wider pore-size range will shed further light on the question of the dimensionality of the transition, which are underway at ISSP Univ. of Tokyo.

REFERENCES
1) D.J.Bishop et al. Phys. Rev. B<u>24</u>, 5047 (1981)
2) T.Minoguchi and Y.Nagaoka, Prog. Theor. Phys. <u>80</u>, 397 (1988)
3) J.Machta and R.A.Guyer, Phys. Rev. Lett. <u>60</u>, 2054 (1988)
4) K.Shirahama et al. Jpn.J.Appl.Phys. <u>26</u>, Suppl. 26-3, 293 (1987)
5) V.Kotsubo and G.A.Williams, Phys. Rev. B<u>33</u>, 6106 (1986)

THE SINGULAR BOUNDARY RESISTANCE OF SUPERFLUID ^{4}He

Fang Zhong, Jim Tuttle and Horst Meyer
Department of Physics, Duke University, Durham, NC 27706

We report further experimental investigations on the singular boundary resistance ΔR_k of ^{4}He with an impurity of 2 ppb. As T_λ is approached, a weakly diverging, heat-independent contribution to ΔR_k is first observed. At a reduced temperature $|\varepsilon| < |\varepsilon_c(Q)|$ where $\varepsilon = (T-T_\lambda)/T_\lambda$, a more strongly diverging, heat dependent anomaly in R_k is recorded. Here T is the bulk fluid temperature. The weakly diverging contribution was first detected by Duncan et al.[1] using ultra high resolution (5 nK) thermometry. In the present work, this contribution was found to be roughly a factor of three stronger than reported in ref. 1 and could be detected with our Ge thermometry (0.3 μK resolution). The heat-dependent boundary resistance was observed by Dingus et al.[2] and in more detail by Duncan et al.,[1] and possibly by Lipa et al.[3] for very low heat current.

A weak singularity of the resistance is predicted to result from a boundary layer, the "healing length", of thickness l comparable to the correlation length ξ. In this layer the heat current converts from the bulk counter-flow to the diffusive heat flow $\kappa \nabla T$ where $\kappa(\varepsilon)$ is the diverging heat conductivity in the normal phase. Frank and Dohm's predictions[5] lead to an effective power law, $\Delta R_k \approx 2 \times 10^{-3} |\varepsilon|^{-0.27}$ K cm^2/W. There is no explanation so far for the heat-dependent singularity, which we have investigated over the range $8 < Q < 46$ μW/cm^2, where Q is the heat current. Most recently, Duncan et al.[4] have studied this singularity over the same range of Q.

Our cell was a parallel-plate conductivity cell with face diameter of 3.4 cm and spacing 0.19 cm between the lapped and gold-plated faces of two copper blocs. The top plate was connected via small holes to a fill space containing a capacitor to measure the density via the dielectric constant. The average temperature of the bulk fluid was taken in a first approximation as $\bar{T} = (T_{TOP} + T_{BOT})/2$ where T_{TOP} and T_{BOT} are the temperatures of the Ge resistors embedded in the top and bottom copper blocs. Also $\Delta T = T_{BOT} - T_{TOP}$ is the measured temperature difference that includes the drop between thermometers across the top and the bottom copper blocs, and the two boundary resistances, $\Delta T/Q = R_b(Q,T) = R_b^0(T) + \Delta R_k(Q,\varepsilon)$.

We have investigated the boundary resistance during two separate series of experiments between which the cryostat was warmed up to room temperature. Data were taken at constant Q by decreasing $|\varepsilon|$ by small steps until at $\varepsilon^*(Q)$ a sudden large increase of ΔT showed the onset of the normal phase. The temperature was then slowly decreased until suddenly ΔT returned to its former value at $\tilde{\varepsilon}^*(Q)$, and measurements were then conducted with further increasing $|\varepsilon|$. No hysteresis in $\varepsilon^*(Q)$ nor in $\Delta R_k(Q,\varepsilon)$ was observed. The principal results from these two series are 1) The observed weak singularity has an exponent consistent with the predicted one, but a larger amplitude. Also it is larger than observed by Duncan et al.[2] (see Fig. 1).
2) The maximum $\Delta R_k = R_b(Q,\varepsilon^*) - R_b^0$ was 0.5 ± 0.05 cm^{2}K/W, independent of Q within our scatter, and of comparable magnitude as found by Duncan et al..[4] 3) The plot of $|\varepsilon_c|$ versus Q over the same range of Q can be represented by a simple power law $|\varepsilon_c| = a Q^n$ with $n \approx 1.04$ and shows small but systematic differences between

series 1 and 2. The results are compared in Fig. 2 with the most recent data by Duncan et al.[4] and earlier work[2]. A detailed description and analysis of these experiments is in preparation. The authors thank R.V. Duncan for providing a copy of his thesis and for comments on this paper. This research was supported by the NSF grant DMR 8516156.

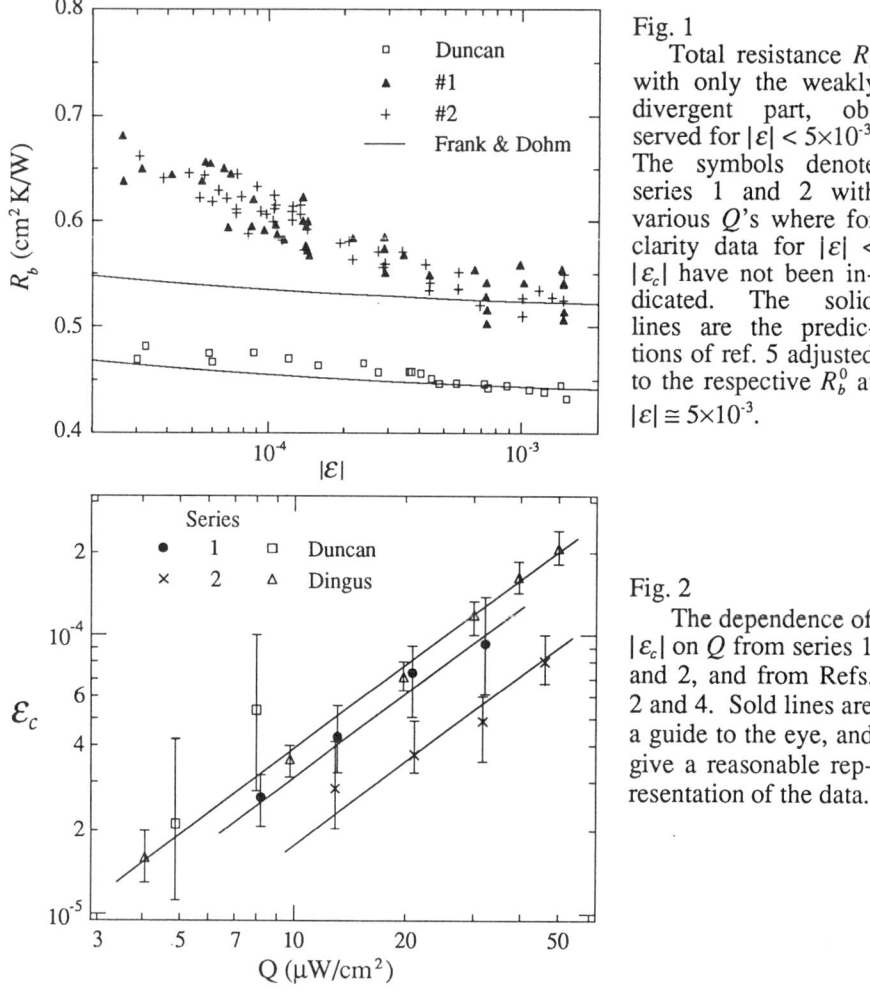

Fig. 1
Total resistance R_b with only the weakly divergent part, observed for $|\varepsilon| < 5\times 10^{-3}$. The symbols denote series 1 and 2 with various Q's where for clarity data for $|\varepsilon| < |\varepsilon_c|$ have not been indicated. The solid lines are the predictions of ref. 5 adjusted to the respective R_b^0 at $|\varepsilon| \cong 5\times 10^{-3}$.

Fig. 2
The dependence of $|\varepsilon_c|$ on Q from series 1 and 2, and from Refs. 2 and 4. Sold lines are a guide to the eye, and give a reasonable representation of the data.

REFERENCES

1) R.V. Duncan, G. Ahlers and V. Steinberg, Phys. Rev. Lett. **58**, 377 (1987).
   R.V. Duncan and G. Ahlers, *Jpn. J. Appl. Phys. Suppl.* **26-3**, 363 (1987).
2) M. Dingus, F. Zhong and H. Meyer, *J. Low Temp. Phys.* **65**, 185 (1986).
3) T.C.P. Chui, Q. Li and J.A. Lipa, *Jpn. J. Appl. Phys. Suppl.* **26-3**, 371 (1987), *Bull. Am. Phys. Soc.* **33**, 1373 (1988).
4) R.V. Duncan, thesis, UCSB (1988), R.V. Duncan and G. Ahlers to be published
5) D. Frank and V. Dohm, *Phys. Rev. Lett.* (in print).

WETTING OF HELIUM FILMS TO SILVER SUBSTRATES

R.J. Dionne and R.B. Hallock

Laboratory for Low Temperature Physics
Department of Physics and Astronomy
University of Massachusetts, Amherst, MA 01003

Dash and his co-workers have reported[1] data based on quartz microbalance techniques and vapor pressure isotherms which have been interpreted as evidence that ^{4}He film growth is not uniform as a function of temperature on either Ag or Au surfaces. The results were surprising since ^{4}He has a strong substrate interaction and has generally been assumed to wet those substrates (and hence grow smoothly and continuously as a film). Subsequent work by Taborek and Senator[2] utilized graphite and platinum bent-fiber oscillators to study the mass loading of the fiber both above and below T_λ with the conclusion that ^{4}He wets the studied substrates below T_λ but not above.

In an effort to address this problem by means of a simple technique, we have conducted capacitance experiments[3]. We have built parallel plate capacitors of different gap sizes, located them in an experimental chamber and continuously observe the capacitance as the temperature is very slowly changed from below T_λ to above or the reverse. In the presence of complete wetting for the case of a saturated helium film (ie., one where a puddle of bulk liquid is present in the apparatus) we expect to see a continuously rising capacitance as the vapor pressure increases with temperature. Should the wetting become incomplete at some temperature we expect to see deviations from the smoothly increasing capacitance as a function of temperature.

We have run versions of the apparatus with various rates of temperature change in the vicinity T_λ. An illustration of the sorts of results we obtain[3] is given in figure 1. The smooth curve of overlapping data points is the expected result for the change in the capacitance due *entirely* to the effects of the vapor pressure change with temperature. We see no large-scale anomalous behavior in the capacitance. Thus, we believe that for the surfaces we have used for these experiments (silver evaporated on borosilicate glass and exposed to room air for a number of days) there is no anomalous situation and the helium film wets the surface as expected on the basis of simple binding energy arguments.

We are able to induce[3] irregularities in the capacitance signal by creating a temperature gradient in the apparatus by, for example, sweeping the temperature too fast. For example, if we sweep the temperature at the rate of 5 μK/sec by the application of heat in the experimental chamber, we observe the capacitance to decrease sharply by an amount suggestive of a thinning of the helium film by about 25 atomic layers upon warming through the lambda point. Whether or not the behavior we can induce by temperature gradients is related to the anomalous apparent wetting seen by others is not clear to us; we have different surfaces. What is clear to us is the fact that we do not appear to see anomalous behavior so long as we are careful to keep the temperature gradients in the apparatus very small.

The general behavior we have seen has also recently been observed by Zimmerli and Chan[4]. In that case vibrating fibers were used and no anomalous behavior was seen under conditions of very small temperature gradients. There too, the induction of temperature gradients resulted in apparent anomalous wetting. Similar behavior has been seen by Lea et al.[5]

Very recent work by Graham and Taborek[6] with a copper cell has seen apparent anomalous wetting for the case of ^{3}He − ^{4}He mixture films.

Although we believe that the basic question of the wetting of a helium film to a substrate has been resolved with the conclusion that helium films wet both above and below T_λ for the surfaces we have studied, there is an intriguing aspect of the film thickness which remains in our work. We observe in figure 2 an apparent nearly symmetric deviation in the film thickness of several atomic layers associated with the transition. Recently there have been interesting predictions concerning the effect of criticality on wetting layers. According to the theory[7] as applied to helium, when the helium coherence length becomes comparable to or larger than the helium film thickness, the thickness of a thick helium film in equilibrium with the saturated vapor pressure will be reduced. For saturated films of the type we have studied, about a 1% thinning of the film is predicted[7] to occur within one or two mK of the lambda transition. There is the intriguing possibility that we may have seen evidence for this but additional work will be necessary to be certain.

This work was supported by the NSF through DMR 85-17939 and DMR 88-20517.

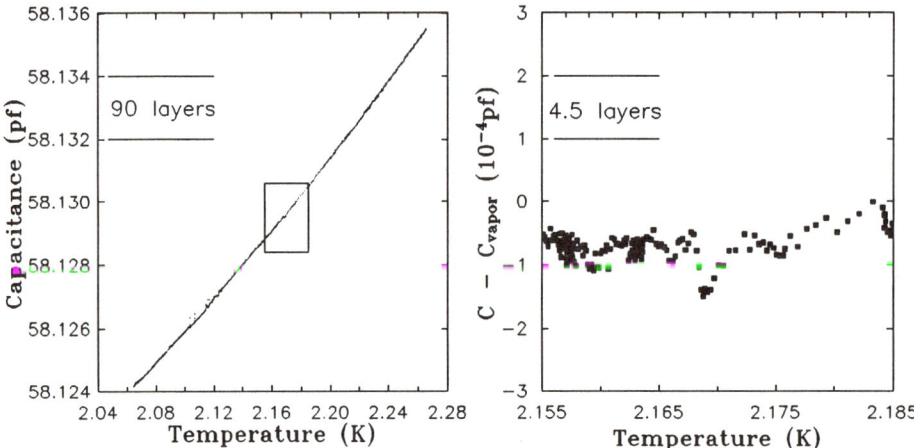

Fig. 1 Capacitance of the saturated film–vapor system. The smooth evolution of data points is as expected due entirely to the vapor.

Fig. 2 The difference between the observed capacitance and that expected due to the vapor pressure for the boxed data of fig. 1. Note the anomaly near T_λ.

REFERENCES

1. A.D. Migone, J. Krim, J.G. Dash and J. Suzanne, Phys. Rev. B **31**, 7643(1985).
2. P. Taborek and L. Senator, Phys. Rev. Lett. **57**, 218(1986).
3. R. Dionne and R.B. Hallock, Bull. Am. Phys. Soc. **33**, 442(1988).
4. G.A. Zimmerli and M.H.W. Chan, Bull. Am. Phys. Soc. **32**, 433(1987).
5. M.J. Lea, D.S. Spencer and P. Fozooni, Phys. Rev. B **35**, 6665(1987).
6. M. Graham and P. Toborek, Bull. Am. Phys. soc. **33**, 442(1988).
7. J.O. Indekeu, J. Chem. Soc. Faraday Trans. II **82**, 1835(1986).

EVIDENCE FOR MAGNETIC ORDERING IN THE BOUNDARY LAYERS OF ³He ON GRAFOIL

L.J. Friedman, A.L. Thomson[a], C.M. Gould, H.M. Bozler
Department of Physics, University of Southern California
Los Angeles, CA 90089-0484

P.B. Weichman, and M.C. Cross
Department of Physics California Institute of Technology
Pasadena, CA 91125

ABSTRACT

The boundary layer of ³He on exfoliated graphite (grafoil) is an interesting model system for two dimensional magnetism. The low field NMR spectra of ³He boundary layers on exfoliated graphite show collective modes for $T < 1$ mK. We measure the amplitude and frequency for these modes with the static H_0 field applied parallel to the graphite planes and varying continuously between 0 and 15 gauss. One of the modes extrapolates to a nonzero frequency and amplitude as the field is dropped to zero. We interpret these nonzero intercepts as an indication of zero field magnetic order in the ³He boundary layers. The possible role of the weak nuclear dipolar interactions will be described.

INTRODUCTION

Boundary layers of ³He have shown low temperature ferromagnetic properties for a large number of substrates.[1] The system is unique in that the ferromagnetism is exhibited by the nuclear spins as opposed to the ferromagnetic films that are electronic in origin.[2] Also, physical interchanges between the adsorbed atoms are thought to govern the exchange interaction responsible for the ferromagnetic tendency.[3] Exfoliated graphite (grafoil) is a particularly well suited substrate for studying the magnetic properties of ³He boundary layers. Experiments reveal[4] that the structural coherence of the first layer is on the order of 100 Å. In addition, the substrate planes are oriented within ± 15⁰ so as to create a well defined geometry. Experiments performed with ³He bulk liquid[5,6] filling the pores of grafoil and with a few monolayers[7,8] adsorbed on the surface have both resulted in low temperature polarizations of the boundary layer which far exceed Curie values. The film experiments[8] have shown that the magnetization of one layer (presumably the second) is well approximated by a two dimensional Heisenberg model in the relatively large fields used there (~ 100 gauss). Local dipole fields from the ³He are only of order a few gauss for maximum polarization of the boundary layers. It is therefore desireable to study this system in applied fields of comparable magnitude in order to see effects of the ordering on the spin dynamics. In our present work[9] we use very low field SQUID NMR in the simplest geometry (field parallel to the surface planes) to investigate deviations from the ideal Heisenberg model and the zero field ordering.

© 1989 American Institute of Physics

The strict two-dimensional isotropic Heisenberg model has a phase transition only at T = 0: No finite magnetization can exist for T > 0 and zero applied field.[10] In the present experimental system, however, there exist long-ranged dipolar interactions between the ^{3}He nuclei. The dipole energy plays two distinct roles. First, it is anisotropic with the spins preferring to point in the plane of the substrate. This would be true even if the dipole force were not long ranged, and is simply a consequence of its purely antiferromagnetic character when the spins point perpendicular to the plane. The dipole energy therefore breaks the Heisenberg symmetry, reducing it to at most planar (i.e. XY). This effect was seen as a spin flop transition[6] in a prior experiment with H_0 perpendicular to the grafoil planes. The dipolar anisotrpy energy competed with Zeeman energy, resulting in a reorientation of the surface domains as the applied field was decreased below a few gauss.

The second and more interesting effect of the dipole energy is to allow long range order to exist at sufficiently low temperatures. Yafet et al.[11] have shown that the first (low temperature) spin wave correction to an assumed ferromagnetically ordered state is finite in two dimensions when the long range part of the dipole energy is taken into account. The dipole term alters the character of the long wavelength spin wave stiffness from E ∝ k² to E ∝ k. This eliminates the usual infrared divergence and allows a violation of the Mermin-Wagner result.[10]

The 1 mK measured onset temperature is of order the exchange energy, J. Since the dipolar coupling constant, δ, is of order 0.1 μk, and since the transition temperature must vanish with δ, this seems paradoxical at first sight. However, a quick estimate reveals otherwise: At low temperatures and zero field the pure Heisenberg correlation length, ξ, diverges exponentially in J/k_BT.[12] Dipolar forces become important when the energy, $\delta\xi^2$, associated with a cluster of aligned spins of linear size ξ is roughly k_BT. Hence an estimate for T_c is given by the solution to $k_BT_c \sim \delta\xi^2$, which yields

$$k_BT_c \approx \frac{2c_1 J}{\ln(c_2 J/\delta)} \qquad (1)$$

to leading logarithmic order in J/δ (here c_1 and c_2 are constants of order unity). The transition temperature therefore vanishes very slowly with δ, and is determined primarily by J, consistent with our experiments.

In this work we report measurements that strongly suggest the presence of a zero field magnetic order in ^{3}He surface layers. When the applied field is below a few gauss, our NMR spectra display two lines in addition to the absorbtion at the Larmor frequency of the ^{3}He nucleus. We monitor the amplitude and frequency of these additional lines as the applied field is varied between 0 and 15 gauss. Extrapolation to zero applied field shows a nonzero intercept for one of the lines, indicating collective behavior in very low magnetic fields.

EXPERIMENT

Our grafoil substrate has about 1 m² of surface area and an open volume of 0.05 cc which is filled with liquid ³He during the experiment. A column of liquid ³He provides a thermal link between the NMR cell and the Lanthanum diluted Cerium Magnesium Nitrate (LCMN) susceptibility thermometer. Both the NMR and LCMN thermometer tower are mounted on the nuclear stage of a copper demagnetization cryostat. The grafoil used in this work had been previously heated in H_2 at 1600° C and was loaded into the experimental tower in an N_2 atmosphere. The low field NMR uses a SQUID based detection scheme that has been described elsewhere[6,13] but has been modified for the present work. In previous work, the applied H_0 field was trapped in a superconducting aluminum cylinder. The heat required for retrapping prevented us from making any changes in H_0 during a demagnetization cycle. In the present apparatus, the aluminum cylinder is replaced by a 400 turn superconducting solenoid powered from a room temperature current source. The noise that would otherwise interfere with the SQUID operation is shorted out with a superconducting shunt. The shunt can be driven normal with an attached heater which is thermally anchored to the still of our dilution refrigerator. This scheme allows us to sweep the field with no discernable heating of the NMR cell even at the lowest temperatures. The H_0 field in this work is applied parallel[14] to the grafoil planes.

RESULTS

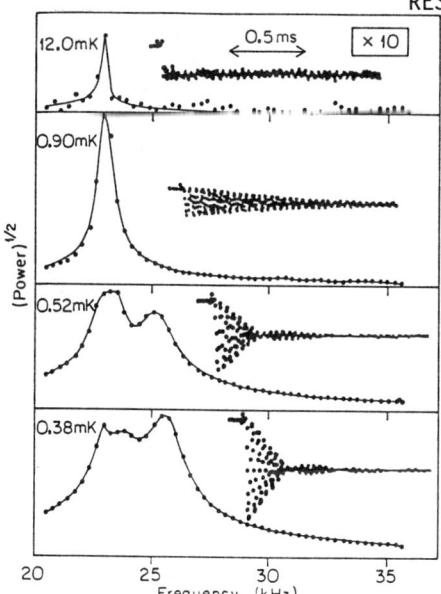

Fig. 1 Free induction decays with their associated Fourier transform power spectra with a H_0 = 7.1 gauss field applied parallel to the grafoil planes.

We first illustrate several qualitative features that are evident directly from the raw data. In Figure 1 we display a series of free induction decay signals and their associated fourier power spectra as the temperature is lowered with H_0 = 7.1 gauss. At high temperatures (T = 12 mK) we see a single narrow line from the ³He liquid in the cell. As the temperature is lowered the surface signal begins to dominate. There appears a distinct beating in the FID accompanied by additional peaks in the Fourier transforms. In the 0.52 mK scan, we can resolve two broad peaks. The lower frequency peak, however, is further resolved into two separate lines as the temperature reaches 0.38 mK. The sharp peak at the lowest frequency then appears to coincide with the high temperature liquid signal frequency. Notice how the high frequency peak has grown at a frequency already displaced from the Larmor frequency f_0. We therefore refer to this as the displaced mode and denote its

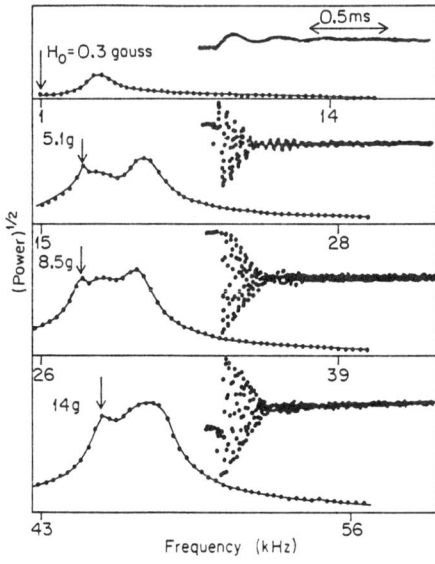

Fig. 2 Free induction decays and Fourier transform power spectra for T = 0.38 mK and varying applied fields. The arrows indicate the Larmor frequency for each field.

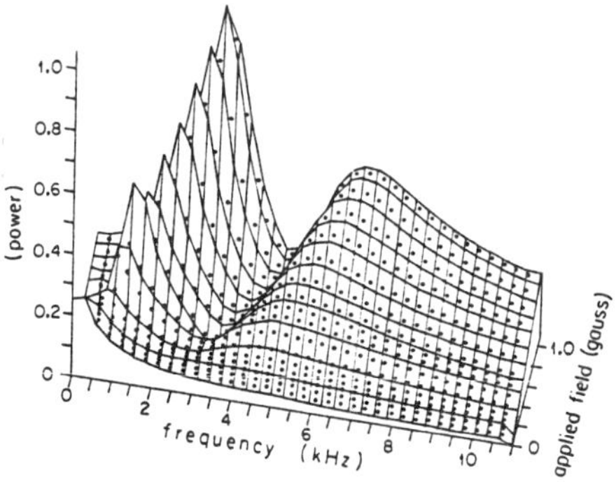

Fig. 3 A family of transforms accumulated at T = 0.59 mK and $H_0 \leq 1$ gauss. Notice how one mode appears to have a zero field intercept around 3 kHz.

frequency by f_d. However, the center peak in the 0.38 mK data has first appeared close to f_0 and subsequently shifted away from it. We then refer to this as the shifted mode with frequency f_s. These features will be further clarified in the analysis we will subsequently discuss.

Figure 2 displays a different series of FIDs together with their power transforms. In this instance T is a constant 0.38 mK and the field is varied as indicated next to each curve. The arrows identify the location of the Larmor frequency for these fields. In the 5.1 gauss to 14 gauss data it again appears that the f_s (shifted) mode shifts away from the f_0 Larmor peak and merges with the higher frequency f_d displaced mode at the highest field. Notice that the frequency shift ($f_d - f_0$) for the displaced mode does not seem to change appreciably as H_0 assumes different values.

In Figure 3 we display a series of transforms accumulated at T = 0.59 mK and H_0 below one gauss. The purpose of this three dimensional plot is to illustrate the behavior of the displaced mode frequency as the field approaches zero. The narrow peak at the f_0 Larmor frequency (which at these low fields is indistinguishable from f_s) tends toward zero frequency at zero field. The remarkable feature in this figure is that the peak we associated with the displaced mode clearly has a finite frequency intercept around 3 kHz at $H_0 = 0$. The f_d mode also broadens considerably at these low fields.

We may extract more quantitative information from our data by fitting our free induction signals to a trial function. In Figure 4 we reproduce the earlier T = 0.38 mK, H_0 = 7.1 gauss Fourier transform that is typical of our low temperature signals. Since we consistently see three peaks in our low temperature Fourier transforms, we use three lines in our fitting function corresponding to the Larmor, shifted and displaced modes mentioned above. In Figure 4 we show the total fit (solid line) to the spectrum in addition to the three separate component lines. For comparison, we also show a high temperature liquid signal at the Larmor frequency which is more than an order of magnitude smaller than the solid signal. The solid lines appearing in Figures 1 and 2 were also derived from our fits to the FIDs there.

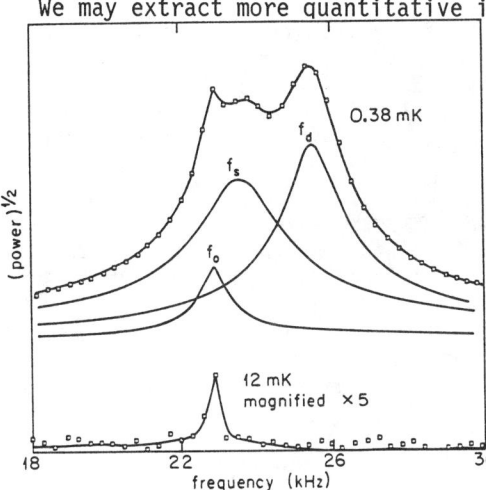

Fig. 4 The Fourier transform power spectra of the data (open squares) and our fits (solid lines). We show a H_0 = 7.1 gauss line broken into its three components along with a magnified liquid line that was recorded at 12 mK.

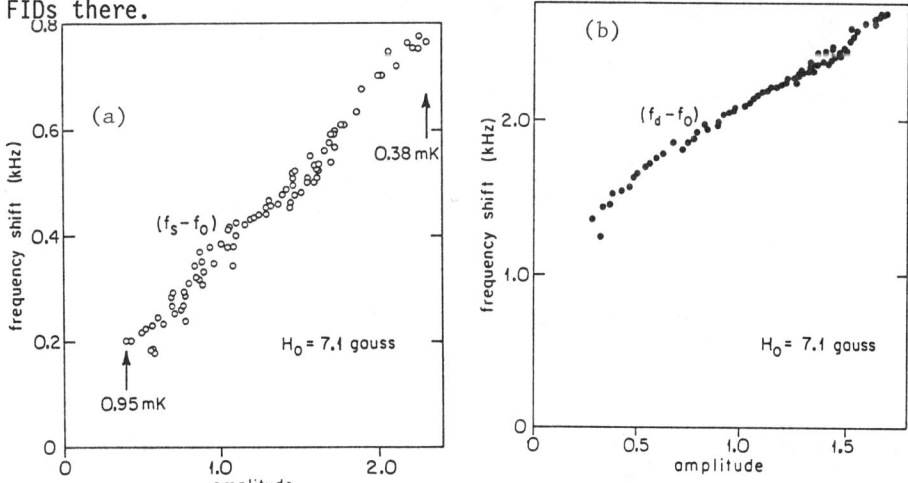

Fig. 5 The difference frequencies for the shifted mode ($f_s - f_0$) and displaced mode ($f_d - f_0$) plotted against their respective amplitudes. The plots cover the 0.95 mK to 0.38 mK temperature range. (a) The shifted mode frequency difference is linear in the amplitude and extrapolates to a zero intercept at high temperatures. (b) The displaced mode shift extrapolates to a nonzero value at zero amplitude (high temperature) indicating that the mode grows at a frequency displaced from the Larmor frequency.

The frequencies and amplitudes of the three components of the NMR response show different temperature dependences. The Larmor line itself remains fixed in frequency f_0 as the temperature is decreased. In Figure 5a we show the frequency shift $(f_s - f_0)$ for the center line shifted mode vs. its own amplitude. The plot covers the 0.95 mK down to 0.38 mK temperature range. The line is first resolved with a small frequency shift which grows along with its amplitude. Within the scatter of the data the $(f_s - f_0)$ shift is proportional to the amplitude of the mode and approaches a zero shift intercept at the (high temperature) zero amplitude extreme. However, the displaced mode $(f_d - f_s)$ vs. amplitude plot in Figure 5b is qualitatively different. In this instance it is evident that the mode grows from a frequency that is already displaced above the Larmor line. i.e. $(f_d - f_0)$ approaches a nonzero value as the line amplitude goes to zero. These plots therefore agree with our qualitative observations made from Figure 1 and justify the "shifted" and "displaced" mode nomenclature.

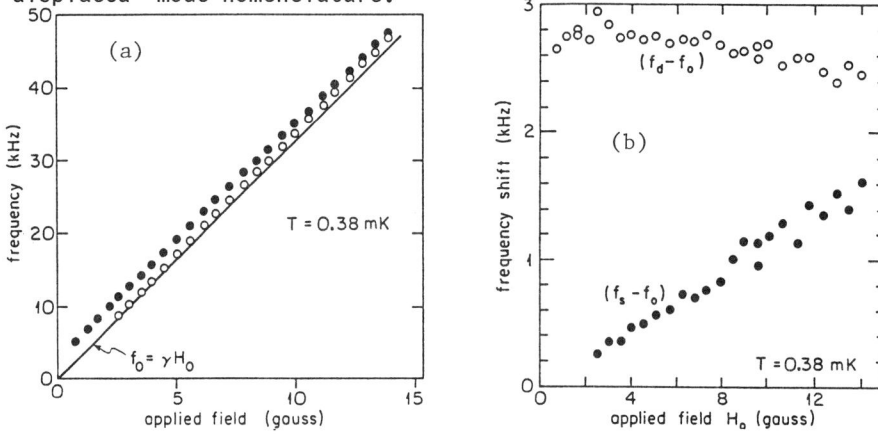

Fig. 6 (a) The frequencies f_0, f_s, and f_d are potted against the applied field at T = 0.38 mK. The open circles record the shifted mode frequency and the closed circles record the displaced mode frequency. The solid line indicates the Larmor frequency. (b) The shifted mode difference frequency $(f_s - f_0)$ is linear in the applied field while the displaced mode frequency difference $(f_d - f_0)$ shows an almost constant displacement of 2.7 kHz above the Larmor frequency.

We may also use our fits to track the field dependence for the three lines. In Figure 6a we display the low temperature field dependence for the frequencies f_0, f_s and f_d. The shifted and displaced mode are always shifted above the Larmor frequency. In our prior work[6] with H_0 perpendicular to the grafoil planes and $H_0 > 3$ gauss there was a mode observed with a 2.5 kHz negative shift relative to the Larmor frequency. This was due to having the applied field of a magnitude sufficient to align the magnetization perpendicular to the grafoil planes. The demagnetizing field then produced a negative frequency shift. With H_0 parallel to the planes this negative shift does not occur.

In Figure 6b we subtract off the Larmor frequency and show only the $(f_d - f_0)$ and $(f_s - f_0)$ difference frequencies. We then see that the shifted mode frequency difference (f_s-f_0) is linear in the applied field. The displaced mode frequency difference (f_d-f_0) in contrast approaches a nonzero value as the applied field is decreased to zero. At higher fields we see a small decrease in (f_d-f_0). Though the difference frequencies (f_d-f_0) and (f_s-f_0) appear to be approaching one another, our system cannot reach the frequencies necessary to trace the behavior through this interesting regime.

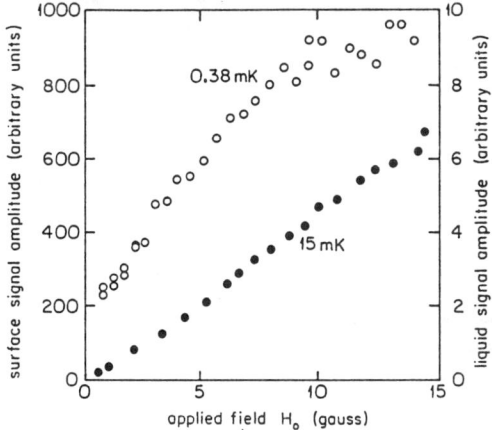

Fig. 7 Total amplitude of the NMR signal at 0.38 and 15 mK (same arbitrary units for each). The high temperature liquid signal amplitude extrapolates to zero amplitude for zero field while the surface signal displays an offset.

Our signals from the dc SQUID amplifier are proportional to the sample magnetization, independent of the applied field. Thus, in contrast to a resonant NMR scheme, we can directly compare the relative amplitudes at varying applied fields. In Figure 8 we show the recorded amplitudes versus H_0 for liquid ^{3}He measured in our cell at 15 mK. As expected, the amplitude is linear in the applied field. We also show in Figure 7 the amplitude versus H_0 for the total solid signal at T = 0.38 mK. In contrast to the case of the liquid signal, we see that the amplitude does not extrapolate to zero for zero applied field. At the highest fields we see a slight bending over of the curve that may be due to saturation effects. Additional field sweeps reveal zero field offsets that grow as the temperature is further reduced. However, no offset may be resolved in a field sweep taken at 0.85 mK.

Our fits to the data reveal that the finite amplitude offset present at zero applied field arises from the contribution of the displaced mode to the total amplitude. This is illustrated in Figure 8a. The displaced mode amplitude extrapolates to a nonzero value at zero field while the shifted mode amplitude passes directly through zero for zero applied field. The field for the Figure 8a data was swept from negative to positive values. The slight asymmetry in the zero field displaced mode offsets may be due to some hysteresis in the system.

Figure 8b shows several of the free induction decays that have been analyzed for display in figure 8a. The numbers adjacent to each FID in Figure 8b correspond with those points labelled in Figure 8a. The long time behavior is due only to the narrower line of the shifted mode (which is again is not separable from the Larmor line at

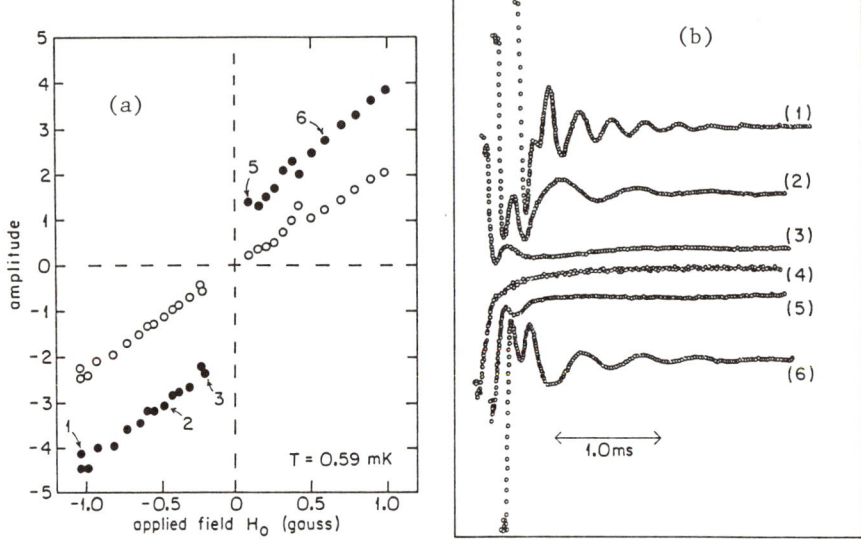

Fig. 8 (a) With T = 0.59 mK we show the amplitude vs. field dependence for the two modes contributing to the surface signal. The shifted mode (open circles) extrapolates to zero amplitude for zero field while the displaced mode (closed circles) shows an offset as the field passes through zero. (b) A series of free induction decays for the data displayed in (a). The numbers labelling each FID correspond to the points labelled in (a). The vertical scales are not the same for the six FIDs displayed.

these fields). The beating present at short times arises from the displaced mode signal. Notice how the shifted mode frequency passes through zero as the field is swept from -1.0 to +1.0 gauss. The overall phase of the FID inverts between FID #3 and #4. The displayed FIDs #3 and #5 are at the lowest fields included in the Figure 8a. The FID #4 occurs for a field intermediate between that for #3 and #5, but the displaced mode has broadened so much that there is only a small hint of an oscillation at short times where the curve appears to break slightly.

Figure 9 illustrates the extent of broadening for the displaced mode as the field approaches zero. This is again from the same field sweep data displayed in Figure 8. At higher fields the linewidth approaches a constant value. We saw evidence for this broadening in the family of Fourier transforms displayed in Figure 3 and the free induction decays in Figure 8b. Fixing the linewidth in these low fields would reduce the size of the amplitude jump in Figure 8a, but this results in very poor fits to the free induction decay signals. Allowing the linewidth to vary as a free parameter results in the linewidths displayed in Figure 9 and the jump seen in Figure 8a.

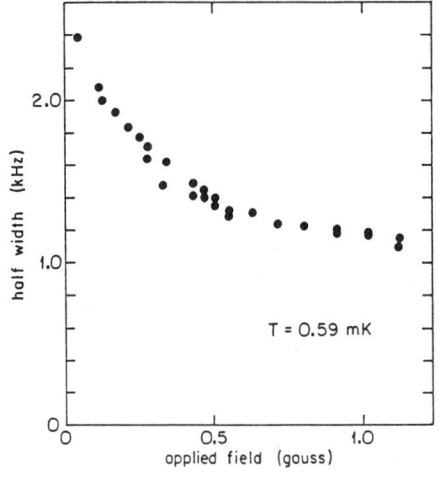

Fig. 9 Half linewidths of the displaced mode for $H_0 \leq 1$ gauss. The line broadens by a factor of 2 above its high field constant value.

DISCUSSION

The frequency shifts can be used to estimate polarization. For a two dimensional array of spins (with H_0 parallel to the plane) we expect an NMR line at:[6]

$$\omega^2 = \gamma^2 (H_0^2 + \lambda M H_0) \qquad (2)$$

with λM giving the dipole demagnetizing field. For full polarization of a triangular lattice of spacing 3.78 Å (second layer of ^{3}He on graphite) the computed λ implies a shift of 1.77 kHz. Our kHz shifts imply polarizations on the order of 30% or greater. We may compare this implied polarization with the measured amplitude of the signal through two methods. The high temperature (T ~ 15 mK) ^{3}He liquid signal may be compared with the low temperature (T < 1 mK) solid signal to arrive at a polarization for the solid. Alternatively, the SQUID output voltage may be expressed in terms of input flux or in this instance solid polarization. This second method correctly predicts the size of the liquid signal at high temperatures. Both of these methods imply solid ^{3}He polarizations around 0.4%, in contrast to the 30% estimate for the frequency shift data.

The measurements at a few monolayers[8] in higher fields were consistent with nearly full polarization of the entire surface area. Restricted domain size on our grafoil surface may partially explain the polarization discrepancy in the present instance. If there are domains of size comparable to the regions of structural coherence[4] (~100 Å), then for a single domain one finds that the ratio of magnetic field to thermal energy, $MH_0/k_B T$, is about 0.2 for

T = 0.5 mK, H_0 = 1 gauss and 100% polarization. Thermal fluctuations will therefore decrease the sample magnetization by roughly this same factor. However a further reduction by a large factor is still needed to explain the 0.4% polarization mentioned above.

We should note that the thermally excited domains would precess about the applied field at a frequency quite different from that given by equation (1) in which near alignment is presumed. With M = constant for a precessing domain we may numerically integrate the equations of motion to arrive at the orbit area and frequency. This procedure reveals that at low fields (H_0 < 3 gauss) there may in fact be two positive shifted peaks in the density of states $\rho(f)$. This predicted behavior does not, however, persist up to the highest fields where we still experimentally observe two modes ('shifted' and 'displaced'). While thermal excitations of the domain magnetization may help explain the small magnetization that we observe, it apparently does not account for the observation of the two shifted peaks in the high field NMR spectra.

Aside from restricting the domain size, surface quality may still play an additional important role in governing the observed frequency shifts and polarization. Free electron spins (dangling surface bonds) occurring at roughly the same density as ^{3}He structural domains (~100 Å) would produce an internal field sufficient to account for the 3 kHz shift of the displaced mode. The ^{3}He spins lying next to such an electron spin feel a much larger field, of order 1 kG, but this is still much less than the effective field between ^{3}He spins due to the exchange coupling. A single domain should then still precess as a single rigid moment, with the electron field averaged over the whole domain determining the shift. The dipolar interaction energies between electrons and between otherwise uncoupled domains are about two orders of magnitude smaller than the electron-domain coupling, and are therefore unimportant. One expects thermal effects and random placement of electron spins should lead to a broadened spectrum of zero field frequency shifts and domain orientations. Such effects may then be responsible for the small recorded amplitudes as well.

The frequency shift and amplitude behavior of the shifted mode is qualitatively consistent with Equation (2). Putting a M = χH_0 dependence (from Figure (8a)) into Equation (2) we obtain $f_s = \gamma H_0 (1 + \lambda\chi)^{1/2}$ which is consistent with the shifted mode frequencies of Figure (6b) with $(1 + \lambda\chi)^{1/2}$ = 1.035. The shifted and displaced modes may arise from two separate regions on the surface. One region would be characterized by an enhanced linear susceptibility, M α H_0, and the other by an actual ordering that leads to the offsets seen for the displaced mode.

The main theoretical task is a quantitative theory of the 2-d ordering. The work of Ref. (11) demonstrates that an ordered state exists at sufficiently low temperatures. If one varies the power law rate, r^{-p} with which the dipole potential decays at large distances a Kosterlitz-Thouless phase can intervene between the ordered and paramagnetic phases for $4 > p > 3\frac{3}{4}$.[15] However in the physical case (p = 3) one finds that the entire Kosterlitz-Thouless phase is unstable to the addition of a small dipole interaction, and therefore the transition occurs directly from the paramagnetic to the ferromagnetic phase. For $p > 4$ the force is effectively short ranged, and the Kosterlitz-Thouless phase survives right to T=0, consistent with the Mermin-Wagner theorem[10].

Although dipolar ferromagnets have been treated in higher dimensions within the ϵ-expansion about 4 dimensions, they have not been considered in the geometries necessary to describe the present experiments. The present hope is that some kind of expansion in the decay rate of the dipole force (perhaps about $p = 3\frac{3}{4}$), analogous to similar expansions near 4 dimensions, will yield information about the critical exponents.[16]

In summary, our observations on the field dependence for the frequency and amplitude of the displaced mode point to a collective behavior in fields less than one gauss, consistent with correlated domains of at least 10^4 Å^2. Our experiments seem to be probing regions of the ^{3}He spin dynamics whose interpretation is limited not by experimental resolution, but by surface irregularity. A more uniform substrate such as ZYX may be needed to achieve a cleaner realization of the dipole driven ferromagnetic ordering transition. The experiments described in this paper have approached the limit of useful low temperatures and magnetic fields for ^{3}He ordering on grafoil because the size of a magnetic coherence area is approaching the size of the structural coherence allowed by the substrate.

This work has been supported by the National Science Foundation through grants DMR88-00291 (LJF,HMB), DMR85-19970 (CMG), and DMR87-15474 (PBW,MCC), the Weingart Foundation through The California Institute of Technology (PBW), and by the Science and Engineering Research Council (U.K.) (ALT).

REFERENCES

a) Permanent address: Department of Physics, University of Sussex, Falmer, Brighton BN1 9QH, England.
1) D.F. Brewer, J.S. Rolt, Phys. Rev. Lett. 29, 1485 (1972); A.I. Ahonen, T. Kodama, M. Krusius, M.A. Paalanen, R.C. Richardson, W. Schoepe, Y. Takano, J. Phys. C 9, 1665 (1976); A.I. Ahonen, T.A. Alvesalo, T. Haavasoja, M.C. Veuro, Phys. Rev. Lett. 41 494 (1978); Y. Okuda, A.J. Ikushima, H. Kojima, Phys. Rev. Lett. 54 130 (1985); P.C. Hammel, PhD Thesis Cornell University 1984 (unpublished); B.N. Engel, G.G. Ihas, G.F. Spencer, Jap. Jour. Appl. Phys. 26 315 (1987); Y. Okuda, A. Fukushima, A. Ikushima, Jap. Jour. Appl. Phys. 26 269 (1987).

2) W. Durr, M. Taborelli, O. Paul, R. Germar, W. Gudat, D. Pescia, M. Landolt, Phys. Rev. Lett. 62 206 (1989); T. Beier, H. Jahrreiss, D. Pescia, T. Woike, W. Gudat, Phys. Rev. Lett. 61 1875 (1988).

3) J.M. Delrieu, M. Roger, J.H. Hetherington, J. Low Temp. Phys. 40, 71 (1980); M. Roger, Phys. Rev. B 30, 6432 (1984); M. Roger, J.M. Delrieu, Jap. Jour. Appl. Phys. 26, Suppl. 26-3, 267 (1987); H. Jichu, Y. Kuroda, Prog. Theor. Phys. 67, 715 (1982); H. Jichu, Y.

Kuroda, Prog. Theor. Phys. 69, 1358 (1983).

4) H.J. Lauter, H.P. Schildberg, H. Godfrin, H. Wiechert, R. Haensel, Canadian Journal of Physics 65 1435 (1987); J. K. Kjems, L. Passell, H. Taub, J. G. Dash, A. D. Novaco, Phys Rev. B 13 1446 (1976); C. Bouldin, E. A. Stern, Phys. Rev. B 25 3462 (1982).

5) H.M. Bozler, T. Bartolac, K. Luey, A.L. Thomson, Phys. Rev. Lett. 41, 490 (1978); H. Godfrin, G. Frossati, D. Thoulouze, M. Chapellier, W.G. Clark, J. Physics (Paris), Colloq. 39, C-287 (1978); H.M. Bozler, D.M. Bates, A.L. Thomson, Phys. Rev. B 27, 6992 (1983).

6) L.J. Friedman, S.N. Ytterboe, H.M. Bozler, A.L. Thomson, and M.C. Cross, Phys. Rev. Lett. 57, 2943 (1987); L.J. Friedman, S.N. Ytterboe, H.M. Bozler, A.L. Thomson, and M.C. Cross, Canadian Journal of Physics 65 1351 (1987); H.M. Bozler Jap. Jour. Appl. Phy. 26, Suppl. 26-3 1849 (1987).

7) H. Franco, H. Godfrin, D. Thoulouze, Phys Rev. B 31, 1699 (1985). H. Franco, R.E. Rapp, H. Godfrin, Phys. Rev. Lett. 57, 1161 (1986); H. Godfrin, Can. J. Phys. 65, 1430 (1987).

8) H. Godfrin, R.R. Ruel and D.D. Osheroff, Phys. Rev. Lett. 60, 305 (1988).

9) L.J. Friedman, A.L. Thomson, C.M. Gould, H.M. Bozler, P.B. Weichman, M.C. Cross, Phys. Rev. Lett. 62 1635 (1989).

10) N.D. Mermin, H. Wagner, Phys. Rev. Lett. 17, 1133 (1966).

11) Y. Yafet, J. Kwo, E.M. Gyorgy, Phys. Rev. B 33, 6519 (1986).

12) P. Kopietz, P. Scharf, M.S. Skaf, and S. Chakravarty, State University of New York at Stony Brook preprint (1988).

13) L.J. Friedman, A.K.M. Wennberg, S.N. Ytterboe, H.M. Bozler, Rev. Sci. Instrum. 57, 410 (1986).

14) The effect that the spread in grafoil plane angles might have on observed NMR signals is discussed in references 6.

15) P.B. Weichman, to be published.

16) The possibility of this approach was suggested by D.S. Fisher, private communication.

NUCLEAR SPIN HEAT CAPACITY OF ^{3}He ADSORBED ON GRAPHITE

Dennis S. Greywall
AT&T Bell Laboratories, Murray Hill, New Jersey 07974, USA

ABSTRACT

The heat capacity of ^{3}He adsorbed on graphite has been measured for films between one and five atomic layers and for temperatures between 2 and 200 mK. These results are compared with recent magnetization data which also show several anomalies in this coverage regime. Prior to third layer promotion the second layer is found to solidify into a registered structure with unusual properties. This contradicts the model proposed to explain the NMR measurements.

It has been appreciated for nearly two decades that the first few layers of liquid ^{3}He adjacent to a solid surface exhibit unusual magnetic behavior at very low temperatures. This phenomenon, however, is still not completely understood even though significant progress has been made in recent years, particularly via measurements made as a function of adsorbed coverage.[1-3] The most detailed of the current experiments are the ^{3}He-on-graphite magnetization measurements of Franco et al.[1] which exhibit several intriguing features for film thicknesses between roughly 2 and 5 layers. The main feature is a hugh ferromagnetic peak which is located by Franco et al. at 2½ layers and attributed to the freezing of the second ^{3}He layer into a solid with an extremely large exchange energy.

We discuss here the first heat capacity measurements for ^{3}He adsorbed on graphite in the low millikelvin regime. These measurements provide information which is complimentary to the magnetization data,[1] but contradict the proposed model. The heat capacity data indicate that the second layer freezes prior to third promotion into a registered solid with unusual nuclear spin properties. This phase appears to be intimately related to the ferromagnetic anomaly observed at higher coverages.

Figure 1 shows the heat capacity results as a function of coverage for several different isotherms and also makes comparison with the magnetization data.[1] The lowest portion of the figure shows results at 2.5 and 5 mK where the heat capacity is dominated by the nuclear spins. The

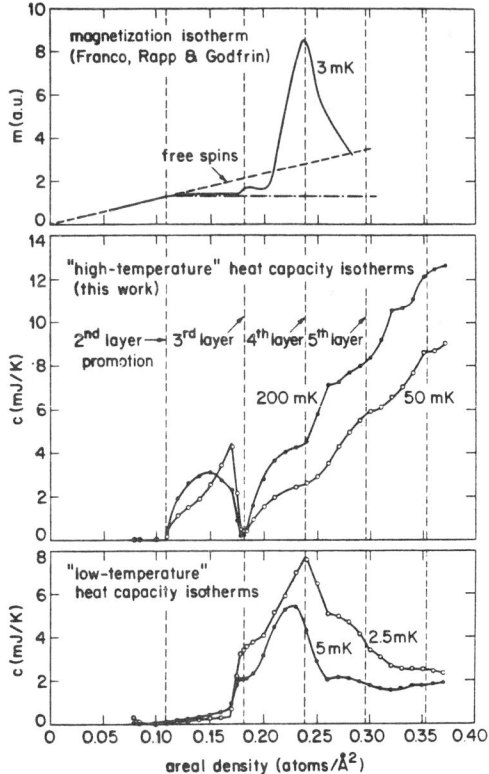

Fig. 1. Heat capacity as a function of coverage for several isotherms. The heat capacity at 2.5 and 5 mK is dominated by the nuclear spin contribution of the second layer atoms. The curves at 50 and 200 mK are determined by the fluid contributions from the various layers. Comparison is made with the magnetization results of Franco et al.

middle portion shows isotherms at 50 and 200 mK. At these much higher temperatures the heat capacity is mainly due to the fluid contributions from the various layers.

The abrupt increases in the 200 mK isotherm at 0.109, 0.182, and at 0.239 atoms/Å^2 correspond to promotion of atoms into the second, third, and fourth layers respectively. The abrupt decrease in this isotherm at

0.169, which is accompanied by a sharp increase in the 2.5 mK curve, is a consequence of the solidification of the second layer. Note that the peak in the spin heat capacity and in the magnetization coincide, perhaps coincidentally, with fourth layer promotion. The ferromagnetic anomaly therefore occurs in a rather complicated system consisting of several layers. A further complication is that the second layer at this coverage is most likely in a transition region between registered and incommensurate solid.

The claim that the freezing taking place prior to third layer promotion is into a registered phase is based mainly on the low density of the second layer when the transition occurs. At a corresponding first layer density the solid is registered with respect to the graphite substrate, but for the second layer the registry is presumably with respect to the first ^{3}He layer which exists at high density as an incommensurate solid.

Assuming that the registered solid has a triangular lattice suggests that three particle ring exchange should be the dominant exchange process,[4] and this leads to an effective ferromagnetic interaction which can cause ordering only at $T = 0$ in two dimensions. This is contradicted by the magnetization data, however, which show a magnetization less than the free spin value and also by the heat capacity results which exhibit a sharp peak at 2.5 mK which is suggestive of a finite temperature phase transition.

Another puzzling property of the registered phase data is the spin entropy computed from the heat capacity measurements. The entropy is only half the expected value of $k_B \ln 2$ per atom. The explanation must be that below the lowest experimental temperature of 2 mK, the heat capacity has a second anomaly corresponding to the ordering of the remaining spin degrees of freedom.

This behavior could be explained if there were zero point vacancies in the registered structure since there vacancies would presumably induce spin polarons.[5] The spins forming part of the polaron itself would order at a much higher temperature than those spins more distant from a vacancy.

Elser[6] has proposed a very different model which applies to the perfect registered structure. He proposes that the second layer atoms are arranged as indicated in Fig. 2. The darker circles represent second layer atoms which form a close pack lattice with an areal density 4/7 of that of the first layer. This is in agreement with the experimental finding. There are two types of lattice sites for the second layer atoms. Three quarters of the atoms (B) sit centered above four first layer atoms at a position of low potential energy while the remaining atoms (A) are located directly above first layer atoms and therefore at potential maxima. These atoms have a large zero point motion. As a consequence, the exchange of two B atoms

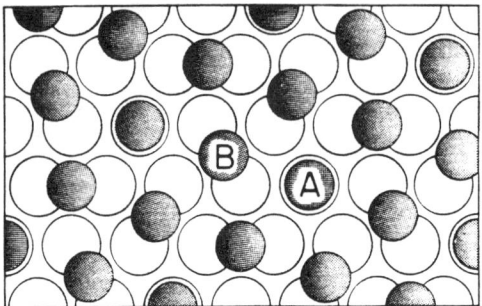

Fig. 2. Structure of the second layer registered solid proposed by Elser. Note that one fourth of the atoms are located directly above first layer atoms and therefore at substrate potential maxima.

occurs more often than at A-B exchange, since the A atom hindering the first process has a high probability of simply being out of the way. If the A-B exchange rate is extremely small, then the A spins order at a very low temperature. This explains a portion of the entropy discrepancy. A large B-B exchange would lead to a dominant antiferromagnetic spin interaction in agreement with the NMR results.

Although more detailed calculations are necessary, this model looks extremely hopeful and may provide a basis for our understanding of the much more complicated region near the ferromagnetic peak.

References

1. H. Franco, R. E. Rapp, and H. Godfrin, Phys. Rev. Lett. **57**, 1161 (1986).
2. Y. Okuda, A. J. Ikushima, and H. Kojima, Phys. Rev. Lett. **54**, 130 (1985).
3. D. S. Greywall and P. A. Busch, Phys. Rev. Lett. **60**, 1860 (1988).
4. M. Roger, Phys. Rev. Lett. B **30**, 6432 (1984).
5. R. A. Guyer, Phys. Rev. Lett. **39**, 1091 (1977).
6. Veit Elser, to be published.

FIELD DEPENDENCE OF THE ³He–SUBSTRATE SPIN COUPLING

T.J. Gramila, F. Van Keuls, Y. Hu, R.C. Richardson
Cornell University, Ithaca, N.Y. 14853

Although the magnetic relaxation for normal ³He has been known to be dominated by surface processes for some time, the dynamics of the relaxation are not understood.[1] The temperature dependence of the relaxation was accounted for in 1983 by Hammel and Richardson in a model which assumes that the relaxation takes place in a solid-like surface layer.[2] Particularly puzzling, however, is the linear dependence of the relaxation time T_1 on field. This has been observed for an extremely broad range of frequencies - from under a hundred kilohertz to over a hundred megahertz.

In order to try to further understand the relaxation process, we have made N.M.R. measurements of both the ³He and the substrate ¹⁹F relaxation times for ³He adsorbed on a powder of CaF_2 crystallites (average particle size about 0.5 microns). Our measurements fall into two groups. The first of these is made as a function of ³He coverage at a magnetic field of 3.2 KGauss and at a temperature of 96 mK. Especially at the lowest coverages, it is important to anneal the films at about 8 K for a couple of hours after each addition of ³He. The second group of these measurements is made at a fixed coverage and temperature but with varying magnetic fields.

The coverage dependence data are shown in figure 1. The dashed curve is a rough indication of monolayer coverage as indicated by B.E.T. In these measurements, the ¹⁹F relaxation rate is completely dominated by surface processes (the bare substrate T_1 is greater than two days). At low coverages, both the ³He and the ¹⁹F have identical functional dependences on coverage, then exhibit a pronounced peak and drop off at coverages near one monolayer. These data are evidence that both spin species relax via the same channel : magnetic fluctuations in the surface layer of ³He atoms.

The field dependence data are shown in figure 2 for a coverage of 3 layers and for full cell. The three layer ³He data displays two distinct linear regions. The high field region has a a pronounced negative intercept, while the low field region has a very nearly zero intercept. The three layer ¹⁹F data has a field dependence very similar to that of the ³He data, again indicating that both have the same relaxation process. Low field ¹⁹F measurements are complicated by finite spin diffusion times in the substrate.

These data are most useful when they are compared to the full cell measurements. The ¹⁹F full cell data is virtually identical to the three layer data. Since the ¹⁹F spins can "see" only the first layer of the ³He atoms, this means that the dynamics of the first layer are unaltered by the bulk liquid overlayers. The ³He full cell data, however, is substantially different from the three layer data. There are two distinct linear regions,

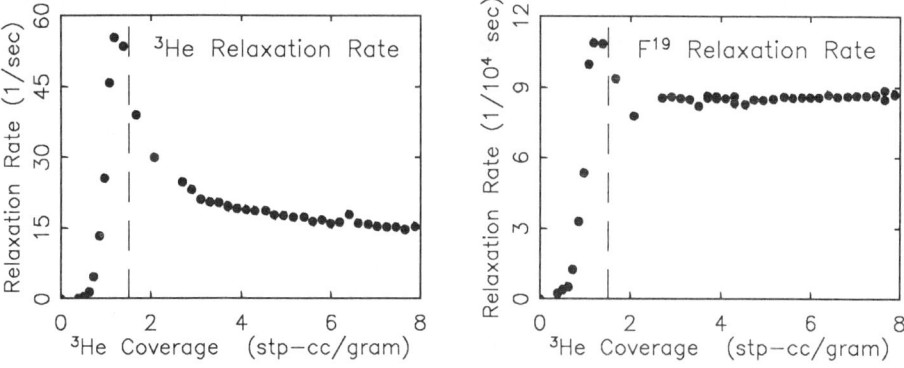

Figure 1. Coverage dependence of the ³He and ¹⁹F relaxation rates

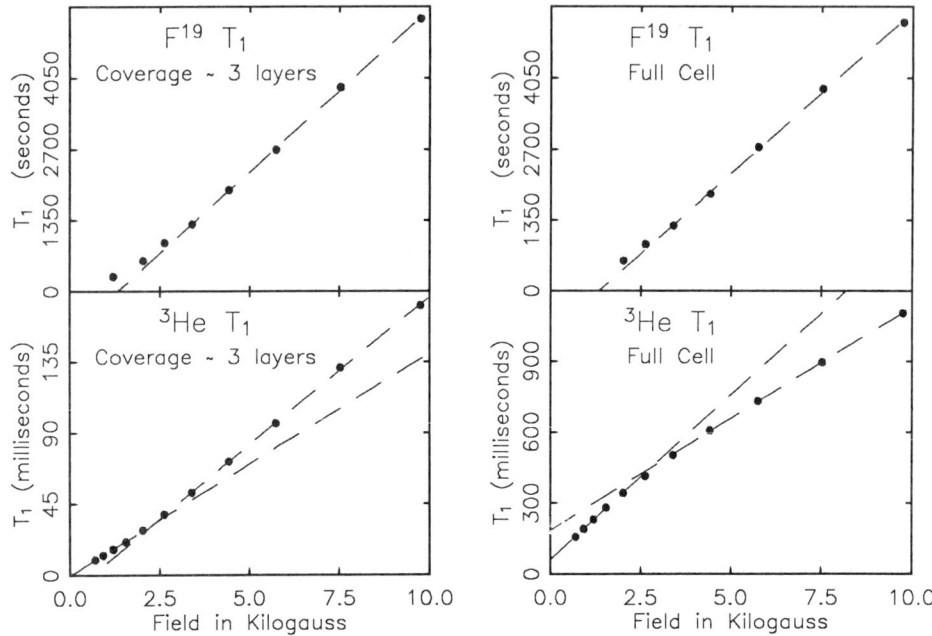

Figure 2. Field dependence of the relaxation times for 3 layers and full cell

but now both have positive intercepts. The dynamics of the first layer are unchanged, so the data show that more than the first layer participates in the relaxation process.

This result is indicated more dramatically in figure 3. The fundamental assumption of recent models is that the relaxation takes place only in the solid-like surface layer. The effect of the liquid above that then is simply to provide more magnetization to be relaxed at the surface. The measured T_1 can be expressed as :

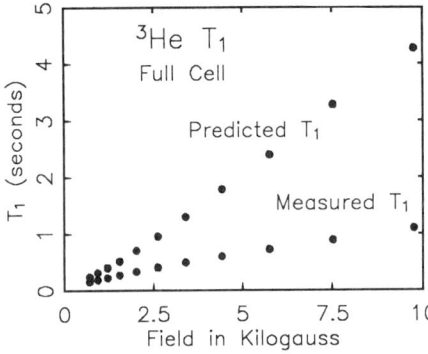

Figure 3. Comparison of predicted and measured full cell relaxation times

$$T_1 = T_{1\text{surface}} * \frac{\text{Total Helium Magnetization}}{\text{Surface Helium Magnetization}}$$

This accounts for the observed temperature dependences of both the substrate and the ^{3}He relaxation times. We have observed that dynamics of the first layer are unaltered by the addition of liquid beyond three layers (i.e. $T_{1\text{surface}}$ is unchanged), so this expression predicts that the full cell T_1 should just equal the three layer T_1 times the ratio of their magnetizations. This is plotted in figure three (upper curve) as well as the measured values for T_1. The measured relaxation is four times as fast as expected! We conclude that more than just the first layer is participating in the relaxation process; there must be significant relaxation in the liquid overlayers.

REFERENCES

1. See the review by R.C. Richardson, Physica , 126B (1984) 298.
2. P.C. Hammel, R.C. Richardson Phys. Rev. Lett 51 (1983) 2124

OSCILLATIONS OF CRITICAL PARAMETERS IN THIN FILMS OF ^{3}He

L.S. Borkowski[1], G. Harań[2], L. Jacak[3]
1 Virginia Polytechnic Institute and State University, USA
2 Institute for Low Temperature and Structure Research, Polish Academy of Sciences, Wrocław, Poland
3 Technical University of Wrocław, Poland

In this paper we consider some aspects of thin films ($p_F^{-1} \ll d \leq \xi_0$) of ^{3}He. d is the film's thickness and ξ_0 - the correlation length of the Cooper pair. We investigate the influence of surface roughness on the superfluid transition. It was suggested that in thin films of ^{3}He on a rough substrate there may exist a gapless superfluid phase[1] due to a diffusive scattering at the boundary. The peculiar feature of the gapless state is that we may have both an energy gap and a nonzero density of states at the Fermi surface.

We consider the ABM phase which was found to be stable in thin film in this thickness range[2]. The boundary conditions imposed on a single-particle wave function are

$$\Psi(x,y,d) + u(x,y)\frac{\partial}{\partial z} \Psi(x,y,z)\Big|_{z=d} = 0, \quad \Psi(x,y,0) = 0. \quad (1)$$

Here $u(x,y)$ represents height of the irregularities on the surface liquid-solid and is a gaussian random function with a zero mean and a δ-type correlation. The second of eqs. (1) means an infinite potential wall at the other boundary.

Following standard methods of the Abrikosov-Gorkov theory we obtain the gap equation as a function of film thickness and degree of surface roughness. Details of this derivation can be found elsewhere[3]. Solving the gap equation for the parameter of the critical surface roughness ζ giving the transition to a nonzero density of states at the Fermi surface, we arrive at

$$1 = \zeta \sum_\nu \frac{\cos^4\theta_\nu}{\cos\theta_\nu \left[\sin^2\theta_\nu + \zeta^2 U^2 \cos^4\theta_\nu\right]^{\frac{1}{2}}}, \quad (2)$$

where $\cos\theta_\nu = \frac{\nu}{\nu_0}$, ν being the number of one of the parallel Fermi circles, $\nu_0 = \left[\frac{2md^2}{\pi^2}\mu\right]^{\frac{1}{2}}$ and μ is the chemical potential. We look for the smallest value of ζ leading to a nonzero solution for U. Figs. 1-4 present results of numerical calculations.

The final result is the phase diagram for the transition: superfluid-gapless superfluid phase, Figs. 3 and 4, where we use the formula

$$\frac{p_F^4 w^2}{\Delta_0/\mu} = \frac{(p_F d)^6}{\pi^4 \nu_0^5} \frac{\Delta}{\Delta_0} \zeta . \quad (3)$$

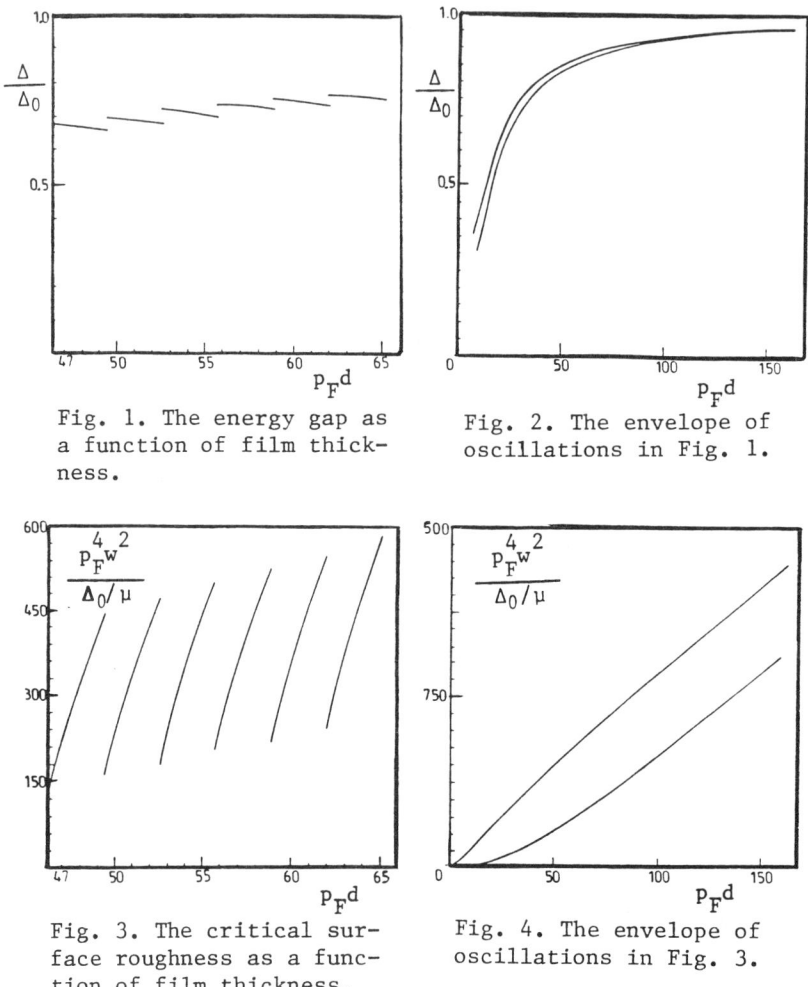

Fig. 1. The energy gap as a function of film thickness.

Fig. 2. The envelope of oscillations in Fig. 1.

Fig. 3. The critical surface roughness as a function of film thickness.

Fig. 4. The envelope of oscillations in Fig. 3.

Here p_F, μ, Δ_0 and T_c are bulk values. $p_F^4 w^2$ is a direct measure of the critical surface roughness in units of interatomic distance, w denotes the average height of 'bumps'. Our calculations are performed for $T = 0.01\ T_c$. Let us note that Fig. 3 resembles typical characteristics of the de Haas - van Alphen effect. It is worth mentioning in this context that this and other interesting effects also appear in thin films of ^{3}He in magnetic fields, and we plan to report on it in near future.

REFERENCES

1. Z. Tešanović, O.T. Valls, Phys. Rev. B <u>34</u>, 7610 (1986).
2. M.R. Freeman et al., Phys. Rev. Lett. <u>60</u>, 596 (1988).
3. G. Harań, L.S. Borkowski, L. Jacak, submitted to Physica B.

SUPERFLUID ^{3}He FILM FLOW — A PHASE TRANSITION AND A SUBSTRATE EFFECT[†]

J.P. Harrison, A. Sachrajda[*], S.C. Steel and P. Zawadzki
Queen's University, Kingston, Ontario, Canada K7L 3N6

INTRODUCTION

Experiments have shown that superfluid ^{3}He films will flow when driven by a gravitational potential.[1,2] The present study, described briefly here, has focused on film flow over a substrate with no sharp curvature which was preplated with 0, 1 or 3 monolayers of ^{4}He (0, 1, or 3 M/L).

EXPERIMENTAL DETAILS

Our technique is to measure the rate at which a beaker of superfluid ^{3}He empties as the Rollin's film which adheres to the walls of the beaker flows over the rim of the beaker. The present beaker was machined with a rounded lip of radius 0.5 mm and the surface was electropolished. Trial polishes gave a very smooth surface with isolated pits ~1 μm in diameter. However the actual beaker surface had lines of pits perhaps due to a variance in the final polish. The flow rate was studied as a function of temperature from 0.4 to 0.9 mK, as a function of film thickness from 100 to 200 nm, and as a function of ^{4}He preplating on the beaker surface.

RESULTS AND DISCUSSION

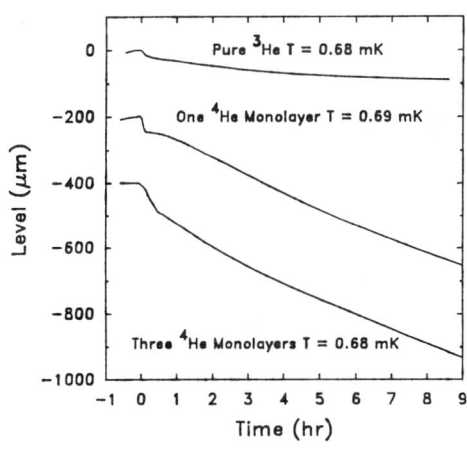

Figure 1 shows the level inside the beaker as a function of time for cooldowns with 0, 1 and 3 M/L. The level is static until the temperature is reduced below 1 mK. The ^{3}He then becomes superfluid, the film flows, and the level drops. All cooldowns showed two flow regimes. There was an initial fast flow until the level had dropped ≃20 μm for 0 M/L or ≃100 μm for 1 and 3 M/L. After an abrupt change, there followed a slow flow that continued for up to 60 hours with the flow rate decreasing as the level dropped and therefore as the film at the rim thinned.

[*] Division of Physics, N.R.C. Ottawa
[†] Research supported by NSERC.

In the slow flow regime the flow rate was significantly higher for 1 and 3 M/L than for 0 M/L. A detailed analysis, to be described elsewhere, showed that the 0 M/L film behaved as a slab with one diffuse and one specular surface (the free surface) and the 1 M/L film as a slab with two specular surfaces. That is, the 0 M/L film of thickness d showed a suppressed transition temperature T_c which agreed with that predicted for a slab with thickness 2d and with two diffuse boundaries (ie. $2d/\xi(T_c) = \pi)^3$. By contrast, the 1 M/L film showed no suppression of T_c even for films with $d/\xi(T) \simeq 1$. Freeman et al made the similar observation that 2 M/L of ^{4}He changed the boundaries of 350 nm ^{3}He slabs from diffuse to specular.[4]

It is believed that the abrupt feature in the level versus time curves is the expected phase transition between the B-like and A-like phases of thin ^{3}He films.[3,5] For a slab of thickness w and diffuse boundaries, this should occur at $w/\xi(T) = 7$. For 0 M/L where w = 2d a detailed analysis of our results shows $w/\xi(T) \sim 5$ for this feature. However for this first order phase transition a certain amount of superthinning can be expected.[6] Unexpectedly, for the 1 M/L films with two specular boundaries, where w = d, we find the transition at $w/\xi(T) \simeq 1.5$; the theoretical prediction is for a higher value of $w/\xi(T)$.[5]

CONCLUSION

New results with a rounded lip beaker have demonstrated diffuse and specular boundary ^{3}He films for substrates with 0 and 1 or 3 ^{4}He monolayers of preplating. Further as the film thins there is an apparent phase transition which we identify as the B-like to A-like transition. Finally, as a note of caution, very recent experiments with a smoother substrate (copper and stainless steel) show no flow of the A-like 0 M/L film. There is flow of the B-like 0 M/L and 1 M/L films and of the A-like 1 M/L.

ACKNOWLEDGEMENTS

We appreciate assistance with electron microscopy from P.Y. Timbrell and discussions with A. Tyler.

REFERENCES

1. J.G. Daunt R.F. Harris-Lowe, J.P. Harrison, A. Sachrajda, S.C. Steel, R.R. Turkington and P. Zawadzki, J. Low Temp. Phys., 70, 547 (1988)
2. J.C. Davis, A. Amar, J.P. Pekola and R.E. Packard, Phys. Rev. Lett., 60, 302 (1988)
3. A.L. Fetter and S. Ullah, J. Low Temp. Phys., 70, 515 (1988)
4. M.R. Freeman, R.S. Germain, E.V. Thuneberg and R.C. Richardson, Phys. Rev. Lett., 60, 596 (1988)
5. Y-H Li and T-L Ho, Phys. Rev. B, 38, 2362 (1988)
6. T-L Ho, private communication.

MAGNETIC RELAXATION OF NORMAL ^{3}He ON A SURFACE COATED WITH ^{4}He

Yu.M.Bunkov, V.V.Dmitriev, Yu.M.Mukharskii, D.A.Sergatskov
Institute for Physical Problems, Kosygin St.2, Moscow, 117334, USSR

Measurements on surface magnetic relaxation in a bulk sample of normal ^{3}He at temperatures below 10 mK and under different pressures are described. Linear dependence of T_1 on the temperature is observed. The ^{4}He film on the surface does not essentially affect the relaxation. Solidification of this film changes the slope of the dependence of T_1 on the temperature.

It is well known that the magnetic relaxation of a normal ^{3}He takes place on walls of the experimental cell and the magnetization of the sample is transferred to the walls by a spin diffusion (see for example the review[1]). The longitudinal magnetic relaxation in this case obeys a linear dependence on the temperature and on the value of the external magnetic field. Hammel and Richardson[2] supposed the model which explains the temperature dependence of T_1. In this model the relaxation is connected with the solid ^{3}He layer on the surface. The rate of the relaxation is proportional to the magnetic moment of the layer, which obeys Curie-Weiss low. Experiments which was done in confined geometry are in agreement with the theory[2,3]. The coating the surface by few monolayers of ^{4}He results in sufficient decrease of the relaxation rate[4].

We have carried experiments with a bulk sample of ^{3}He. The walls of the experimental cell were covered by lavsan (material analogous to mylar) foil. The experimental volume has a form of a cylinder (ø6mm, length - 6 mm) and is connected with other parts of the cell by a channel ø1mm. T_1 is measured by a standard two-pulse method in a magnetic field of 142 Oe. At temperatures below 5 - 10 mK the recovery of the longitudinal magnetization after the first tipping pulse is exponential and we may not take into consideration the restriction of the relaxation by the spin diffusion (this is also confirmed by numerical estimations where values of the spin diffusion coefficient have been taken from[5]).

Results of our experiments when the cell was filled by pure ^{3}He are shown in Fig.1 by open symbols. As it is seen from the figure experimental points obeys the following dependence:

$$T_1 = A(P) \left[T + \Delta(P) \right] \qquad (1)$$

where $A(P)$ some coefficient, depending on pressure and $\Delta(P)$ is close to 0.

After warm up to the room temperature we have repeated the experiments, but now we have condensed in a cold cell 2 liters (at normal conditions) of ^{4}He and then added the mixture (15% ^{4}He, 85% ^{3}He). So at superlow temperatures we have two phases: pure ^{3}He in the upper part of the cell and 6-10 % ^{4}He-^{3}He mixture in the lower part. The experimental volume of our cell is placed in the upper part, so in this case we measure T_1 of ^{3}He. The walls of the experimental volume should now be covered by a thick (100 A) film of ^{4}He (ref.6). The concentration of ^{3}He in this film should be 0 near the wall and he absence of the solid layer of ^{3}He on the surface should essentially change the relaxation. Solid points in Fig.1 correspond to measurements with ^{4}He film. As it is seen from the figure the change of the relaxation rate is not large. It is worth to mark that value of Δ for pressures 11 and 24 bars now is negative. We can not explain the obtained results, but it is clear that linear temperature dependence of T_1 at least for bulk samples of ^{3}He is not connected with the solid ^{3}He layer. At 27 bars the ^{4}He film should be solid. In this case the slope of the dependence of T_1 on the temperature increases.

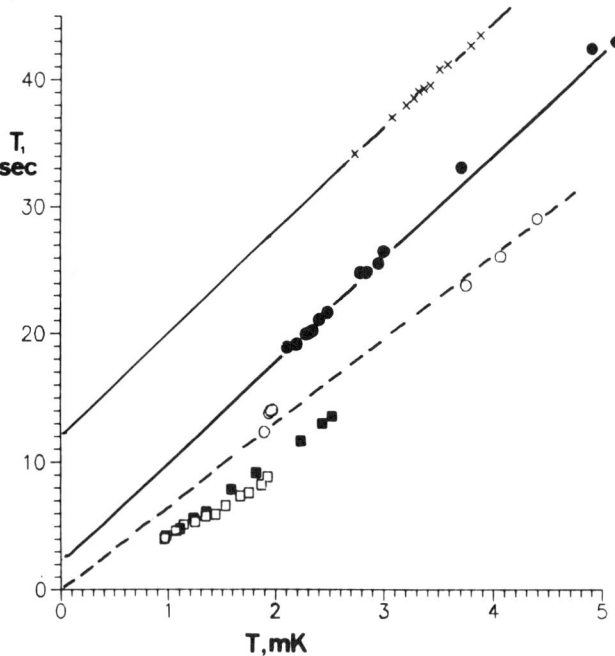

Fig.1 T_1 in normal ^{3}He at 0 bar (□,■), 11 bar (o,●), 24 bar(x).
Closed symbols and crosses correspond the experiment with ^{4}He film.

1. R.C.Richardson, Physica, **126B+C,** 298 (1984)
2. P.C.Hammel, R.C.Richardson, Phys.Rev.Lett., **52,** 1441 (1984)
3. H.Franco, H.Godfrin, H.Lauter, D.Thoulouse, in Proc. of 17th Int. Conf. of Low Temp. Phys., North Holland,(1984),p.755
4. H.Godfrin, G.Frossati, B.Hebral, D.Thoulouse, J.Phys, **C7,** 275 (1980)
5. A.S.Sachrajda, D.F.Brewer, W.S.Truscott, J.L.T.P.,**56,** 617 (1984)
6. V.P.Peshkov, JETP Lett., **21,** 162 (1975)

STRUCTURE IN THE MAGNETIZATION OF THIN ^{3}He–^{4}He MIXTURE FILMS

R.H. Higley, D.T. Sprague and R.B. Hallock
Laboratory for Low Temperature Physics
Department of Physics and Astronomy
University of Massachusetts, Amherst, MA 01003

Several studies of the properties of ^{3}He-^{4}He films have shown mixture films to be a remarkably rich and interesting system. At low temperature, small concentrations of ^{3}He behave as a nearly ideal Fermi gas in two dimensions atop the ^{4}He film with $T_F = \pi \hbar^2 N / k_b m_3 A$ where A is the surface area and N is the number of ^{3}He atoms. For increasing ^{3}He coverage, one can study the evolution of interactions among the ^{3}He.

We report measurements of the magnetization of ^{3}He as a function of the coverage, d_3, and temperature for $0.005 \leq d_3 \leq 4$ atomic layers and $40 \leq T \leq 250$mK for the case of a ^{4}He film of thickness 2.14 atomic layers. The experiments utilize the substrate Nuclepore in an experimental arrangement which has been described previously[1]. Pulse echo techniques are employed in a static field $H_0 = 2$T. The 200nm diameter Nuclepore pores are oriented perpendicular to the static field. The total surface area is $A = 1.77$ m^2.

For the case of an ideal Fermi gas in two dimensions (2DIFG) the magnetization is given by $M = M_{30}(1 - \exp(-T_F/T))$ where $M_{30} = N\mu_m^2 H_0 / k_b T_F$. For an interacting Fermi liquid (2DFL) with $T \ll T_F$ we expect $M = m^{**}M_{30}(1 - \exp(-T_F/T))$ with $m^{**} = m^*/(1 + F_0^a)$, $m^* = m_h(1 + (F_1^S)/2)$ where m_h is the hydrodynamic mass, F_0^a and F_1^S are (two dimensional) Fermi liquid parameters and $T_F = \pi \hbar^2 N / k_b m^* A$.

An example of our magnetization data[2] extrapolated to $T = 0$ is shown in figure 1. Several regions of coverage are apparent. For $0.1 \leq d_3 \leq 0.6$ layers we have a 2DFL with a linear increase of $M(T=0)/M_{30} \sim m^{**}$ with coverage.

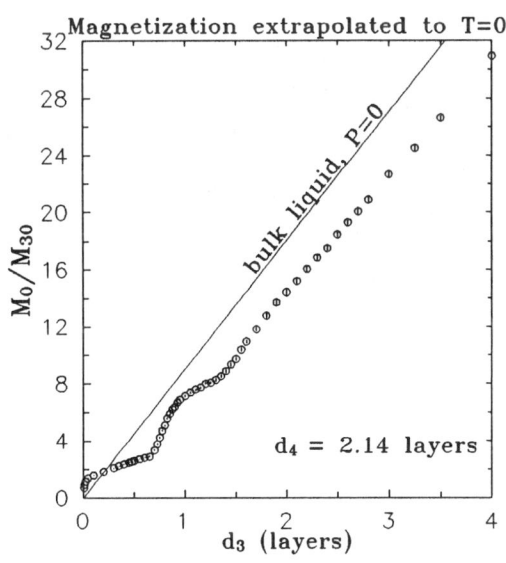

Fig. 1 Magnetization extrapolated to T=0 as a function of the ^{3}He coverage, at a fixed ^{4}He coverage of 2.14 layers. The solid line is the magnetization of the corresponding amount of 3-d bulk liquid ^{3}He.

For $d_3 \sim 0.75$ layers, M(T=0) doubles. At $d_3 \sim 1.5$ layers we see a second step in M(T=0). For $d_3 \geq 2$ layers the evolution of M(T=0) appears bulk-like. Over the entire d_3 range we have $T_1 \sim 1$ sec and $T_2 \sim 7$ msec, essentially constant.

The steps in $M(T=0)/M_{30}$ at $d_3 \sim 0.75$ and 1.5 layers suggest the population of higher potential energy states for the ^{3}He. Supporting this interpretation, we observe dM/dT > 0 at $d_3 \sim 0.65$ layers for temperatures near 0.2K; this positive derivative suggests that a process of thermal activation is present.

We have examined a simple two-level model[2] with the separation between levels ϵ_{12} = constant for $d_3 < 1.3$ layers. We extend the 2DFL model by summing states over both levels and fit the data under several simplifying assumptions: (1) the same interactions in both levels, (2) there is a constant ratio between F_1^s and F_0^a, (3) we require $T_F = \epsilon_{12}$ at $d_3 = 0.75$ layers and (4) there is a small distribution in ϵ_{12}, $\delta\epsilon_{12}$ brought about by the substrate. The results of a fit to the data at 40 mK are shown in figure 2. For this fit we find $\epsilon_{12} = 1.40$K and $\delta\epsilon_{12} = 55$ mK.

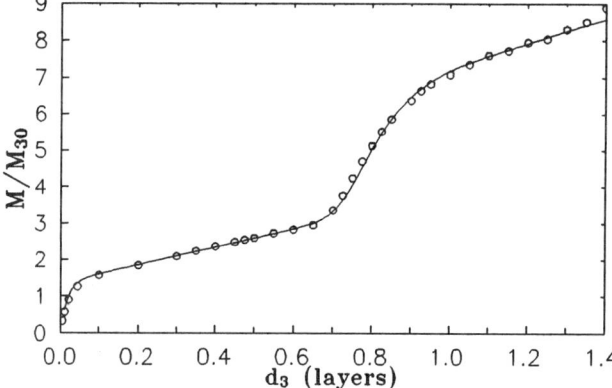

Fig. 2 Magnetization at T=40mK versus coverage of ^{3}He, fit by a simple 2-level Fermi liquid model, with gap energy distributed about 1.40K with width 0.055K.

Simultaneous measurements of the Q and frequency of third sound are collected with a third sound resonator located in the apparatus. There is evidence that the third sound Q has a minimum near 0.4 atomic layers and a maximum in the vicinity of 0.7 atomic layers; the frequency is smooth. Earlier data[3] taken with a different resonator for $d_4 = 5.6$ atomic layers also showed substantial structure in the Q with local maxima in the Q at $d_3 \sim 0.7$, 1.4 atomic layers. The steps in the magnetization appear to be directly connected to the observed structure in the Q.

It is apparent that thin ^{3}He-^{4}He mixture films represent fertile ground for the study of two (or more) level Fermi systems in two dimensions.

This work was supported by the NSF through DMR 85-17939 and DMR 88-20517.

REFERENCES

1. J.M. Valles Jr., R.H. Higley, B.R. Johnson and R.B. Hallock, Phys. Rev. Lett. **60**, 428(1988).
2. R.H. Higley, D.T. Sprague and R.B. Hallock, Bull. Am. Phys. Soc. **34**, 678(1989).
3. F.M. Ellis and R.B. Hallock, Phys. Rev. B **29**, 497(1984).

QUASIPARTICLE INTERACTION OF ^{3}He IMPURITITES ON ^{4}He FILMS

M. Saarela
Department of Theoretical Physics
University of Oulu
90570 Oulu 57
Finland

E. Krotscheck and J. L. Epstein
Center for Theoretical Physics
Texas A&M University
College Station
Texas 77843

At low temperatures, ^{3}He atoms adsorbed to the surface of liquid ^{4}He film form a dilute quasi-two-dimensional Fermi system. If the ^{4}He film is more than a monolayer thick ^{3}He impurities "float" on top of the ^{4}He film. Experimentally ^{3}He density can be varied over a wide density range which makes this system a very interesting Fermi fluid[1]. We have formulated a microscopic theory for the quasiparticle interaction between ^{3}He atoms on the surface of the ^{4}He film based on the variational Jastrow-type wave function[2].

$$\Psi^{II}_{A+2}(\vec{r}_1,\vec{r}_2,\vec{r}^B_1,\ldots,\vec{r}^B_A) = exp\frac{1}{2}\{u^I_1(\vec{r}_1) + u^I_1(\vec{r}_2) + u^{II}_2(\vec{r}_1,\vec{r}_2) + \sum_{1\leq i\leq A}[u^{IB}_2(\vec{r}_1,\vec{r}^B_i) + u^{IB}_2(\vec{r}_2,\vec{r}^B_i)]\}\Psi_A(\vec{r}^B_1,\ldots,\vec{r}^B_A)\Phi(1,2), \quad (1)$$

where $u_1(\vec{r})$ and $u_2(\vec{r}_i,\vec{r}_j)$ are the one- and two-particle correlation functions that describe the spatial structure and the short range correlation of the system, $\Psi_A(\vec{r}^B_1,\ldots,\vec{r}^B_A)$ is the wave function of the ^{4}He film, and $\Phi(1,2)$ is a two-particle Slater determinant which takes care of the antisymmetry of two ^{3}He particles.

We consider a ^{4}He film adsorped on a substrate which is translationally invariant in the x-y plane. In this geometry, all two-body quantities depend on the distances z_i of both particles from the substrate, and their separation $\vec{r}_\parallel$ parallel to the surface. We assume that the ^{3}He impurities occupy the lowest state with respect to their motion perpendicular to the surface. The single- particle part of the impurity wave function is then $exp[\frac{1}{2}u_1(z)+i\vec{q}_\parallel\cdot\vec{r}_\parallel]|\sigma>$, where $\vec{q}_\parallel$ is the momentum of the particle parallel to the surface. Since the z dependence of the single-particle wave function is the same for all impurities, $\Phi(1,2)$ needs to contain the spin degrees of freedom of the ^{3}He impurities and the motion parallel to the surface.

The correlation energy between two impurity particles is

$$\Delta E = E^{II}_{A+2} - 2E^I_{A+1} + E_A \quad (2)$$

where E^{II}_{A+2}, E^I_{A+1}, and E_A are the total energies of the systems containing two, one, and zero impurities, respectively. Using the hypernetted-chain approximation, one derives the following expression for the energy normalized to unit surface area.

$$\Delta E(1,2) = \int dz_1 dz_2 d^2 r_\parallel \rho^I_1(z_1)\rho^I_1(z_2)V_{eff}(z_1,z_2,r_\parallel)\{1 - \delta_{\sigma_1,\sigma_2}exp[i(\vec{q}_1-\vec{q}_2)\cdot\vec{r}_\parallel]\}, \quad (3)$$

with the quasiparticle interaction

$$V_{eff}(\vec{r}_1,\vec{r}_2) = g(\vec{r}_1,\vec{r}_2)[v(|\vec{r}_1-\vec{r}_2|) + w_I(\vec{r}_1,\vec{r}_2)] + \frac{\hbar^2}{2m_3}\{|\nabla_1[g(\vec{r}_1,\vec{r}_2)]^{1/2}|^2 + |\nabla_2[g(\vec{r}_1,\vec{r}_2)]^{1/2}|^2\}. \quad (4)$$

The one-body densities $\rho^I_1(z_i)$ are normalized to unity, $g(\vec{r}_1,\vec{r}_2)$ is the impurity-impurity distribution function, m_3 their mass, and $v(r)$ the two particle potential[3]. The "induced interaction" $w_I(\vec{r}_1,\vec{r}_2)$ describes the interaction that is induced by the exchange of density fluctuations through the ^{4}He background film.

The distribution function $g(\vec{r}_1, \vec{r}_2)$ is obtained by minimizing the energy $\Delta E(1,2)$. The technique for deriving the Euler equation and the "induced potential" $w_I(\vec{r}_1, \vec{r}_2)$ is described in ref. 4. The solution of the Euler equation determines the quasiparticle interaction V_{eff}. Knowing V_{eff} we can calculate the Fermi-liquid parameters[5] F_1^s, F_0^a, and the magnetic susceptibility $\chi(0)$.

$$\chi(0)/\chi_{30} = (m_H/m_3)(1 + F_1^s/2)/(1 + F_0^a). \qquad (5)$$

It is given in units of the magnetic susceptibility χ_{30} of the two-dimensional free Fermi gas. The only remaining unknown quantity in this equation is the "hydrodynamic mass" for which we used the experimentally determined value $m_H = 1.26 m_3$. The ^{3}He coverage dependence of the magnetic susceptibility, shown in Fig. 1. is related to the the momentum dependence of the quasiparticle interaction because the phase space integration involves integrals over the quasiparticle interaction ranging from momentum q=0 to q=2k_F. Increasing ^{3}He densities probe then higher momenta of the quasiparticle interaction. Our theoretical curves agree reasonably well with recent measurements[1].

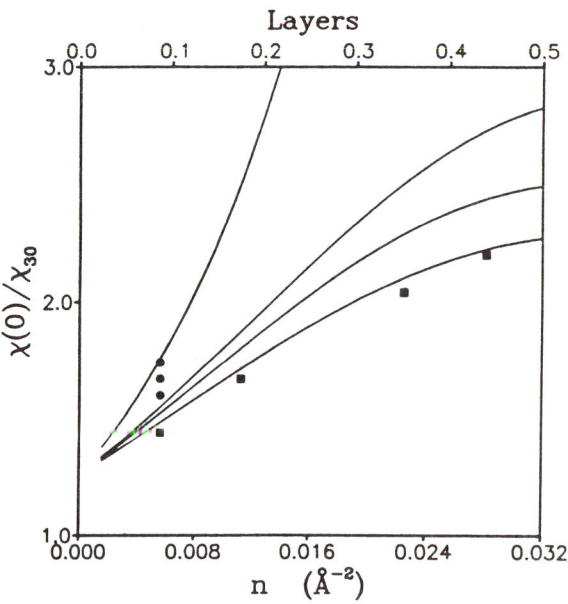

FIG. 1. The magnetic susceptibility of the ^{3}He film as a function of areal density n in atom/Å^2 and coverage in layers. The uppermost curve corresponds to a ^{4}He coverage of 0.15 atom/Å^2, and the lowest one to the largest coverage of 0.30 atom/Å^2. The squares are the experimental results of Ref. 1 for a ^{4}He film of about 34 Å thick. The circles are susceptibility data of Ref. 1 for a ^{3}He density of 0.088 layers on ^{4}He films of 10, 12.2, and 17.1 Å thick.

REFERENCES

1. J. M. Valles, Jr., R. H. Higley, R. B. Johnson, and R. B. Hallock, Phys. Rev. Lett. **60**, 428 (1988).
2. E. Krotscheck, M. Saarela, and J. L. Epstein, Phys. Rev. Lett. **61**,1728,(1988).
3. R. A. Aziz, V. P. S. Nain, J. C. Carley, W. L. Taylor, and G. T. McConville, J. Chem. Phys. **70**, 4330 (1979).
4. E. Krotscheck, M. Saarela, and J. L. Epstein, Phys. Rev. B **38**, 111 (1988), and references therein.
5. J. Bardeen, G. Baym, and D. Pines, Phys. Rev. **156**,207 (1967).

POLARIZED QUANTUM SYSTEMS

CHAIRMAN

Maurice Chapellier
Laboratoire de Physique des Solides
Université de Paris-Sud

KINETIC PHENOMENA IN SPIN-POLARIZED QUANTUM SYSTEMS

A.E.Meyerovich
Department of Physics, Northwestern University, Evanston, IL 60208

Abstract

I present a brief review of recent theoretical and experimental achievements concerning the kinetics of spin-polarized quantum systems. Recently serious attention has been paid to generalized and more accurate schemes of derivation of kinetic equations for such systems. Within different approaches the exchange and non-local effects, virial corrections and diagrammatic methods were studied in detail. The first calculations of transverse relaxation time responsible for attenuation of spin waves have also been performed. The new experimental results on transport phenomena are in reasonable agreement with theoretical predictions. The future studies of spin-polarized quantum systems are hampered by the lack of adequate and concise description of the particles' interaction with the walls such as boundary slip effects and magnetic relaxation. The latter is especially interesting and important at low temperatures because of possible magnetic ordering in the boundary ^{3}He layers.

1. Introduction

The considerable progress has been achieved recently in theoretical and experimental studies of spin-polarized quantum systems. Below I give a brief review of these new developments, and try to outline the remaining problems which seem to me to be the most intriguing. Some of the data are summarized in my reviews (Meyerovich, 1987; 1989a) and in the Proc.3-d Int. Conf. on Spin-Polarized Quantum Systems (Torino, Italy, July 1988).

Mostly I will consider kinetics of different ^{3}He↑ systems though some of the ideas and results can be applied to the spin-polarized atomic hydrogen as well (the most important kinetic studies for H↑ and D↑ cover the depolarization and recombination processes and do not concern the polarization changes in transport and collective phenomena which I am mostly interested in). I will discuss both the longitudinal kinetics (transport) and transverse efffects (spin waves). What is more, now it is possible to discuss the kinetic peculiarities of spin-polarized systems in a more general context of kinetics in systems with arbitrary internal degrees of freedom. Some very recent results on kinetic equation for such systems provide one with more general understanding of the situation and show the interrelations between spin-polarized and other systems.

In the next Section I discuss new achievements in the general kinetic description of polarized systems. Then, in Sec.3, I proceed with more concrete recent theoretical and experimental results. In the concluding section (Sec.4) I list some unsettled problems which I believe to be rather important.

2. Spin-Polarized and Other Systems of Particles with Internal Degrees of Freedom

There is an important question. Is there anything very special about spin-polarized quantum systems that makes them different from systems of particles with other (non-spin) internal degrees of freedom? The general theory of such systems seems to be very well developed at least for dilute (gaseous) phases. Why not to substitute general commutation relations for operators corresponding to arbitrary internal degrees of freedom by commutation rules for spin operators and to introduce some asymmetry in the populations of internal states (the analog of spin polarization)? In principle, it is possible and must lead to complete and correct results. Especially this should be true for dilute (gaseous) systems for which there exists thoroughly studied general Boltzmann equation called for systems with the internal degrees of freedom the Waldmann-Snider equation (see, e.g., the review by Moraal, 1975). From this point of view the only peculiarity of systems in question is their extreme quantum nature as a result of rather low temperatures which makes various macroscopic quantum effects very pronounced and easily observable.

Fortunately, the studies of spin-polarized systems were performed independently. Otherwise we probably would not learn of the existence of the most exciting effects such as spin waves in Boltzmann gases and giant growth of transport coefficients with polarization.

The reason is that the existent Boltzmann equation for particles with internal degrees of freedom (e.g., molecules; the so-called Waldmann-Snider equation) does not include the effects of quantum identity of particles. Thus one should improve the Waldmann-Snider equation in order to describe the exchange effects properly.

Such a generalization has been recently performed (Meyerovich, 1989 b). The resulting kinetic equation is too cumbersome to be given here explicitly. I will only mention, that one gets some additional (exchange) terms in both the mean field and dissipative parts of the collision operator. The corrections to dissipative terms are not very important - within the standard Waldmann-Snider approach one can always take into account these corrections by proper symmetrization of scattering probabilities in the collision integral (it is not always easy, but at least possible). But I do not know any general procedures which one may use to obtain properly symmetrized mean field terms.

The resulting equation describes all the variety of kinetic phenomena (including quantum collective ones) in Boltzmann gases of particles with arbitrary internal degrees of freedom. If the only internal degrees of freedom are the spin ones (and the interaction is purely exchange and spin independent), then this equation reproduces all known effects in Boltzmann spin-polarized gases. But this equation is more general, and provides one with reliable approaches to some new phenomena.

Non-exchange effects. Usually the studies of spin-polarized quantum gases are performed within the exchange approximation neglecting all spin non-conserving interactions. Waiving of this assumtion permits one to study the influence of weak non-exchange (dipole) effects on properties of ^{3}He↑ (for polarized hydrogen these terms may describe the influence of slow recombination on kinetics). Small non-exchange terms being accounted for perturbatively provide one with the description of the frequency dispersion effects for longitudinal transport and relaxation processes. Moreover, without the exchange approximation there is no splitting of kinetic equation into independent longitudinal and transverse parts, and one can study the coupling between transport and spin-wave phenomena.

Collective coherent effects. The possibility of propagation of different (damped) collective modes in gases of particles with arbitrary internal degrees of freedom is ensured by the existence of strong molecular field with non-zero internal frequencies Ω which may be schematically written as

$$\Omega = \mathrm{Tr}\,\{\,\hat{P}\,\hat{T}\,[\hat{f}^{(0)},\,\hat{\delta f}]\,\}\,/\,\mathrm{Tr}\,\{\,\hat{P}\,\hat{\delta f}\,\}$$

where $\hat{T}$ is the scattering T-matrix, $\hat{f}^{(0)}$ is the equilibrium Wigner distribution function (which is a matrix in the space of internal variables), $\hat{\delta f}$ describes the deviation of the desired symmetry from equilibrium, $\hat{P}$ is the projection operator which characterizes the mode in question, and [...,...] is the commutator. The frequencies Ω are non-zero when (i) $\hat{\delta f}$ is non-diagonal in internal states and (ii) $\hat{f}^{(0)}$ is not a δ-matrix, i.e. there is some asymmetry in the equilibrium populations of internal states. With non-zero Ω there is a possibility of propagation of collective modes with quadratic spectrum, $\omega - \omega_0 \sim k^2 T/m\Omega$, where T is the temperature, **k** is the corresponding wave vector, and m is the mass of particles. The situation becomes the most simple in the case of equidistant internal energy levels when

$$\omega = \omega_0 + (k^2 T/m\Omega)(1 - i/\Omega\tau)/(1 + 1/\Omega^2\tau^2),$$

where ω_0 is the distance between the levels and τ is the usual collision relaxation time (certainly, a gas of two-level particles always corresponds to this case). Only in this case the situation is completely analogous to the spin waves in spin-polarized gases.

Non-local effects. Usually one assumes the locality of interaction taking into account only the lowest order (uniform) terms in the gradient expansion of the collision integral. One may waive this assumption and consider the non-local corrections to both coherent (mean field) and non-coherent (dissipative) interaction terms. The non-locality does not lead to any striking new effects, but it is very interesting and enlightening technically. The non-local effects for dilute polarized Boltzmann gases lead (Meyerovich, 1989b) to small corrections in density to the spin waves spectrum of the order of $\hbar\Omega/T$ (Ω is proportional to a density of a gas). These terms become important when one introduces the density (virial) corrections to the Boltzmann equation (Tastevin *et al*, 1988; Laloe, 1988). Moreover, the accurate study of non-local terms permits one to compare the classical kinetic (Lhuillier and Laloe, 1982) and Fermi-liquid (Bashkin and Meyerovich, 1979, 1981; Bashkin, 1981; Meyerovich, 1985) approaches to kinetic phenomena in polarized Boltzmann gases. The kinetic equations within these approaches differ by small non-local terms. The detailed analysis demonstrated that the corresponding high-order terms within both approaches are incomplete and should be modified. Farthermore, this analysis confirmed the earlier phenomenological prediction (Meyerovich, 1983) of the existence of important large non-local contributions for dense highly polarized systems such as liquid ^{3}He↑.

There appeared some other generalizations and improvements of kinetic equation. Laloe (1988) and Tastevin *et al* (1988) corrected the Waldmann-Snider equation making it possible to introduce virial corrections and to describe exactly the short-range two-body correlations. This was achieved by introducing, instead of usual Wigner transformation of density matrices, the so-called free Wigner transform which is somewhat analogous to interaction representation in the many-body theory. As a result, it became more easy to describe exactly the two-body collision. But the price was a considerable complification of kinetic equation governing the succession of two-particle collisions and describing the dynamics of corresponding distributions. Neverthless, it became possible to reproduce the Beth-Uhlenbeck (1937) second virial correction directly from the kinetic equation thus confirming the validity and the importance of the approach. Within this approach it is also possible to study the non-local terms, but the comparison with the more traditional approach outlined above is still rather difficult. Maybe it is worthwhile to try to rewrite the final equation of Tastevin *et al* (1988) through the usual Wigner distributions and their derivatives (such a translation seems to be possible, but not very easy) in order to achieve better understanding of shortcomings of more conventional approaches.

There is another possibility of rigorous derivation of kinetic equation starting not from the master equation or BBKGY hierarchy, but exploiting one of the types of diagrammatic techniques such as, for example, Keldysh or Kadanoff-Baym formalisms. The latter approach has been used recently by Jeon and Mullin (1988) and by Ruckenstein and Levy (1989). But it is still early to make definite conclusions about possible advantages of such an approach, since the first publications only reproduced the already known results within this approach.

3. New Experimental and Theoretical Results on Transport in Spin-Polarized Systems

To an extent the situation with transport in spin-polarized bulk helium systems is now rather clear although there are only few experimental results.

In dense liquid ^{3}He there appeared first observations of change of viscosity with polarization (Kopietz *et al*, 1986; Vermeulen *et al*, 1988; Kranenburg *et al*, 1988). It is clear, that at not very high polarization $p=(N_+-N_-)/(N_++N_-)$ (N_+ and N_- are the densities of particles with up and down spins) the change in viscosity is quadratic in polarization,

$$\delta\eta/\eta = ap^2$$

Of course, at higher polarizations there should be deviations from this first term of the expansion in p. Usually in helium most of dimensionless parameters are of the order of unity. So when it turned out that, according to Kopietz *et al* (1986), the coefficient a is large and negative, $a \sim -25$, it was regarded as a sign of closeness to some possible phase transition (Bedell and Sanchez-Castro, 1986; Hess and Quader, 1987). But the later experiments gave more "normal" (and positive) values of a; $a \sim 2 - 4$ (Vermeulen *et al* 1988; Kranenburg *et al*, 1988). This last value seems to be more reliable since the experimental conditions corresponded to more equilibrium ones (especially in the case of Vermeulen *et al*,1988,when helium was polarized by the brute force technique).

The "normal" absolute value of the coefficient a makes it more difficult to present consistent quantitative theoretical description. The microscopic models do not have any definite accuracy, and all (except for those with some phase transitions) give reasonable values of a. Since up to now we do not have any consistent microscopic description of non-polarized ^{3}He, one can hardly expect reliable microscopic calculations of polarization effects.

On the other hand, the phenomenological calculations (Meyerovich, 1983; Anderson *et al*, 1987) are also not very helpful quantitatively - the phenomenological scattering probabilities of quasiparticles are unknown, and their polarization dependences cannot be extracted from the pressure dependences of transport coefficients of non-polarized ^{3}He. Qualitatively it is reasonable to believe that a should be positive and of the order of unity since the effective density of states increases and the main (s-wave) scattering channel becomes suppressed with polarization (Bashkin and Meyerovich, 1981).

The situation with dilute phases of ^{3}He↑ (^{3}He↑ -^{4}He liquid mixtures and ^{3}He gas) is very different. The large growth of transport coefficients with polarization in such phases was reliably predicted long ago for both degenerate (Bashkin and Meyerovich, 1978) and non-degenerate (Meyerovich, 1978, 1982; Lhuillier and Laloe, 1982) cases, but except for qualitative observations of Greywall and Paalanen (1981) up to now there were no experimental results. The first quantitative experiment have appeared only recently for both ^{3}He gas (Leduc *et al*, 1987) and ^{3}He↑ -^{4}He solutions (Bowley *et al*, 1988). In both experiments the observed increase of thermal conductivity and viscosity was in very good agreement with numerical calculations of Lhuillier (1983) for ^{3}He↑ gas and of Hampson et al (1988) for ^{3}He↑ -^{4}He liquid mixtures.

There are also several recent calculations of transport parameters for dilute polarized sytems. There appeared numerical calculations of transport coefficients for intermediate (between Boltzmann and degenerate) temperature region (Jeon and Mullin,1987; Hampson et al, 1988) in the s-wave scattering approximation. These variational results in the high-temperature limit coincide exactly with the first order Chapman-Enskog calculations of Lhuillier and Laloe (1982) and Meyerovich (1982), and in low-temperature limit - with variational results of Bashkin and Meyerovich (1978) and Meyerovich (1982). Such an exact fit of the results in both limits is not accidental and is due to the fact, that in all mentioned approximate approaches the trial functions were constants. In higher order approximations such exact correspondence would be rather unlikely: the Chapman-Enskog expansion and the exact solution of kinetic equation in degenerate case are based on trial functions of very different symmetries. These new calculations are applied to dilute ^{3}He↑ -^{4}He solutions.

Another recent result concerns the calculation (Mullin and Miyake, 1986) of transport coefficients in semidegenerate ^{3}He↑ - ^{4}He liquid mixtures - systems (Meyerovich, 1978, 1985), in which due to the very high polarisation p, $N_+ \gg N_-$, the spin-up component is dgenerate while the spin-down component is a Boltzmann one. Unfortunately, due to the difference in parametrizations it is rather difficult to compare the corresponding numerical results for viscosity with earlier calculations of Meyerovich (1978) for exactly the same case.

The most interesting is the situation with the so-called transverse relaxation time, $\tau_\perp$, and transverse spin diffusion coefficient, $D_\perp$, introduced by Meyerovich (1985) for exchange spin-polarized systems. The problem is very important (Meyerovich, 1985, 1987, 1989 a): though all usual continuous and spin echo NMR experiments study the relaxation and diffusion of transverse components of magnetization ($\tau_\perp$ and $D_\perp$), the standard spin diffusion calculations address only the diffusion of longitudinal component of magnetization (the diffusion coefficient D and the corresponding diffusion exchange relaxation time $\tau_{||}$). Of course, at low polarizations the transverse and longitudinal coefficients are the same. Moreover, the calculations by Lhuillier and Laloe (1982) show, that there is practically no difference between these coefficients in the Boltzmann limit when the energies of particles do not depend on polarization. However, Meyerovich (1985) qualitatively demonstrated that in degenerate case $\tau_{||} \gg \tau_\perp$, and that the ratios $D_\perp / \tau_\perp$ and $D / \tau_{||}$ being the universal functions of temperature coincide in high-temperature Boltzmann region and become more and more different with lowering temperature. It was shown, that the experimental results of Gully and Mullin (1984) on $\tau_\perp$ and $D_\perp$ are inconsistent with any calculations of $\tau_{||}$ and D near and below degeneracy temperature, but, at the same time, these data on the ratio $\tau_\perp / D_\perp$ perfectly coincide with the theoretical temperature dependence of $\tau_\perp / D_\perp$ (Meyerovich, 1985) for all temperatures. Now the situation became more clear since there appeared direct kinetic calculations for $\tau_\perp$ by Jeon and Mullin (1988), McHale (1988), which confirmed the prediction $\tau_{||} \gg \tau_\perp$ (Meyerovich, 1985) for low temperatures. It is worth mentioning, that there are some plans (Nunes et al, 1988) of measuring for the first time the longitudinal diffusion coefficient D directly. If these attempts succeed, the picture will be complete.

It seems that nearly all possible analytical calculations for spin-polarized systems have already been done, and the remaining problems should be approached numerically or, at least, seminumerically. One of the few exceptions concerns the region where the chemical potential is nearly zero, and all the characteristics may be expanded in powers of this small chemical potential (Kumar, 1988).

The last problem I want to mention here is the problem of (spin) pressure diffusion. In usual hydrodynamic limit the flows (or currents) are proportional to the driving forces - gradients of macroscopic variables. In unpolarized (and one-component) systems there are two currents (mass and heat flows) and two driving forces. For spin-polarized sytems there is an additional flow (spin diffusion current) and an additional driving force - the gradient of polarization. Of course, the matrix of diffusion coefficients relating the currents to the driving forces contains not only the diagonal coefficients but also the off-diagonal ones. Thus the spin current contains the term proportional to the gradient of pressure P. This term corresponds to the pressure diffusion in ordinary binary mixtures, and in this context may be called the spin pressure diffusion current (Meyerovich, 1982, 1983). Usually the calculation of pressure diffusion is quite trivial since the diffusion current is proportional to the gradient of chemical potential, and the contribution to the (spin) diffusion current proportional to ∇P is equal to $D\ [(\partial \mu/\partial P)/(\partial \mu/\partial p)]\nabla P$. However, it is known (Zhdanov et al, 1962) that the calculations of pressure diffusion currents must include the terms proportional to the second spatial derivatives of mass velocity - due to the Navier-Stokes law these derivatives are of the same order as the first gradient of pressure in the stationary state. Therefore, the corresponding terms make the contribution to the (spin) pressure diffusion current of the same order as usual pressure diffusion terms thus leading to considerable "viscous" renormalization of pressure diffusion coefficient. Such calculations were performed by Ivanova and Meyerovich (1988) for different types of spin-polarized quantum systems. It turned out, that for all dilute binary (or polarized) systems the (spin) pressure diffusion ratio is equal to

$$k_{sp} = P\left[\frac{1}{m_+}\frac{\partial \mu_+}{\partial P} - \frac{1}{m_-}\frac{\partial \mu_-}{\partial P} - \frac{\eta_+}{m_+ N_+ \eta} + \frac{\eta_-}{m_- N_- \eta}\right] / \left[\frac{1}{m_+}\frac{\partial \mu_+}{\partial p} - \frac{1}{m_-}\frac{\partial \mu_-}{\partial p}\right]$$

where the indices + and - correspond to the components with up and down spins, m_+ and m_- are the (effective) masses of (quasi)particles, and η_+, η_- are the partial viscosities of (spin) components. The final expressions for k_{sp} through the particles' thermodynamic functions and cross-sections can be obtained from this equation in a rather straightforward manner. Note, that above the partial viscosities represent some sort of auxiliary quantities; they may be observed as real independent physical quantities only in high-frequency experiments studying, for example, the sound absorption in spin-polarized ^{3}He↑ systems.

Some progress has also been achieved in the study of spin waves in ^{3}He↑ systems. The most interesting development was the confirmation by Ishimoto et al (1987) of the existence of some definite ^{3}He concentration (between 3% and 5%) at which Ω being proportional to $F^{(a)}_0 - F^{(a)}_1/3$ is equal to zero ($F^{(a)}_i$ are the harmonics of the antisymmetric part of the Fermi liquid function). At lower concentrations of ^{3}He in ^{3}He-^{4}He solutions the internal frequency $\Omega > 0$, and at higher concentrations $\Omega < 0$. The presence of such a point could also be deduced from the previous observations by Owers-Bradley et al (1984) when Ω was negative at high ^{3}He concentrations (all the low concentration data lead to positive values of $F^{(a)}_0$ and

Ω). When Ω is equal to zero, the propagation of spin waves with quadratic spectrum becomes impossible (see Sec.2), and the spin wave spectrum obtains a very different structure (Bedell, 1989).

4. Remaining Problems

There are still many unsolved problems most of which concern dense phases such as liquid ^{3}He$\uparrow$and concentrated ^{3}He$\uparrow$-^{4}He solutions. But the real progress in this area is hindered not by some complicated peculiarities of polarized systems, but by the lack of comprehensive non-model many-body theory for dense quantum systems. On the other hand, the phenomenological approaches to these systems seem to be rather exhausted. Of course, since experimentally the field is not yet studied thoroughly enough, there is always a possibility of unexpected developments.

I would rather make comments on two other problems. The first concerns one of the most important current problems of ultralow temperature physics - the problem of superfluidity of ^{3}He in ^{3}He-^{4}He mixtures. At present it is the only one remaining system with considerable orbital entropy at ultralow temperatures. Therfore the solutions (along with other nuclear magnetic systems) inevitably will play more and more important role in future ultralow-temperature developments. The problem is to estimate reliably the superfluid transition temperature for ^{3}He in solutions. The existing estimates vary from much less than 10^{-3} mK up to about 1 mK. The reasons are very simple. Within all reasonable theories the pairing temperature is exponentially small in ^{3}He concentration. Even a very small uncertainty in the exponent leads to a large scatter (sometimes in orders of magnitude) for T_c. The scatter in the exponent is caused either by choice of different models or by uncertainity in interaction parameters. Note, that the reasonable description of transport and/or thermodynamic properties within some models does not mean that such models provide reasonable predictions for T_c. Below I illustrate this on the example of application of the usual BCS formula for solutions with 3% ^{3}He. In the Table I give the values of scattering length (extracted from different experimental data) and the corresponding values of T_c.

Data	Scattering length, A	T_c, mK
viscosity; Fisk and Hall,1972	- 0.75	0.03
thermal conductivity; Abel *et al*,1967	- 0.83	0.06
spin diffusion; Anderson *et al*, 1966	- 0.54	0.0015
Murdock *et al*, 1981	- 0.52	0.001
$\Omega\tau$ /D ; Gully and Mullin,1984	- 0.5 --- - 0.7	0.0006 --- 0.02

What can be done to improve the accuracy? The problem for theory is to estimate the deviations from BCS formula at these concentrations, and the problem for experimentalists is to eliminate uncertainty in the interaction parameter (scattering length). From my point of view, the best way to achieve it is to perform more accurate spin echo experiments in highly polarized ^{3}He$\uparrow$-^{4}He mixtures - such an experiment provides one simultaneously with the data on $D_\perp$,$\Omega\tau_\perp$ and their ratio $D_\perp$

$/\Omega\tau_\perp$ which being the universal function of temperature (see above) gives the most accurate and model-independent data on scattering length. Up to now there were no signs of the superfuid transition in the ^{3}He subsystem down to the temperatures below 0.2 mK (the latest attempt has been made by Ishimoto *et al*, 1987).

There is an additional difficulty in observation of this superfluid transition. The superfluid transition with s-wave pairing may occur only if T_c is large enough in comparison with $\hbar/\tau^*$, where τ^* is the spin non-conserving collision relaxation time. The ^{3}He - ^{3}He magnetic nuclear dipole-dipole relaxation time is too long, and such processes are not dangerous. But at ultralow temperatures the (quasi)particles' mean free paths are very large, and the collisions with the walls become important. As a result, if T_c is of the order of several μK or lower, the collisions with the walls may prevent pairing. At such temperatures one should coat the walls by some non-magnetic materials (e.g., H_2) though such coating certainly does not improve the magnetic and temperature equilibration.

Unfortunately, there are no comprehensive theories and experimental data to obtain unambiguos values and low-temperature dependencies of τ^* for quasiparticles collisions with the walls. I feel that at present the absence of reasonable theory of magnetic relaxation at the walls is one of the major shortcomings. It hinders both theoretical and experimental progress in studies of polarized systems at ultralow temperatures. As a result, we practically do not understand the depolarization and some other processes, and cannot interpret and predict several important experiments. Of course, the wall relaxation strongly depends on the structure and the material of the wall, but still it is possible to understand some general features. Below I try to indicate some problems and outline some of the approaches.

Boundary slip. In most of experiments with spin-polarized systems one deals not with equilibrium, but stationary states of systems often in the presence of strong gradients (e.g., of temperature) and corresponding flows. In the presence of strong gradients the usual hydrodynamic boundary conditions (zero tangential velocity on the wall) are insufficient, and even small slip effects being multiplied by large gradients lead to considerable renormalization of system parameters. In this case one must use slip boundary conditions and to introduce a matrix of slip coefficients (in our case, 3x3; see, e.g., Ivanova and Meyerovich, 1988) relating three boundary currents (mass, heat and magnetic moment) to three driving forces - gradients of chemical potential, temperature, and tangential velocity. Standard stationary experimental situations correspond to zero mass flow, small known magnetization flow (slow depolarization), and known temperature gradient. Thus using the matrix of slip coefficients one can find the heat flow, and polarization and pressure gradients. At present the values of most slip coefficients are unknown. For calculations at low temperatures one may use the approach developed by Jensen *et al* (1980) and Onsager relations between off-diagonal coefficients derived by Ivanova and Meyerovich (1988).

Magnetic relaxation on the walls. This problem is crucial for ultralow temperature studies of polarized ^{3}He↑ systems. One cannot make any reliable predictions without information on the time of longitudinal spin relaxation on the walls. The problem is complicated by the presence of peculiar solidified and/or liquid helium boundary layers (in the presence of ^{4}He these layers consist mostly of ^{4}He, but even a small admixture of ^{3}He in the boundary layers has a very large influence on the wall relaxation). Therefore the magnetic relaxation on the walls is a two-step process - exchange interaction of ^{3}He from the bulk with ^{3}He particles in the boundary layers, and the interaction of ^{3}He boundary particles with magnetic subsystem of substrate.

The latter processes depend on the stucture of the wall and may correspond to magnetic dipole interaction with conduction electrons (metals) and paramagnetic atoms, or to exchange interaction with some nuclei (e.g., ^{19}F). These processes may be also affected by possible diffusion of ^3He into the wall. As a result, this step in depolarization differs from case to case.

The depolarization of bulk $^3He\uparrow$ via the exchange interaction with helium particles in the boundary layers is a more general and strongly temperature dependent process. One of the origins of such dependence is the T^{-2} dependence of mean free paths and spin diffusion coefficients in normal dense $^3He\uparrow$ or $^3He\uparrow$-^4He; in superfluid ^3He this factor may be masked by superfluid spin currents. Another temperature dependent factor is the (quasi)particle cross-section by the wall in the channel with the change of spin state. This cross-section is analogous to that of depolarization cross-section for polarized neutrons by heavy nuclei, and depends on the structure of localized ^3He levels near the wall. If there are resonance or quasi-resonance levels near the Fermi energy, then depolarization goes through the formation of long-lived quasi-bound state (analog of compound nuclei), and the depolarization probability in the scattering act is nearly 1/2. On the other hand, if there are no qasi-bound states near the Fermi energy, then the depolarization probability is much smaller, especially for ^3He-^4He solutions where the probability of a purely specular elastic reflection is practically equal to 1 due to the properties of one-dimensional scattering of low energy particles.

The situation becomes very different below the temperature of a possible magnetic phase transition in ^3He boundary layers. This transition may take place for many different substrates, and below this transition ^3He is probably ferromagnetic. Below the transition the depolarization rate will begin to decrease exponentially as $exp(-J/T)$ (J is the characteristic magnetic interaction energy in the boundary layers) down to the the temperatures about $T \sim \hbar J^{1/2}/l\, m^{1/2}$ when this exponetial factor reaches the value $exp[-l\,(mJ)^{1/2}/\hbar]$ corresponding to tunneling of inverse spin through the ferromagnetically ordered boundary layer of the thickness l. In this case the depolarization process is somewhat analogous to that considered for magnetically ordered ^3He liquid-solid interface (Meyerovich and Spivak, 1981).

Summarizing, I want to emphasize, that the wall depolarization is characterized by several unknown parameters, and our real understanding of the situation will improve only after systematic experiments on depolarization time.

References

Abel W.R., Johnson R.T, Wheatley J.C., and Zimmerman W. 1967,
  Phys.Rev.Lett.,**18**,737
Anderson A.C., Edwards D.O., Roach W.R., Sarwinski R.E., and Wheatley J.C.
  1966,Phys.Rev.Lett., **17**,367
Anderson R.H., Pethick C.J., and Quader K.F. 1987, Phys.Rev. B, **37**, 1620
Bashkin E.P. 1981, JETP Lett., **33**,8
  -- 1987, Sov.Phys.JETP,**66**,482
Bashkin E.P., and Meyerovich A.E. 1978, Sov.Phys.JETP,**47**,992
  -- 1979, Sov.Phys.JETP,**50**,196
  -- 1981, Adv.Phys., **30**, 1
Bedell K.S. 1989, Phys.Rev.Lett., **62**, 167
Bedell K.S., and Sanchez-Castro C.1986, Phys.Rev.Lett., **57**, 854
Beth E., and Uhlenbeck G.E. 1937, Physica, **4**, 915
Bowley R.M., Owers-Bradley J., and Main.P.C. 1988, in: Proc.3d Int.Conf. on

Spin-Polarized Quantum Systems (July 1988, Torino, Italy)
Fisk D.J., and Hall H.E. 1972, Proc.LT-13, eds.K.D.Trimmerhaus,
   W.J.O'Sullivan, E.F.Hammel, Plenum, N.Y., v.1
Greywall D.S., and Paalanen M.A. 1981, Phys.Rev.Lett.,**46**,1292
Gully W.J., and Mullin W.J. 1984, Phys.Rev.Lett., **52**,1810
Hampson T.M.M., Bowley R.M., Brugel D., and McHale G. 1988, J.Low
   Temp.Phys, **73**, 333
Hess D.W., Quader K.F. 1987, Phys.Rev. B, **36**, 756
Ishimoto H., Fukuyama H., Nishida N., Miura Y., Takano Y., Fukuda T.,
   Tazaki T., and Ogawa S. 1987, Phys.Rev.Lett., **59**,904
Ivanova K., and Meyerovich A.E. 1988, J.Low Temp.Phys, **72**,461
Jensen H.H., Smith H., Wolfle P., Nagai K., and Bisgaard T.M. 1980, J.Low
   Temp.Phys.,**41**,473
Jeon J.W., and Mullin W.J. 1987, J.low temp.Phys., **67**, 421
-- 1988, J.Phys.(Paris),**49**, 1691
Kopietz P., Dutta A., and Archie C.N. 1986, Phys.Rev.Lett., **57**, 1231
Kranenburg C.C., Wiegers S.A., Roobol L.P., van de Haar P.G., Johemsen R.,
   and Frossati G. 1988, Phys.Rev.Lett., **61**, 1372
Kumar P. 1988, privite communication
Leduc M., Nacher P.J., Betts D.S., Daniels J.M., Tastevin G., and Laloe F.
   1987,Europhys.Lett.,**4**,59
Laloe F. 1988, On Snider Equation, preprint
Lhuillier C. 1983, J.Phys.(Paris),**44**,1
Lhuillier C., and Laloe F. 1982, J.Phys.(Paris),**43**,197;225
McHale G. in: Proc.3d Int.Conf. on Spin-Polarized Quantum Systems (July 1988,
   Torino, Italy)
Meyerovich A.E. 1978, Phys.Lett. **A69**,279
-- 1982, J.Low Temp.Phys.,**47**,271
-- 1983, J.Low Temp.Phys.,**53**,487
-- 1985, Phys.Lett. **A107**,177
-- 1987 in: Prog.Low Temp.Phys. (ed.D.F.Brewer, North Holland,
   Amsterdam), v.11, p.p. 1-73
-- 1989 a, in: Anomalous Phases of ^{3}He (eds. W.P.Halperin,
   L.P.Pitaevski, North Holland, Amsterdam)
-- 1989 b, Phys.Rev. B, in print
Meyerovich A.E., and Spivak B.Z. 1981, JETP Lett.,**34**,551
Moraal H. 1975, Phys.Rep.,**17**,225
Mullin W.J., an Miyake K. 1986, J.Low Temp.Phys., **63**, 479
Murdock E.S., Mountfield K.R., and Corruccini L.R. 1981,
   J.Low Temp.Phys.,**31**,581
Nunes G.,Jr., Jin C., and Lee D.M. 1988, in: Proc.3-d Int.Conf. on Spin-Polarized
   Quantum Systems (July 1988, Torino, Italy)
Owers-Bradley J.R., Chocholacs H, Mueller R.M., Buchal Ch., Kubota M., and
   Pobell F. 1984, Phys.Rev.Lett., **51**, 2120
Proc. 3-d Int. Conf. on Spin-Polarized Quantum Systems. 1989 (July 1988, Torino,
   Italy)
Ruckenstein A.E., and Levy L.P. 1989, Phys.Rev.B, **39**, 183
Tastevin G., Nacher P.J., and Laloe F. 1988, in: Proc. 3-d Int.Conf. on
   Spin-Polarized Quantum Systems (July 1988, Torino, Italy)
Vermeulen G.A., Schuhl A., Rasmussen F.B., Joffrin J., Frossati G., and
   Chapellier M. 1988, Phys.Rev.Lett., **60**, 2315
Zhdanov V., Kagan Yu., and Sazykin A. 1962, Sov.Phys.-JETP,**15**,596

NUMERICAL STUDY OF NONLINEAR SPIN WAVES IN A FERMI LIQUID

D. Candela
University of Massachusetts, Amherst MA 01003

The Problem. Leggett[1] derived the following equation for the space-dependent Magnetization **M(r)** in a degenerate Fermi liquid:

$$\partial \mathbf{M}/\partial t = \gamma \mathbf{M} \times \mathbf{H}(\mathbf{r}) - \nabla_i \mathbf{J}_i, \qquad (1a)$$

where the spin-current tensor is

$$\mathbf{J}_i = -[D_0/(1+\mu^2 M^2)]\{\nabla_i \mathbf{M} + \mu(\mathbf{M} \times \nabla_i \mathbf{M}) + \mu^2(\mathbf{M} \cdot \nabla_i \mathbf{M})\mathbf{M}\}. \qquad (1b)$$

D_0 is the spin-diffusion coefficient, and the dimensionless quantity μM becomes much larger than one in the collisionless (low temperature, high field) limit. The same equation applies to a spin-polarized but non-degenerate quantum gas.[2]

Although it depends on several approximations, equation (1) has been well verified in two limits: 1. For an infinite medium, the response of the system to a ϕ-τ-π pulse train results in a reduced spin-echo amplitude (the Leggett-Rice effect); this has been seen experimentally.[3] 2. For very small tipping angles of **M** away from its equilibrium direction, (1) reduces to the Schrödinger equation; thus spin waves are predicted and they also have been observed.[4]

Experiments have not yet addressed the validity of (1) in situations where **M(r)** is tipped far from its equilibrium direction in a spatially varying manner. The predictions of equation (1) in this highly nonlinear situation are not easily derived.

Nonlinear Spin-Wave Theory. Lévy[5] has applied methods developed for nonlinear wave equations to this problem. In the zero-temperature limit, he finds the direction l of **M** obeys the "Heisenberg ferromagnet equation",

$$\partial \mathit{l}/\partial t = \gamma \mathit{l} \times \mathbf{H}(\mathbf{r}) - \mu M D_\perp \mathit{l} \times \nabla^2 \mathit{l} \qquad (2)$$

where $D_\perp = D_0/(1+\mu^2 M^2)$. This equation was previously known to have soliton solutions for an unbounded situation, and cnoidal-wave solutions for a periodic or bounded sample (cnoidal waves are described by the Jacobian elliptic function cn(x,k)). Lévy addressed the general bounded-sample problem, which has an infinite hierarchy of quasiperiodic solutions. These solutions are strictly valid only at zero temperatures and with a perfectly uniform magnetic field **H**.

If the components of l are used to form the complex number $w = (l_x + i l_y)/(1+l_z)$, then the effect of non-zero temperature ($\mu M < \infty$) may be included yielding[5]

$$\partial w/\partial t = -i\gamma H(\mathbf{r})w - iD\{\nabla^2 w - 2w^*(\nabla w)^2/(1+w^*w)\} \qquad (3)$$

where $D = D_0/(1-i\mu M)$ is a complex diffusion coefficient.

Numerical Simulations. We intend to test these ideas using a dilute ^{3}He-^{4}He solution in an 8-10T field at low temperature. As a guide to the experiments we have numerically integrated Equation (3) in one space dimension. The simulation is checked against the exact theory for zero temperature and field gradient and special initial conditions. The temperature, field gradient, and initial conditions are varied to match achievable experimental situations.

These simulations were carried out on a small desktop computer (IBM-AT type) using a finite-difference method with between 100 and 200 space points. Stable implicit methods (e.g. Crank-Nicholson) do not easily generalize to the nonlinear case, so a simple explicit method which is not strictly stable is used. The instability to rapid spatial oscillations does not cause a problem over the time scale of the simulations provided sufficiently small time and space steps are used. However, the nonlinear problem requires much more computer time than the corresponding linear problem.

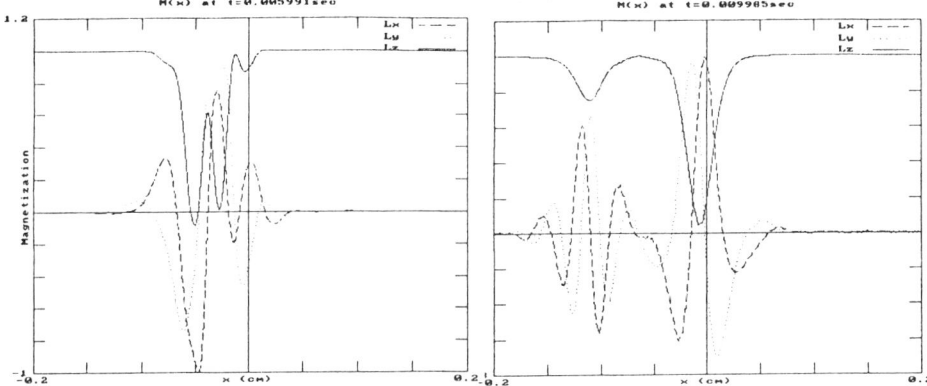

Fig. 1. This simulation shows "soliton" behavior: two localized traveling disturbances pass through each other unscathed despite the complex and highly nonlinear interaction at the collision point.

REFERENCES

1. A.J. Leggett, J. Phys. C **3**, 448 (1970).
2. E.P. Bashkin, JETP Lett **33**, 8 (1981); C. Lhuillier and F. Laloë, J. Phys. (Paris) **43**, 197, 225 (1982).
3. L.R. Corruccini, D.D. Osheroff, D.M. Lee, and R.C. Richardson, J. Low Temperature Phys. **8**, 229 (1972).
4. J.R. Owers-Bradley et al., Phys. Rev. Lett. **51**, 2120 (1983); P.J. Nacher et al., J. Phys. Lett. (Paris) **45**, L441 (1984); B.R. Johnson et al., Phys. Rev. Lett. **52**, 1508 (1984); D. Candela et al., J. Low Temperature Phys. **63**, 369 (1986).
5. L.P. Lévy, Phys. Rev. B **31**, 7077 (1985).

TRANSVERSE SPIN DIFFUSION IN POLARIZED FERMI GASES

J.W. Jeon and W.J. Mullin
Laboratory for Low Temperature Physics, University of Massachusetts,
Amherst, MA 01003, U.S.A.

ABSTRACT

We report the solution of a recently derived kinetic equation for a dilute quantum system at arbitrary degeneracy and polarization. From this equation we have developed a spin hydrodynamic equation that allows the generalization of the treatment of spin waves and the Leggett-Rice effect to polarized degenerate Fermi gases. We find a transverse spin-diffusion collision time $\tau_\perp$ that is often shorter than the corresponding longitudinal collision time, $\tau_\parallel$. As $T \to 0$, $\tau_\perp$ approaches a T-independent value, in contrast to $\tau_\parallel \sim 1/T^2$. Spin waves should thus continue to be damped even to $T = 0$ K.

DISCUSSION

Lhuillier and Laloë (LL)[1] introduced a kinetic equation, describing the time evolution of a matrix distribution function, and valid for arbitrarily spin-polarized gases. Their collision integral produces two types of terms — the usual dissipative terms that determine transport coefficients, and a reactive term which plays a role similar to that occurring in the Landau-Silin equation for a degenerate Fermi fluid. This latter feature describes what is called "identical-particle spin rotation" and is the basis for the existence of spin waves in this system. The Landau-Silin equation for Fermi liquids, first presented by Silin[2], in 1958, includes a spin-rotation term and the generalization of Landau's mean-field terms to the non-diagonal spin case. However, no form of the dissipative collision integral was presented at that time and Silin limited his discussion to "collisionless" spin waves. We have been able to derive, by Green's function methods, a kinetic equation equivalent to that of LL valid for degenerate quantum systems at arbitrary polarization, and including an explicit expression for the (matrix) collision integral. We have applied this equation to transverse spin phenomena in dilute Fermi systems to examine the effects of polarization and degeneracy on spin waves, and have also considered the generalization of the Leggett-Rice effect[3] to such situations. We find that the collision time governing transverse spin diffusion can be much smaller than that for the longitudinal case leading, for one thing, to the remarkable result that the damping of transverse spin waves becomes a constant at low temperatures, even to $T = 0$ K. These results confirm speculations made by Meyerovich in 1985.[4] The results, valid in Born approximation, are given in two recent papers[5].

We find[5] that longitudinal and transverse spin diffusion coefficients are given by the forms $D_\parallel = \alpha_\parallel \tau_\parallel$ and $D_\perp = \alpha_\perp \tau_\perp$ where $\alpha_\parallel$ and $\alpha_\perp$ are polarization dependent parameters. The spin-rotation parameter is given by $\mu = -nV(0)\tau_\perp/\hbar$, where n is the density and $V(\mathbf{p})$ is the Fourier transform of the potential. Meyerovich[4] reasoned that a relaxation approximation to the collision integral ought to contain two separate relaxation times, $\tau_\parallel$ and $\tau_\perp$, and our results verify his surmise. We have found explicit expressions for $\tau_\perp$ and $\tau_\parallel$ from variational solution of our kinetic equation. These are shown in Fig. 1. In the Boltzmann limit, the transverse and longitudinal quantities

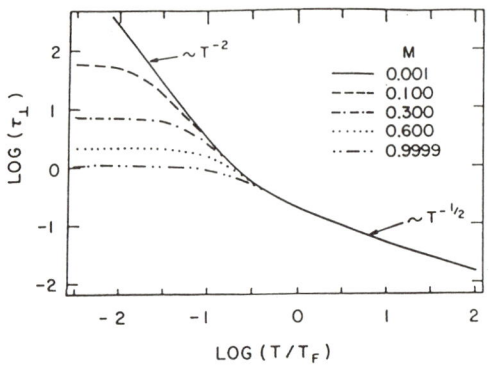

Fig. 1. Transverse spin diffusion collision time $\tau_\perp$ [in units of $A = [3/(8\pi^2 V(0) kT_F)]^2 (\hbar^7/m^{*3})$] vs. T/T_F.

coincide. In the degenerate limit, for small polarization M, $\tau_\perp$ appears to coincide with $\tau_\parallel$, and $D_\perp$ with $D_\parallel$. For larger M and $T < T_F$, $\tau_\perp$ falls below $\tau_\parallel$. In fact, as $T \to 0$, $\tau_\perp$ ultimately *always* approaches a finite limit that depends on M, although for very small M this limit may be approached only at unreachably low temperatures.

To understand our result consider a case of longitudinal spin diffusion in a polarized degenerate system. The Fermi spheres corresponding to two regions at x and x+dx are not quite the same size. The one at x has an up-spin sphere that is a little larger than that at x+dx, and the down-spin sphere at x is a little smaller that that at x+dx. Consider the diffusion of an up spin from x to x+dx. If that spin is in the narrow annular region of up spins that constitutes the difference between the two up-spin fermi spheres, it is out of equilibrium when it reaches x+dx and must scatter somehow to get equilibrated. Up spins that are farther down in the fermi sphere may not be able to move from x because their momentum states at x+dx are already occupied; or perhaps such a spin has a large wave packet so that it is really the same spin as in the momentum state at x+dx. Thus the scattering occurs just in a little layer around the fermi sphere, just as the usual Boltzmann equation tells us about longitudinal diffusion.

The case of transverse diffusion (spin echo) is quite different. The fermi spheres are the same size at x and x+dx; but they have slightly different *directions* of magnetization. Thus a spin migrating from x to x+dx in *any momentum state between the up and down fermi spheres* is out of equilibrium and must scatter to return to local equilibrium. On the other hand an up spin or a down spin in momentum state below momentum p_{F-}, the fermi momentum of the smaller down-spin sphere, comes from a region of zero magnetization into a zero magnetization region and "senses" no lack of equilibrium and does not need to scatter. Thus in the transverse case, we require scattering to restore equilibrium throughout the region between the up and down-spin fermi spheres, which our new Boltzmann equation formalism tells us. The phase space for this scattering is much larger than in the longitudinal case so $\tau_\perp$ can be smaller than $\tau_\parallel$.

This research was sponsored by the NSF through grant INT-8715042.

REFERENCES

1. C. Lhuillier and F. Laloë, J. Phys. (France) **43**, 197; 225 (1982).
2. V.P. Silin, Soviet Phys. JETP **6**, 945 (1958).
3. A.J. Leggett, J. Phys. C **3**, 448 (1970).
4. A.E. Meyerovich, Phys. Lett. **107A**, 177 (1985).
5. J.W. Jeon and W.J. Mullin, J. Phys. France **49**, 1691 (1988); preprint (1989).

VISCOSITY CHANGE IN LIQUID ³He ON POLARIZATION

G.A. Vermeulen, A. Schuhl, F.B. Rasmussen, J. Joffrin and M. Chapellier
Laboratoire de la Physique des Solides, Universite Paris–Sud, Orsay, France

ABSTRACT

The relative increase in the viscosity of liquid ³He caused by a magnetic field has been observed, for the first time at conditions close to equilibrium. At the field 10 Tesla, temperature 50 mK and pressure 3 MPa, corresponding to a polarization of 3.9 %, the observed viscosity increase was positive and of relative magnitude
0.3 ± 0.15 %.

INTRODUCTION

In the decade following the first polarization experiments on liquid ³He, very few experimental studies on properties of the strongly polarized liquid have been published. Concerning a possible viscosity change, two conflicting experiments exist, yielding results of opposite signs. Both measurements suffer from the far-from-equilibrium conditions that exist in liquid ³He when polarized by the Castaing-Nozieres method of rapid melting.

To avoid this complication, our measurement was done with liquid polarized under equilibrium conditions. The price we had to pay then, was the smallness of the polarization that we could reach under the best conditions available to us (10 Tesla, 50 mK): about 4 %.

PRINCIPLE OF THE METHOD

Under the conditions of interest, the dependence on polarisation m and temperature T for the viscosity η may be expressed as $\eta = \eta_1(1 + \alpha m^2) / T^2$, where η_1 and α are constants. With α of order unity and m only about 4 %, the requirements on reproducibility of thermometer and viscometer would be impossible to fulfill in a measurement where one simply compares η in zero field with η in a strong magnetic field. Our method employs a constant external field B_0, a vibrating wire viscometer and NMR techniques. The polarization is reduced from its equilibrium value m_0 to a smaller value m by RF irradiation, and by analyzing the subsequent viscometer response we can distinguish viscosity changes due to the inevitable temperature variations from those due to variations in polarization.

Using the model depicted in fig. 1, we have calculated the (almost instantaneous) viscometer response δQ_{sat} due to the saturating RF pulse, and the integrated change δQ_{rel} during the subsequent relaxation. Their ratio is:

$$\frac{\delta Q_{sat}}{\delta Q_{rel}} = \frac{1 + \gamma \alpha T^2 / T^{**}}{(m_0 - m)/(m_0 + m) - \gamma \alpha T^2 / T^{**}} \quad (1)$$

Here, γ is defined from the molar specific heat C_V by $C_V = \gamma RT$, T^{**} means the "magnetic temperature" (characterizing the magnetic susceptibility), and T is the temperature of the experiment. All temperature changes are assumed small compared with T, which again is considered small compared with the Fermi temperature. From a measurement of the ratio (1), temperature T, and m/m_0, we can determine α. A value of α may also be obtained from the time development of δQ, as relaxation proceeds (the result is not a simple exponential).

Data from a typical measurement are shown in figs. 2 and 3. From a number of measurements like these we derive $\alpha = +2 \pm 1$. This value is in direct conflict with a prediction based on the nearly metamagnetic model ($\alpha = -30$) and does not support paramagnon calculations, either.

© 1989 American Institute of Physics

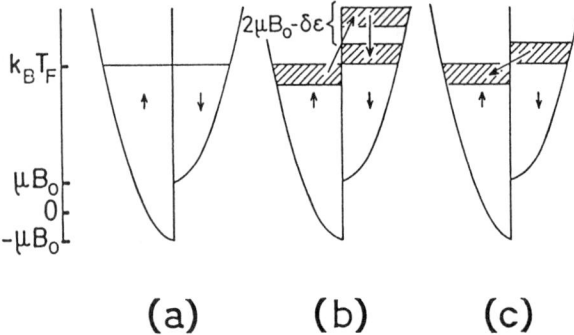

Figure 1. Coupling between the reservoirs of nuclear spin magnetic energy and quasiparticle kinetic energy in liquid ^{3}He.

(a) equilibrium in field B_0 (μ is the nuclear moment).

(b) spins in a band $\delta\varepsilon = \mu B_0 (m_0 - m)/m_0$ are excited by RF irradiation. The quasiparticles rapidly release their surplus kinetic energy as heat.

(c) the spin system relaxes back to (a) and more heat is generated from quasiparticle kinetic energy; this proces, proceeding with the spin-lattice relaxation time, is slow compared with (b). In the case of a complete saturation (m = 0), the amounts of heat released in (b) and (c) are equal.

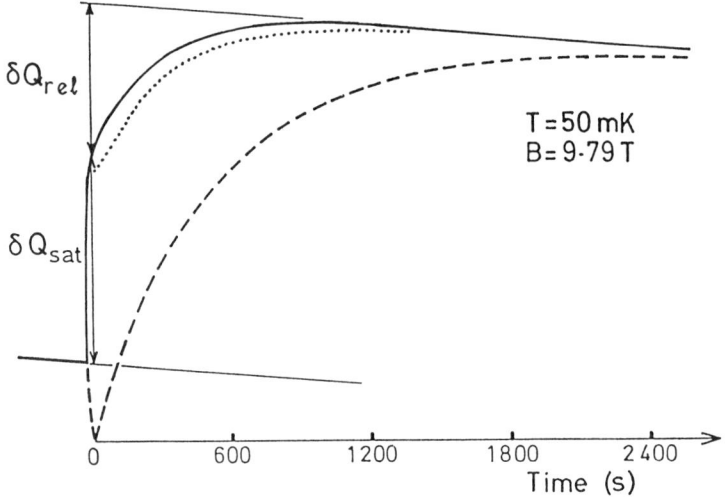

Figure 2. Viscometer quality factor (amplitude) as observed during saturation and subsequent relaxation.

Full line: recorder trace; the ratio $\delta Q_{sat}/\delta Q_{rel} = 1.36$, from which we derive $\alpha = 2$.

Dotted line: curve calculated with $\alpha = 0$.

Dashed line: curve calculated with $\alpha = -30$.

A detailed account and discussion of our work will appear shortly (*J. Low Temp. Phys.* **76**, 43 (1989)).

BOSE EINSTEIN CONDENSATION: COMPRESS OR EXPAND?

Isaac F. Silvera
Harvard University, Cambridge, MA 02138

ABSTRACT

One of the great challenges of low temperature physics is to produce spin-polarized hydrogen (H↓) in a Bose-Einstein condensed (BEC) state. This has been hampered by recombination to H_2 which both limits the density and prevents achievement of low temperatures. In this paper, six ideas which use compression and/or expansion of H↓ to achieve BEC are discussed.

INTRODUCTION

Since the development of techniques to produce spin-polarized hydrogen (H↓) as a quasi-stable gas,[1] the door has been opened to the exciting possibility of observing Bose-Einstein condensation (BEC) in a weakly interacting Bose gas.[2] The expression for the critical temperature, plotted in fig. 1, is

$$T_c = 3.31 \frac{\hbar^2}{mk_B} n^{2/3}, \qquad (1)$$

where n is the density. The initially produced density of hydrogen of 10^{14} atoms/cm^3 ($T_c=3.4\times10^{-5}$ K) was soon increased to 10^{17} cm^{-3} ($T_c=3.4\times10^{-3}$ K). These critical temperatures were well out of the range of experiment. Higher densities or lower temperatures could not be reached due to increased recombination. Subsequent experiments have employed compression of H↓ to reach densities in the 10^{18} cm^{-3} range[3,4,5] however, temperatures were limited to about 550 mK, whereas recent experiments in a magnetic trap have demonstrated cooling of hydrogen to 800 microkelvin,[6] but with densities of about 2×10^{12} cm^{-3}. In both cases conditions were far from BEC as indicated in fig. 1. In this article I would like to discuss six methods that experimental groups are currently considering to achieve BEC. Before doing so it is useful to consider some of the important properties of spin-polarized hydrogen.

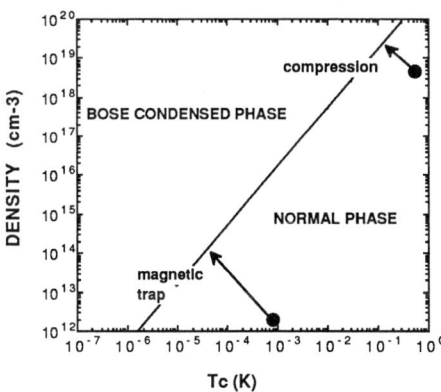

Fig. 1. The phase boundary line for BEC. The solid circles show how close current experiments have approached this line.

II. REVIEW OF PROPERTIES OF H↓

A single hydrogen atom in the ground 1s state has four spin states corresponding to the various couplings of the electron and nuclear spins, as shown in fig. 2. In a strong magnetic field the two lower states labelled a and b have their electron spins oriented preponderantly against the field. The b-state is a pure spin state, whereas the a-state has a small admixture of the reversed spin, due to the hyperfine interaction. Atoms in both of these states can lower their energies by moving to stronger magnetic fields; these atoms are called high field seekers. The c and d-state atoms have their spins parallel to the field and are low field seekers. In a gas of hydrogen, intrinsic transitions between the hyperfine levels can take place due to magnetic dipole-dipole interactions or spin-exchange.

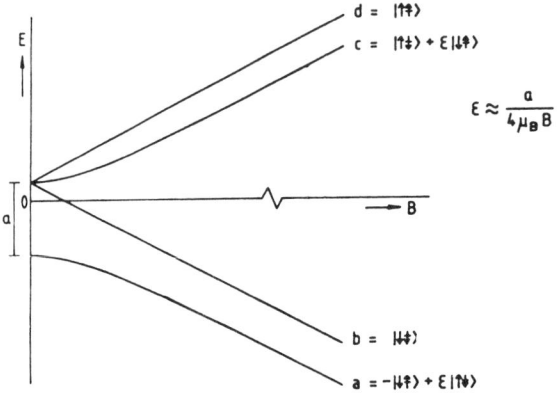

Fig. 2. The hyperfine diagram of hydrogen as a function of magnetic field. a is the hyperfine interaction parameter ($H = a\vec{i} \cdot \vec{s}$). The states are shown in terms of the projection of the electron spin m_s (↓) and the nuclear spin m_i (↓).

When a gas of spin-polarized hydrogen is in the presence of a surface, atoms will adsorb on the surface due to the attractive surface potential. If this potential is too strong, at low temperatures the density of hydrogen on the surface will become so large that the atoms will rapidly recombine to H_2. By covering surfaces with a thin film of helium the surface adsorption is strongly inhibited. The relationship between the surface coverage σ and the density n is

$$\sigma = n\lambda \exp(\varepsilon/k_B T), \qquad (2)$$

where $\lambda = (2\pi\hbar^2/mk_B T)^{1/2}$ is the thermal deBroglie wavelength and ε is the binding energy of an H atom to the surface, with $\varepsilon/k_B \approx 1$ K for ^{4}He and about 0.35K for ^{3}He. This expression is only valid in the low-density high temperature limit; near the conditions for BEC the surface saturates at a significantly large value of σ.[2] The atoms on the He surface form an almost perfect 2-dim gas and it is believed that they will exhibit a Kosterlitz-Thouless transition to a superfluid state.[2]

The interaction of isolated hydrogen atoms with the helium surface is interesting and important. First we note that low energy hydrogen atoms will not penetrate or dissolve into helium. When an atom scatters off of the surface, there are three possible channels: 1) specular reflection; 2) inelastic, non-sticking reflection; and 3) sticking. It has been found experimentally that at low temperature, channel 2 is insignificant and that channel 3 has the sticking probability[7]

$$S = 0.3T \qquad (3)$$

where T is the gas temperature in degrees kelvin, so that most scattering is specular. Thus energy exhange with the helium takes place by an atom sticking and thermalizing with the surface or an atom desorbing into the gas.

In the gas phase hydrogen recombines at a density reducing rate

$$\frac{dn}{dt} = -K_v^{(3)} n^3, \qquad (4)$$

since three bodies are required in a collision to conserve energy and momentum. On the surface the rate is

$$\frac{d\sigma}{dt} = -K_s^{(2)} \sigma^2 - K_s^{(3)} \sigma^3. \qquad (5)$$

The two-body term is allowed since the helium atoms in the surface can act to conserve the energy and momentum. These expressions are grossly simplified and one should actually write down rate equations for each of the hyperfine states.[2]

When two atoms of hydrogen recombine they release $D_0/k_B = 52{,}000K$ of energy. As a consequence, recombination generates heat at a rate:

$$\dot{Q} = (\dot{n} \times \text{volume} + \dot{\sigma} \times \text{area}) D_0/2. \qquad (6)$$

Let us consider the obstacles which have been in the path of BEC until now. The highest densities of hydrogen have been achieved by compressing tiny bubbles (diameter $\gtrsim 150$ microns) up to about 5×10^{18} cm^{-3}. However at this density the recombination energy heats the gas, preventing achievement of low temperatures. Indeed the gas can get hot enough (T $\gtrsim$ 1K) for the reversed electron spin states to be populated. This leads to faster recombination and ultimately the explosion of the bubble. If an attempt is made to achieve BEC at low densities, substantially lower temperatures are required. However, this leads to population of the surface states (cf. eq. 2) so that the surface recombination becomes important. The resulting heating prevents achievement of the required low temperatures. For this reason there have been substantial efforts in developing magnetic traps to isolate the atoms from the surface.

III. SIX PATHS TO BEC

In the past few years several ideas have emerged which may well open the path to BEC in hydrogen. These proposals use some form of compression or expansion or combinations of both. In the following we shall briefly discuss all of these ideas, referring the reader to the original literature, when it exists, for details.

A. Bubble compression

The technique of compressing a hydrogen bubble submerged in a sea of helium was developed by Sprik et al[8] and exploited by several groups. The idea is to prepare a sample of hydrogen gas with the atoms in the b-state (this state is a pure spin state and is much more stable than the a-state or mixtures[2]). The gas is enclosed in a dome shaped chamber above a sea of liquid helium as shown in fig. 3. By raising the level of the helium the H gets compressed, forming a sessile bubble in the dome. The bubble's volume can be measured by capacitance techniques. As the atoms decay the volume shrinks. Measurement of this decay rate can be used to identify the decay process as $\dot{n}_b = K^{(3)}_{bbb} n_b^3$, with a like term for the decay of the surface density. Here n_b is the density of b-state atoms and $K^{(3)}_{bbb}$ is the three-body dipole-dipole-dipole recombination rate constant. As mentioned above, due to the rapid recombination, high density bubbles can overheat and can be terminated by an explosion. Earlier experiments were performed in magnetic fields up to 10 T. Gillaspy et al[9] extended these studies to 20 T to see if the high fields would reduce the constant $K^{(3)}_{bbb}$. They found that $K^{(3)}_{bbb}$ rose and fell sharply at 20 T. Increasing the field to 25 T will probably reduce it further. The rate constant for the surface atoms has not yet been measured.

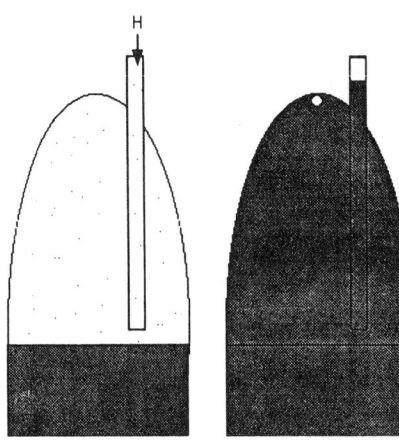

Fig. 3. Schematic of a bubble compression experiment. On the left, a cell is filled with H↓ with the helium level down. On the right, the level is raised to compress the H↓ into a bubble in the dome of the cell.

All together one might expect a decrease of 8-100 in the decay rate by going to high fields. This is far from adequate to reduce the thermal gradients to observe BEC.

Silvera et al[10] have shown that the thermal gradient in a bubble is proportional the square of the radius, R. By studying bubbles with a diameter of order 10 microns, the gradients could be reduced by several hundred from those of earlier work, which might bring one into the range of BEC. However a few problems (and advantages) arise. The equation of state of a bubble is

$$P = P_h + 2\gamma/R \qquad (7)$$

where P is the pressure of the internal gas, P_h is the hydrostatic pressure of the helium, and $2\gamma/R$ is the surface tension pressure of the bubble with surface tension γ. We consider a pressure $P=P_c$, the critical pressure for BEC, which corresponds to n_c. Since for a given P_c we want the smallest radius, we choose $P_h=0$. Larger hydrostatic pressures result in larger values of R for this critical pressure. For a diameter of 10 microns the density is about 7×10^{19} cm^{-3}. The recombination rate will be very high and the lifetime of the bubble very short. It would be advantageous to work at lower densities and with lower decay rates. Silvera et al[10] show that one can use negative hydrostatic pressures in helium to reduce the density by a factor of three for the same radius, and thereby reduce the heating by a factor of 27. A number of other advantages of this geometry which may further reduce the heating are discussed and it is shown that BEC should be observable in these tiny bubbles.

B. Compression with a potential gradient

Kagan and Shlyapnikov[11] (KS) have proposed a clever geometry and compression method for observing BEC. Hydrogen can interact with external potentials due to magnetic ($\vec{B}$) or electric ($\vec{E}$) fields: $U = -\vec{\mu}\cdot\vec{B}$ or $-\frac{1}{2}\alpha E^2$, where $\vec{\mu}$ is the magnetic dipole moment and α is the atomic polarizability. If the density of hydrogen is n_0 for a value U_0, then for a value U

$$n = n_0 e^{(U-U_0)/k_B T} \qquad (8)$$

The resulting compression n/n_0 can be very large for $(U-U_0) \gg k_B T$. KS suggest using a sharp needle (magnetic or conducting) with radius of curvature of order 1 micron to achieve large gradients. The needle would be situated in a large volume filled with atoms, compressing a small number to a high density in a "potential bag" in the vicinity of the needle as shown in fig. 4. The bag and large volume would be in quasi-thermodynamic equilibrium. For modest densities in the large volume the density in the bag can rise above n_c. This is accompanied by a sharp increase in the recombination rate, which can be used to detect BEC. The advantage of this scheme is that the large volume continues to feed atoms into the bag so that BEC is sustained for a relatively long period of time, and that part of the recombination heat is dissipated in the large volume. A prime difficulty is that to obtain large gradients, a needle with a radius of curvature of order 1 micron is required. Such a needle may have difficulty carrying away that

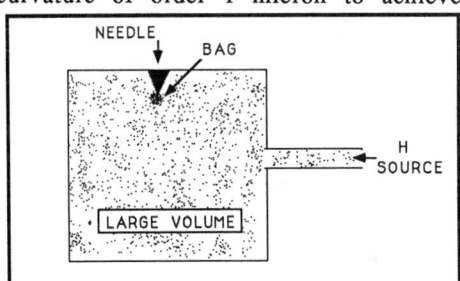

Fig. 4. Schematic of the potential gradient compression experiment.

recombination heat and the helium could be "burned" off of its surface so that it rapidly recombines the hydrogen.[1,2]

C. The static magnetic trap

An alternate route to BEC is to isolate the atoms from the helium walls with a magnetic trap to inhibit wall recombination and then to cool the atoms in the trap to achieve BEC. Ideally one would like to use a magnetic field maximum and trap the a or b-state atoms shown in fig. 2. Unfortunately in free space a magnetic field cannot have a maximum, however a minimum is allowed.[13] Hess[14] proposed to use a magnetic field minimum and trap the c and d-state atoms in the trap shown in fig. 5.

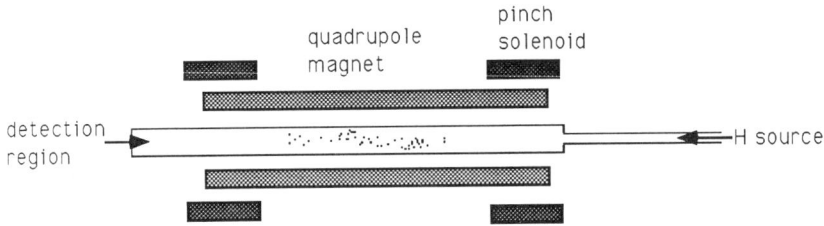

Fig. 5. Schematic of the static magnetic trap. Atoms are detected by lowering the pinch field so that the gas can move to the detection region.

The trap consists of a magnetic quadrupole field plus two pinch fields which define the potential well. The trap is limited to low densities (implying low T_c) since the c and d-state atoms (which fill it) have density dependent relaxation rates to the a and b-states which are ejected from the trap (anti-trap states). At a density of about 10^{13} cm^{-3} the decay rate becomes manageable and the remaining trapped atoms are in the d-state. When a hot d-state atom in the trap escapes, or evaporates, it reduces the energy of the remaining atoms, which subsequently thermalize to a lower temperature. This evaporative cooling can be continued by lowering the field of the pinch coils to get to lower and lower temperatures. Such a trap has been realized in the laboratory by Hess et al[15] as well as by van Roijen et al.[16] More recently Doyle et al[6] have evaporatively cooled to 800 microkelvin, but the densities were in the 10^{12} cm^{-3} region, a few orders of magnitude away from the critical density.

One of the inherent difficulties in this trap is that it currently starts the cooling at a low initial density. Tommila[17] has suggested that it may not be possible to cross the BEC line starting with such initial conditions. The basic problem is that as the temperature of the trapped atoms are lowered the collision rate for thermalization, $\Gamma_{th} \propto$ velocity $\propto \sqrt{T}$ decreases until the spin-flip rate, which is independent of T at low temperatures, dominates.[14] Since the spin-flip rate is more rapid for higher density, it is the losses of the <u>lowest energy</u> atoms in the middle of the trap which dominate and result in evaporative <u>heating</u>. The cooling limit is estimated to be about 30 microkelvin. Another important problem which needs theoretical attention is the question of the time required to develop a zero-momentum component, i.e. will the gas thermalize into a Bose distribution if the conditions are otherwise proper for BEC?

We note that the static trap uses expansion to cool the gas, whereas the field gradient compresses the atoms to the field minimum.

D. The microwave trap

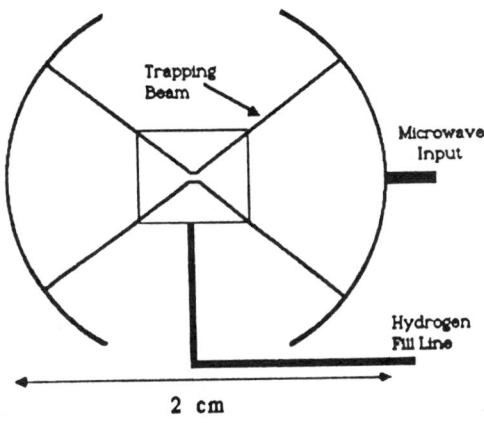

Fig. 6. Schematic of a microwave trap using a concentric resonator to produce an absolute maximum in the microwave field.

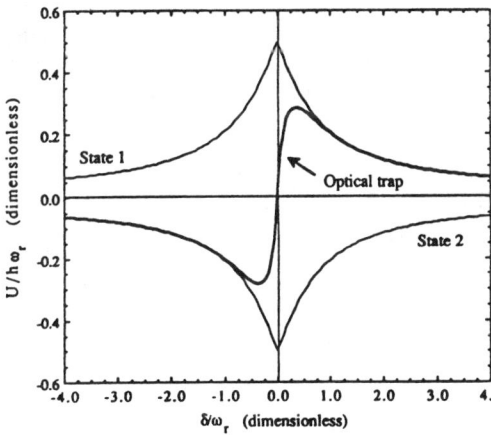

Fig. 7. The microwave trap potential as a function of detuning for states |1> and |2>. The potential for an optical trap is also shown for comparison. Due to the rapid spontaneous emission the laser trap depth is zero at $\delta=0$.

An ac magnetic field can have a maximum in free space. Agosta et al[18] have proposed a microwave trap resonant for the hyperfine states (b and c-states). A possible geometry for such a trap is shown in fig. 6 in which a microwave field is sharply focussed in a cavity to yield a high field strength with a maximum. By simulataneously diagonalizing the atomic states and the radiation field, one finds the dressed atom states

|1> = $\cos\theta$ |b,N-1> + $\sin\theta$|c,N>   (9a)
|2> = − $\sin\theta$ |b,N-1> + $\cos\theta$ |c,N>.   (9b)

Here N is the number of photons in the field and $\tan 2\theta = \omega_r/\delta$; δ is the detuning from resonance, $\omega - \omega_0$, where ω_0 is the resonance frequency and ω is the microwave frequency, and the rabi frequency ω_r is proportional to the microwave field. States |2> and |1> are trap and anti-trap states, respectively, which have the potentials shown in fig 7. We note that exactly on resonance, |2> is an equal admixture of b and c-states. Due to the c-admixture, collisions can result in relaxation to state |1> and ejection. By detuning to negative values of δ, the c-admixture in state |2> is reduced and the state becomes rather stable. The advantage of the microwave trap is that it should be possible to fill it to high densities, limited by recombination. The principal difficulty is technical in that microwave fields of several hundred gauss at $\nu \sim 50$ Ghz are required to have a reasonable trap depth, relative to the initial temperature of the gas of H↓.

The microwave field gradient compresses the atoms, whereas evaporative cooling can be realized by detuning to lower the walls of the trap.

E. Laser cooling in a trap

In the past several years cooling and trapping of atoms with lasers has been extensively demonstrated.[19] In order to cool H to achieve BEC, a tuneable laser source at 1216 Å is required to excite the 1s-2p transition (Lyman-alpha). Such lasers are currently becoming available as pulsed sources, using non-linear harmonic and mixing techniques. However, the power levels are very low (~0.01 microwatt) and the pulse repetition rate is very slow (10-20 hz) so that there is no possibilty of trapping the atoms with the laser. Thus, although the laser could be used to cool, if the gas were confined in a chamber with helium covered walls then the heat exchange with the walls would render the cooling insignificant. There are possibilities of increasing the laser power by many orders of magnitude in the near future.[20]

Hijmans et al[21] have proposed using a weak Lyman-alpha laser source to cool atoms in a static magnetic trap. The quantum limit of cooling with a laser for a two-level system is

$$T_{min} = \Gamma/2 \qquad (9)$$

where Γ is the spontaneous decay rate. For the 1s-2p transition T_{min} is 2.2 mK. In their scheme the trapped gas would first be cooled with the laser and subsequently be evaporatively cooled to lower temperatures. The advantage is that initial densities for evaporative cooling are much higher which may overcome the concerns for the existing static trap. Detection of the atomic density and temperature could be achieved by probing the distribution with the laser and analyzing the spontaneous emission radiation.

F. Trap-free laser cooling of H↓ to BEC

With technological advances one may expect to have Lyman-alpha sources available in the not-too-distant future with power levels in the milliwatt regime. Although these power levels would not be sufficient to trap a gas of H↓, I would like to present a heretofore unpublished proposal for Bose condensing hydrogen in the lower hyperfine states at relatively high densities.

Let us first consider the order of magnitude of laser power that is required (it is more convenient to consider the number of photons/sec: 6×10^{17} photons/sec=1 watt). For $T_c = T_{min} = 2.2$ mK, $n_c = 5.1 \times 10^{16}$ cm^{-3}. If one starts with a gas at T=100 mK, then simple considerations show that it requires about 15 photons to reduce the average momentum of an atom to the average at 2.2 mK. We shall show that roughly 10^{14} atoms are sufficient to attain the critical density, so that a minimum of order 1.5×10^{15} photons are required in a short period of time (of order 1 s).

The basic idea is to start with a tube of atoms in an inhomogeneous magnetic field, as shown in fig. 8. For initial temperatures such that $\mu B/k_B T \gtrsim 1$ the density is rather uniform. Energy is then removed from the atoms with the laser so that they compress towards the field maximum. The atoms collide with each other to thermalize as they collapse towards the center (recovering some kinetic energy due to the magnetic field gradient forces). They also collide with the walls, but, according to eq. 3, as the atoms cool off, fewer and fewer exchange energy with the walls, so that, for example at 1 mK only 3 in 10,000 wall collisions exchange energy. Thus the atoms become Kapitza resistance isolated from the walls. Indeed if the gas Bose condenses, then the condensate fraction does not exchange energy with the walls at all, as has been pointed out by Kagan and Shlyapnikov[22].

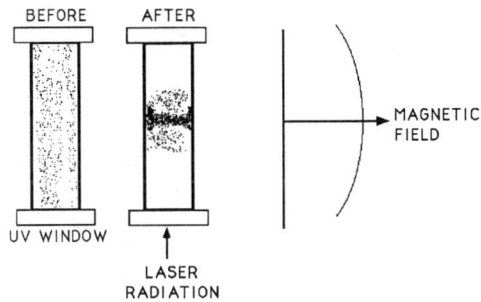

Fig. 8. Laser compression experiment. The static axial magnetic field profile is shown on the right. On the left is the cell before laser cooling, showing the rather uniform density distribution. In the middle panel, the cell is shown after laser cooling, demonstrating the field compression of the condensate to the field center.

The flux of heat from the walls is given by

$$\dot{Q}_W = \left(\frac{1}{4} n \bar{v} A\right) S (T_w - T_g) k_B \qquad (10)$$

where the factor in the parenthesis is the atomic flux hitting area A, S is given in eq. 3, T_w is the wall temperature and T_g is the gas temperature. To cool the gas the laser cooling rate must be much faster than the heating rate of eq.(10).

The problem of achieving BEC in the above geometry has been analyzed in detail by Goldman et al[23] who show that $T_c \propto (N/A_0)^{1/2}$ where N is the total number of atoms in a tube of area A_0. If we use a tube with a diameter of 1 mm, then for $T_c = 5$ mK (somewhat above T_{min}) $N \approx 2.7 \times 10^{14}$. For a tube 1 cm long, the initial density at 100 mK is $2.7 \times 10^{16}/cm^3$, which rises to $6 \times 10^{17} cm^{-3}$ at the center when Bose condensed. The axial width of the condensate fraction is about 340 microns and the normal component is about 2 mm wide.

IV. CONCLUSIONS

Several promising experimental proposals for achieving BEC have been discussed along with the advantages and disadvantages. These techniques use various approaches of expanding or compressing the gas, or combinations of both. It is not easy to predict which scheme is most likely to work, as the experiments are very demanding. However it seems that one of the most attractive (and difficult) approaches is the microwave trap because of the possible high densities and low temperatures.

I would like to thank M. Reynolds for commenting on this manuscript. This research has been supported by the Department of Energy, Grant no. DE-FG02-85ER45190.

REFERENCES

1. I.F. Silvera and J.T.M. Walraven, Phys. Rev. Lett. 44, 164 (1980).
2. See for example the review by I.F. Silvera and J.T.M. Walraven, Progress in Low Temperature Physics, Vol. X, D.F. Brewer, ed. (Elsevier Science Publishers, B.V., Amsterdam, 1986), p. 139, or T.J. Greytak and D. Kleppner, New Trends in Atomic Physics, Vol. 2, Les Houches 1982, eds. G. Grynberg and R. Stora, (North Holland, Amsterdam), p. 1125.
3. R. Sprik, J.T.M. Walraven, and I.F. Silvera, Phys. Rev. B32, 5668 (1985).
4. D.A. Bell, H.F. Hess, G.P. Kochanski, S. Buchman, L. Pollack, Y.M. Xiao, D. Kleppner, and T.J. Greytak, Phys. Rev. B34, 7670 (1986).
5. T. Tommila, E. Tjukanov, M. Krusius, and S. Jaakkola, Phys. Rev. B36, 6837 (1987).
6. J.M. Doyle, J.C. Sandberg, N. Masuhara, I.A. Yu, D. Kleppner, and T.J. Greytak, preprint.
7. J.J. Berkhout, E.J. Wolters, R. van Roijen, and J.T.M. Walraven, Phys. Rev. Lett. 57, 2387 (1986).
8. R. Sprik, J.T.M. Walraven, and I.F. Silvera, Phys. Rev. Lett. 51, 479 (1983), erratum 51, 492 (1983).
9. J.D. Gillaspy, I.F. Silvera, and J.S. Brooks, Phys. Rev. B38, 9231 (1988).
10. I.F. Silvera, J.D. Gillaspy, and J.G. Brisson, Proc. Int. Conf. on Spin Polarized Quantum Systems, Torino, Italy 1988, in press.
11. Yu. Kagan and G.V. Shlyapnikov, Phys. Lett. 130A, 483 (1988).
12. S.A. Vasilyev, A.Y. Katunin, I.I. Lukasevich, and G.V. Shlyapnikov, preprint.
13. W.H. Wing, Prog. in Quant. Electr. 8, 181 (1984).
14. H.F. Hess, Phys. Rev. B34, 3476 (1986).
15. H.F. Hess, G.P. Kochanski, J.M. Doyle, N. Masuhara, D. Kleppner, and T.J. Greytak, Phys. Rev. Lett. 59, 672 (1982).
16. R. van Roijen, J.J. Berkhout, S. Jaakkola, and J.T.M. Walraven, Phys. Rev. Lett. 61, 931 (1988).
17. T. Tommila, Europhys. Lett. 2, 789 (1988).
18. C.C. Agosta, I.F. Silvera, H.T.C. Stoof, and B.J. Verhaar, Phys. Rev. Lett., 62, 2361 (1989).
19. S. Chu, J.E. Bjorkholm, A. Ashkin, and A. Cable, Phys. Rev. Lett. 57, 314 (1986); A.L. Migdall, J.V. Prodan, W.D. Phillips, T.H. Bergeman, and H.J. Metcalf, Phys. Rev. Lett. 54, 2596 (1985).
20. T. McIlrath, private communication.
21. T.W. Hijmans, O.J. Luiten, I.D. Setija, J.T.M. Walraven, Proc. Int. Conf. on Spin Polarized Quantum Systems, Torino, Italy 1988, in press.
22. Yu. Kagan and G.V. Shlyapnikov, Phys. Lett. 95A, 309 (1983).
23. V.V. Goldman, I.F. Silvera and A.J. Leggett, Phys. Rev. B24, 2870 (1981).

THE ROLE OF MAGNETIC FIELD GRADIENT IN THE SPIN WAVE SPECTRUM OF SPIN POLARIZED ATOMIC HYDROGEN

N. P. Bigelow, J. H. Freed and D. M. Lee
Cornell University, Ithaca, NY

The transport properties of spin polarized atomic hydrogen are strongly affected by identical particle quantum exchange effects. In particular these effects give rise to the observation of nuclear spin wave modes in the NMR spectrum[1]. Related effects have been observed in spin polarized ^{3}He gas as well as dilute mixtures of ^{3}He in ^{4}He[2]. The consequence of exchange effects on the transport properties of a spin polarized quantum gas have been considered theoretically by Bashkin, Lhuillier and Laloë, and Lévy and Ruckenstein[3]. These theories derive a set of coupled nonlinear differential equations which describe the time evolution of the spin current and density in the hydrodynamic limit.

For the case of small-tipping-angle pulsed NMR experiments the equations of motion can be linearized, and when expressed in a frame rotating at the Larmor frequency the result resembles a one dimensional Schrödinger equation with damping:

$$i\dot{M}_+ = \{\gamma\delta H_o + iD_o(1+i\mu P)/(1+\mu^2 P^2)\nabla^2\}M_+ \qquad (1)$$

where $M_+ = M(\mathbf{r},t) \equiv M_x + iM_y$ is the complex transverse magnetization density, D_o is the self diffusion coefficient, δH_o is the residual static field gradient, γ is the effective nuclear gyromagnetic moment, P is the sample polarization and μ is a figure of merit which relates the importance of exchange scattering as compared to diffusive momentum scattering. In these experiments we consider the case of a linear static field gradient applied along the axis of our cylindrical NMR resonator. The experimental apparatus has been discussed elsewhere [1]. In this case the solutions to equation (1) with reflecting boundary conditions at the cell walls [1] are standing spin wave modes. These modes are Airy functions bound at one extreme by the cell wall nearest the static field minimum. At the other extreme the mode is bound by the static field gradient δH_o. The eigenvalues for these modes define the mode frequencies and the eigen spectrum is given by:

$$\omega_n = -\alpha_n\{(\gamma\delta H_o)^2 D_o(i+\mu P)/(1+\mu^2 P^2)\}^{\frac{1}{3}} \qquad (2)$$

where n is the mode index and α_n is found from the zeroes of the derivative of the nth Airy function Ai. It is useful to note that these frequencies are measured relative to the Larmor frequency at the field maximum. In this solution, we neglect the perturbation of the boundary at the high side of the field gradient, with the assumption that the modes decay sufficiently rapidly beyond the field gradient boundary that any perturbation is small. This assumption is shown to be particularly true for the low index modes.

The experiments were performed by loading a sample of hydrogen and allowing the sample to achieve high nuclear polarization by preferential recombination[4]. The sample cell was regulated at 373 mK and small ($\approx 10°$) tipping angle pulses were applied at each field gradient setting. To check for systematic effects certain gradient settings were repeated at the end of the experiment. The total duration of the experiment was less than 20 minutes so that there were no significant changes in the sample density, which was about $1.2 \times 10^{16} cm^{-3}$. The NMR free induction decays were detected using a two phase heterodyne spectrometer, digitized and stored for computer analysis.

The frequencies of the modes were determined using a linear prediction with singular value decomposition technique (LPSVD)[5]. The typical fourier spectra of the free induction decays for two different gradients are shown in figure 1. Because the exact Larmour frequency at the reflecting wall is not well defined, we considered only frequency shifts of the higher modes relative to the first mode. The splittings of the three modes immediately following the first (most prominent mode) are shown in the upper the three curves in figure 2. A direct consequence of the dispersion relation (eqn. 2) is that, if the relative splittings of the modes are divided by the differences in the coefficients $(\alpha_n - \alpha_m)$, the resulting

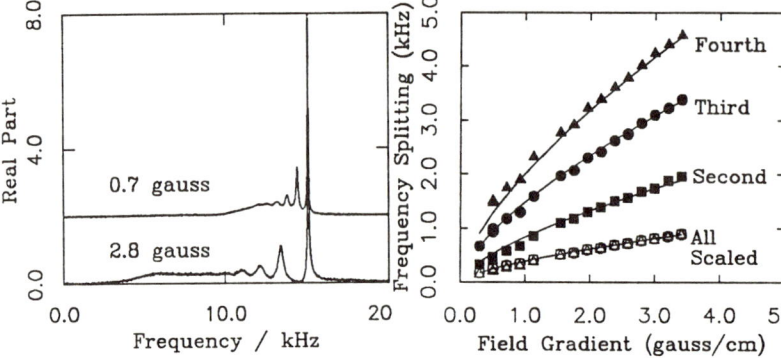

Figure 1. Typical nuclear-spin-wave spectrum for two different static magnetic field gradients. The gradient is applied along the axis of the cylindrical sample cell/NMR resonator and the inhomogeneously broadened background lineshape reflects the rectangular geometry in this direction.

Figure 2. Frequency splitting of the second, third and fourth spin wave modes relative to the first mode. The solid lines are the theoretical predictions for the splittings of simple Airy function modes. The lowest curve is the result of scaling the individual splittings to a mode independent splitting as described in the text. The solid line is the same scaling of the analytic solution.

number is mode independent. The result of this scaling is shown in the lowest curve of figure 2. The collapse of these three upper curves onto a single line demonstrates the validity of our analytic solution of the theory and of the accuracy of the experimental conditions. As a further check, the ratios of the splittings for different gradients were calculated. The result is a direct measurement of the ratios of the field gradients. When compared to the inhomogeneously broadened lineshape the results were found to agree within 8%. We would like to point out that these results are independent of values of μ and D_o. The measurement of these values is presented elsewhere[6]. The authors would like to thank Prof. B. W. Statt for experimental assistance and useful discussions. This work was supported by the National Science Foundation under grant DMR-8616727.

REFERENCES
1. B. R. Johnson, J. S. Denker, N. Bigelow, L. P. Lévy, J. H. Freed and D. M. Lee, Phys. Rev. Lett. **52**, 1508 (1984).
2. P. J. Nacher, G. Tastevin, M. Leduc, S. B. Crampton and F. Laloe, J. Physique Lett. **45**, L-441 (1984).; W. J. Gully and W. Mullin, Phys. Rev. Lett. **52**, 1810 (1984).
3. E. P. Bashkin JETP Lett. **33**, 8 (1981).; C. Lhuillier and F. Laloë, J. de Phys. **43**, 197 and 225 (1982), L. Lévy and A. E. Ruckenstein, Phys. Rev. Lett. **52**, 1412, (1984).
4. R. W. Klein, T. J. Greytak and D. Kleppner, Phys. Rev. Lett. **47**, 1195 (1981); B. W. Statt and J. A. Berlinsky, Phys. Rev. Lett. **45**, 2105 (1980).
5. N. P. Bigelow, D. M. Lee, J. H. Freed and B. W. Statt, *Proc. Conf. on Spin Pol. Quant. Sys.*, (to be published, World Scientific).
6. N. P. Bigelow, J. H. Freed, D. M. Lee (submitted for publication).

SOLID HELIUM

CHAIRMAN

Douglas D. Osheroff
Department of Physics
Stanford University

This session was made possible, in part, through
a generous donation provided by

P.O. Box 2006 Bradenton, Florida 33508 • Tel: 813 746-3515

Since 1976

Harra Technical Sales
P.O. Box 6454
Clearwater, FL. 33518
813/797-7527

- SEMICONDUCTOR & VACUUM EQUIPMENT
- LEAK DETECTORS • QUARTZWARE CLEANERS
- MASS ANALYZERS & MASS FLOW INSTRUMENTS

MEASUREMENTS OF MAGNETIC SUSCEPTIBILITY OF hcp SOLID ^{3}He

T.Mamiya, H.Yano, T.Kato, Y.Minamide, Y.Miura, S.Inoue and T.Uchiyama
Department of Physics, Nagoya University, Chikusa-ku, Nagoya 464
Japan

ABSTRACT

We have precisely measured ac magnetic susceptibility of hcp solid ^{3}He down to tens of μK with an improved SQUID technique, and observed ferromagnetic tendency in hcp solid ^{3}He.

INTRODUCTION

The three-atom exchange interaction is deemed to be dominant in the hcp phase of solid ^{3}He because of crystal symmetry. The interaction is expected to lead to the ferromagnetic order in the nuclear spin system at ultra low temperatures. On the other hand, in the bcc phase the four-atom exchange has turned out to be dominant and an antiferromagnetic phase transition is observed. The direct exchange interaction in solid ^{3}He should be evidenced by means of the comprehensive studies on both phases.

Takano et al[1]. reported that the magnetization measured down to 43 μK on hcp solid ^{3}He showed ferromagnetic nature. We aim at measuring the magnetization in hcp phase more precisely with improved SQUID technique. We installed a SQUID probe at the bottom of the cryostat so that a flux locked loop was not exposed to the strong magnetic field for demagnetization cooling. With this configuration the magnetization measurements became possible both in static and in ac methods at ultralow temperatures. Another point is to devise a ^{3}He cell for the measurements of magnetic susceptibility at ultralow temperatures.

EXPERIMENTAL

The ^{3}He sample is cooled by means of two stage demagnetization. The first stage is composed of 2.6 mole PrNi$_5$ and the second one, 17 mole copper, which is cut from an OFHC piece. The experimental space is shown in Fig.1. The resistance ratios of the copper nuclear bundle and the silver material in the nuclear stage were 2800 and 1700, respectively. The boundary electrical resistance between a platinum piece for an NMR thermometer and a silver piece at 4.2 K becomes as low as 0.01 μΩ when screws of 5 mm in diameter are used for fastening two pieces. We applied a torque up to 2 N·m to stainless steel screws having 3 mm in diameter to assemble the silver material in the experimental space and the copper bundle. The ^{3}He sample cell is shown in Fig.2. We made a hybrid cell, which was composed of the inner material of pure silver containing sintered silver powder and the outer tube of coin silver. This device guaranteed high thermal conductivity in the inner part and negligible eddy current heating with a large penetration depth of 1 cm at 16 Hz because of the low conductive outer material. The amplitude of the magnetic field produced by the primary coil was 0.004 gauss at 16 Hz with an order of 1 pW heat leak. The unbalance of the astatic pair, 12 turns each, of the secondary coil was 0.01 for one of the pair. The noise of the ac susceptibility is $3\times10^{-4}\phi_0/\sqrt{Hz}$.

Installing the SQUID probe at the bottom of the cryostat needed some special techniques.[2] Tuning of the tank circuit was achieved when the total length of the coaxial cable between the SQUID and the rf head was 6 m, which corresponded to half a wavelength at 19 MHz. In this situation the height of the signal(triangular wave) was 60 mV peak to peak, which was compared to 120 mV in the original commercial SQUID. The noise for the dc magnetization is lower than $1\times10^{-2}\phi_0/\sqrt{Hz}$.

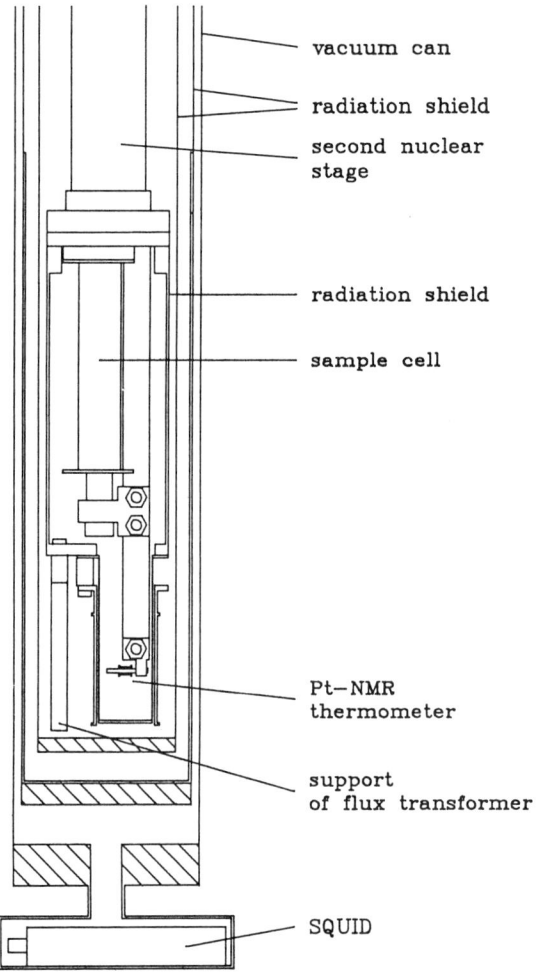

Fig. 1. Experimental space in the nuclear stage.

Flux jumps occurred sometimes. In order to get the dc magnetization, one has to connect smoothly the data before and after the jumps. Moreover dc magnetization measurements are influenced by the change of flux due to sweeping fields of various magnets. Compared to dc magnetization measurement, ac susceptibility measurement is more reliable. Ac suseptibility is immune even from the sweeping field of the demagnetization magnets because of lock-in detection. This is a marked advantage over dc measurement.

Temperatures were measured with NMR Pt thermometers which were installed at the top of the bundle and in the experimental space at the bottom of the bundle. NMR thermometers were calibrated above 0.91 mK (antiferromagnetic ordering temperature in bcc solid) against a ^{3}He melting curve thermometer, which was installed at the top of the copper bundle, based on the temperature scale of Fukuyama et al[3]. In this temperature range the two NMR thermometers and the melting curve thermometer are at the same temperature with a negligible temperature difference less than 6μK. In regards to temperature difference, the top thermometer showed 159 μK when the bottom thermometer read 126 μK. This indicates a downward heat flow along the nuclear stages. Temperatures are usually measured once an hour at 125 kHz at the lowest temperatures. A personal computer(NEC PC-8801) used for recording data generated large heat leak. We put the computer outside the shielded room and used optical GPIB links(OMRON Z3G) to connect the electronics and the computer, so that the heat leak due to the computer was reduced to undetectable level.

Demagnetization of PrNi$_5$ starting at 9 mK in 5.4 T down to 0.053T

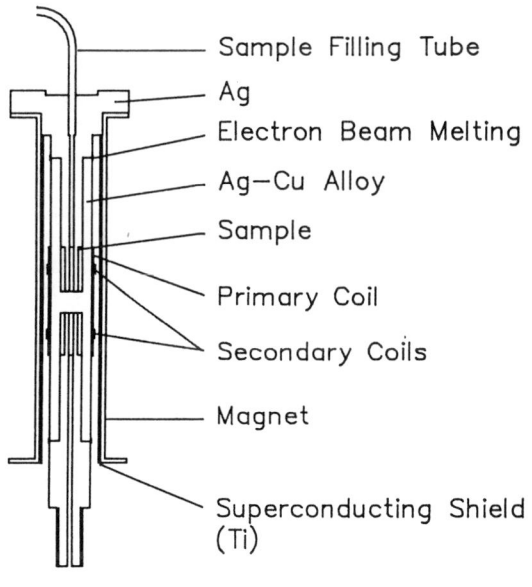

FiG. 2. Sample cell for magnetic susceptibility measurements of solid ^{3}He.

precools the effective 7 mole of Cu to 5.3 mK in 7.2 T. Demagnetization of Cu from 7.2 T to 0.10 T results in the final temperature of 70 µK. While taking data during warm up, we sometimes applied heat pulses of the order of 100 µJ and also applied heat of 2 nW in addition to natural heat leak of 1 nW using a heater wound on the upper part of the copper bundle. Thus warm up took 14 days from 70 µK to 8 mK.

hcp solid ^{3}He at a molar volume of 19.29 cm^3 was formed by the blocked capillary method. The sample was annealed 12 hours at temperature just below the solidification point of liquid to bcc solid and the bcc to hcp transformation point, respectively. Pressure equilibrium was monitored by means of a pressure gauge of the Straty and Adams type which was mounted at the top of the copper bundle. The molar volume was also determined with this pressure gauge. A magnetic field of 19 gauss for the dc magnetization measurements was trapped in a titanium superconducting cylinder during cooling down with a dilution refrigerator. Measurement of ac susceptibility was done with an impedance bridge by using the SQUID as a null detector.

RESULTS AND DISCUSSION

The relaxation time for susceptibility equilibrium after switching on and off heat pulses is a few minutes at 200 µK and 20 minutes at

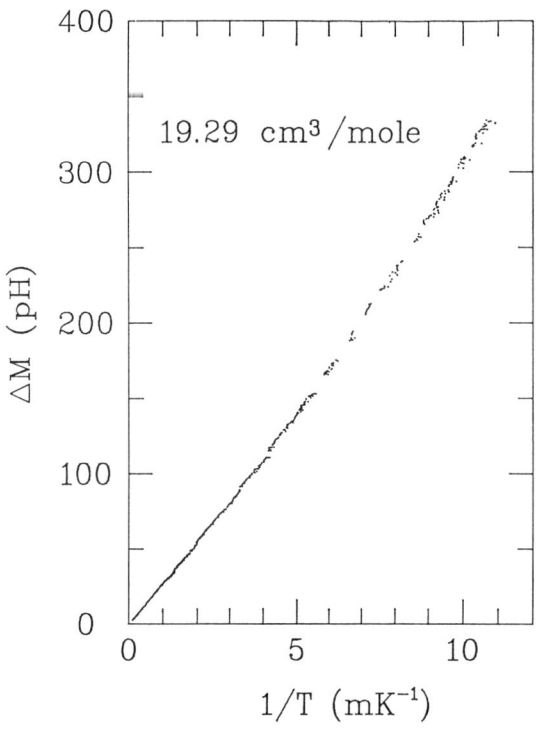

Fig. 3. Mutual inductance which is proportional to the susceptibility versus inverse temperature.

100 µK, respectively. The relaxation time is considered to come mainly from the relaxation between the cell wall material and the ^{3}He sample. The temperature lag of the sample during the continuous heating was checked to be less than 0.1 µK even at the lowest temperatures. The real part of mutual inductance of the sample, which is proportional to the susceptibility, is shown as a function of inverse temperature in Fig.3. The quantity ΔM in the ordinate indicates the change of mutual inductance from that at high temperatures. The upward curvature in this figure clearly shows a positive Curie Weiss temperature i.e. ferromagnetic property in hcp solid ^{3}He.

Next we deduce the Curie temperature. The measured mutual inductance M was fitted to the following Curie Weiss type equation:

$$\Delta M = M - M_0 = C/(T - \Theta + B/T)$$

If we take B = 0 and assume errors of 0.5 pH for M and 1 % for T, which expresses the status of the present experiment, a weighted least square fitting yields M_0 = 59.94 ± 0.04 pH, C = 26.15 ± 0.04 pH mK, and Θ = 14.08 ± 0.19 µK. Taking into account various least square fittings, with weighted and unweighted, yields Θ = 14.7 ± 0.7 µK. The positive Θ is clearly seen as the intercept of the extrapolated experimental data with T axis in Fig.4. This value is a factor two less than the previous value[1] at the corresponding molar volume. Our

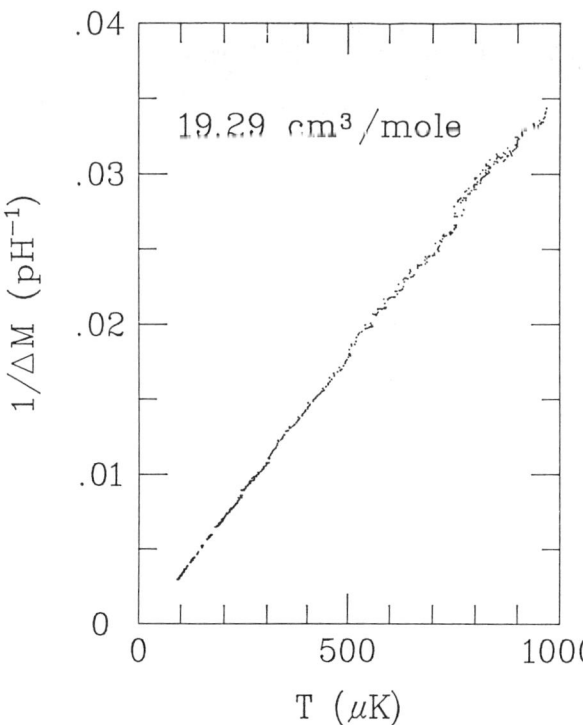

Fig. 4. Inverse mutual inductance versus temperature.

Θ was obtained with improved precision and improved thermometry. The detail will be more clearly studied since the lowest temperature attainable in the present apparatus is reduced as demagnetization is being repeated. In regards to the entire variation of mutual inductance over the present range of temperature, estimated inductance associated with susceptibility of solid ^{3}He is $2.3 \times 10^{-14}/T$ (henry) for the sample volume 0.045 cm^3. The variation of mutual inductance in the range from 100 μK to 1 mK is 210 pH. This estimate is in agreement with the experiment. Incidentally the variation of mutual inductance of the empty cell, i.e. the background is 2 pH at most over the entire temperature range. For the time being we neglect the variation of the background. The imaginary part of the mutual inductance and dc magnetization were also measured and these give results consistent with the variation of the real part of the mutual inductance. Studies for other molar volumes and at lower temperatures are going on.

We acknowledge experimental help of H.Kondo, T.Suzuki, H.Shibayama, W.Itoh, T.Kurokawa, M.Katsuki and the technical division of Institute of Plasma Physics, and are grateful to K.Iwahashi, Y.Nagaoka and K.Adachi for helpful discussion. One of us (T.M.) wishes to thank Yamada Science Foundation for financial support.

REFERENCES

1. Y.Takano, N.Nishida, Y.Miura, H.Fukuyama, H.Ishimoto, S.Ogawa, T.Hata and T.Shigi, Phys.Rev.Letters **55**, 1490 (1985)
2. T.W.Bradshaw and W.P.Pratt,Jr., Cryogenics **19**, 553 (1979)
3. H.Fukuyama, H.Ishimoto, T.Tazaki, and S.Ogawa, Phys.Rev.**B36**, 8921 (1987)

PRESSURE STUDY OF hcp SOLID ^{3}He IN MAGNETIC FIELDS

Y. Miura, S. Abe, S. Sugiyama, and T. Mamiya
Department of Physics, Nagoya University, Chikusa-ku,
Nagoya 464, Japan.

and

R. C. Richardson
Laboratory of Atomic and Solid State Physics, Cornell
University, Ithaca, New York 14853.

ABSTRACT

Pressure measurements have been made in hcp solid ^{3}He down to 0.93 mK in magnetic fields up to 0.2 T. An increase in pressure by the field is observed contrary to bcc solid, indicating ferromagnetic behavior in hcp solid ^{3}He.

INTRODUCTION

In solid ^{3}He the magnetic properties of the nuclear spin system are described in terms of the multiple-exchange interactions arising from the large zero-point motion of ^{3}He atoms. In the bcc phase, many thermodynamic and nmr experiments[1] have revealed much about the spin interactions. In the hcp solid, however, only magnetization measurements[2] have been reported at low temperatures. Thermodynamic pressure measurements would be helpful in understanding the magnetic properties of hcp ^{3}He; but the pressure due to the exchange interactions is estimated to be much smaller in hcp solid than in bcc solid. In this paper, we report the first measurements of thermodynamic pressure P(T,H) in hcp solid ^{3}He in magnetic fields H and compare the results to the model independent high-temperature expansion,

$$P(T,H)-P_0 = A/T + B/T^2$$
$$= R/8T(\partial \tilde{e}_2/\partial V) + R/2(\gamma \hbar H/2k_B T)^2(\partial \Theta/\partial V) \qquad (1)$$

where R is a gas constant, V is the molar volume, and Θ is a Weiss temperature. The results show that the dominant interactions in the hcp solid are ferromagnetic, in agreement with the result of magnetization measurements[2] and with the theory[3].

EXPERIMENTAL

The pressure of hcp solid ^{3}He was measured by means of a

capacitive strain gauge cell combined with capacitance bridge. The pressure resolution of the cell is given by[4]

$$\Delta P = 2E/3r(15P_m/s_y)^{1.5}d \cdot (\Delta C/C) \qquad (2)$$

where r, E, P_m, s_y, d, and $\Delta C/C$ denote the diaphragm radius, modulus of elasticity, the maximum pressure, the yield stress, the capacitor gap, and the relative sensitivity of the capacitance bridge. In spite of a large P_m, it is essential to reduce the resolution ΔP in the pressure measurement in hcp solid more than in the bcc solid because the exchange energy in the hcp phase is much smaller. The resolution and the stability of the capacitance bridge were improved by using a large r and large capacitor plates of the cell, and by reducing the stray capacitance in the capacitance bridge.

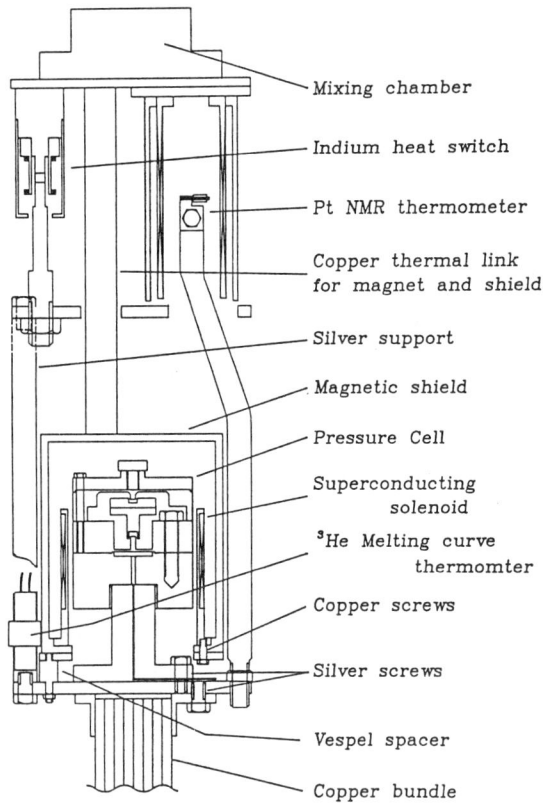

Fig. 1. Schematic drawing of apparatus used for P(T,H) measurement. The pressure gauge cell, the ^{3}He melting curve thermometer, and the copper bundle are screwed into the same silver flange for good thermal contact.

Figure 1 shows a schematic diagram of the experimental setup. The diaphragm, made of beryllium copper, is 16 mm in diameter by 1.6 mm thick, the central post which supports the lower capacitor plate is 2.0 mm in diameter. The capacitor plates, made of silver are 18 mm in diameter, and were polished to 1 μm. The upper capacitor plate is movable and is rigidly fixed at an optimum position before cooling, corresponding to a capacitor gap of 18 μm at 180 bar. At low temperatures the capacitance is 53 pF with the cell empty, and 140 pF at 180 bar. The stray capacitance of the cable connecting the cell with the null detector, which decreases the off-balance output of the bridge, is reduced to 250 pF, resulting in high capacitor resolution $\Delta C/C = 5 \times 10^{-9}$. In order to reduce the drift of the bridge a vacuum capacitor anchored to the mixing chamber of a dilution refrigerator was used for a reference capacitor, and the coaxial cables from the cell to the null detector and the ratio-transformer at room temperature were grouped together and were inserted into the vacuum line. The stray capacitance of the cables which connect the cell with the ratio-transformer was also reduced to 340 pF to stabilize the bridge. Using the values of $E = 1.3 \times 10^6$ bar, $r = 8.0$ mm, $P_m = 170$ bar, $s_y = 1.5 \times 10^4$ bar, $d = 20$ μm, and $\Delta C/C = 5 \times 10^{-9}$, Eq. (1) gives a value of $\Delta P = 1$ μbar which agrees with the experimental value at 4.2 K.

The base of the sample cell was made of the coin silver (92.5% silver, 7.5% copper) and screwed into a thick silver link for good thermal contact with refrigerant and thermometers. The cell chamber was packed with fine silver powder to provide a large surface area of 2.3 m^2 for thermal contact with the sample. The thickness of bulk ^{3}He between the silver powder and the strain gauge diaphragm was chosen to be about 24 μm at 150 bar so as to minimize the thermal relaxation time. Cooling was achieved by adiabatic demagnetization of copper nuclei. The cell temperature was determined from the melting pressure of the ^{3}He. The melting pressure gauge was calibrated against the superfluid ^{3}He-A, ^{3}He-B transition, and the magnetic transition of bcc solid[5] for every cooling run. To isolate the ^{3}He melting pressure thermometry and the platinum nmr-thermometer from the magnetic field applied to the hcp solid, the cell and the cell magnet were enclosed in magnetic shields made of iron and μ-metal which were in thermal contact with the mixing chamber.

An applied magnetic field to the empty cell was found to change the capacitance of the cell, corresponding to a pressure change of less than 170 μbar in fields up to 0.2 T at 25 mK. No temperature dependence of the capacitance of the empty cell, however, has been observed from the temperature 8 mK down to 1 mK at 0.2 T, therefore only the constant value has been subtracted from the experimental values in magnetic fields to determine the pressure due to nuclear magnetism of solid ^{3}He.

The sample of ^{3}He was slowly solidified and annealed at a temperature slightly below its melting temperature for 8 hours. After cooling down to 0.9 K in order to avoid supercooling of the bcc solid into the hcp phase region, the sample was again annealed for 8 hours at a temperature slightly below that of the bcc-hcp transformation. The molar volume of the sample was determined from the pressure of the annealed hcp solid[6] within an accuracy of ±0.01 cm^3/mol.

The pressure of the ^{3}He was measured by changing the sample cell temperature in steps. The thermal time constant τ was measured by observation of the relaxation of the ^{3}He pressure when the cell temperature was changed stepwise. τ was found to be less than 20 min. in zero field down to 0.93 mK. τ at 1 mK in 0.2 T, for example, was estimated to be 30 min. due to the spin diffusion process[7] in solid ^{3}He and 50 min. due to the Kapitza resistance[8], while a value of about 250 min. was observed. Since such a long τ was observed only in magnetic fields, it may be explained in terms of the large heat capacity and the low thermal conductivity of the massive diaphragm made of beryllium copper. In order to achieve a thermal equilibrium between the ^{3}He and the thermometer the cell temperature was held constant at least for a period of 3τ between every step.

RESULTS AND DISCUSSION

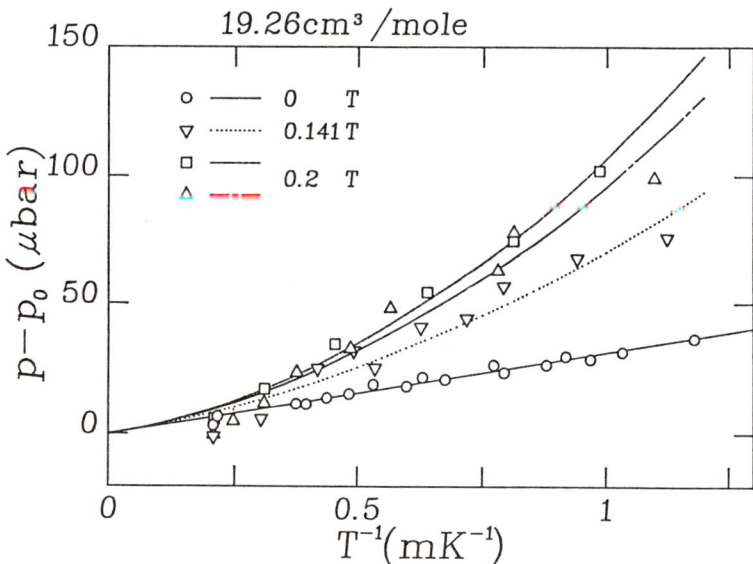

FIG. 2. Pressure differences versus T^{-1} for V=19.26 cm^3/mol in fields of 0.0, 0.14, and 0.2 T. Various symbols are for the different cooling runs. The curves represent least-squares fits to the data with Eq.(1).

Figure 2 shows the pressure difference versus T^{-1} for a molar volume of 19.26 cm^3/mol and in fields of 0.0, 0.14, and 0.2 T. The zero of the pressure P_0 was chosen so that each curve extrapolated through zero at the high temperature limit. The scatter in the data is indicative of the long-time stability of our pressure gauge. An increase in pressure due to magnetic fields is observed in hcp solid ^{3}He, indicating ferromagnetic behavior or $\partial\Theta/\partial V>0$. This is the first time the sign of $\partial\Theta/\partial V$ or Θ has been determined from non-susceptibility studies. The positive sign of Θ means, in terms of the multiple-exchange model, that the exchange among odd number of atoms, most likely three nearest-neighbor atoms, are dominant. Our result is in qualitative agreement with the recent magnetization measurement[2] and the theoretical prediction[3].

The solid and broken lines represent least-squares fitting to the data with a model-independent high temperature expansion Eq.(1), which contains three fitting parameters, A, B, and P_0. The values of B and P_0 in fields were determined using the fixed value of A obtained in zero field. Figure 3 shows that the values of B are proportional to H^2 as expected from Eq.(1). A least-squares analysis of the data shown in Fig. 2 yields the values 30±3 μbar·mK for A and (17±5)×10^2 μbar·mK^2/T^2 for B/H^2.

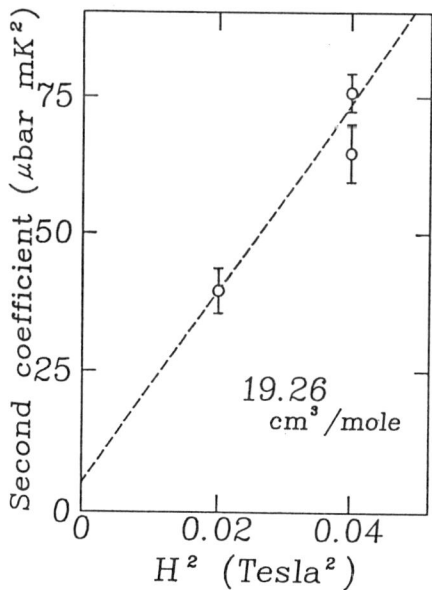

Fig. 3. Field dependence of the coefficient of the second term B in Eq.(1). The straight line represent least-squares fit to the data.

Since the values of $\tilde{e}_2$ and Θ could be obtained by integrating the fit coefficients A and B with respect to volume over a wide range in molar volume, the pressure measurement has been extended to in other molar volumes to determine the volume dependences of A and B. A volume correction should also be necessary to convert our data to constant volume results in order to obtain the quantitative results because a pressure change of 50 μbar, for example, changes the sample volume by 0.1 ppm which is not negligible in the pressure measurement in hcp solid. We found the correction factor to be 1.25 at 24.08 cm^3/mol and 1.66 at 19.26 cm^3/mol by comparing our pressure data at 24.08 cm^3/mol with the reported isochoric data[9]. This implies that the isochoric data at 19.26 cm^3/mol is equal to the factor 1.64 times the raw data shown in Fig. 2. While our pressure measurements are limited to single molar volume of 19.26 cm^3/mol, our result seems to be consistent with an effective exchange parameter of nearly 10 μK. In Fig. 2 we cannot resolve within our experimental error a deviation from the high-temperature expansion of the ferromagnetic Heisenberg Hamiltonian involving the exchanges among only nearest-neighbor atoms.

ACKNOWLEDGMENTS

We wish to thank N. Matsushima, S. Inoue, H. Shibayama, W. Itoh, and T. Kurokawa for help with the experiment. We are also grateful to Y. Nagaoka, and K. Adachi for useful discussions.

REFERENCES

1. See a review by M. Cross, Jpn. J. Appl. Phys. 26, Suppl.26-3, 1855 (1987).
2. Y. Takano, N. Nishida, Y. Miura, H. Fukuyama, H. Ishimoto, S. Ogawa, T. Hata, and T. Shigi, Phys. Rev. Lett. 55, 1490 (1985).
3. M. Roger, Phys. Rev. B30, 6432 (1984).
4. G. C. Straty and E. D. Adams, Rev. Sci. Instrum. 40, 1393 (1969).
5. D. S. Greywall, Phys. Rev. B33, 7520 (1986).
6. G. C. Straty and E. D. Adams, Phys. Rev. 150, 123 (1966). E. R. Grilly and R. L. Mills, Ann. Phys. (N.Y.) 8, 1 (1959).
7. R. D. Giffard, J. P. Stagg, W. S. Truscott, and J. Hatton, J. Low Temp. Phys. 31, 817 (1978).
8. T. Mamiya, A. Sawada, H. Fukuyama, Y. Hirao, K. Iwahashi, and Y. Masuda, Phys. Rev. Lett. 47, 1304 (1981).
9. D. G. Wildes, M. R. Freeman, J. Saunders, and R. C. Richardson, J. Low Temp. Phys. 62, 67 (1986).

CHARGE MOTION IN SOLID HELIUM

A. F. Andreev

Institute for Physical Problems
USSR Academy of Sciences, 117334 Moscow, USSR

INTRODUCTION

Although ionic motion in ^{4}He crystals has been the subject of extensive investigation (see review by Dahm[1] and later work by Golov et al.[2]) the structure and mechanism for ion motion were not understood. Many important experimental results were not explained:

1. Keshishev[3] observed two types of behavior of ion velocities in large electric fields. At low pressure (25.8 atm) the dependence of the positive ion drift velocity on electric field E was less than linear for $E > 7 \times 10^3$ v/cm. For all other cases the ion velocities were fit empirically by the relation $v \propto (E+E_0)^3$ where the parameter $E_0(T)$ is an increasing function of temperature T. The nature of this relation was not clear.

2. Golov et al.[2] report a decrease in ion velocities with further increase of electric field.

3. Results by Lau et al.[4] suggest an anisotropic mobility of positive ions characterized by different activation energies for motion along the c- axis and in the basal plane of hcp ^{4}He crystals.

4. Golov et al.[2] observed an interesting temperature dependence of ion velocities. At higher temperatures the velocities decrease exponentially. In the region of intermediate temperatures a complicated (even nonmonotonic) temperature dependence was observed. At lower temperatures velocities again decrease exponentially with a different activation energy.

This work is based mainly on that of Andreev and Savishchev[5]. The concept of inelastic scattering of delocalized vacancies by ions developed by Andreev and Meyerovich[6] is used and the above mentioned experiments are explained. A new geometrical approach to the problem of ion structures is proposed. Direct measurements of the anisotropy of ion velocities are required for experimental realization of this approach.

GEOMETRY OF ION DISPLACEMENTS AND DRIFT VELOCITY

It is generally accepted that the positive ion complex consists of a singly charged bare helium ion (He^+, He_2^+, or even He_3^+, see ref. 1) localized in the crystal lattice with an enhanced density surrounding the ion core due to electrostriction. The negative ion complex consists of a cavity and an electron localized inside. The characteristic size of the complex is of the order of the interatomic distance for ions of both signs.

The ion transport in helium crystals is caused by processes that involve displacements of the ion to neighboring localized states. The most important structure is the set of the corresponding displacement vectors $\vec{u}_i$, i=1,2,... . To explain the general situation let us consider simple examples. Let the negatively charged cavity be a single vacancy. In the hcp crystals there are two translationally nonequivalent positions of the vacancy, corresponding to two sublattices of the hcp lattice. Figure 1 shows the configuration of lattice sites of the crystal. The sites belonging to the first and the second sublattices are shown respectively by open and full circles. Vectors $\vec{u}_1,...,\vec{u}_6$ also shown in Fig. 1 correspond to the displacement processes of the ion to neighboring sites within the first sublattice. The same vectors correspond to the displacements within the second sublattice. The displacements by the vectors $\vec{u}_7,...,\vec{u}_{12}$ transfer the ion from the first to the second sublattice. The displacements by the vectors $-\vec{u}_7,...,-\vec{u}_{12}$ transfer the ion backward from the second to the first sublattice. Thus, the set of displacement vectors $\vec{u}_i$ for a single vacancy ion complex consists of 18 vectors $\vec{u}_1,...,\vec{u}_{12}, -\vec{u}_7,...,-\vec{u}_{12}$.

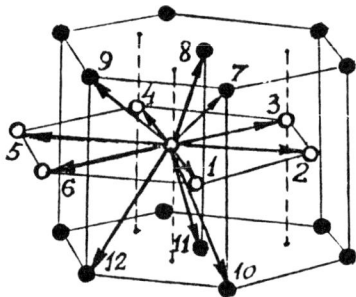

Fig. 1

Let n_1, n_2 be the probabilities to find the ion in the first and the second sublattices, respectively, and $W^{(a)}(\vec{u}_i)$ be the probabilities of the displacements per unit time by the vectors $\vec{u}_i$ of the ion occuping an a-th (a=1,2) sublattice site. We have

$$n_1 + n_2 = 1,$$

$$\partial n_1/\partial t = -n_1 \sum_{i=7}^{12} W^{(1)}(\vec{u}_i) + n_2 \sum_{i=7}^{12} W^{(2)}(-\vec{u}_i). \qquad (1)$$

Under stationary conditions we get from (1)

$$n_1 = \left[\sum_{i=7}^{12} W^{(2)}(-\vec{u}_i)\right]\left[\sum_{i=7}^{12}[W^{(1)}(\vec{u}_i) + W^{(2)}(-\vec{u}_i)]\right]^{-1},$$

$$n_2 = \left[\sum_{i=7}^{12} W^{(1)}(\vec{u}_i)\right]\left[\sum_{i=7}^{12}[W^{(1)}(\vec{u}_i) + W^{(2)}(-\vec{u}_i)]\right]^{-1}, \qquad (2)$$

The mean drift velocity of the ion is determined now by the formula

$$\vec{v} = n_1 \sum_{i=7}^{12} \vec{u}_i W^{(1)}(\vec{u}_i) + n_2 \left[\sum_{i=1}^{6} \vec{u}_i W^{(2)}(\vec{u}_i)\right.$$

$$-\sum_{i=7}^{12}\vec{u}_i W^{(2)}(-\vec{u}_i)\bigg]. \qquad (3)$$

As a second example let us consider negative ion complexes consisting of four vacancies. The most compact configurations of four vacancies correspond to four vertices of tetrahedrons. There are four translationally nonequivalent configurations of this type. These are, for example, the tetrahedrons

1. 0-10-11-12   2. 0-7-8-9
3. 11-0-3-4   4. 8-0-3-4

where 0 is the central site in Fig. 1. Let us enumerate corresponding translationally nonequivalent types of tetrahedrons by the index a=1,2,3,4, respectively.

The most probable transitions between different configurations consist in displacements of one of the vacancies along the hexagonal axis. These are the displacements of the vacancy 0 downwards and upwards, respectively, for the configurations a=1 and a=2 and the displacements 11→8 and 8→11 for the configurations a=3 and a=4. These transitions result in displacements of the ion as a whole by vectors $\pm\vec{u}_1$ where $\vec{u}_1 = (1/4)\vec{a}_3$ and $\vec{a}_3$ is an elementary translation vector along the hexagonal axis.

To provide the possibility of the ion motion over the entire volume of the crystal one should take into account additional processes of displacement by two vacancies simultaneously. A typical example is the displacement of two vacancies from 10,12 into 3,4. The initial configuration a=1 is transformed into a=3 by this process. The corresponding displacement vector for the ion as a whole is $\vec{u}_3 = \vec{u}_8^{(\text{prev})} - \vec{u}_1$ where $\vec{u}_8^{(\text{prev})}$ is the 8-th vector of the previous example. Due to crystal symmetry there are two more analogous vectors $\vec{u}_{2,4} = \vec{u}_{7,9}^{(\text{prev})} - \vec{u}_1$. Similarly, transitions from the configuration a=2 to a=4 displace the ion by vectors $\vec{u}_{5,6,7} = \vec{u}_{10,11,12}^{(\text{prev})} + \vec{u}_1$. The set of the displacement vectors $\vec{u}_i$ for the four vacancy ion complex consists of 14 vectors $\vec{u}_1,...,\vec{u}_7, -\vec{u}_1,...,-\vec{u}_7$.

If n_a is the probability of the a-th configuration, then

$$\sum_a n_a = 1 \qquad (4)$$

and we have under stationary conditions

$$0 = \partial n_1/\partial t = -n_1\sum_{i=1}^{4}W^{(1)}(\vec{u}_i) + n_2 W^{(2)}(-\vec{u}_i) + n_3\sum_{i=2}^{4}W^{(3)}(-\vec{u}_i),$$

$$0 = \partial n_2/\partial t = n_1 W^{(1)}(\vec{u}_i) - n_2\left[W^{(2)}(-\vec{u}_i) + \sum_{i=5}^{7}W^{(2)}(\vec{u}_i)\right] + n_4\sum_{i=5}^{7}W^{(4)}(-\vec{u}_i),$$

$$0 = \partial n_3/\partial t = n_1 \sum_{i=2}^{4} W^{(1)}(\vec{u}_i) - n_3 \sum_{i=1}^{4} W^{(3)}(-\vec{u}_i) + n_4 W^{(4)}(\vec{u}_i),$$

$$0 = \partial n_4/\partial t = n_2 \sum_{i=5}^{7} W^{(2)}(\vec{u}_i) + n_3 W^{(3)}(-\vec{u}_i) - n_4 \left[W^{(4)}(\vec{u}_i) + \sum_{i=5}^{7} W^{(4)}(-\vec{u}_i) \right], \quad (5)$$

The probabilities n_a are determined from the last equations. They are zero order homogeneous functions of the probabilities $W^{(a)}(\vec{u}_i)$ of the transition processes. The mean drift velocity of the four vacancy ion is expressed in terms of $W^{(a)}(\vec{u}_i)$ as

$$\vec{v} = n_1 \vec{u}_i W^{(1)}(\vec{u}_i) + n_2 \left[-\vec{u}_1 W^{(2)}(-\vec{u}_1) + \sum_{i=5}^{7} \vec{u}_i W^{(2)}(\vec{u}_i) \right]$$
$$+ n_3 \sum_{i=1}^{4} (-\vec{u}_i) W^{(3)}(-\vec{u}_i) + n_4 \left[\vec{u}_1 W^{(4)}(\vec{u}_1) - \sum_{i=5}^{7} \vec{u}_i W^{(4)}(-\vec{u}_i) \right]. \quad (6)$$

We can formulate now a general approach without making any suggestions concerning the specific structure of the ion. In the general case we have a number of translationally nonequivalent types a=1,2,... of localized states connected by transition processes characterized by a set $\vec{u}_i$ of displacement vectors and by probabilities $W^{(a)}(\vec{u}_i)$ of transitions from the a-th state accompanied by the displacements $\vec{u}_i$. The mean drift velocity of the ion in this general case is

$$\vec{v} = n_a \sum_{a} \vec{u}_a W^{(a)}(\vec{u}_a), \quad (7)$$

where n_a are zero order homogeneous functions of $W^{(a)}(\vec{u}_i)$. We shall consider below specific mechanisms of the transitions of the ions between the localized states. As a result we shall find the dependence of the probabilities $W^{(a)}(\vec{u}_i)$ on the temperature, magnitude and orientation of the electric field. It will be seen that equation (7) can be efficiently used to find the set of vectors $\vec{u}_i$ from experimental data of the anisotropy of the ion velocities.

Let us emphasize that our approach may be applied also for positive ions. Characteristic vectors $\vec{u}_i$ depend on the symmetry of ion positions in the elementary cell of the crystal and particularly on whether they are He^+, He_2^+ or He_3^+ ions.

INELASTIC SCATTERING OF VACANCIES FROM IONS

Suppose that the basic mechanism of ion motion is due to the existence of thermally activated vacancies. The displacements of the ion complexes considered above result here from the following processes (Figs. 2,3).

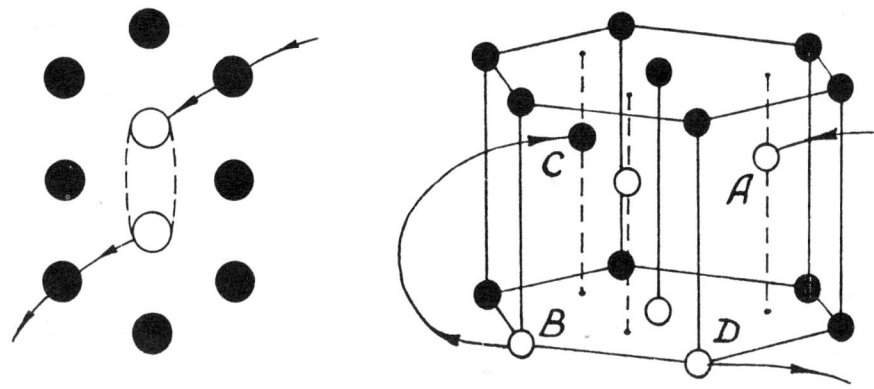

Fig. 2   Fig. 3

When moving through a crystal a neutral vacancy may find itself in a lattice site adjacent to one vacancy ion complex (Fig. 2). A two vacancy ion complex is formed as an intermediate state. It should decay then in a neutral vacancy and one vacancy ion, with the probability 1/2 that this ion will be displaced from its initial position.

The vacancy-induced transition involving the displacement of two vacancies simultaneously is shown in Fig. 3 for the four vacancy ion complex. Moving through a crystal a vacancy comes to the lattice site A, then the B vacancy comes to C and finally the D vacancy goes to infinity. Since such a complicated process follows a tunneling trajectory its probability is exponentially small.

The displacement of a positive ion is a result of any process in which a trajectory of a vacancy goes through a lattice site occupied by the ion.

The key point is that vacancies in ^{4}He crystals become, due to their quantum nature, delocalized quasiparticles. The processes described above can be considered as the quantum-mechanical inelastic scattering of delocalized vacancies from localized ion complexes[6]. The probabilities $W^{(a)}(\vec{u})$ in eq. (7) can be written in the form

$$W^{(a)}(\vec{u}) = \int \frac{d^3\vec{p}}{(2\pi)^3} \int \frac{d^3\vec{p}'}{(2\pi)^3} W^{(a)}(\vec{u};\vec{p},\vec{p}')n(\vec{p}) \qquad (8)$$

where $W^{(a)}(\vec{u};\vec{p},\vec{p}')$ is a differential probability of the inelastic process in which the ion is displaced by the vector $\vec{u}$ from the a-th localized state, $\vec{p}$ and $\vec{p}'$ are initial and final quasimomenta of the vacancy, $n(\vec{p}) = \exp(-\epsilon(\vec{p})/T)$ is the equilibrium distribution function of the vacancies.

The next key point is that, according to experiments[3,7], the characteristic value of the inelastic scattering cross-section is smaller than the characteristic area of the elementary cell of the crystal by a factor $\sim 10^{-2} - 10^{-3}$. The nature of this small factor is connected with the complication of the tunneling transition and the exponential character of its probability. As a result we have

$$W^{(a)}(\vec{u};\vec{p},\vec{p}') \propto \exp[-\phi^{(a)}(\vec{u};\vec{p},\vec{p}')] \tag{9}$$

where ϕ is large. The main contribution to the integral (9) is made by a small region near minimum of the expression

$$\phi^{(a)}(\vec{u};\vec{p},\vec{p}') + \frac{\epsilon(\vec{p})}{T} \tag{10}$$

We have here two important limiting cases.

1. At low enough temperatures $T \ll \Delta/\phi$ where Δ is the width of the energy band of vacancies, the minimum of the expression (10) is realized for quasimomentum $\vec{p} = \vec{p}_0$ corresponding to the bottom of the band. The emission of phonons by the vacancy in the scattering process is impossible in this case.

In the region of weak electric fields $e\vec{E}\vec{u} \ll \Delta$ where e is the ion charge, the transition probabilities (8) are (see ref. 5)

$$W^{(a)}(\vec{u}) = \lambda_a T^2 \exp[-\epsilon(\vec{p}_0)/T]\Phi\left(\frac{|e\vec{E}\vec{u}|}{T}\right)$$
$$\times \begin{cases} 1 & \text{if } e\vec{E}\vec{u} > 0, \\ \exp\left(-\frac{|e\vec{E}\vec{u}|}{T}\right) & \text{if } e\vec{E}\vec{u} < 0, \end{cases} \tag{11}$$

where

$$\Phi(z) = \int_0^\infty e^x[x(x+z)]^{1/2}dx = \begin{cases} 1 & \text{if } z \ll 1, \\ (\pi z/4)^{1/2} & \text{if } z \gg 1, \end{cases}$$

λ_a are constants and $\epsilon(\vec{p}_0) = \min \epsilon(\vec{p})$.

The ion drift velocity is first proportional to E with increasing E at $e\vec{E}\vec{u} \ll T$ and then at $T \ll e\vec{E}\vec{u} \ll \Delta$ increases like $E^{1/2}$.

With a further increase in field the drift velocity decreases and finally for $e\vec{E}\vec{u} > \Delta$ ion motion corresponding to a given $\vec{u}$, is blocked. This fact follows immediately from energy conservation $e\vec{E}\vec{u} = \epsilon(\vec{p}') - \epsilon(\vec{p})$ and the definition of the width of the band $\Delta = \max[\epsilon(\vec{p}') - \epsilon(\vec{p})]$.

Near the threshold $\Delta - e\vec{E}\vec{u} \ll \Delta$ we have

$$W^{(a)}(\vec{u}) = \gamma_a T^2 \exp[-\epsilon(\vec{p}_0)/T]\Psi[(\Delta - e\vec{E}\vec{u})/T] \tag{12}$$

where γ_a are constants and

$$\Psi(z) = z^2 \int_0^1 [x(1-x)]^{1/2}e^{-zx}dx = \begin{cases} (\pi z/4)^{1/2} & \text{if } z \gg 1, \\ (\pi/8)z^2 & \text{if } z \ll 1. \end{cases}$$

The drift velocity decreases with increasing electric field like $(\Delta - e\vec{E}\vec{u})^{1/2}$ in region $T \ll \Delta - e\vec{E}\vec{u} \ll \Delta$ and like $(\Delta - e\vec{E}\vec{u})^2$ for $\Delta - e\vec{E}\vec{u} \ll T$.

To include vacancy scattering with the simultaneous emission of a phonon, it is sufficient to take into account that $\vec{p}$ and $\vec{p}'$ are not connected now by energy conservation used above, and to add the factor $1 + N(\omega)$ to the right-hand side of eq. (8). Here $N(\omega) = (e^{-\omega/T} - 1)^{-1}$ and $\omega = \epsilon(\vec{p}) - \epsilon(\vec{p}') + e\vec{E}\vec{u}$ is the phonon energy.

At high temperatures $T \gg \Delta/\phi$ the second term in the expression (10) can be neglected. The minimum of this expression is realized then for temperature-independent quasimomenta $\vec{p} = \vec{p}_m$ and $\vec{p}' = \vec{p}'_m$. It is important to note that $\vec{p}_m$ and $\vec{p}'_m$ are field-independent for $e\vec{E}\vec{u} \ll U$ where U is the characteristic interaction energy between particles in the crystal. In fact, the function ϕ is characteristic of the tunneling trajectories of neutral vacancies, and is field-independent if $e\vec{E}\vec{u} \ll U$. The energy of the emitted phonon

$$\omega = \epsilon(\vec{p}_m) - \epsilon(\vec{p}'_m) + e\vec{E}\vec{u} \tag{13}$$

is small compared with the Debye temperature θ in region $e\vec{E}\vec{u} < \Delta \ll U, \theta$, which is most interesting for us. In this case the probability of emission is proportional to the third power of the phonon energy. Using eq. (8) we have

$$W^{(a)}(\vec{u}) = A\exp[-\epsilon(\vec{p}_m)/T]\omega^3[1 + N(\omega)] \tag{14}$$

where A is a constant and ω is determined by eq. (13).

The probability of vacancy scattering with the simultaneous absorption of a phonon is simply the probability of the reverse process. Let the b-th state be the result of the displacement of the a-th state by the vector $\vec{u}$. The probability of the reverse transition b→a with the absorption of a phonon is

$$W^{(b)}(-\vec{u}) = A\exp[-\epsilon(\vec{p}'_m)/T]\omega^3 N(\omega). \tag{15}$$

In the intermediate temperature region $T \simeq \Delta/\phi$ the minimum of the expression (10) is realized for temperature dependent $\vec{p}_m(T)$ and $\vec{p}'(T)$. Equations (14) and (15) determine in this more general case only the field dependence of the probability.

For $e\vec{E}\vec{u} \gg T$ the processes with the simultaneous emission of a phonon and $e\vec{E}\vec{u} > 0$ are dominant. We can neglect the phonon distribution function in eq. (14). So we obtain

$$W^{(a)}(\vec{u}) = B\cos^3\theta(E + E_0)^3 \tag{16}$$

where B is a field-independent constant, θ is the angle between $\vec{E}$ and $\vec{u}$, and

$$E_0(T) = [\epsilon(\vec{p}) - \epsilon(\vec{p}')]/eu\cos\theta. \tag{17}$$

All the characteristic features of experimental results mentioned in the Introduction are easily explained[5] on the basis of eqs. (7), (11), and (14)-(17).

It is interesting that there exists a direct possibility to determine the set of displacement vectors $\vec{u}_i$ form experiments on the anisotropy of the ion velocity.

The point is that for $e\vec{E}\vec{u} \gg T$ the velocity changes jump-wise whenever the vector $\vec{E}$ passes through a plane normal to one of $\vec{u}_i$ as the field direction is varied.

REFERENCES

1. A.J. Dahn, in Progress in Low Temp. Phys., v. 10 (North- Holland, Amsterdam, 1986), p. 73.
2. A.I. Golov, V.B. Efimov and L.P. Mezhov-Deglin, Sov. Phys. JETP 67, 325 (1988).
3. K.O. Keshishev, Sov. Phys. JETP 45, 273 (1977).
4. S.C. Lau, A.J. Dahm and W.A. Jeffers, J. de Physique 39, 6-86 (1978).
5. A.F. Andreev and A.D. Savishchev, Sov. Phys. JETP (1989), in press.
6. A.F. Andreev and A.E. Meyerovich, Sov. Phys. JETP 40, 776 (1975).

NUCLEAR MAGNETISM OF BCC SOLID ^{3}HE IN A HIGH MAGNETIC FIELD

H. Ishimoto, Hiroshi Fukuyama, T. Fukuda,
T. Okamoto, T. Tazaki, K. Sakayori and S. Ogawa
The institute for Solid State Physics, The University of Tokyo,
7-22-1 Roppongi, Minato-ku, Tokyo 106, Japan

ABSTRACT

Isochoric pressure measurements have been made for bcc solid ^{3}He with a high density in a magnetic field up to 7 Tesla. The sample has successfully been cooled down to 0.4 mK at 7 T. Temperature dependence of the pressure suggests that the upper critical field for the molar volume of 22.53 cm^3/mol is below 7 T.

INTRODUCTION

Nuclear magnetic order in solid ^{3}He is known to arise from exchanges among ^{3}He atoms. The hard cores of interatomic potential favor a coherent exchange of more than two particles. Exchange of an odd number of particles leads to ferromagnetic interactions and that of an even number of particles does to antiferromagnetic ones. These two competing interactions cause the interesting and peculiar phase diagram as a function of temperature and magnetic field. Three phases have been observed, a para magnetic phase (PARA), the low field ordered phase (U2D2) and the high field ordered phase (HFP). The observed phases qualitatively compare well with those predicted in the mean field approximation of the multiple exchange model with a few exchange parameters.[1,2] There are still discrepancies between the predictions and the experiments. It is not clear whether the discrepancies arise from flaw of the approximation or from the the neglect of the higher order exchange parameters. Exceptionally the upper critical field (H_{c2}) between the PARA and the HFP phase is exactly correct even in the mean field approximation[3] and can be compared with the experimental results. Experimentally the value on the melting curve is estimated to be (22.3±0.4) T from the melting pressure measured below 11 T.[4] Such a high field is difficult to reach in the present state of art of superconducting magnet technology. This difficulty can be removed by increasing the sample density, which decreases the exchange interaction. Then the H_{c2} comes into the field region of our apparatus. On the other hand, the whole phase diagram shrinks and we have to cool the high density sample down to further lower temperatures in the high magnetic field. In the present paper, we report our experimental trial to cool the high density sample in the field of 7 T and its prelimanary results.

EXPERIMENTAL

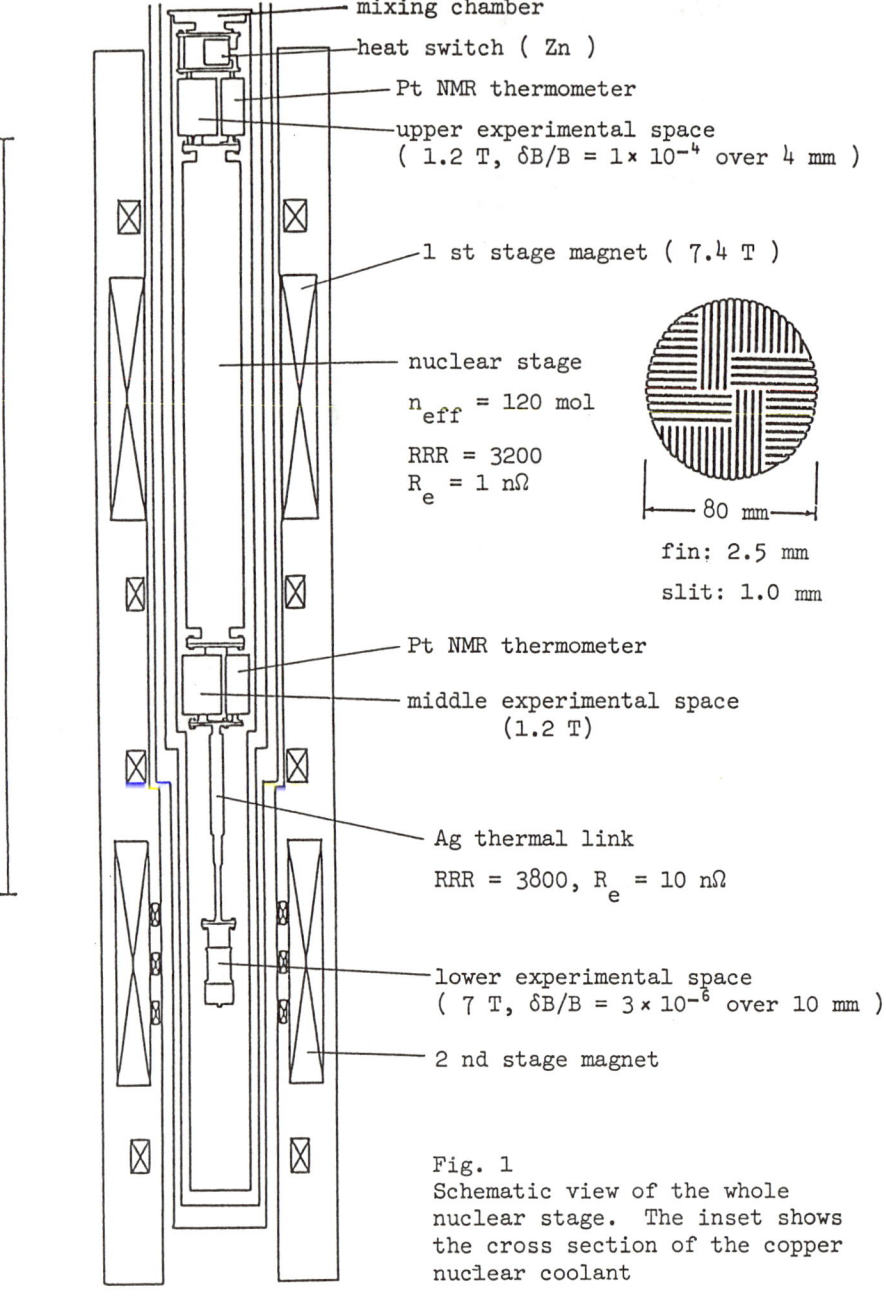

Fig. 1
Schematic view of the whole nuclear stage. The inset shows the cross section of the copper nuclear coolant

a) Cryogenics

For the above purposes, a nuclear refrigerator with a large cooling power and a minimum temperature of microkelvin region has been constructed. The schematic drawing of the nuclear stage is shown in Fig. 1. The first stage consists of 120 mol effective OFHC copper,[5] whose cutaway view is given in the inset. It was annealed in an atomosphere of oxygen (2×10^{-4} Torr) for three days at a temperature of 950° C , RRR of copper being 3200. The second stage is just the sample cell in the magnetic field of 7 T, whose homogeneity is 3×10^{-6} over 1 cm^3. They are thermally linked by a silver rod with several slits to reduce an eddy current heating. The thermal annealing increased RRR of the silver rod up to 3800, which corresponds to elecrical resistance of 10 nΩ. For this value, the temperature difference between both ends of the link at 0.4 mK is estimated to be 10 μK, even if the heat leak of 10 nW enters into the sample cell. The present total heat leak is less than that, and we can know the sample temperature quite well by the platinum wire NMR thermometer outside the demagnetization magnet.

b) Sample cells

In the center of 7 T, we have two sample cells. One is a stain gauge type pressure sensor with a reference capacitor on top of it and the other is an NMR cell. Both are made of pure Ag, whose RRR is 3800, except a diaphragm and screws of Ag-Si alloy. The bulk solid ^{3}He sample is formed in a small gap of 50-100 μm between the cell body and the sintered sponge. The sponge is made of 1100 Å Ag powder, and its increased surface area is helpful to reduce the thermal time constant of the sample in the high magnetic field. The NMR cell consists of a coaxial resonant cavity , whose resonant frequency is variable between 80 and 220 MHz. In the middle space between the two 7 T magnets, the platinum NMR thermometer and a ^{3}He melting curve thermometer are placed. The former is calibrated by the latter based on the latest temperature scale.[6]

RESULTS AND DISCUSSIONS

Before measurements at the high field, the pressure was measured for the molar volume of 22.53 cm^3/mol at 0 T in the dilution refrigerator temperature region. This kind of experiments have been made by Panczyk and Adams,[7] and Van Degrift.[8] As shown in Fig. 2, the present data agree very well with the latter, which means that our sample in the gap is sufficiently thick and no volume correction for the constant volume condition is needed. Assuming that the Gruneisen constant is the same, 18, for various exchange interactions, the first coefficient in the high temperature expansion, e_2, is estimated to be 0.44 mK^2, which compares reasonably with the experimental value of 6.8 mK^2 on the melting curve. Combined with the pressure data at 3.3 T, the Weiss temperature θ is derived to be -0.368 mK, in good agreement with that estimated from our previous magnetization measurements.[9]

Thermal relaxation of the sample is important to cool it to μK region in the high magnetic field. The thermal time constant

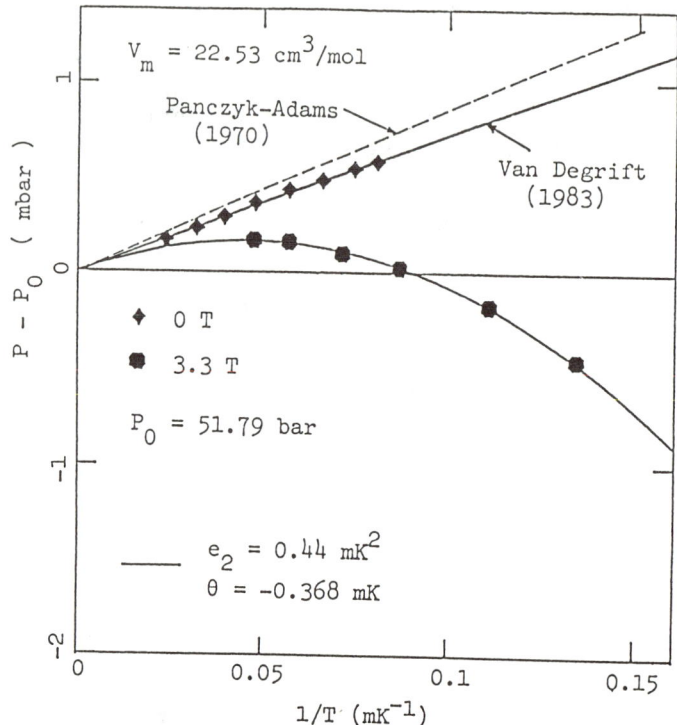

Fig.2 Isochoric pressure for the molar volume of 22.53 cm^3/mol at 0 and 3.3 T in the high temperature region. The solid lines are the fitted one for the high temperature expansion formula.

obtained from pressure measurements at 7 T is shown in Fig. 3. In the figure, the calculated heat capacity of the sample, which is not distinguishable from that of a free paramagnet, is given. Time constant has a maximum around 5 mK, corresponding to the peak of the heat capacity. Once crossing the peak, refrigeration down to 0.5 mK seems to have no great difficulty. The observed temperature dependence of the pressure at 7 T was well fitted to a curve calculated for the PARA phase by the mean field theory. This indicates that the sample was in the PARA phase and it was actually cooled to 0.4 mK. Therefore, it is strongly suggested that H_{c2} exists below 7 T for the molar volume of 22.53 cm^3/mol.

The next step to observe H_{c2} is to sweep the magnetic field very slowly. It usually takes a day or two just to go to the next field, because the sample cell is made of very pure metal. The experiments are going on to determine H_{c2} exactly at various temperatures for several molar volumes, including the magnetization measurements by NMR.

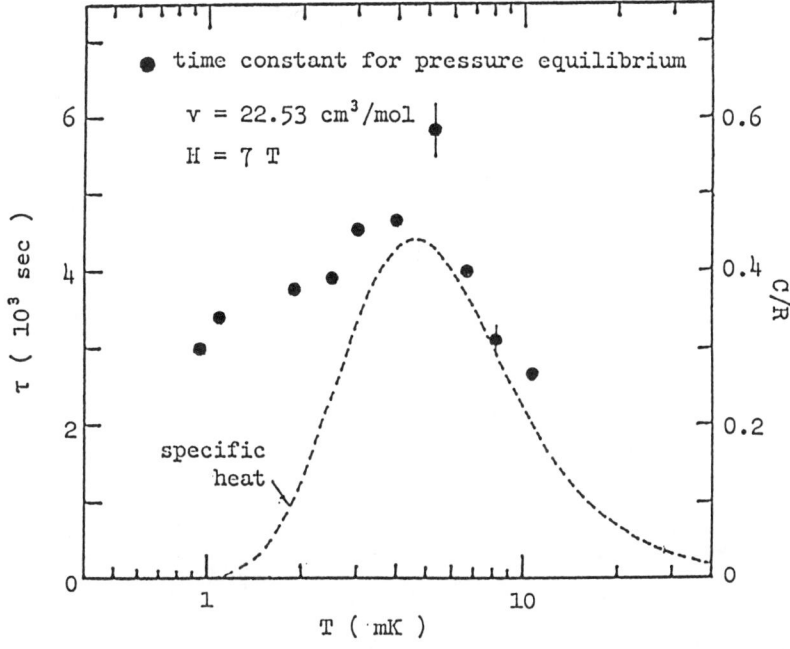

Fig.3 Thermal time constant of pressure equlibrium for 22.53 cm^3/mol in the high magnetic field of 7 T. The dashed line is the calculated specific heat.

REFERENCES

1. M. Roger, J.H. Hetherington and J.M. Delrieu, Rev. Mod. Phys. 55, 1 (1983)
2. H. Stipdonk, and J.H. Hetherington, Phys. Rev. B31, 4684 (1985)
   K. Iwahashi, Y. Miwa and Y. Masuda, J. Phys. Soc. Jpn. 53, 3088 (1984)
3. J.H. Hetherington, Physica 108B, 855 (1981)
4. D.D. Osheroff, H. Godfrin and R.R. Ruel, Phys. Rev. Lett. 58, 2458 (1987)
5. Hitachi Cable, Ltd., 2-1-2 Marunouchi, Chiyoda-ku, Tokyo, Japan
6. H. Fukuyama, H. Ishimoto, T. Tazaki and S. Ogawa, Phys. Rev. B36, 8921 (1987)
7. H.F. Panczyk and E.D. Adams, Phys. Rev. A1, 1356 (1970)
8. C.T. Van Degrift, AIP Proc. No.103, 16 (1983)
9. Y. Miura, N. Nishida, Y. Takano, H. Fukuyama, H. Ishimoto, S. Ogawa, T. Hata and T. Shigi, Phys. Rev. Lett. 58, 381 (1987)

DOMAIN STRUCTURE OF UUDD HELIUM-3 STUDIED BY NMR

Y. Sasaki, Y. Hara, T. Mizusaki and A. Hirai
Department of Physics, Kyoto University, Kyoto 606, Japan

INTRODUCTION

We have been studying non-linear spin dynamics and spin relaxation in uudd solid ^{3}He by using NMR techniques[1,2]. It has been found that our "single crystal" specimen was always composed of multi-domains. In oder to investigate the domain structure of our specimen, we have constructed a small superconducting magnet system with current shims to improve the homogeneity of the magnetic field and to apply linear magnetic field gradients along three principal directions. With this magnet system, we performed a preliminary "magnetic imaging" experiment to observe the magnetic domain structure of uudd solid ^{3}He.

EXPERIMENTAL RESULTS

Our experiments have been performed along the melting curve, and our method of making a single crystal of uudd solid ^{3}He is similar to that used by Osheroff et al.[3]. We first cooled liquid ^{3}He to about 0.5 mK and then applied a heat pulse ($\sim$ 1 erg) to an NMR coil which was wound around the sample cell to initiate the crystal growth. We avoided usage of a heater wire for nucleating the seed inside the space where the solid crystal was made. Temperature, T, of the sample cell raised momentarily above the nuclear spin ordering temperature, T_N, and then a paramagnetic solid was formed, which cooled down gradually below T_N again. Then, usually three resonance lines corresponding to the three possible ℓ-vectors (the direction of uudd sequence) are observed. The quality of the initial crystal was not good, as seen by the fact that each NMR line width was large or each resonance line was splitted into several peaks. Then, we melted a part of the crystal by expanding the volume of the ^{3}He cell in the Pomeranchuk device, and then grew the crystal again by compressing the volume. By repeating these process several times, each NMR line usually became narrower and its fine structure disappeared, but there were always three resonance lines in a single crystal sample. It was interesting to investigate the magnetic domain structure in a single crystal of the uudd solid ^{3}He. For this purpose the magnetic "imaging" technique should be very useful.

A preliminary experiment was performed here. Fig. 1 shows an example of the NMR signals, where we notice one line with small $\cos^2\theta$ (θ being the angle between ℓ-vector and the external magnetic field, B_θ) splits into two, marked by C1 and C2. For this sample, we applied a linear field gradient along three principal axis of the laboratory frame (not directly related with the principal axes of the single crystal sample) and observed the change of the resonance frequency and the line width, and then analysed these data with local oscillator approximation. The result is shown in Fig. 2. It seems that the domain is rather large and the whole sample ($\sim$ several mm^3)

is divided into a few domains. It should be stressed that the two domains corresponding to C1 and C2 in Fig.1 are separated in space. Such a splitting of one NMR line is often observed. This may be caused by the distortion of the crystal due to "magneto-striction". If this is the case, the ratio of c/a (c and a are a lattice constant of the bcc solid ^{3}He along ℓ-vector and that perpendicular to it, respectively.) with $|c/a - 1| \simeq 10^{-3}$ accounts for the observed value of the splitting.

We have also extended our previous NMR line width measurements[2] in a magnetic field with better homogeneity and obtained more reliable data on the temperature dependences of the relaxation parameter.

REFERENCES

1. T.Kusumoto, O.Ishikawa, T.Mizusaki and A.Hirai, J. Low Temp. Phys. **59** (1985) 269.
2. Y.Sasaki, K.Sasayama, T.Mizusaki and A.Hirai, Jpn. J. Appl. Phys. **26-3** (1987) 417.
3. D.D.Osheroff, M.C.Cross and D.S.Fisher, Phys. Rev. Letters **24** (1980) 792.

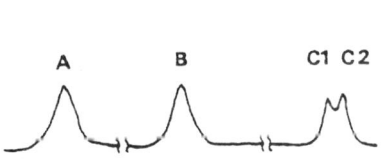

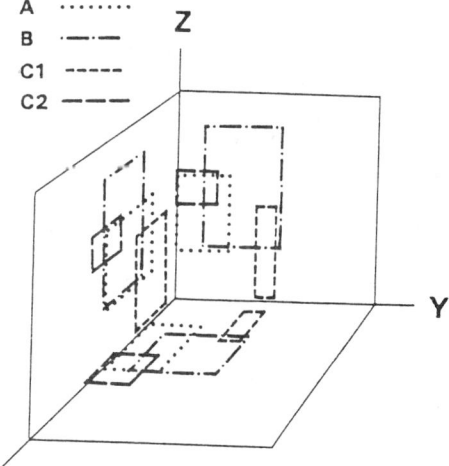

Fig. 1. One example of the NMR spectra from a uudd solid ^{3}He single crystal, whose magnetic domain structure is determined by imaging technique, as shown in Fig.2.

$T/T_N \simeq 0.7$, $B_0 \simeq 40$ mT.

A; $\cos^2\theta = 0.586$
B; $\cos^2\theta = 0.352$
C1; $\cos^2\theta = 0.069$
C2; $\cos^2\theta = 0.067$

Fig. 2. Schematic representaion of the magnetic domain structure in a single crystal uudd solid ^{3}He. There are four domains. Each parallelepiped indicates only the size of the domain, not the actual domain shape.

TOWARDS A FIRST PRINCIPLES THEORY OF ^{3}HE MAGNETISM

J.H. Hetherington and M.Roger

Physics Dept., Michigan State U.,E. Lansing, Mich. 48824, and

SPSRM, CEA Orme des Merisiers, 91191 Gif s/Yvette,France

The coupled cluster approximation (CCA) was developed[1] mainly for application to systems of fermions(e.g. nuclear matter). So far as we know the method has not previously been applied to spin lattice systems despite the long history of attempts to improve on available methods for those problems. We have applied the CCA to the spin Hamiltonian for bcc ^{3}He. This Hamiltonian is now known to contain several types of exchange including 2-spin exchange and 3-spin ring exchange, which lead to Heisenberg type terms, and 4-spin and 6-spin ring exchanges which lead to σ^4 and σ^6 type terms.[2-5] Until now the ground state of this problem has only been studied in mean field theory[2,6,7] or by exact diagonalization of the 16-atom cluster[8], and somewhat tentatavely by the random phase approximation[2,9].

The formal statement of the CCA method is as follows: Assume that the (unnormalized) ground state of the system is of the form $|\Phi> = e^S|0>$, where S involves only raising operators and where $|0>$ is the mean field ground state. This assumption is always possible if $|\Phi>$ is not exactly orthogonal to $|0>$. Then $He^S|0> = E_0 e^S|0>$. Multiplying on the left by e^{-S} one has $\tilde{H}|0> \equiv e^{-S}He^S|0> = E_0|0>$. Introducing the normalized basis built on the mean field ground state we may write:

$$<0|\tilde{H}|0> = <0|e^{-S}He^S|0> = E_0, \qquad (1)$$

$$<n|\tilde{H}|0> = <n|e^{-S}He^S|0> = 0. \qquad (2)$$

The equations (2) are requirements on S so that (1) will be true. There are more equations (2) than parameters in S but if the lower states' equations are satisfied the energy given by (1) is a good approximation.

An important point, which leads to finiteness of the equations 2, is that the transformed Hamiltonian $\tilde{H}$ is a closed expression: Since $\tilde{H} = H + [H,S] + \frac{1}{2!}[[H,S],S] + \cdots$ and since S contains only raising operators and H contains only a finite product of raising and lowering operators, each successive commutation with S reduces the expression until it terminates. Thus the series is finite for any H with a finite number of σ's in each term. As a result the equations 2 are finite and tractable.

The general spin exchange Hamiltonian is written:[2-5]

$$H_{ex} = -\sum_{P^\sigma}(-1)^p J_P P^\sigma \qquad (3)$$

where the sum runs over all permutations P^σ acting on N spin variables. $(-1)^p$ represents the sign of the permutation P. We choose the parameters determined by Monte-Carlo calculations[5] adjusted from $v = 24.12$ to the melting molar volume 24.22 by the usual 'Grüneisen' coefficient. Hence, we take $J_{NN} = 0.50$, $J_{NNN} = 0.07$, $J_T = 0.21$, $K_P = 0.29$, $K_F = 0.03$, $S_1 = 0.04$, $S_2 = 0.02$ as the exchange parameters in mK.

Athough the main qualitative features of the phase diagram are accounted for by the preponderant exchange processes[2,3] (pair exchange between first neighbors, J_{NN}; triple exchange on the most compact triangles, J_T; and cyclic four spin exchange on planar cycles with four first neighbors, K_P), calculations from first principles[4,5] have revealed that the exchange frequencies do not decrease with the number of particles involved as fast as previously conjectured. Some higher processes (six spin ring exchanges, etc.), although smaller, should be taken into account in quantitative theoretical predictions. We apply the coupled cluster approximation to a model including the above three dominating exchanges. 'Mean field corrections' are then added to take into account the smaller exchanges. Thus we consider first the following multiple exchange Hamiltonian:

© 1989 American Institute of Physics

$$H_{ex} = J_{NN} \sum_{i<j}^{(1)} P_{ij} + J_{NNN} \sum_{i<j}^{(2)} P_{ij} - J_T \sum_{i<j<k}^{(T)} \{P_{ijk} + P_{ijk}^{-1}\} + K_P \sum_{i<j<k<l}^{(Pl)} \{P_{ijkl} + P_{ijkl}^{-1}\} \quad (4)$$

where $P_{ij}, P_{ijk}, P_{ijkl}$ represent cyclic permutations of two, three, four spins respectively. The sums are performed on first (1) and second (2) neighbor pairs, the most compact triangles (T) and planar four-spin cycles (P). We apply the CCA to (4) using an S which includes all n-spin flips generated by the Hamiltonian itself: pair flips between first, second and third neighbors, three-spin flips on the most compact triangles, four-spin flips on planar four-site cycles. The contributions of the small parameters K_P, S_1, S_2, ($\approx 0.03mK$) to the ground state energy are taken to be the same as their contributions to the mean field energy[7]

The magnetization curve thus obtained for the 'pf' phase is represented in Fig. 1 (solid curve) and compared to the experimental results. The agreement is excellent. The transition field $B_{c1} \approx 0.76\ T$ between the 'uudd' and 'pf' phase is higher than the experimental value: 0.45 T. This difference is compatible with the errors bars on the exchange parameters. Some small neglected exchange processes might lower this transition field[7].

We finally emphasize that the set of exchange frequencies adopted is compatible with all other thermodynamic data on bcc ^3He. The Curie Weiss temperature, $\Theta \approx -1.54mK$, is in agreement with experimental data, $\Theta \approx -1.8 \pm 0.2mK$. A rough estimate of T_{c1}, the zero field first order transition temperature is:[2] $T_{c1} \approx 1.2mK$ in good agreement with the experimental value $\approx 1.04mK$. The spurious second onder transition of Ref. 6 persists but is probably an artifact of mean field theory as outlined there.

Most experiments on bcc ^3He magnetism have been performed at the melting molar volume. A few experimental studies[10] at lower molar volume have shown that most physical quantities simply scale as v^γ with $\gamma = 18 \pm 2$ (this somewhat surprising result is qualitatively confirmed by exchange-frequency variations calculated from first principles[4]). The physical parameter which is the most sensitive to small variations of the ratios between various exchange frequencies is the transition field B_{c1} between the two almost degenerate 'uudd' and 'pf' phases. We predict on the basis of the actions calculated in Ref. 4 that at $v = 20.07\ cm^3/mole\ B_{c1}/B_{c2} = 0.055$, compared with the value $B_{c1}/B_{c2} = 0.036$ at melting molar volume.

REFERENCES

1. R. F. Bishop and H. G. Kümmel, Physics Today, March 1987, p.52.
2. M. Roger, J. H. Hetherington and J. M. Delrieu, Rev. Mod. Phys. **55**, 1 (1983).
3. M. C. Cross and D. S. Fisher, Rev. Mod. Phys. **57**, 881 (1985).
4. M. Roger, Phys. Rev. B **30**, 6432 (1984).
5. D. M Ceperley and G. Jacucci, Phys. Rev. Lett. **58**, 1648 (1987).
6. H. Stipdonk and J. H. Hetherington, Phys. Rev. B **31**, 4684 (1985).
7. H. Godfrin and D. D. Osheroff, Phys. Rev. B **38**, 4492 (1988).
8. M. C. Cross and R. N. Bhatt, Phys. Rev. B**33**, 7809 (1986).
9. K. Iwahashi, Y. Miwa and Y. Masuda, J. Phys. Soc. Japan **53**, 3088 (1984).
10. T. Mamiya, A. Sawada, H. Fukuyama, Y. Hiro, and Y. Masuda, Phys. Rev. Lett. **47**, 1304 (1981).

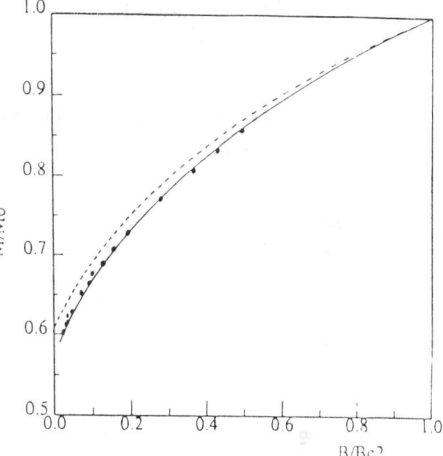

Fig. 1 The reduced magnetization versus reduced field according to mean field (dashed), CCA (solid), and experiment.

LOW TEMPERATURE SUSCEPTIBILITY OF LARGE MOLAR VOLUME bcc SOLID ^{3}He[†]

M.E.R. Bernier, M. Bassou, M. Chapellier, M. Rotter*
DPhG/SPSRM, CEN-Saclay, 91191 Gif-sur-Yvette Cedex, France

The magnetic susceptibility χ provides important informations on the many body phenomena in solid ^{3}He. Precise measurements have been performed in recent years, in good agreement with the multiple exchange description of the solid[1,2,3] and experiments made at large molar volume reported an unexpected decrease of χ at low temperature[4].

We investigated the temperature variation of χ for these large molar volumes by measuring simultaneously the susceptibility of both the ^{3}He sample in an open volume at the top of a cell and the fluorine spins of teflon spheres of 2000 Å in diameter packed at the bottom of the same cell. For each temperature we also recorded the pressure inside the cell using a BeCu strain gauge. The susceptibility of the ^{19}F spins which follows a Curie law down to the millikelvin range was used as a secondary thermometer calibrated by comparison with the melting curve of ^{3}He at low temperature. The samples were formed using the blocked capillary method and annealed before taking data.

Our measurements were made on three samples of molar volumes V, 24.850 cm^3, 24.574 cm^3 and 24.085 cm^3. Two of the samples melt at low temperature and by recording both the pressure inside the cell and the susceptibility of the sample we obtain the melting temperature and can correlate the susceptibility measurements to the amount of liquid formed in the cell. We reduce our data according to a procedure used in reference 2, i.e., we plotted $(\chi T)^{-1}$ as a function of T^{-1}. Our measurements give directly the integrals of the ^{3}He and ^{19}F signals, I_{He} and I_F, two quantities proportional to the susceptibilities of He and F, the susceptibility of F being proportional to T^{-1}. We thus plot I_F/I_{He} as a function of I_F, i.e., $(T\, I_{He})^{-1}$ as a function of T^{-1}. The data are given in figures 1-2-3 for the three molar volumes:

(1) The sample $V = 24.085$ cm^3 remains solid in the whole temperature range and the Curie Weiss constant θ_{cw} deduced from figure 1, $\theta_{cw} = 1.63$ mK, is in good agreement with the most recent measurements[1,2].

(2) The sample $V = 24.850$ cm^3 melts at low temperature (~ 180 mK) and χ shows a decrease due to the lower χ of the liquid phase. The expected susceptibility for the liquid solid mixture in an open geometry is plotted as a solid line in figure 2 but in our cell the liquid is preferentially formed in the confined geometry between the teflon spheres and is not seen by the ^{3}He NMR coil wound around the open volume of the cell. This accounts for an observed decrease of χ smaller than expected.

(3) The same phenomenon is observed at $V = 24.574$ cm^3. When cooling down, the solid melts at a temperature lower than that of the preceding molar volume ($T \sim 75$ mK) and the amount of liquid formed is smaller and localized around the teflon spheres. The quantity of liquid formed in the open geometry is too small to be observed in our susceptibility measurements. We thus obtain $\theta_{cw} \sim 2.10$ mK for V close to 24.574 cm^3, in good agreement with the extrapolation of the most precise measurements[1,2].

The susceptibility of large molar volume bcc ^{3}He samples measured in this experiment is in good agreement with values reported earlier at lower V but no anomaly can be observed[5]. A more detailed analysis of these experiments can be found in reference 6.

[†] Work supported in part by NATO 88/703.
* Permanent address: Charles University, Prague, Czechoslovakia.

REFERENCES

1. C.T. Van Degrift, W.J. Bowers, Jr, P.B. Pipes, D.F. McQueeney, Phys. Rev. Letters 49, 149 (1982).
2. W.P. Kirk, Z. Olejniczak, P. Kobiela, A.A.V. Gibson, A. Czermak, Phys. Rev. Letters 51, 2128 (1983).
3. M.E.R. Bernier, E. Suaudeau, M. Roger, Phys. Rev. B 38, 737 (1988).
4. W.P. Kirk, Z. Olejniczak, P.S. Kobiela, A.A.V. Gibson, Proceedings of LT17, Contributed papers p.273, U. Eckern et al. Editors, Elsevier Science Publishers B.V. 1984.
5. M.E.R. Bernier, E. Suaudeau, Phys. Rev. B 38. 784 (1988).
6. M.E.R. Bernier, M. Bassou, M. Chapellier, M. Rotter, submitted to Phys. Rev. B

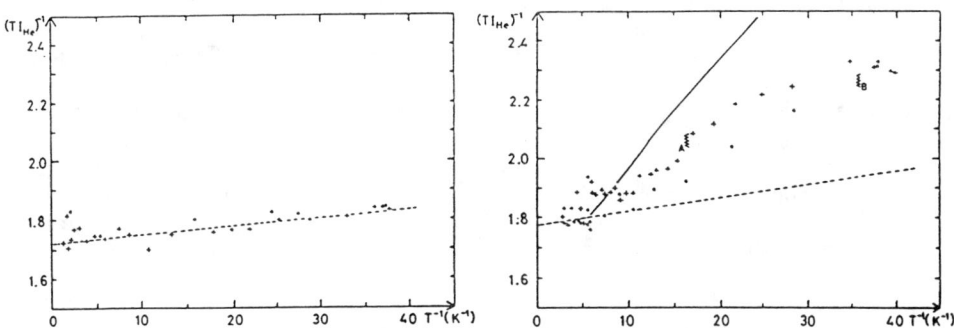

Figure 1 : The ratio of the integrals of the NMR lines of ^{19}F and ^{3}He as a function of T^{-1} at $V = 24.085$ cm^3.

Figure 2 : The ratio of the integrals of the NMR lines of ^{19}F and ^{3}He as a function of T^{-1} at $V = 24.850$ cm^3.

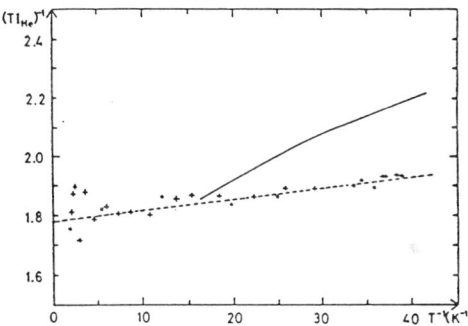

Figure 3 : The ratio of the integrals of the NMR lines of ^{19}F and ^{3}He as a function of T^{-1} at $V = 24.574$ cm^3.

VACANCIES IN bcc SOLID ^{3}He

M.E.R. Bernier and J.H. Hetherington*
DPhG/SPSRM, CEN-Saclay, 91191 Gif-sur-Yvette Cedex, France

The nature of vacancies in solid He is still not satisfactorily explained. Due to the influence of quantum effects in the solid several types of vacancies could exist: thermally activated vacancies exist as in most classical solids but the larger zero point motion of the atoms suggests that they are non-localized forming a band of excitation energies. The existence of zero point vacancies has also been suggested and the possibility of multivacancies has not been ruled out experimentally.

Vacancies can be observed in various experiments: (1) The most direct method is the measurement by X-ray diffraction of the lattice parameter as a function of temperature[1]. At constant volume the thermal monovacancy concentration x is directly deduced and related to f, e and s respectively the free energy, the energy and the nonconfigurational entropy of formation; (2) The observation of a specific heat larger than the phonon contribution in constant volume experiments is interpreted as a contribution of thermally activated vacancies and e can be deduced from the experiments provided the nature of the defect (as monovacancies) is assumed[2]; (3) The unpinning of dislocations in sound attenuation experiments[3] is attributed to thermal vacancies and also gives a measure of e; (4) In NMR, the vacancies are observed by the effect of their mobility on the relaxation. Two temperature regimes can be considered where the vacancies modulate various interactions between the ^{3}He spins[4,5]. The important parameter is then x/τ where τ^{-1} is a characteristic jump frequency of a vacancy to a neighbouring site. The experimental activation parameter measured is then $e + E_b$ where E_b is the energy of the barrier in classical diffusion. If the vacancies tunnel from one site to another $E_b = 0$. If the vacancies are assumed to form a band, the measured parameter also includes the bandwidth Δ; (5) Besides X-ray studies, the most direct way for studying vacancies is the measurement of the pressure for a constant volume sample as a function of temperature. Below melting, the pressure is the sum of the contributions of the various energy reservoirs of the system: exchange, phonons and vacancies and the experiment gives access to the characteristic parameters of these reservoirs[6]. The temperature dependence of the pressure exhibits a minimum between a low T regime where the exchange contribution dominates and a high T regime where the contributions of both the phonons and vacancies dominate. Assuming a Debye spectrum for the phonons the contribution of vacancies to the pressure gives e and its molar volume dependence.

Microscopic model

We have extended a finite bandwidth model formerly discussed by Hetherington[7]. We consider a band of states which represents both the vacancies and the associated local modes. For convenience, we assume that the vacancy and phonon part of the spectrum are separable, the phonon part being just the ordinary phonon spectrum and the vacancy part including local phonon modes.

We write for the vacancy part of the energy: $H = \sum_\alpha m_\alpha E_\alpha$ where m_α is the occupation number of a combined vacancy-local phonon state and is either 0 or 1.

We can therefore separate the phonons and vacancy statistics and we are then interested in a description of the vacancy density, effect on pressure and specific heat in terms of a density of states $\rho(E)$ where $N \rho(E) dE$ is the number of states E_α in the range of energy $(E, E + dE)$.

---

* Physics Department, Michigan State University, East Lansing, MI 48824 USA.

We then calculate the partition function

$$Z = \sum_{\{m_\alpha\}} \exp\left(-\beta \sum_\alpha m_\alpha E_\alpha\right) \qquad (2)$$

where $\{m_\alpha\}$ describes all the configurations of vacancies, and show that it can be simply written $\ell n\, Z = N \int \rho^*(E) e^{-\beta E} dE$. (This expression should not be confused with the slightly more familiar formula $Z = \int \rho(E) \exp(-\beta E) dE$ wherein $\rho(E)$ is the density of all states, not the density of "single particle" states).

The total vacancy energy, the heat capacity, the contribution to the pressure and the number of vacancies are directly deduced.

All thermodynamic results are unified by the simple density $\rho^*(E)$ that we do not need to determine more specifically for thermodynamic purposes. We may simply think of $\rho^*(E)$ as a convenient parametrization of $\ell n\, Z$.

In reference 7 it was shown that the very accurate specific heat data of Greywall can be fitted by the form $\rho^*(E) = A/2(E - \Phi)^2$ for $E > \Phi$ and $\rho^*(E) = 0$ for $E < \Phi$.

Examination of the empirical values of A and Φ as a function of volume given in 7 shows that numerically $A \sim \Phi^{-3}$ for each molar volume and thus $\rho(E) = 1/2\Phi^{-3}(E - \Phi)^2$ can be used to fit Greywall's data. Using this form for $\rho^*(E)$ we obtain $\ell n\, Z = N/(\beta\Phi)^3 e^{-\beta\Phi}$.

We applied this analysis to various types of experiments formerly published[1,2,5,6] and also to our new pressure data. The parameters Φ deduced are compared to one another in figure 1. The detailed analysis and more references can be found in Ref.8.

We notice at once[8] that the data for all non X-ray experiments on vacancies are consistent with each other in the framework of a finite bandwidth vacancy model. However, the most direct method, X-ray measurement of the lattice parameter, gives a larger number of vacancies than any other experiment. If this proves to be a well established experimental fact, one will have to look more precisely into the very microscopic nature of the vacancies to solve the puzzle.

REFERENCES

1. S.M. Heald, D.R. Baer and R.O. Simmons, Phys. Rev. B30, 2531 (1984).
2. D.S. Greywall, Phys. Rev. B15, 2604 (1977) and Phys. Rev. B16, 5129 (1977).
3. J.R. Beamish and J.P. Franck, Phys. Rev. B28, 1419 (1983).
4. N. Sullivan, G. Deville, A. Landesman, Phys. Rev. B11, 1858 (1975).
5. M.E.R. Bernier, J. Low Temp. Phys. 56, 205 (1984).
6. I. Iwasa and H. Suzuki, J. Low Temp. Phys. 62, 1 (1986).
7. J.H. Hetherington, J. Low Temp. Phys. 32, 173 (1978).
8. M.E.R. Bernier and J.H. Hetherington, Phys. Rev. B to be published (1989).

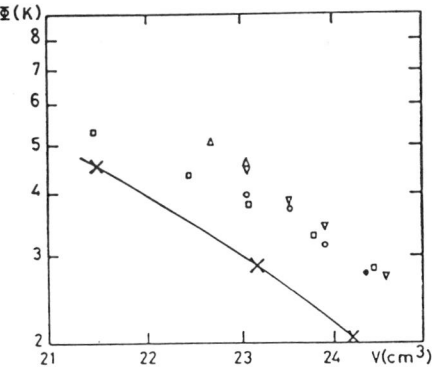

Fig.1: Vacancy band parameter Φ deduced from various experiments: □ Specific heat, o,• Pressure, △ NMR "hight T", ▽ NMR "low T", × X rays, as a function of V.

SOLID ⁴He : SEARCH FOR SUPERFLUIDITY

G. Bonfait[+], H. Godfrin[*], B. Castaing
C.R.T.B.T.-CNRS, BP 166 X, 38042 Grenoble, France
[*]I.L.L., BP 156 X, 38042 Grenoble, France

ABSTRACT

Theoreticians agree to say that the superfluidity of a Bose solid can exist. The experiments to detect such an effect have failed up to now. We present here an argument that justifies the exploration of the 1 mK-20 mK temperature range. We have made an experiment to detect a mass flow through the solid ⁴He down to 4 mK. No anomalous effect occured, this result still lowers the upper limit for the superfluidity of solid ⁴He.

INTRODUCTION

While the cryogenic technics allowed to reach lower and lower temperatures, exotics phenomena appeared in quantum fluids and solids : superfluidity of the liquids ³He and ⁴He and nuclear order for solid ³He. Only the solid ⁴He escapes such a phenomenon. However, twenty years ago, Andreev and Lifshitz[1] suggested a Bose-Einstein condensation for the vacancies existing in the solid ⁴He, this leading to a solid superfluid. During a decade, models have been proposed[1-17], and today, the theoreticians agree to state that a Bose condensation must occur, and not to exclude phenomena associated with the superfluidity. They are all very prudent on predicting the transition temperature T_c and the superfluid fraction ρ_s/ρ. Compiling the different results, one gets :

$$0.1 \text{ mK} < T_c < 100 \text{ mK}, \quad 10^{-6} < \rho_s/\rho < 0.5$$

Up to now, the experiments[18-22] performed in order to detect anomalous behaviour of ⁴He failed.

MOTIVATIONS

A calculation, ab nihilo, of the transition temperature is rather difficult but one can be helped by results from solid ³He, a rather well understood quantum solid[23,24]. In this solid, a long range order, other than crystalline, occurs at $T_N = 1$ mK (nuclear antiferromagnetic order). This order is the consequence of interatomic exchange (2, 3, 4,...particles exchange), which also exists in the ⁴He solid. We think that the Bose statistic will be revealed thanks to the atomic exchange and so, the degeneracy temperature should be the exchange temperature.

The nuclear order-temperature is obtained by a summation of the energies of the different exchange cycles. Due to the fermionic character of the ^{3}He atoms, those energies are added or subtracted dependent of the cycle parity. For ^{4}He -boson- all the energies[24] are to be added. Then, "if ^{3}He was a boson", the order would occur at 17 mK on the melting curve (v_m = 24.22 cm^3/mole). The highest molar volume of the solid ^{4}He is 21 cm^3/mole, and, if we apply the law $T_N \propto v^{18}$ found for ^{3}He, we find that the order transition T_c would be around 1.5 mK. This crude model shows that the 1 mK-20 mK temperature range has to be explored.

EXPERIMENTAL SET-UP

We have performed an experiment in order to observe a possible anomalous behaviour in the creep properties of solid ^{4}He at low temperature : if the solid becomes superfluid, we hope that creep will be accelerated (or allowed) by atoms or by vacancies transport through the solid. The principle of the experiment is shown on figure 1. The solid is grown by adjunction of atoms from an external tank. When the interface reaches the height h_o, the cell becomes separated in two parts, isolated by the solid plug. One can then obtain an height difference, thus a pressure difference δP or a chemical potential difference $\delta \mu$ between these two parts. If superfluidity appears, this $\delta \mu$ should vanish -it will be our criterium for superfluidity- and that can be done only by flow of atoms through the solid until equal level on each side is obtained. The cell (Fig. 2) is located in the mixing chamber of a dilution refrigerator. It is made of three concentric copper tubes, electrically isolated, and separately cooled down by sintered silver exchanger. We had check earlier, in a similar arrangement that the ^{4}He is thermalised down to 4 mK through the electrodes surface. The solid level detection is obtained by a capacitance measurement between two neighbour cylinders ; we are able to detect a 0.1 mm change.

We have made the superfluidity test with two crystal types. First, we grew the solid in the bcc phase and cooled it down to 4 mK. Such a procedure was performed in order to obtain an highly dislocated crystal due to the traumatism of the phase transition bcc → hcp (1,46 K)[25]. We thought so that this could help transport through defects lines. During 3.5 days at 4 mK, we have not detected any level change.

On the contrary, the other crystal type was a very "pure" one, grown very slowly (8 hours) at 4 mK. We then obtained a crystal with a very small dislocations and defects density and we hoped to repell the ^{3}He atoms (~ 10^{-7}) in the liquid phase. With this crystal, we hope to have avoided any crystal wedging due to dislocations.

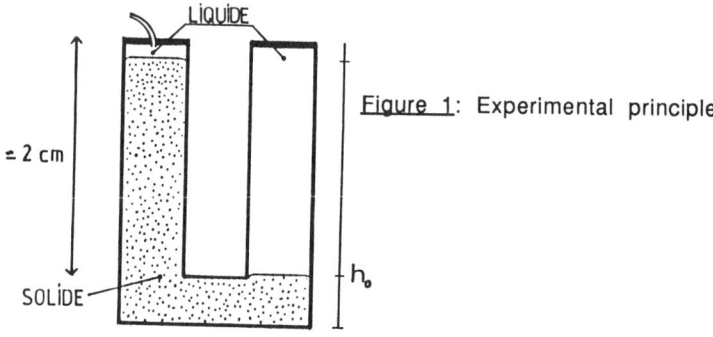

Figure 1: Experimental principle

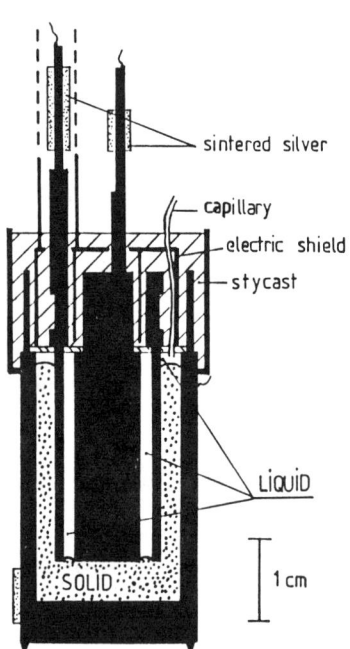

Figure 2 : Experimental cell

During 6.5 days at 4 mK, we do not have detected any level change. That gives an upper limit for the atoms velocity in the crystal below 0.2 Å/s.

CONCLUSION

Are we able to chose between the two following explanations ?
i) solid ^{4}He is not superfluid down to 4 mK
ii) solid ^{4}He is superfluid but the superfluid fraction and the critical velocity are related by :

$$\frac{\rho_s}{\rho} * v_c < 0.2 \text{ Å/s}$$

which is the same indetermination that Bishop et al[22] and Greywall[21] had obtained down to 25 mK.

It is difficult to settle between these two explanations. An order of magnitude determination[26], based on liquid ^{4}He theory, gives for the critical velocity of solid ^{4}He $v_c \cong \hbar/md \sim 5.10^4$ Å/s (d : hydrodynamic diameter). Such a critical velocity would lead to a superfluid fraction below 4.10^{-6} which is close to the lowest fraction predicted by the theoreticians. Even considering warnings from the theoreticians that signs of superfluidity in solid ^{4}He would be difficult to detect, we think that our experiment indicates that tests at lower temperature are to be performed.

REFERENCES

*Present address : Departamento de Quimica, LNETI, P-2686 Sacavem, Portugal
1. A.F. Andreev and I.M. Lifshitz, Zh. Eksp. Teor. Fiz. 56, 2057 (1969) [Sov. Phys. JETP 29, 1107 (1969)].
2. J. Sarfatt, Phys. Lett. A 30, 300 (1969).
3. H. Matsuda and T. Tsuneto, Prog. Theor. Phys. Suppl. 46, 411 (1970).
4. A.J. Leggett, Phys. Rev. Lett. 25, 1543 (1970).
5. G.V. Chester, Phys. Rev. A 2, 256 (1970).
6. D.A. Kirzhnits and Yu.A. Nepomnyashchii, Zh. Eksp. Teor. Fiz. 59, 2203 (1970) [Sov. Phys. JETP 32, 1191 (1971)].
7. R.A. Guyer, Phys. Rev. Lett. 26, 174 (1971).
8. W.J. Mullin, Phys. Rev. Lett. 26, 611 (1971).
9. K.S. Liu and M.E. Fisher, J. Low Temp. Phys. 10, 655 (1973).
10. J.F. Fernandez and M. Puma, J. Low Temp. Phys. 17, 131 (1974).
11. Y. Imry and M. Schwartz, J. Low Temp. Phys. 21, 543 (1975).
12. Y.C. Cheng, Phys. Rev. B 14, 1946 (1976).

13. A. Widom and D.P. Locke, J. Low Temp. Phys. $\underline{23}$, 335 (1976).
14. W.M. Saslow, Phys. Rev. Lett. $\underline{36}$, 1151 (1976).
15. W.M. Saslow, Phys. Rev. B $\underline{15}$, 173 (1977).
16. M. Liu, Phys. Rev. B $\underline{18}$, 1165 (1978).
17. Y.C. Cheng, Phys. Rev. B $\underline{23}$, 157 (1981).
18. A. Andreev, K. Keshishev, L. Mezov-Deglin, and A. Shalnikov, Zh. Eksp. Teor. Fiz. Pis'ma Red. $\underline{9}$, 507 (1969) [JETP Lett. $\underline{9}$, 306 (1969)].
19. H. Suzuki, J. Phys. Soc. Jpn $\underline{35}$, 1473 (1973).
20. V.L. Tsymbalenko, Zh. Eksp. Teor. Fiz. Pis'ma Red. $\underline{23}$, 709 (1976) [JETP Lett. $\underline{23}$, 653 (1976)].
21. D.S. Greywall, Phys. Rev. B $\underline{16}$, 1291 (1977).
22. D.J. Bishop, M.A. Paalanen, J.D. Reppy, Phys. Rev. B $\underline{24}$, 2844 (1981).
23. M. Roger, J.H. Hetherington, J.M. Delrieu, Rev. Mod. Phys. $\underline{55}$, n° 1 (1983).
24. H. Godfrin, D.D. Osheroff, Phys. Rev. B $\underline{38}$, 4492 (1988).
25. E.R. Grilly, J. Low Temp. Phys. $\underline{11}$, 33 (1973).
26. J. Wilks, "*The properties of Liquid and Solid Helium*", Clarendon Press, Oxford 1967.

ROUGHENING AND WETTING TRANSITIONS IN DILUTE ^{3}He—^{4}He MIXTURE CRYSTALS

Yoash Carmi
Case Western Reserve University, Cleveland, OH 44106

Stephen Lipson and Emil Polturak
Technion-Israel Institute of Technology, Haifa 32000, Israel

In a previous letter we reported on a two wetting transitions of solid ^{4}He on the Cu walls of the experimental cell[1]. Recently we found that the cause for these transitions is a very small ^{3}He contamination. Crystals grown from commercial ^{4}He do not show any wetting behavior, while crystals with a small concentration of ^{3}He, as low as 4.10^{-7}, partially wetted the Cu walls when the temperature was raised very slowly. These crystals did not wet a MICA crystal which had been cleaved in situ. The role of the ^{3}He atoms in these transitions is not clear yet. The dramatical change in the behavior of the crystals surface can be explained only if substantial amount of ^{3}He is absorbed on it. To invertigate this we performed a systematic observation on the growth shape of HCP crystals with ^{3}He concentration ranging from 8.10^{-7} to $1.5 \cdot 10^{-4}$.[2] During a slow growth the disappearance of the c-facets indicats a lowering of the roughening transition Tr from 1.28K to 1.08K (Fig. 1). In order to explain this we used a model which described successfully the analog system of the liquid-gas interface of dilute mixtures. In this model the ^{3}He atoms are described as 2D Fermi gas bound to the interface.[3] From the change of Tr[4] we could calculate the absorption of ^{3}He to be up to a quarter of a monolayer with binding energy of about 10K. The ^{3}He atoms can lower their zero point energy by migrating to the interface where the density profile has a deep minima because of the large structural mismatch between the solid and the liquid[5]. From the lowering of Tr one can conclude that the discontinuous wetting transition at 1.04K[1] is due to approaching Tr, where the c-facet becomes mobile, from below. In addition, the strange star-shapes of the growing and melting crystals (Fig. 2) indicate the existance of growth instabilities near 1K and that the ^{3}He concentration in the solid is not uniform at the time of growth.

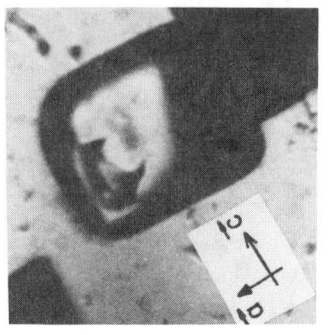
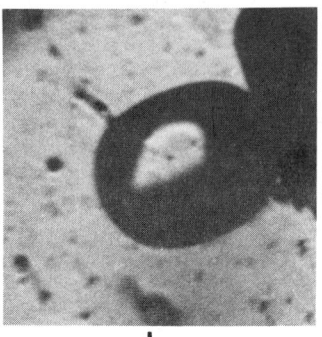

Fig. 1. Growth shapes of crystals. $x = 3 \times 10^{-5}$ near the roughening temp. Showing the disappearance of the c-facet. (a) 1.00K (b) 1.04K.

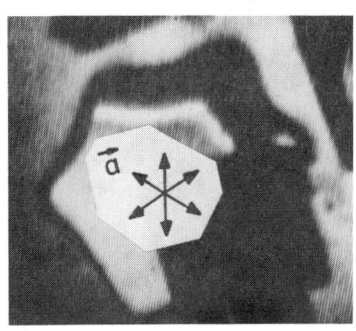
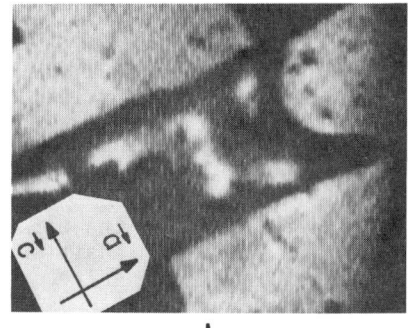

Fig. 2. Growth and melting shapes near 1K, showing growth instabilities. (T>T_R of the a facet). $x = 8.10^{-6}$. (a) Growth shape (b) Melting shape.

1. Y. Carmi, S. Lipson, E. Polturak, Phys. Rev. Lett. 54, 2042, 1985.
2. Y. Carmi, S. Lipson, E. Polturak to be published in Phys. Rev. Lett.
3. D. Edwards, W.F. Saam, Prog. in Low Temp. Phys. VIIA, D. D. Brewer, ed., 285, (North Holland, 1978).
4. R. Kariotis, H. Suhl, B. Yang, Phys. Rev. B32, 4551, 1985.
5. A. Bonissent, B. Mutaftschiev, in Chemistry and Physics of Solid Surfaces Vol. III., R. Vanselow and W. England. eds., (CRC Press) 163, 1982.

THE DYNAMICS OF THE ISOTOPIC PHASE SEPARATION IN SOLID HELIUM

N.Alikacem and M.Richards
School of MAPS, Sussex University, Brighton,England.

Isotopic phase separation in solid helium is an interesting process that has not been studied very extensively. The dynamics of the separation are probably governed by quantum diffusion; for dilute ^3He in solid ^4He, the ^3He rich phase probably consists of small crystallites (where size effects may be important) embedded in a ^4He rich matrix. Finally,there is a small range of pressures wherein the ^3He rich phase will be liquid, ie: one has the possibility of studying liquid ^3He droplets suspended in a ^4He rich solid matrix.

The isotopic phase separation of ^3He-^4He solid mixture containing 0.5% ^3He was studied at temperatures between $1K$ and $0.028K$ by pulsed N.M.R at 1MHz. The solid mixture was formed from the liquid at constant pressure. Two starting pressures were used: firstly, high pressure ($P = 32 \pm 0.2bar$) where the new ^3He rich phase was solid. Secondly,low pressure ($P = 28 \pm 0.3bar$) where the new phase was liquid. It is important to note that our sample was in the form of a long cylinder of diameter of 1 cm and the N.M.R coil occupied the middle third of it.

I- HIGH PRESSURE CASE: $P = 32 \pm 0.2bar$.

a) Growth Of The Signals: A typical run consists of solidifying the liquid at $1.9K$ for approximately 2 hours . N.M.R signals were studied for calibration purposes at $1K$ and the crystal was then cooled slowly to $0.1K$. The onset of phase separation for a 0.5 % mixture was at $0.2K$. The N.M.R signals coming from the ^3He and ^4He rich phase could be monitered separately because of their widely different spin lattice relaxation time T_1. Fig. 1 show a typical plot of signal growth over an 8 hour period . We have other such plots and all shows the following features:

(i) A growth over about 2 hours of the ^3He rich phase till it provides about ($60 \pm 10\%$) of the signal. This is consistent with the ratio of the ^3He atoms in the two phases predicted[1] at $0.1K$.

(ii) The total number of ^3He atoms in the region inside the r.f coil grows with the same time constant.

(iii) Following the increases in (i) and (ii) there is a slower growth or decay of the signals from both phases. Apparent equilibrium is reached after about 6 hours .

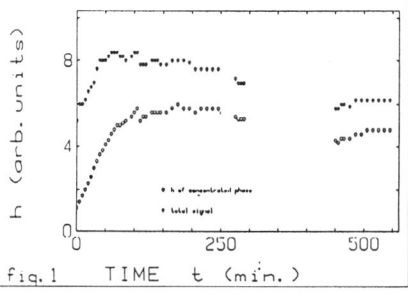
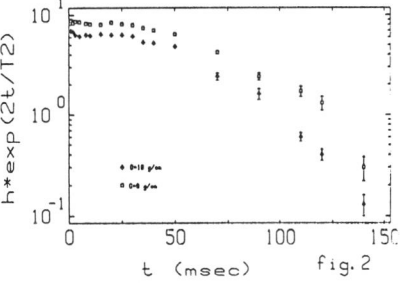

b) The Relaxation Times: (i) T_1 and T_2 (transverse relaxation time) were measured at $0.4K$ and found to be $35 sec$ and $15 msec$ respectively.
(ii) spin-lattice relaxation was measured during the isotopic phase separation and after apparent equilibrium was achieved. In both cases, T_1 was made up of two different components. The short component was assumed to relate to ^3He rich phase and stayed constant at $T_{1S} = 0.35 \pm 0.05 sec$ during the separation process. The long component, comming from the ^4He rich phase, increased during the phase separation, reflecting the depletion of ^3He in this phase.
(iii) The transverse relaxation time, T_2, was measured in the ^3He rich phase only, and it was found to be close to T_{1S} ($T_{2S} = 0.26 \pm 0.05 sec$). these values were appropriate to pure solid ^3He with a molar volume of 24 cm^3 [2].
c) Spin Diffusion: At apparent equilibrium, the damping of the spin echo signal (from the ^3He rich phase) in the presence of a magnetic field gradient was measured. Fig.2 shows data for two different gradients. The data can be fitted to a model of diffusion bounded by a length a in the direction of the gradient[3]. A value of the spin diffusion coefficient, D, of $4 \times 10^{-7} cm^2 sec^{-1}$ and $a = 2 \times 10^{-4} cm$ result, the former being of the order of magnitude found in bulk solid ^3He[2].
The above results can be contrasted with those of Mikheev[4] who worked on a similar sample but at a higher field. He studied the concentrated phase only and observed its growth. However, his diffusion data were markedly different.

II- LOW PRESSURE CASE: $P = 28 \pm 0.3 bar$

a) The Magnetic Susceptibility: At this pressure, the solid mixture was formed from the liquid at $T = 1.65K$. Then, it was cooled slowly to $T = 0.028K$. The spin echo signal of the mixture , being proportional to the magnetic susceptibility $\chi(T)$, was measured from $1K$ to $0.028K$. $\chi(T)$ shows the characteristic degerency effects of a Fermi fluid.
b) Relaxation Times at $0.028K$: Both (T_1) and $(T2)$ were found to be non exponential, but this is attribued the relaxation being at the surface of the ^3He liquid droplets since at this temperature there should be a negligeable number of spins in the dilute (solid) phase.
c) Diffusion coefficient below $0.1K$: The damping of spin echoes was studied in various field gradients, and cannot be described either by bounded diffusion or unbounded diffusion. for instance, there is a little variation with temperature despite D in liquid ^3He being proportional to T^{-2}. It may be that the decay of the spin echoes is governed by ^3He spins leaking out the droplets into dilute phase.

References:
1) W.J.Mulin,Phys.Rev.Lett.,20,254,(1968).
2) R.C.Richardson,D.A.Guyer,and L.I.Zaner,Rev.Mod.Phys.,143, 532,(1971).
3) B.Robertson,Phys.Rev.,151,273,(1966).
4) V.Mikheev et al.,Sov.J.Low.Temp.Phys.,12(6),376,(1986).

NUCLEAR MAGNETIC ORDERING

CHAIRMAN

Robert C. Richardson
Department of Physics
Cornell University

NUCLEAR MAGNETISM IN COPPER AT NANOKELVIN TEMPERATURES

H. E. Viertiö and A. S. Oja*
Low Temperature Laboratory, Helsinki University of Technology,
SF-02150 Espoo, Finland

ABSTRACT

The antiferromagnetic phases of nuclear spins in copper are investigated theoretically. We have found in our Monte-Carlo simulations two ordered spin structures as a function of the external magnetic field applied in the [110] direction and three phases for the field in the [001] direction, in agreement with experiments. Our spin-wave analysis indicates a competition between quantum and thermal fluctuations: the $T = 0$ spin state in zero field is predicted to be a four-sublattice structure while a two-sublattice state is stabilized at $T \lesssim T_c$. The recent experiments on the phase diagram in a magnetic field are briefly reviewed as well as progress in calculating the indirect spin-spin interactions from first principles

INTRODUCTION

During the past few years nuclear magnetism in metals has become a subject of considerable progress. In the quest for ever lower temperatures adiabatic demagnetization[1] is the best procedure: the only remaining entropy in a metal which can be extracted is due to the nuclear spins. At the lowest temperatures these systems show complex ordering phenomena. Yet, the underlying microscopic properties are relatively simple: the nuclear spins are well localized and the spin-spin interactions can be accurately determined. Nuclear magnets, with the possibility of isotopic substitution for introducing disorder, could provide the best experimental realizations of many tractable mathematical models.

The pure metals which have been investigated for nuclear ordering are copper,[2,3] silver,[4] thallium,[5] and praseodymium.[6] Nuclear spins interact via the dipolar and the conduction-electron-mediated exchange interactions, the latter being the dominant coupling mechanism in silver[7] and, particularly, in thallium.[8] This is in a marked contrast to the situation in insulators, in which the nuclear spins are coupled only through the dipolar interaction.[9] Nuclear spins in copper form a unique system in which the antiferromagnetic total-spin-conserving exchange interaction is of the same order of magnitude as the ferromagnetic total-spin-nonconserving dipolar interaction. Competition between these two profoundly different coupling mechanisms results in diverse behavior.

In this paper, we first review the main results of experimental studies of nuclear magnetic ordering in copper.[10] We present the phase diagram in a magnetic field, as was determined in the early susceptibility measurements.[11] We then discuss the recent neutron diffraction experiments,[3] in which the translational period of the magnetic structure was determined. In the theoretical section, the interactions between nuclei are described. The spin configurations of the ordered phases are then discussed.

*Present address: Physics Department, Risø National Laboratory, DK-4000 Roskilde, Denmark.

© 1989 American Institute of Physics

We present some new results of Monte-Carlo simulations and a spin-wave analysis, deriving the macroscopic behavior from microscopic properties.

SUSCEPTIBILITY MEASUREMENTS

Nuclear ordering in copper has been extensively investigated by means of magnetic susceptibility measurements[2], using the Helsinki double stage nuclear demagnetization cryostat.[12] In a typical experiment, the second nuclear stage, i.e., the sample, is demagnetized in a time much shorter than the spin-lattice relaxation time. Therefore, only the nuclear spins cool down while the conduction electrons stay at a much higher temperature on order of 100 µK, the temperature of the first nuclear stage. After the sample has been demagnetized, its low-frequency magnetic susceptibility is monitored while the nuclei warm up towards the electronic temperature. The nuclear antiferromagnetic order shows up in the initial increase of the susceptibility. Later, when the nuclei have warmed into the paramagnetic phase, the susceptibility decreases monotonically with temperature.

The ordered state of copper nuclei in a magnetic field was investigated on a single crystal specimen by measuring the susceptibility in all three Cartesian directions.[11] The external magnetic field was applied in a direction almost parallel to the [001] crystalline axis. The measurements indicated three characteristically different regions as a function of the external field.

The experimental phase diagram in the magnetic field vs. entropy plane is shown in Fig. 1. The three antiferromagnetic phases AF1, AF2, and AF3 are separated by first-order transition regions. Transitions from AF1 and AF2 to the paramagnetic state (P) are also of first order. The critical temperature has been measured in zero field: $T_c = 58 \pm 10$ nK.[13]

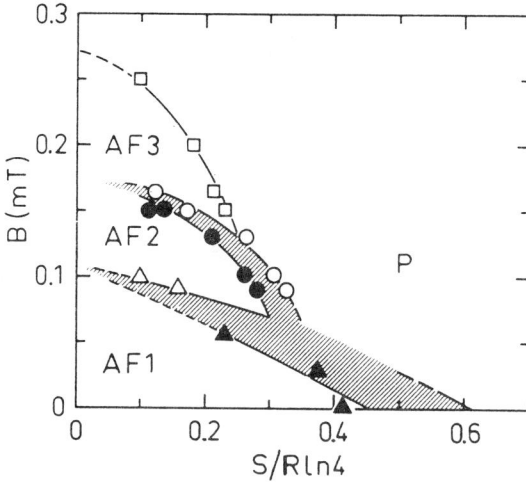

Fig. 1. The external magnetic field vs. entropy diagram[11] of copper nuclear spins. AF1, AF2, and AF3: antiferromagnetic phases. P: paramagnetic phase. Shaded areas indicate first-order transition regions.

NEUTRON DIFFRACTION EXPERIMENT

The neutron diffraction technique provides a microscopic tool for probing the magnetically ordered structure of nuclear spins. The neutron-nucleus scattering length is

$$a = b_o + b\mathbf{I}\cdot\mathbf{s} \qquad (1)$$

The last term depends on the direction of the nuclear spin **I**. In an elastic scattering experiment, the spin-independent scattering length b_0 gives rise to structural reflections while the spin-dependent term proportional to b, will result in additional Bragg reflections if $<I_i> \neq 0$, i.e., the nuclei are magnetically ordered.

The neutron diffraction study of the ordered structure of copper nuclei was undertaken in a collaboration between Risø National Laboratory (Denmark), Hahn-Meitner Institut (West-Berlin), and the Helsinki Low Temperature Laboratory. Fig. 2 shows the measured (001) neutron intensity after the nuclei had been cooled by the final demagnetization.[3] In zero field (Fig. 2(a)) a pronounced neutron reflection is observed for the first five minutes. The simultaneously measured static susceptibility, presented in the inset, displays almost a plateau at the beginning, also indicating an antiferromagnetic order.

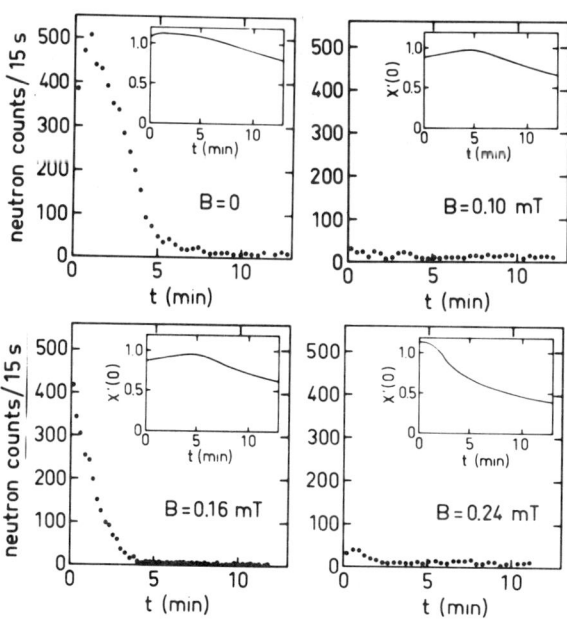

Fig. 2. The measured (001) neutron intensity as a function of time after demagnetization, at various external magnetic fields in the [110] direction.[3] The insets show the behavior of the simultaneously measured static susceptibility.

Similar experiments[3] were carried out in external fields in the [110] direction (Figs. 2 (b)-(d)). After demagnetization to 0.10 mT, hardly any neutrons were observed above the background. However, the initial increase of the susceptibility suggests antiferromagnetic order. The neutron reflection was observed again at 0.16 mT. Finally, at 0.24 mT, only a small neutron intensity was observed. The neutron signal vanishes above the critical field $B_c = 0.25$ mT.

These results show that at least two antiferromagnetic phases, with a characteristic (001) reflection, exist in copper. The translational period at the intermediate fields remains unknown. Other neutron reflections, such as (0 0 1/2) and (1/2 1/2 1/2), have been searched for but not observed.[14]

INDIRECT INTERACTIONS

Nuclear spins in a metal interact via the dipolar force, which is known with great accuracy, and via the indirect conduction-electron-mediated force. The most important indirect coupling mechanism in copper is the Ruderman-Kittel (RK) interaction,[15] which arises in the second-order perturbation theory from the contact term of the hyperfine Hamiltonian. The RK interaction is isotropic in spin space, viz.,

$$H_{RK} = \tfrac{1}{2} \sum_{ij} J_{ij} \mathbf{I}_i \cdot \mathbf{I}_j. \tag{2}$$

The coupling constant J_{ij} oscillates with the distance between the spins. The value of ΣJ_{ij} could be determined from NMR measurements at high spin polarizations.[16] It was found to be positive, indicating an antiferromagnetic spin-spin interaction. The RK force is competing with the ferromagnetic dipolar force: in copper both are of the same order of magnitude.

The indirect interactions between nuclear spins can be derived from the electronic energy bands and wave functions. The RK interaction in copper has recently been calculated by several groups.[17-19] The results by Lindgård, Wang, and Harmon[17] are illustrated in Fig. 3, together with the free-electron approximation of the RK interaction. The antiferromagnetic

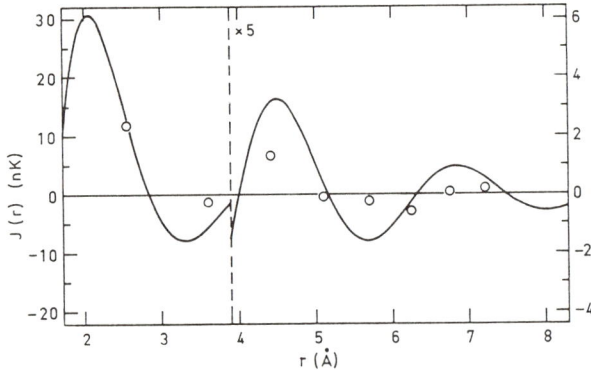

Fig. 3. The Ruderman-Kittel interaction in copper. The full line denotes the free-electron RK force, the magnitude of which was determined by NMR measurements.[16] Circles are the values calculated by Lindgård, Wang and Harmon.[17] Note the change in the vertical scale at $r = 3.9$ Å.

nearest neighbor coupling clearly dominates over the more distant interactions.

Besides the isotropic Ruderman-Kittel interaction between nuclei, the electron-nuclear dipolar and orbital interactions give rise to significant anisotropic coupling terms.[19-21] The symmetry of the anisotropic terms is not dipolar in form but is determined by the symmetry of the crystal.[19] The anisotropic interactions in copper tend to destroy the (001) order by stabilizing a spiral structure. Numerical calculations have not been precise enough to determine accurately the extremely small energy difference between these structures.

ORDERED SPIN CONFIGURATIONS

The ordered structure of nuclear spins in copper was first studied by using the mean field approximation. Kjäldman and Kurkijärvi[22] correctly predicted the (001) ordering. However, the mean field theory is inadequate to identify the ground-state spin configuration: instead it results in a continuum of degenerate states, corresponding to combinations of antiferromagnetic spin modulations in the x-, y-, and z-directions.

The degeneracy of the ground state can be expected to be lifted by thermal and quantum fluctuations. These can be accounted for by employing different theoretical models depending on the temperature regime: we have applied the Ginzburg-Landau expansion at $0 \ll T \lesssim T_c$,[23] the spin-wave (SW) theory at $0 \leq T \ll T_c$,[24] and the classical Monte Carlo (MC) method [25] at $T > 0$.

The antiferromagnetic order parameter describing the (100), (010), and (001) ordering in copper is six-dimensional. We have classified[23] the possible stable spin configurations in zero field by analyzing the Ginzburg-Landau expansion of the free energy. Depending on the expansion coefficients, several local minima, corresponding to metastable states, can be found. Unfortunately, the values of the coefficients are presently unknown.

Our model system for MC simulations consisted of $12^3/2$ classical spins in an fcc lattice with periodic boundary conditions. The dipolar and the indirect Ruderman-Kittel interactions between nuclear spins were truncated to the two nearest neighbors; the coupling constants J_{ij}/k_B were chosen as 12.5 nK and -5.6 nK, respectively, consistently with NMR measurements.[16] The critical temperature of the model was found to be $T_c^{MC} = 65 \pm 2$ nK, in excellent agreement with the experimental result.

The ordered structures were determined by simulated annealing from the paramagnetic to the antiferromagnetic state at $T = 0.2\, T_c^{MC} = 16$ nK as a function of the external magnetic field. At every studied field value several runs of 4000 - 20000 Monte-Carlo steps per spin were performed. In Fig. 4 we present the resulting spin configurations in external magnetic fields in the [001] and [110] directions, corresponding to the experiments in Helsinki and Risø, respectively. The calculated phases are indicated outside the coordinate axes and the experimental phases inside. The field values of the Monte-Carlo simulations have been scaled down in order to make the critical fields given by the simulations and the experiments coincide. The long range of the dipolar interaction makes B_c depend on the shape of the sample. In our MC simulations the sample is a sphere; then the Lorentz and the demagnetization fields exactly cancel each other.[9] The samples used in the experiments[3,11] had the shape close to a parallelepiped with one of the edges much longer than the other two. The field was applied along the long edge; then the demagnetizing field almost vanishes. In order to obtain the prediction for B_c for this geometry, the Lorentz field 0.41 mT has to be subtracted from $B_c\text{MC} = 0.82$ mT found in our simulations for a spherical MC sample: this yields $B_c\text{MC} = 0.41$ mT, which is 50 % larger than the experimental B_c of 0.27 mT. The discrepancy cannot be explained by the fact that the interactions in the simulations have been truncated to the two nearest neighbors. If the range of the dipolar and the free-electron RK interactions is extended up to infinity, B_c increases from 0.41 mT to 0.61 mT. The latter value for B_c is obtained by using the MF expression by Kumar et al.[26]; at $T = 0$ one has for classical spins $B_c\text{MC} = B_c\text{MF}$.

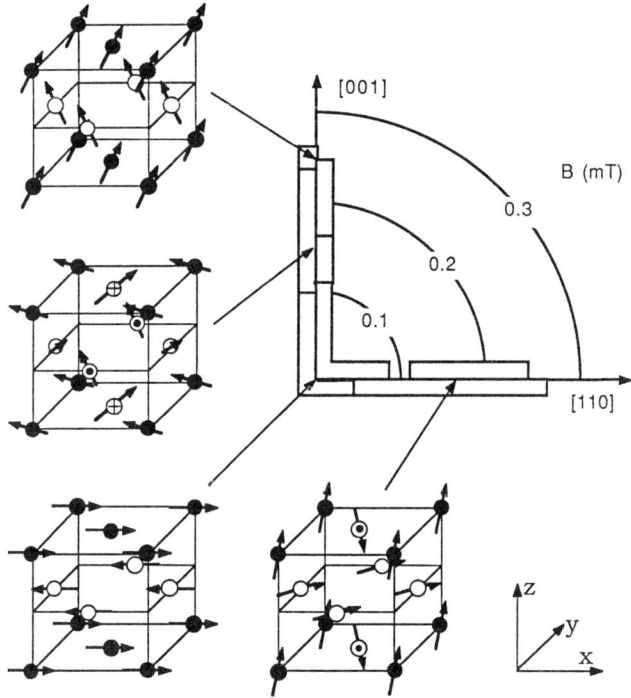

Fig. 4. Spin configurations of the nuclear spin system of copper at $T = 16$ nK as given by our Monte-Carlo simulations. The applied magnetic field in the static susceptibility measurement[11] was in the [001] direction while in the neutron diffraction experiment[3] the field was in the [110] direction. The calculated phases are indicated outside the coordinate axes and the experimental phases inside.

In zero field, the spin configuration is a simple two-sublattice structure with spins parallel to a crystalline axis. With increasing field in the [001] direction, the structure first changes continuously as the spins tilt towards the field direction, until a discontinuous transition to a more complex four-sublattice structure occurs. With further increase in the field, another first-order transition back to the simple two-sublattice state takes place.

In the [110] direction of the field, two phases were found. The high-field structure is a three-sublattice configuration with half of the spins parallel to the field and the rest forming antiferromagnetic pairs tilted towards the field. At low fields we again find the two-sublattice configuration. The number of phases in both field directions is in agreement with the experiments.[3,11]

Some experimental results[3] cannot be understood on the basis of our MC simulations. In addition to the too high theoretical B_c, the disappearance of the neutron reflection between the low-, and high-field phases in the [110] field direction remains unexplained. Another problem arises when a small field in the [110] direction is applied to the zero-field two-sublattice structure. The antiferromagnetic modulation then locks in either x-, or y-direction. Therefore, a single domain state should not give rise to the observed (001) Bragg peak.

Fig. 5 (a) shows in more detail our MC diagram for copper nuclei. In addition to the [001] and [110] axes, three intermediate directions in the field plane $B_x = B_y$ are presented. The empty regions correspond to field values at which the statistics of the calculations performed was insufficient to choose the stable structure among several metastable ones.

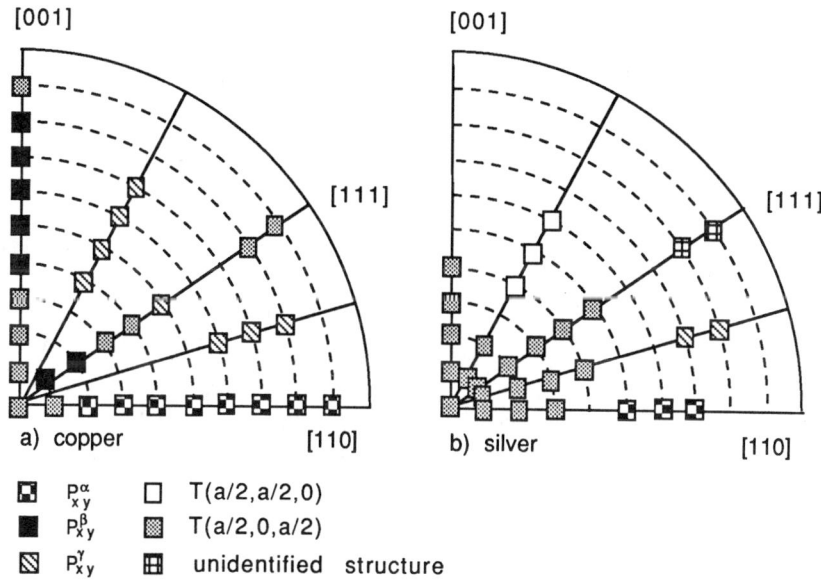

Fig. 5. The phase diagrams of nuclear spins in (a) copper and (b) silver as functions of the magnetic field, determined by simulated annealing to T = 0.2 T_cMC. The magnetic field is in the plane $B_x = B_y$, which contains the high-symmetry directions [001], [111], and [110]. The phases $P\alpha_{xy}$, $P\beta_{xy}$ and $P\gamma_{xy}$ correspond to four-sublattice structures with the plane x = y as the mirror plane of reflectional symmetry.[24] The T(R)-configurations are simple two-sublattice structures which are invariant under a translation by a vector **R**.

Frisken and Miller[27] were the first to report Monte-Carlo simulations of nuclear magnetic ordering in copper. They investigated only the case of the field in the [001] direction, for which they found four antiferromagnetic phases. The highest-, and lowest-field configurations are the same as those obtained in our simulations. The differences in the intermediate fields are presumably caused by the somewhat different Hamiltonians.

For comparison, Fig. 5 (b) presents the corresponding phase diagram for silver. The strength of the indirect interaction, in comparison to the dipolar force, is about four times greater in silver than in copper. This gives rise to major differences in the phase diagrams. The salient feature is the great portion occupied by simple two-sublattice phases in the diagram of silver. The transition temperature of the model was found to be as low as 0.5 nK.

The tendency of the isotropic interaction to stabilize simple collinear structures was first discovered in our spin wave analysis,[24] and later found also in a cluster calculation by Lindgård.[28] The zero-point energies of spin waves lift the degeneracy of the mean-field ground state, resulting in phase diagrams of Fig. 6 (a) for copper and 6 (b) for silver. In the spin-wave calculation the interactions were summed up to infinity; for the RK interaction the free-electron approximation was employed. Despite the different temperatures and interaction ranges in the MC and SW studies, the resulting phase diagrams are very similar: the only essential difference is the larger coverage of the T(a/2,0,a/2) phase in the MC diagram.

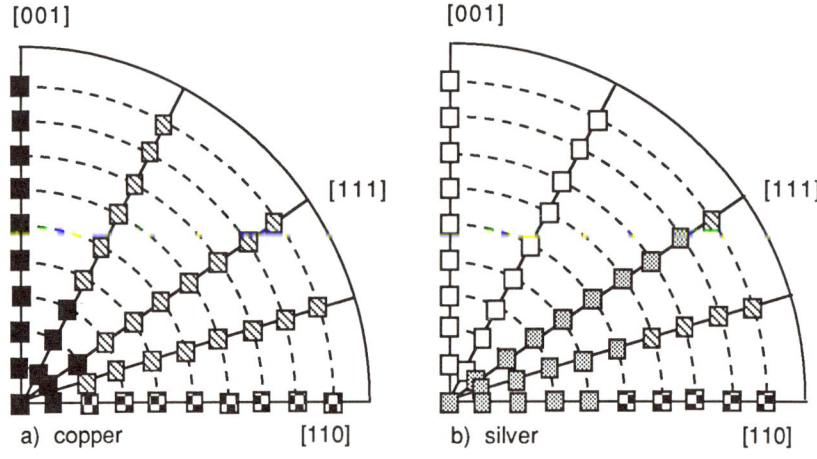

Fig. 6. The phase diagrams of nuclear spins in (a) copper and (b) silver at zero temperature as functions of magnetic field in the plane $B_x = B_y$, calculated by the theory of noninteracting spin waves.

In a recent paper, Lindgård[28] introduces a four-spin cluster model for describing the nuclear spin system of copper. By employing the second order perturbation theory, Lindgård calculates how the isotropic and dipolar nearest neighbor interactions lift the degeneracy of the previously obtained[22,26] MF ground state. A rich phase diagram is found as functions of the magnetic field and the ratio of the two interactions, J/D. For a narrow range of values of J/D, three phases are again found when the field is in the [001] direction, and two phases for the field in the [110] direction. The spin configurations in these phases are different from those found in our MC simulations.

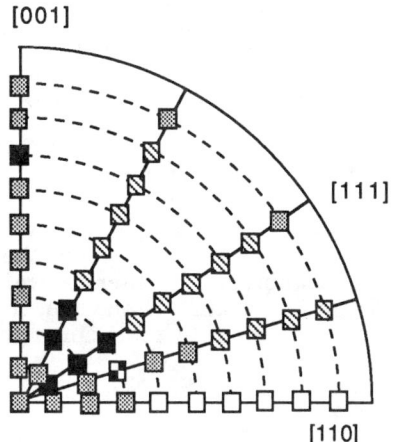

Fig. 7. The phase diagram of nuclear spins in copper at T = 140 nK as a function of the magnetic field in the plane $B_x = B_y$, calculated by the theory of noninteracting spin waves.

With increasing temperature, the SW diagram becomes even more similar to the MC diagram. In Fig. 7 we show the phase diagram of copper, calculated by the SW model at T = 140 nK. The T(a/2,0,a/2) phase now appears essentially in the same regions as in Fig. 5(a).

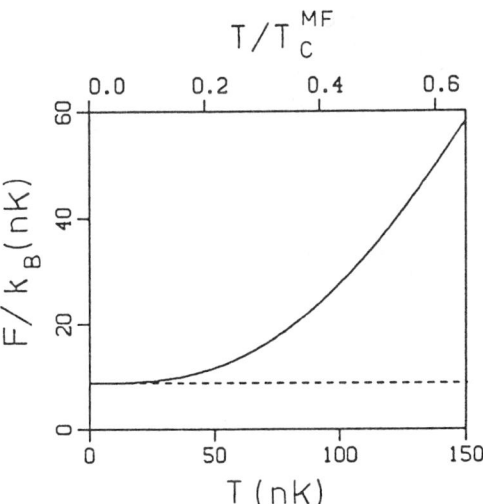

Fig. 8. The free energy F per copper nuclear spin as a function of temperature T for the ground-state structure, calculated by the theory of noninteracting spin waves. The classical ground-state energy of the system has been chosen as the zero of free energy. The solid curve presents the total free energy, while the dashed line indicates the contribution of quantum fluctuations, i.e. the zero-point energy.

The SW theory can also be employed to estimate the relative importance of thermal and quantum fluctuations. In Fig. 8 we present the free energy per copper spin as a function of temperature in zero field, calculated for the ground-state structure. At T = 0 free energy equals the zero-point energy of spin waves, but already at T = 80 nK = 0.35 T_c^{MF} thermal fluctuations make an equally large a contribution. If the temperatures of the spin-wave calculation are scaled down according to the observed T_c of 60 nK, thermal fluctuations are found to dominate above 21 nK. Because the lowest temperature that has been measured in the copper experiments is around 25 nK, one can conclude that the phases observed in the experiments are mainly stabilized by thermal rather than quantum fluctuations. Classical Monte Carlo simulations are, therefore, expected to give results which are relevant to the interpretation of the experiments.

CONCLUSION

Considerable progress has been made in the understanding of the ordered state of copper nuclei, both in experiment and theory. Several antiferromagnetic phases as a function of the external magnetic field have been discovered in static susceptibility and neutron diffraction measurements. The rich phase diagram results from the strongly competing dipolar and indirect interactions as was shown by our Monte-Carlo simulations and spin-wave analysis and other theoretical work. [27,28] Different theoretical approaches yield, however, different phase diagrams and spin configurations. The discrepancies are partly due to the different indirect exchange interactions used in the calculations. The recent first-principles electronic structure calculations[17,19] of indirect interactions are couraging in the sense that the exchange constants in copper are now in excellent agreement with various NMR measurements.[16,29] Hence an accurate microscopic Hamiltonian is available. It is possible that the ordered structures can soon be identified reliably with the theoretical predictions, as more neutron diffraction data become abailable in different directions of the external field.

In future experiments, it would be interesting to extend the measurements on copper to lower temperatures, where quantum fluctuations select the stable state. Studies of nuclear magnetism in other metals are, of course, interesting as well. The strongly exchange-dominated silver is a rather good realization of the extensively studied theoretical model in which spins in a fcc lattice interact via isotropic antiferromagnetic nearest-neighbor forces. The ground-state properties of this model are still unsolved.

ACKNOWLEDGMENTS

We would like to thank B.N. Harmon, P. Kumar, P.-A. Lindgård, O.V. Lounasmaa, M. Steiner, Y. Takano, and X.-W. Wang for useful discussions, and especially K. Siemensmeyer for pointing out the need for the Lorentz field correction to the critical field when the Monte-Carlo results are compared to the experiment. This work has been supported by the Academy of Finland. One of us (HEV) thanks the Jenny and Antti Wihuri and the Magnus Ehnrooth Foundations.

1. K. Andres and O.V. Lounasmaa in Progress in Low Temperature Physics, edited by D.F. Brewer (North-Holland, Amsterdam, 1982), Vol. VIII, p 222.
2. M.T. Huiku, T.A. Jyrkkiö, J.M. Kyynäräinen, M.T. Loponen, O.V. Lounasmaa, and A.S. Oja, J. Low Temp. Phys. $\underline{62}$, 433 (1986).
3. T.A. Jyrkkiö, M.T. Huiku, O.V. Lounasmaa, K. Siemensmeyer, K. Kakurai, M. Steiner, K.N. Clausen, and J.K. Kjems, Phys. Rev. Lett. $\underline{60}$, 2418 (1988).
4. A.S. Oja, A.J. Annila, and Y. Takano, unpublished.
5. G. Eska and E. Schuberth, Jap. J. Appl. Phys., Suppl. $\underline{26\text{-}3}$, 435 (1987).
6. S. Kawarazaki, N. Kunitomi, J.R. Arthur, R.M. Moon, W.G. Stirling, and K.A. McEwen, Phys. Rev. $\underline{B37}$, 5336 (1988).
7. J. Poitrenaud and J.M. Winter, J. Phys. Chem. Solids $\underline{25}$, 123 (1964).
8. Yu. S. Karimov and I.F. Shchegolev, Zh. Eksp. Teor. Fiz. $\underline{41}$, 1082 (1961). [Sov. Phys. JETP $\underline{14}$, 772 (1962)].
9. A. Abragam and M. Goldman, Nuclear Magnetism: Order and Disorder, Clarendon Press, Oxford (1982).
10. For more exhaustive reviews, see A.S. Oja, Physica Scripta $\underline{T19}$, 462 (1987); Ref 2; O.V. Lounasmaa, to be published in Physics Today.
11. M.T. Huiku, T.A. Jyrkkiö, J.M. Kyynäräinen, A.S. Oja, and O.V. Lounasmaa, Phys. Rev. Lett. $\underline{53}$, 1692 (1984).
12. G.J. Ehnholm, J.P. Ekström, J.F. Jacquinot, M.T. Loponen, O.V. Lounasmaa, and J.K. Soini, J. Low Temp. Phys. $\underline{39}$, 417 (1980).
13. M.T. Huiku, T.A. Jyrkkiö, M.T. Loponen, Phys. Rev. Lett. $\underline{50}$, 1516 (1983).
14. T.A. Jyrkkiö, M.T. Huiku, K. Siemensmeyer, and K.N. Clausen, J. Low Temp. Phys. $\underline{74}$, 435 (1989).
15. M.A. Ruderman and C. Kittel, Phys. Rev. $\underline{96}$, 99 (1954).
16. J.P. Ekström, J.F. Jacquinot, M.T. Loponen, J.K. Soini, and P. Kumar, Physica $\underline{98B}$, 45 (1979).
17. P.-A. Lindgård, X.-W. Wang, and B.N. Harmon, J. Magn. Magn. Mater. $\underline{54\text{-}57}$, 1052 (1986).
18. S.J. Frisken and D.J. Miller, Phys. Rev. Lett. $\underline{57}$, 2971 (1986); Phys. Rev. $\underline{B37}$, 10884 (1988).
19. A.S. Oja, X.-W. Wang, and B.N. Harmon, Phys. Rev. $\underline{B39}$, 4009 (1989).
20. N. Bloembergen and T.J. Rowland, Phys. Rev. $\underline{97}$, 1679 (1955).
21. A.S. Oja and P. Kumar, J. Low Temp. Phys. $\underline{66}$, 155 (1987).
22. L.H. Kjäldman and J. Kurkijärvi, Phys. Lett. $\underline{71A}$, 454 (1979).
23. A.S. Oja and H.E. Viertiö, Jap. J. of Appl. Phys., Suppl. $\underline{26\text{-}3}$, 441 (1987).
24. H.E. Viertiö and A.S. Oja, Phys. Rev. $\underline{B36}$, 3805 (1987).
25. H. Metropolis, A.W. Rosenbluth, M.N. Rosenbluth, A.H. Teller, and E. Teller, J. Chem. Phys. $\underline{21}$, 1087 (1953).
26. P. Kumar, J. Kurkijärvi, and A.S. Oja, Phys. Rev. $\underline{B33}$, 444 (1986).
27. S.J. Frisken and D.J. Miller, Phys. Rev. Lett. $\underline{61}$, 1017 (1988).
28. P.-A. Lindgård, Phys. Rev. Lett. $\underline{61}$, 629 (1988).
29. E.R. Andrew, J.L. Carolan, and P.J. Randall, Phys. Lett. $\underline{37A}$, 125 (1971).

NONLINEAR SPIN DYNAMICS AND NUCLEAR ORDERING: Tl METAL AS AN EXAMPLE

G. Eska
Physikalisches Institut, Universität Bayreuth, D-8580 Bayreuth, FRG

ABSTRACT

The spin dynamics of highly polarized nuclear spins is considered for the case of indirect exchange coupling between two spin species. The equations of motion are solved numerically in the molecular field approximation, showing that for strong excitations nonlinear response is expected in NMR experiments. In addition, the minority spin is also treated in the spin wave approximation as an impurity embedded in an almost ferromagnetic host. The models are compared with experimental results available on Tl.

INTRODUCTION

To investigate nuclear ordering phenomena in solids, an understanding of the spin dynamics of highly polarized systems is necessary, especially when nuclear magnetic resonance is used as the experimental technique. In insulators the dipolar interaction between the nuclear spins is the dominant interaction responsible for nuclear ordering which was extensively studied by Abragam's group[1]. In pure metals the indirect exchange interaction of the Rudermann-Kittel type may compete (Cu, Ag) or may even exceed the dipolar interaction as in heavy metals (Tl). So far the only pure metals investigated for nuclear ordering are $Cu^{2,3}$, Ag^4 and Tl^5. All of them have two stable isotopes and their coupling must be included in the equations of motion. This problem was solved[6,7] for the coupling between electronic and nuclear moments (μ) in the case of negligible nuclear polarization (p) and small deviations of the nuclear moments from their equilibrium position in the direction of the external applied field (B_0). These results, valid for FMR or AFMR, can not generally be adapted for NMR at very low temperatures. At large tipping angles (θ) the equations of motion for a system of two nuclear spin species must be solved numerically. This was done in the molecular field approximation assuming an isotropic exchange interaction, strong enough to neclect the dipolar coupling as is the case in Tl.

The exchange frequency for Tl was first determined by Bloembergen and Rowland[8] to $J = A/h = 17.5$ kHz and later by Karimov and Shchegolev[9] to $J = 37.5$ kHz. With this strong exchange nuclear ordering should occur in the μK temperature range as first pointed out by Bloembergen and Rowland[8].

To search for this ordering pulsed NMR experiments were performed below 100 μK by Eska and Schuberth[5,10]. Tl is well suited for NMR work because the two isotopes ^{203}Tl and ^{205}Tl of natural abundance $\alpha_{203} = 0.295$ and $\alpha_{205} = 0.705$ have spin $I = 1/2$ and gyromagnetic factors $\gamma/2\pi$ of 24.33 MHz/T and 24.57 MHz/T. Therefore

one can work at convenient frequencies of around 0.5 MHz in fields of the order of 20 mT.
As Tl is a superconductor with $T_c^{sc} = 2.39$ K and $H_c^{sc} = 17.1 \times 10^{-3}$ A/m the static magnetic field for the NMR was always above 18 mT to keep the samples in the normal state, and it was expected that mainly effects caused by high spin polarization rather then by nuclear ordering will be observable. AF nuclear ordering should be suppressed in such high fields because in the molecular field approximation the critical field for spontaneous nuclear ordering can be estimated to be of the order of $B_c \cong zJh/\mu \cong 3.7$ mT, 5 times less the applied field ($z = 12$ is the number of nearest neighbors). Nevertheless, the applied field is not too far from B_c and one can speculate that Tl may behave somewhat similar to solid ^{3}He where some kind of ordering phenomena show up even at fields far above $B_c (\cong 0.4$ T) and at temperatures above T_c ($\cong 1$ mK).

This would make Tl a unique material to study a metal under equilibrium conditions where the nuclear Zeeman energy equals the thermal energy at a polarization of $p \cong 0.7$ (corresponding to $T_c \cong 20$ μK and $B_c \cong 18$ mT) and where these two energies compete with the Cooper pairing energy of the conduction electrons which mediate the ordered nuclear state, evidently a fascinating situation.

Indeed, several unusual features were observed experimenmentally [5,10], and one of them is of interest here: For polarizations $p > 0.4$ changes in the NMR line width and amplitudes occured; especially a line splitting showed up which did not depend on temperature and only weakly on field. Normally two lines (sometimes three) were observed which had 30 % less line width but, depending on the field gradient, up to 5 lines could also be seen. As this behavior can not be explained in the model worked out by Oja et al.[11] two other approaches will be discussed in this paper: first, as mentioned, spin dynamics within the molecular field approximation, and second, ^{203}Tl as "impuriton" in the ^{205}Tl matrix. Depending on spin concentration and impurity distribution this may cause spin glass behaviour as well as/or spin wave phenomena. The easiest case is considered. The impurity spin points in the direction of the host magnetization and deviations from this direction are small. As Cu at high polarizations should behave similar to Tl, many of the numerical results are given for Cu because first, this metal is easier to handle for the experimentalist and the predictions can easier be verified, and second, many of the nonlinear effects in the spin dynamics could not be seen in low field experiments on Tl due to the four times smaller g-factor difference between the two isotopes and the 65 times stronger exchange interaction which causes a merging of the two isotopes into one NMR line[9].

SPIN DYNAMICS

The equations of motion for a system of two spin species coupled by indirect exchange and dipolar interaction are given in the molecular field approximation by

$$\frac{d}{dt}\vec{m}_{1,2} = \gamma_{1,2}[\vec{m}_{1,2} \times \vec{B}_{eff}] - \vec{m}_{1,2}/\tau + \overleftrightarrow{\beta} [\vec{m}_{1,2}\times[\vec{m}_{1,2} \times \vec{B}_{eff}]]. \quad (1)$$

The first two terms on the RHS are the usual used terms in the Bloch equation for the magnetization $\vec{m}_1$ and $\vec{m}_2$ and τ is the relevant relaxation time T_2 or T_1 depending on which components of the magnetizations are considered. The third term is another relaxation term of the Landau-Lifschitz type and β gives the strength of this relaxation channel. The effective field is

$$B_{eff} = \begin{pmatrix} B_1\cos^2\omega t \\ -B_1\cos\omega t\sin\omega t \\ B_0 - \frac{\omega}{\gamma} - \mu_0 M_{bulk} \end{pmatrix} - (\lambda-d)(m_1+m_2) + \delta\begin{pmatrix} (m_1^z-m_2^z) - (m_1^y-m_2^y) \\ (m_1^x-m_2^x) - (m_1^z-m_2^z) \\ (m_1^y-m_2^y) - (m_1^x-m_2^x) \end{pmatrix}. \quad (2)$$

The usual convention is used with B_{eff} in the rotating frame and the static field in z-direction. The rf field B_1 is linearly polarized but as in this paper only pulsed NMR spectra are simulated, $B_1 = 0$ after the pulse. For experiments on bulk samples where the skin depth is much smaller than the dimensions of the sample, $\mu_0 M_{bulk}$ can not be neglected if $p \cong 1$. The third term in eq. 2 includes all the interactions of the type $[I_1 \times I_2]$ and the second term comes from $(I_1 \cdot I_2)$ contributions like dipolar ($\overleftrightarrow{d}$) and indirect exchange ($\overleftrightarrow{\lambda}$).

The exchange interaction is dominating in Tl metal. Therefore only this interaction will be considered. The molecular field is defined by $\overleftrightarrow{\lambda} \vec{m}$ with λ in the nearest neighbor approximation

$$\lambda = \frac{\sum A_{ij}}{h^2\gamma_1\gamma_2\rho} = \frac{\alpha_{1,2}zJ}{h\gamma_1\gamma_2\rho_{1,2}}. \quad (3)$$

$\rho_{1,2} = \alpha_{1,2}$ N/V is the particle density. For simplicity isotropic exchange is assumed, and as $J_1 = \left(\frac{\gamma_1}{\gamma_2}\right)^2 J_2$ and $\frac{\gamma_1}{\gamma_2} \cong 1$ for Tl, the same strength of the inter- and intra isotopic exchange can be assumed.

Eqs. 1 were solved numerically to obtain the transverse component of the magnetizations for arbitrary tipping conditions which define the starting conditions for the differential eqs. 1. The results are plotted after Fourier transformation in figs. 1-5.

Fig. 1 shows the temperature dependent simulation of Cu-NMR spectra taken for $0.04 < p \le 1$. Beside the decreasing resonance frequencies with increasing polarization the change of intensities is also demonstrated showing that the less abundant isotope has almost all intensity at $p = 1$. Both effects, frequency shift and intensity change, would change direction with a sign change in λ.

The influence of the dipolar interaction is demonstrated in fig. 2 where for $\lambda/\mu_0 = -0.7$ different demagnetizing factors are used, which changes the intensity and position of the lines.

For small deviations from equilibrium eqs. 1 can be solved analytically and the frequencies of the isotopes with spin I_1 and I_2 can be calculated[8] and read for nearest neighbor exchange

Fig. 1:

Simulations of Cu - NMR spectra taken in 10 mT and a tipping angle $\Theta = 1°$. Dipolar interaction was set to zero, the exchange parameter $\lambda = -0.8\ \mu_0$, and the temperatures were 100 μK (a), 50 μK (b), 20 μK (c), 10 μK (d), 5 μK (e), 2 μK (f) and 1 μK (g). Lorentz field included ($\mu_0/3$). Dashed lines indicate position of ^{63,65}Cu at p=0.

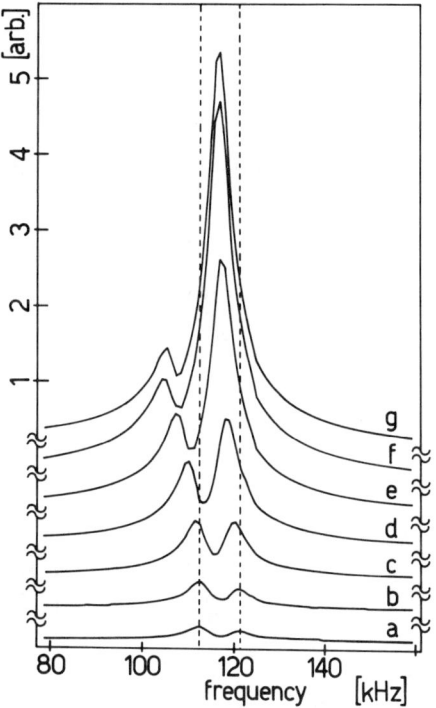

Fig. 2:

Simulations of Cu - NMR similar to fig. 1 but dipolar interactions included. Demagnetizing factors change sign between (a) and (b); full dipolar interaction (c). In curve (d) the dipolar strength was increased by a factor 5 above the usual strength in Cu. Curve e demonstrates the influence of an additional relaxation channel of the Landau-Lifschitz form. Curve c simulates quite well what was measured by Ekström et al.[12]

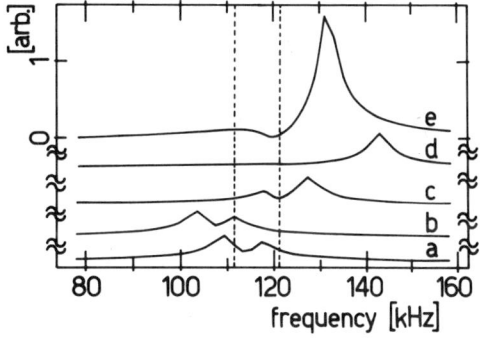

$$\nu_{1,2} = \tfrac{1}{2}\left[\nu_1{}^0 + \nu_2{}^0 \pm \left((\nu_1{}^0-\nu_2{}^0)^2 + 4\alpha_1\alpha_2 I_1 I_2(zJ)^2 p^2\right)^{\tfrac{1}{2}}\right] \quad (4)$$

with

$$\nu_{1,2}^0 = \gamma_{1,2} B_0/2\pi + (zJ) I_{2,1} \alpha_{2,1} p . \quad (5)$$

For small exchange and polarization eq. 4 reduces to eq. 5, and obviously the direction of the frequency shift is given by the sign of J and the magnitude by J, α and p. At high polarization and for very strong exchange interaction, which means

$$(\nu_1{}^0-\nu_2{}^0)^2 \ll 4\alpha_1\alpha_2 I_1 I_2(zJ)^2 p^2, \quad (6)$$

and for $I = I_1 = I_2$, as it is the case for Cu and Tl, one obtains

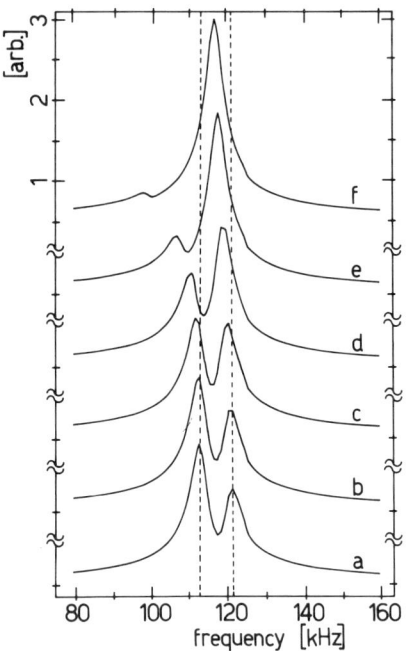

Fig. 3:

Demonstration of an increase of the exchange interaction. (Parameters as in fig. 1) except λ/μ_0 = -0.1 (a), -0.2 (b), -0.4 (c), -0.8 (d), -1.6 (e) and -3.2 (f). p = 0.43.

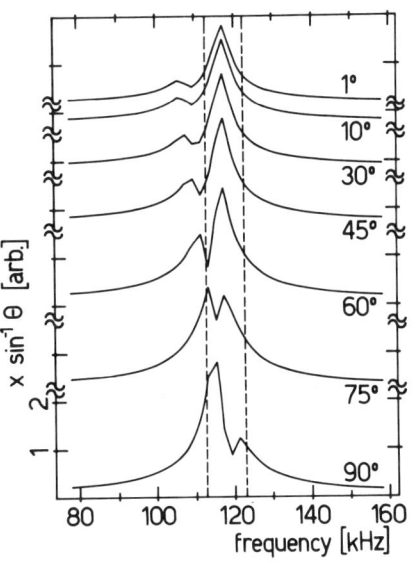

Fig. 4:

Cu-NMR spectra at various tipping angles, 10 mT and 1 μK (p ≅ 1). Both isotopes are tipped by the same amount. λ = -0.7μ_0, Lorentz field $\mu_0/3$ included but without dipolar terms. Intensities normalized to tipping angle.

$$\nu_{1,2} = (\gamma_1+\gamma_2)B_0/4\pi + (zJ) \, I \, p \, (\sqrt{\alpha_1} \pm \sqrt{\alpha_2})^2/2. \qquad (7)$$

Thus, in case of Tl (or Cu) where $\alpha_{1,2} \cong 0.7, 0.3$, the frequency of one mode is just between the two Larmor frequencies and the second mode is shifted away by $(zJ)Ip$. But, as can be seen from fig. 3, there is no intensity left in this second mode.

In fig. 3 the same parameters as in fig. 1 were used except that λ/μ_0 was varied from -0.1 to -3.2 and p was fixed to p = 0.43. With increasing strength of the exchange interaction, only one mode survives and this mode is sitting almost in the middle of the Larmor frequencies and is shifted only weakly in the high exchange limit.

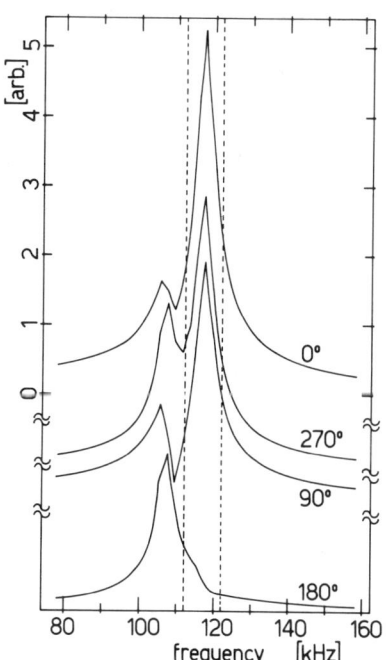

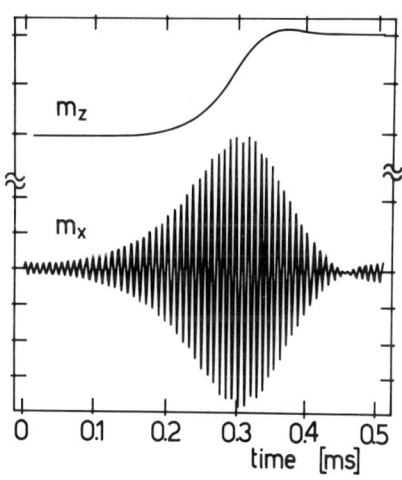

Fig. 5:

Cu-NMR spectra when the phase angle between the two isotopes is varied. Both isotopes are tipped by $\theta = 1°$. Other parameters as in fig. 4. $\varphi = 0$ (a), $\varphi = \pi/2$ (b), $\varphi = \pi$ (c) and $\varphi = 3\pi/2$ (d).

Fig. 6:

Time spectra for the transverse (m_x) and longitudinal (m_z) component of the magnetization after $1°$ tipping for ^{63}Cu and $179°$ for ^{65}Cu. $\varphi = 0$. Other parameters as in fig. 4.

But, beside the polarization dependent shifts there are also tipping dependent frequency and intensity variations as can be seen from fig. 4 showing Cu at p = 1 in B_0 = 10 mT. Both isotopes are tipped by the same amount. A mode crossing seems to occur at around Θ = 75° and the second mode disappears at tipping angles around Θ = 150°. Non linear behavior is also found for small tipping angles ($\Theta \cong 1°$) when the phase angle ϕ between the two magnetizations is changed as demonstrated by fig. 5. For certain conditions even longitudinal oscillations can be excited as shown in fig. 6.

From the above discussion it is obvious that NMR at high polarization can not be done in bulk samples because of the phase and tipping angle shifts within the skin depth. Thus, with this respect, low temperature NMR should preferably be done on foils which are thin compared to the rf penetration depth.

Analyzing the "strong exchange" condition (eq. 6) for the case of Tl only one resonance line would be expected for $B_0 < 1.7 \cdot 10^{-6} zJp = 0.8$ p Tesla. But, as an exchange merged line is found experimentally for p < 0.01 in fields as high as 0.25 T J has to be increased by a factor of 60 if the merging of the Tl lines would be caused by this effect. This is shown in fig. 7 where λ/μ_0 is varied from 100 to 1600 for Tl in 0.2 T at 10 mK. The one mode

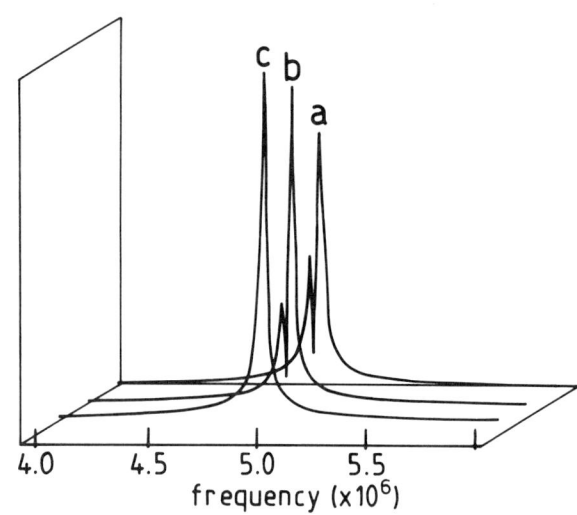

Fig. 7:
Tl-simulation. Θ = 90°, T = 10 mK, B_0 = 0.2 T, λ/μ_0 = 100 (a), 400 (b) and 1600 (c).

case occurs at around 1600 whereas λ/μ_0 should be 26 according to ref. 9. The simulations showed that if nothing else would be changed but the polarization, this strong exchange mode ($\lambda/\mu_0 \cong 1600$) would never resplit into two lines. At around $p \cong 0.8$, λ/μ_0 can be lowered in this model down to 10 and still only a single mode would be visible. This failure of the theory indicates that the molecular field approach is probably not adequate to describe the observations, even at high temperatures.

The merging of the two NMR lines in Tl can be explained by a second order perturbation calculation within the linear response theory which was first done by Karimov and Shchegolev [9] at $p \cong 0$ and later by Oja et al.[11] for arbitrary p. Their result for the frequencies of the two modes is

$$\nu_{1,2} = \nu_{1,2}^0 \mp \frac{\frac{\alpha_{2,1}}{z}(1-p^2) - \alpha_2\alpha_1 p^2}{4(\nu_1^0 - \nu_2^0)} (zJ)^2 \tag{8}$$

where the $(1-p^2)$ term is responsible for the merging which becomes weak at high polarization because of decreasing thermal fluctuations. The external field below which merging would occur at low p is given by

$$B_{merge} \lesssim \frac{\sqrt{zJ}\sqrt{|\alpha_2-\alpha_1|}}{2|\gamma_2-\gamma_1|} \tag{9}$$

which yields about 0.2 T for merging in Tl, a value higher than used in the experiments [5,10]. Therefore, at least as long as the $(1-p^2)$-term dominates the p^2-term, which is valid in Tl for $p \ll 0.5$, only one line should be observed in the NMR spectrum. According to Oja et al.[11] this line should resplit for higher polarizations and these lines should then be shifted approximately according to eq. 7 because the condition eq. 6 is fulfilled for Tl at $p \cong 0.5$ and $B_0 \cong 0.2$, the highest field used in the experiments. But, no temperature dependent shift at all is observed for the splitted line [10]. Unfortunately, nothing is said about the intensities of these "resplitted" lines in ref. 11. The model of ref. 11 seems not to be adequate because the experimentally observed splitted lines in Tl at high p are not shifted by p and because within this model not more than the occurence of two lines can be explained whereas up to 5 could be seen.

SPIN WAVES

As pointed out by I.A. Fomin[13] the occurence of sharp lines at frequencies below the Larmor frequency in low temperature Tl-NMR could be due to a bound state below the spin wave band of total width $W = 2I(zJ)$ This could happen in Tl at high polarizations because the minority isotope ^{203}Tl could be treated as an "impurity" in the almost ferromagnetically oriented ^{205}Tl. The impurity problem in ferromagnets is described in the literature[14]. The ^{203}Tl nucleus with the smaller gyromagnetic value acts on spin waves as an attracting center. In the model discussed here this bound state is only formed if the external field is above a threshold field B_K. For $B_0 > B_K$ this bound state splits off from the Larmor frequency (= lowest frequency of the spin wave band). The splitting Δ should not depend on temperature if the temperature is low enough. Both, B_K and Δ are functions of W. A first estimate showed that for the limit of small concentrations

$$B_K \geq \frac{\pi}{2} \frac{W}{\gamma_{205} - \gamma_{203}} \cong 0.47 \; [T] \qquad (10)$$

is much too high to be relevant for the present problem of Tl. (experiments: $B_0 < 0.2$).

The above one impurity calculation can be generalized to a small but finite concentration of ^{203}Tl.[15] One finds in the "T-matrix approximation" the following susceptibility

$$\chi(\omega,k) = \frac{-2 <I_z>}{\omega - (\omega_0 + \xi(k)) - \alpha_{203} T(\omega)} \qquad (11)$$

with $\xi(k)$ the wave vector dependent spin wave frequency, and

$$T(\omega) = \frac{-\kappa}{1 + \kappa \, I(\omega-\omega_0)} . \qquad (12)$$

κ is defined by $\kappa = (\gamma_{205} - \gamma_{203})B_0$ and

$$I(\omega-\omega_0) = \frac{1}{N} \sum_{k'} \frac{1}{\omega - (\omega_0 + \xi(k'))} . \qquad (13)$$

$I(\omega-\omega_0)$ can be calculated using a square root model for the density of spin wave states

$$\rho(\varepsilon) = \frac{8}{\pi W^2} \sqrt{\varepsilon(W-\varepsilon)} \qquad (14)$$

and one finds

$$I(\omega-\omega_0) = \int_0^W d\varepsilon \, \frac{\rho(\varepsilon)}{\varepsilon - \omega - \omega_0} = \frac{8}{W^2} \left(\omega - \omega_0 - \frac{W}{2} - (\omega-\omega_0)^{\frac{1}{2}}(\omega-\omega_0-W)^{\frac{1}{2}} \right) \qquad (15)$$

The NMR absorption spectra can be derived by taking the imaginary part of eq. 11 and setting k = 0. To include the line width Γ in eq. 11, ω must be replaced by $\omega + i\Gamma$.

For vanishing concentration the frequency splitting Δ of the bound state is given by the pole of eq. 12, for $B_0 > B_K$

$$\Delta = -\kappa \left(1 - \frac{W}{4\kappa} \right)^2 \qquad (16)$$

leaving more or less all the intensity at the Larmor frequency of the host. For very high fields the bound state splitting equals the Larmor splitting of the two isotopes.

The situation changes with increasing impurity concentration, then the bound state splitting occurs already at fields below B_K. Thus, with parameters valid for Tl, the mode frequency is shifted away from the Larmor position even in fields as low as 20 mT as demonstrated in fig. 8. For the simulations on Tl W = 450 kHz,

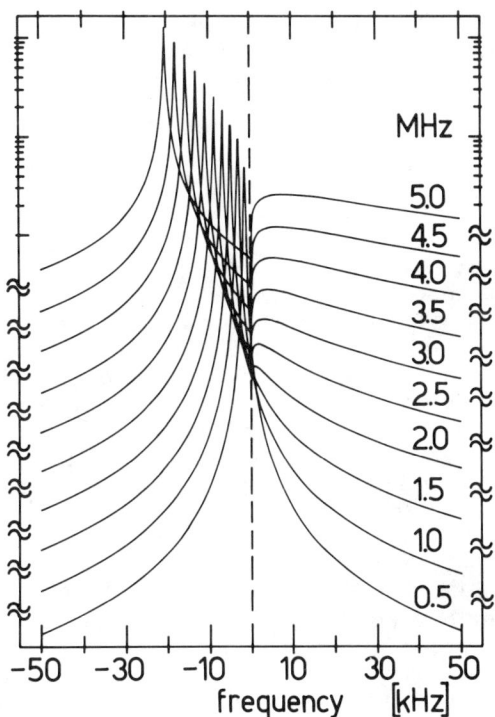

Fig. 8

Spin wave model : $\chi(\omega)$ as function of field. Parameters for Tl (see text). Dashed line indicates the Larmor position of the host given in MHz . Notice: LOG scale, even at half of the critical frequency (11.5 MHz) almost no intensity in the spin wave band.

κ/B_0 = 240 kHz and a fictious Γ = 200 Hz was used. At the lowest field (0.02 T) only a single mode occurs shifted by roughly -2 kHz. In fields around 0.2 T almost all intensity is still kept in the bound state and the shift is -20 kHz only, half of the Larmor shift, and of the order of what was seen in the experiments. For fields much higher than B_K, where $\Delta = \kappa$, the intensities in the bound state and the spin wave band reflect the isotopic abundance.

Obviously, the features of this bound state splitting are similar to what is observed experimentally in Tl at p > 0.4. Firstly, some threshold polarization is necessary to make the Tl ferromagnetically like looking. Secondly, the increase of polarization by lowering the temperature stabilizes this ferromagnetic alignement of the moments in bigger parts of the sample and consequently increases the intensity in the bound state line. Thirdly, the splitting does not depend on temperature and only weakly on field. Certainly, the bound state model is still too crude in its present version and must be refined in order to cope with ^{203}Tl concentrations as large as 30% . Results will be discussed elsewhere [16]. This model could apply to other two spin systems at high polarizations, for instance it could possibly explain the line splitting which we have seen in high fields ($B_0 > 1$ T) in NMR on $AuIn_2$ at temperatures above 10 mK [17] and which was recently found in low fields at around 100 μK [18].

CONCLUSION

The observed line splitting of the merged Tl NMR line at p > 0.4 can neither be explained within the linear response theory using the second order calculations of Oja et al.[11] nor with the molecular field approach where a single mode will stay a single mode down to the lowest temperatures.

The computer simulation of the molecular field equations showed a variety of nonlinear effects in the spin dynamics of highly polarized samples ranging from frequency and intensity shifts with tipping angle as well as with the phase angle between the two isotopes. Even longitudinal oscillations can be excited. This makes it necessary to work with metallic foils of a thickness less than the skin depth if high spin polarization is of interest in NMR experiment.

In a first attempt to include spin wave effects it was found that at higher concentrations of the minority spins a bound state splitting occurs even at low fields. Spin wave effects, particularly spin waves bound to the minority spins, could be the origin for the observations. However, more experimental and theoretical work is necessary to decide definitely whether high spin polarization phenomena or some precursor of nuclear ordering is the origin of what was seen in Tl at low temperatures.

ACKNOWLEDGEMENT

Many stimulating discussions with Igor Fomin and Dierk Rainer are gratefully acknowledged. It was/is a pleasure to participate a little bit in their deep insight into physical problems. Discussions with Aarne Oja were also very fruitful. Last, but not least, I really like to mention all the valuable help I was rendered by my co-worker Erwin Schuberth during the experiments.

REFERENCES

1. For details see: A. Abragam and M. Goldman: "Nuclear Magnetism-Order and Disorder", Clarendon Press, Oxford (1982).
2. M.T. Huiku, T.A. Jyrkkiö, J.M. Kyynäräinen, M.T. Loponen, O.V. Lounasmaa, and A.S. Oja, J. Low Temp. Phys. 62, 433 (1986).
3. T.A. Jyrkkiö, M.T. Huiku, K. Siemensmeyer, and K.N. Clausen, J. Low Temp. Phys. 74, 435 (1989)
4. A.S. Oja, A.J. Annila, and T. Takano, Helsinki Symp., June 88
5. G. Eska, and E. Schuberth, Jap.J.Appl.Phys. 26, Sup 26-3, 435 (1987)
6. see: F. Keffer: "Spin waves", Handbuch der Physik XVIII/2, Springer (1966), 1

7. see: E.A. Turov, and M.P. Petrov: "NMR in Ferro- and Antiferromagnets", Halsted Press, New York (1972)
8. N. Bloembergen and T.J. Rowland, Phys.Rev. $\underline{97}$, 1679 (1955)
9. Y.F. Karimov, and I.F. Shchegolev, Sov.Phy. JETP $\underline{14}$, 772 (1962)
10. G. Eska, and E. Schuberth, J.Low Temp.Phys., to be published
11. A.S. Oja, A.J. Annila, and T. Takano, Phys.Rev. $\underline{B38,}$ 8602 (1988)
12. J.P. Ekström, J.F. Jacquinot, M.T. Loponen, J.K. Soini, and P. Kumar, Physica $\underline{98B}$, 45 (1979)
13. I.A. Fomin, private communication
14. Yu.A. Izygumov, and M.V. Medvedev, "Magnetically ordered Crystals containing Impurities", Consultants Bureau, New York (1973)
15. D. Rainer, private communication
16. D. Rainer, I.A. Fomin, and G. Eska, to be published
17. G. Eska and N. Masuhara, unpublished data
18. K. Gloos, R. König, F. Pobell and P. Smeibidl, this conference

ANTIFERROMAGNETIC RESONANCE OF HYPERFINE-ENHANCED NUCLEAR ANTIFERROMAGNET HoVO₄

H.Suzuki, M.Ono and N.Mizutani
Dept.of Phys. Faculty of Science,Tohoku University,Sendai 980,Japan

Hyperfine-enhanced nuclear spins of Ho in HoVO$_4$ orders antiferromagnetically below about 4.9 mK. The nuclear spin ordering, produced by adiabatic demagnetization cooling of the specimen, was observed in magnetic sysceptibility, nuclear orientation and neutron diffraction experiments. For a full understanding of the antiferromagnetic state, measurement of the antiferromagnetic resonance (AF resonance) is essential. However, AF resonance in hyperfine-enhanced nucler spin system is hard to observe, partly because the resonance frequency is rather high,i.e., in the range of ~ 10 to $\sim 10^2$ MHz, and partly because the rf measurement field itself warms up the nuclear spins so quickly. Therefore Bleaney[1] et al studied NMR of ^{51}V in both paramagnetic and ordered nuclear state of Ho and discussed the ordered state of enhanced nuclear spins of Ho from their results.

In the present paper we report the of the AF resonance of ^{165}Ho nuclear spins in HoVO$_4$ The nuclear spin of ^{165}Ho (I=7/2) in the Van Vleck paramagnet HoVO$_4$ couples with the induced enhanced electronic moment through the hyperfine interaction, and acquires an enhanced magnetic moment. The gyromagnetic ratio of the enhanced nuclear spin has anisotropic values of $\gamma/2\pi$=1523 MHz/T along the a and a'-axis and $\gamma/2\pi$=15 MHz/T along the c-axis of the tetragonal structure. The coefficient of electric quadrupole interaction P/h is +25.6MHz. Then when the axis of the quantization is taken along the c axis, the ground state of the nuclear spin of Ho are $|\pm 1/2\rangle$. In the antiferromagnetic state of HoVO4, the a and a' axes are equivalent . In zero magnetic field the nuclear spins are equally divided between antiferromagnetic domains directed along the a-axis and along the perpendicular a'-axis. On increasing the applied field parallel to the a axis, the a' domains, whose spins are in a spin-flop configuration, predominate at about 80 Oe. At higher fields than 400 Oe, all spins are polarized along the field direction.

A single crystal of HoVO4 was roughly rectangular in shape, 4x4x13 (mm), with the long dimension along the c axis. The crystal was attached to the mixing chamber of a ^{3}He-^{4}He dilution refrigerator with thin Cu wires and GE 7031 vanish. The crystal was mounted with the a-axis parallel to the direction of the applied magnetic field. The cw NMR measurement was made by sweeping the magnetic field through resonance at fixed rf frequencies. When we apply a rather small magnetic field less than 400 Oe parallel to the a-axis, i.e., perpendicular to the quantized axis, the energy splitting of the nuclear spin is rather complicated as shown in Fig.1. And the resonance spectrum in low magnetic field region can be expected to be complicated. The spin temperature of the specimen was determined by measurement of the susceptibility of the enhanced nuclear spin itself. The magnetic susceptibility was measured by use of a low-level resonant circuit. By this method we could measure the temperature dependence of the AF resonance field below T_N. In Fig.2 resonance spectra for the measuring frequency of 40.6 MHz at about 4.3 mK ($T<T_N$) and at about 6.0 mK ($T>T_N$) are shown; the shift of the resonance field at T = 4.3mK from the resonance in the paramagnetic state was about 100 Oe. We also measured the AF resonance at 106.9 MHz. The shift of the resonance field at T=4.43mK was 14 Oe.

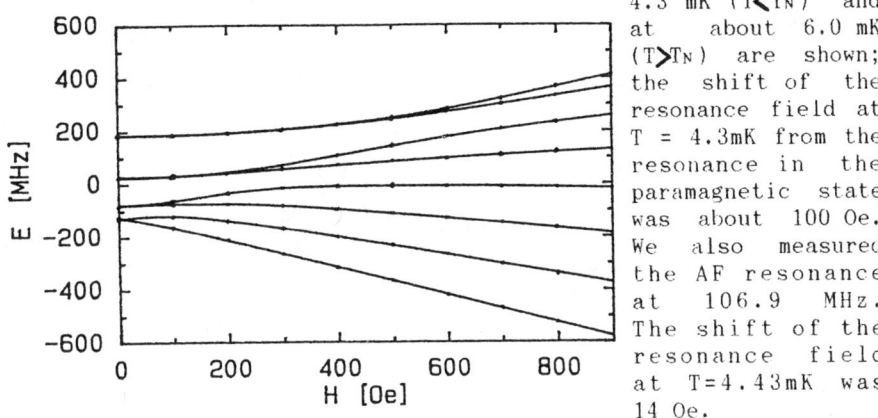

Fig. 1. Energy splitting of Ho nuclear spin I_z of $HoVO_4$. Magnetic field along a-axis.

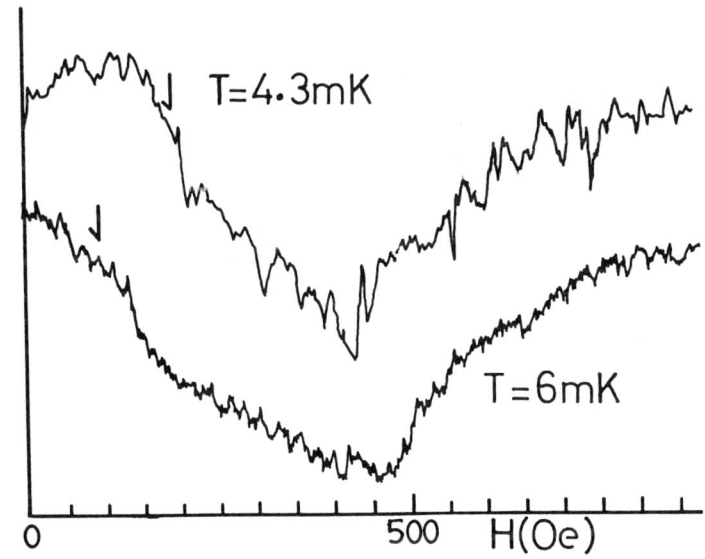

Fig. 2. Resonance spectrum at measuring frequency of 40.6MHz. Arrows show the resonance fields

REFERENCES

1. B.Bleaney,R.G.Clark,J.F.Gregg,Y.Roinel and N.J.Stone,J.Phys.C 20,3175 (1987)

EXOTIC STUDIES

AT

LOW TEMPERATURES

CHAIRMAN

William M. Fairbank
Department of Physics
Stanford University

HUNTING FOR MOST OF THE UNIVERSE—DARK MATTER

David O. Caldwell
Physics Department, University of California, Santa Barbara, CA 93106

ABSTRACT

Probably > 90% of the mass of the universe is known to us only through its gravitational effects. The material is almost surely non-baryonic and then has to be due to a particle or particles not in the standard model of particle physics. The search for such particles, of great importance to astrophysics and particle physics, has so far been conducted with detectors developed for other purposes. Important limits on this dark matter (DM) have been set by searching in proton decay detectors for neutrinos from the annihilation of DM trapped in the sun and by looking for the scattering of DM from the nuclei of double beta decay detectors. The former experiments eliminate sneutrinos and Majorana neutrinos from 10–20 GeV/c^2, and recent results from the UCSB/LBL/UCB/Saclay experiments of the latter type limit DM massive Dirac neutrinos to 4–11 GeV/c^2, Cosmions (which also would solve the solar neutrino problem) to 4–6 GeV/c^2, and generally eliminate DM Dirac-type weakly interacting particles between 11 GeV/c^2 and 6 TeV/c^2.

INTRODUCTION

While it is obviously interesting to know of what most of the universe is made, there is a more specific reason for introducing the subject of dark matter to a Symposium on Quantum Fluids and Solids: so far the search for this elusive material has been done with detectors developed for other purposes, and there is a clear need for low temperature detectors designed for this work and hence for the expertise of the low temperature community. Possible low temperature devices are specifically addressed by Cabrera,[1] while here the subject of dark matter will be introduced and the present status of the search for it reviewed.

THE NEED FOR DARK MATTER

In 1933 Fritz Zwicky reported[2] that the random peculiar velocities of galaxies in the Virgo cluster sometimes exceeded the escape velocity, if all the mass in the cluster were luminous. In other words, the galaxies should not be clustered as observed if the only mass holding them together were that which could be seen. Since that time observations of many astronomical objects, particularly galaxies of all types, have confirmed the need for missing mass, generally labelled dark matter (DM).

Galactic rotation curves provide particularly clear and simple evidence for DM. Using the familiar relation for the rotation of a mass m at a radius r about a mass M, $GMm/r^2 = mv^2/r$, one would expect the tangential velocity, v, to decrease as $r^{1/2}$ as v is measured for m's at increasing values of r outside of M. Instead one observes v to remain approximately constant with r, even out to a radius an order of magnitude larger than that containing the visible mass. Here M is the luminous mass, but this observation must mean that since $v = \sqrt{GM/r}$, really M must increase linearly with r, even if it cannot be seen. That is, there has to be a spherical halo of dark matter pervading the galaxy. Generally most

© 1989 American Institute of Physics

of the visible mass of the galaxy has collapsed to a disk because of the strong interactions of the baryonic matter. The fact that the DM is in a spherical halo is an indication that the material of which it is made is dissipationless; i.e., it is weakly interacting.

While the DM halo needs to contain an order of magnitude more mass than that which is visible in a typical galaxy, there is evidence for an even larger ratio of dark to luminous mass on larger scales. Even on this basis the mass density of the universe which one can account for is only about 0.2 that of the critical density needed for a "flat" universe. That is, one can account for $\Omega = 0.2$, where $\Omega < 1$ gives an open universe (expanding forever) and $\Omega > 1$ gives a closed universe, which will eventually stop expanding and collapse.

It is very likely that $\Omega = 1$ on three grounds. First, unless $\Omega = 1$, incredible fine tuning of the parameters is needed to have $\Omega = 0.2$ today, since unless $\Omega = 1$, Ω is not constant in time. Second, $\Omega = 1$ is required by the inflationary scenario[3] of a period of extremely rapid expansion of the universe, which explains so many issues of cosmology which are otherwise mysteries. Third, there are at least some observations[4] suggesting that Ω is this large.

THE NATURE OF DARK MATTER

Despite the apparent need for a dissipationless halo in galaxies, one may ask whether baryonic matter, our familiar protons and neutrons, could be the DM and in some way not make its presence known to us except by gravitation. If $\Omega = 1$, the answer is no, because nucleosynthesis theory, which accounts so well for the abundance of particularly the light elements, requires that the baryonic contribution to Ω be probably 0.1 and certainly no more than 0.2.

If the restriction on $\Omega = 1$ is relaxed, then one can still eliminate most of the possible forms of baryonic matter (stars, comets, dust, gas, etc.) from various observations or arguments.[5] It is very difficult to make matter with strong and electromagnetic interactions unobservable. Two possible candidates remain: Jupiters (stars of $\sim 10^{-3}$ solar masses, too small for nuclear burning) and black holes of $\gtrsim 10^6$ or $\lesssim 10^{-6}$ solar masses. Black holes having a sufficiently large Ω and with the more likely masses between those extremes are ruled out by their effects not having been seen. However, there are no known mechanisms for forming either sufficient Jupiters or suitable black holes, and such a large number of either one would require a peculiar stellar mass distribution, with enormous (invisible) peaks at one end of the mass scale or the other.

In short, the case is very strong for non-baryonic dark matter. What non-baryonic particles could be responsible? Candidates are usually classified by whether or not they were relativistic at "freeze-out", when in the expansion of the universe the temperature dropped below their rest energy and they could no longer find enough partners with which to annihilate and form other particles, their density then being essentially the current density for long-lived particles.

The main class of hot dark matter candidates is the light neutrino, with a mass of ~ 30 eV/c^2 being appropriate for $\Omega = 1$, since there are about 10^2 neutrinos per cm^3. The standard model requires massless neutrinos, but plausible mechanisms exist to give a neutrino mass. The electron neutrino is no longer a candidate, since neutrinos from supernova 1987a and recent measurements of the ^{3}H endpoint limit the mass to < 20 eV/c^2. Probably more relevant, since such a light neutrino is most likely a Majorana particle (i.e., it is its own antiparticle), and hence could induce neutrinoless double beta decay, is the limit of $\sim 1 - 2$ eV/c^2 on the effective mass from a lifetime limit for that process of 10^{24} years from the UCSB/LBL experiment.[6] The mu and tau neutrinos are still viable candidates, but there are some problems. Unless cosmic strings exist, hot dark matter does not explain galaxy formation, since the neutrinos readily free stream from regions of high density to those of low density, washing out the needed density fluctuations. It is very likely that light neutrinos could not constitute all of dark matter, because dwarf galaxies do not have sufficient phase space to hold enough of such light neutrinos to provide the mass they need.[7]

Cold dark matter (CDM) is currently more popular as an aid to galaxy formation, although there are currently problems with larger scale structure. One unusual form of CDM is the axion, which is discussed in detail by Sikivie,[8] and hence will not be addressed here. The class of CDM considered here goes by the generic name of WIMP (Weakly Interacting Massive Particle). WIMPs could be of two types, Dirac or Majorana particles. The former have spin-independent vector interactions, and the latter have spin-dependent axial vector couplings. Either could be from a presently unknown fourth family. Extra neutrinos exist in broken E_6 as a derivative of string theories, in left-right symmetric particle theories, and in extra U(1) symmetries, in which the neutrino's interaction is less than weak because of the higher mass of the extra Z boson associated with it.

There is a class of WIMPs called Cosmions, which would not only be DM, but also serve to solve the solar neutrino problem.[9] The Cosmion DM could scatter in the sun, being captured and falling in toward the center, where its orbiting would serve to cool the solar core by the $\sim 10\%$ needed to reduce by the observed factor of ~ 3 the flux of the ^{8}B solar neutrinos produced near the center. To do the job, the Cosmion must have a mass > 4 GeV/c^2 in order to avoid too much evaporation from the surface of the sun, but < 15 GeV/c^2 so that the orbit can extend to a large enough radius for sufficient cooling. The latter need also imposes the requirement that the interaction with protons be about 10^2 times the weak interaction strength, yet the annihilation rate must be low and the lifetime very long to maintain enough Cosmions.

The long lifetime needed is provided by some quantum number which is conserved, or nearly so, such as the lepton flavor of the fourth family. The annihilation rate is made low in four Cosmion models having Dirac neutrinos by having a particle-antiparticle asymmetry. A 10 GeV/c^2 neutrino giving $\Omega = 1$ would have about the same asymmetry as that between protons and antiprotons,[10] and it is likely that the asymmetry would arise from similar causes. The main problem is providing the extra interaction strength. One model[11] involves the exchange of an as yet unknown colored scalar triplet, another[12] has a magnetic interaction provided by the existence nearby in mass of a charged partner and a Higgs boson, a third[13] utilizes the extra U(1) mentioned above, but in this case with a lighter Z' which does not couple to the usual leptons, and the last two of the five models use the interaction which would be provided by having a very light Higgs boson. One[14] of the last two has a Dirac neutrino, but the other[15] uses a Majorana particle, which requires more explanation.

Supersymmetry is a popular theoretical concept which provides an explanation of the coexistence of the weak interaction mass scale (the 10^2 GeV/c^2 of the W and Z masses) and the Planck scale (10^{19} GeV/c^2). It requires that every boson have a fermion partner and every fermion have a boson partner. Thus the neutrino's partner is a sneutrino, that of the photon, a photino, that of the Higgs, a Higgsino, that of the Z, a zino, etc. Supersymmetry is clearly broken, since the partners do not have the same mass. In most versions of the theory there is a conserved quantum number, called R parity, so that a supersymmetric particle must decay to another one. Clearly, then, the lightest supersymmetric particle (LSP) must be stable. This might be any of those neutral particles named above, or it might be a mixture of the fermions, photino + Higgsino + zino, which is called the neutralino. If a suitable light Higgs exists, such a neutralino could be a Cosmion, and in that case (since there can be no particle-antiparticle asymmetry for this Majorana particle) annihilation is suppressed because it must occur at low energy through a p-wave interaction. Whether or not it is a Cosmion, the LSP is a very popular DM candidate, although there is as yet no experimental evidence for supersymmetry.

Three arguments favor the Cosmion hypothesis. First, the 10^{-11} admixture to the 10^{57} baryons in the sun is just provided by the flux of DM and the age of the sun, since 10^{46} Cosmions would have accumulated by now. Second, standard solar models do not agree well with the solar core sound speed deduced from helioseismology, whereas Cosmions can help this problem.[16] Third, globular star clusters appear to have ages longer than other determinations of the lifetime of the universe, and Cosmions can speed up the evolution of those stars so as to improve the agreement.[17]

EXPERIMENTAL RESULTS ON DARK MATTER WIMPS

Two quite complementary techniques exist for detecting the existence of WIMPs: annihilation and scattering. Majorana neutrinos and Dirac neutrinos with no particle-antiparticle asymmetry can annihilate in pairs to produce γ-ray lines, positrons, antiprotons, and neutrinos. Searches for the first three have not produced useful limits, but the fourth, utilizing proton decay detectors, has. In the neutrino case, solar capture is again invoked but now in order to produce sufficient annihilation to heavy quarks, the cascade decays of which emit energetic neutrinos detectable terrestrially. Higher energy neutrinos (at least > 2 GeV) interacting in the proton decay detectors have some chance of being detected above the background of atmospherically produced neutrinos, particularly when directionality is utilized. The ratio $r = \bar{S}/\bar{A}$ of the integrated flux of neutrinos from the solar direction to those spread over all directions from atmospheric interactions is shown in Figure 1.[18] It can be seen that the limit on r from the IMB,[19] Frejus,[20] and Kamioka[21] detectors can probably be used to exclude some range of Majorana neutrino masses, although there are large uncertainties in the calculations. However, there is not enough sensitivity to exclude the supersymmetric particles shown, although sneutrinos (not shown) are eliminated as DM.

The Dirac neutrino case requires more discussion. If a Dirac neutrino has no particle-antiparticle asymmetry, then the mass has to be[22] 4 GeV/c^2 (or $\sim$ 2 TeV/c^2) for it to give $\Omega = 1$. In the no-asymmetry case shown in Figure 1 any greater mass requires $\Omega < 1$, but if $\Omega < 0.1$, there would be insufficient DM to provide that needed in our own galaxy, so the curve stops at that point.

The second technique, direct detection,[23] is especially sensitive to Dirac neu-

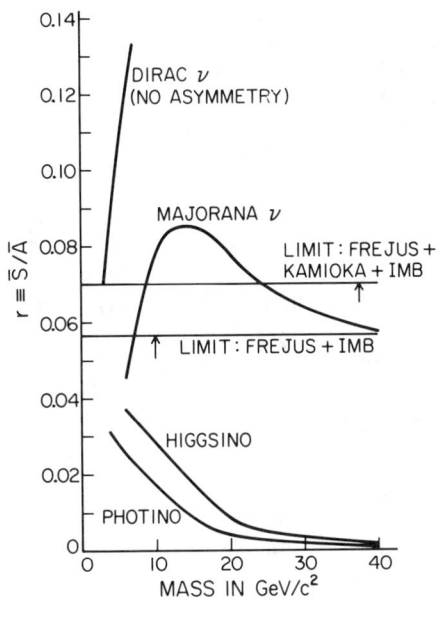

Fig. 1. Ratio, r, of the integrated flux of neutrinos > 2 GeV from annihilations in the sun of the dark matter candidates shown to ν's from background interactions in the atmosphere, as compared to experimental observations.

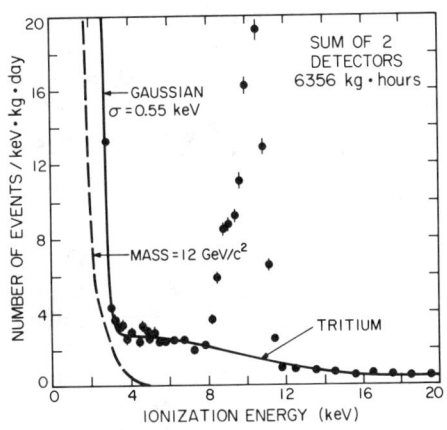

Fig. 2. Ionization energy detected in Ge counters of the UCSB/LBL/UCB experiment. Below the X-ray peaks the data are fitted by gaussian electronic noise ($\sigma = 0.55$ keV) and ^{3}H from spallation when the Ge was above ground, ruling out the expected signal (shown) of a 12 GeV/c^2 Dirac ν at the 95% C.L.

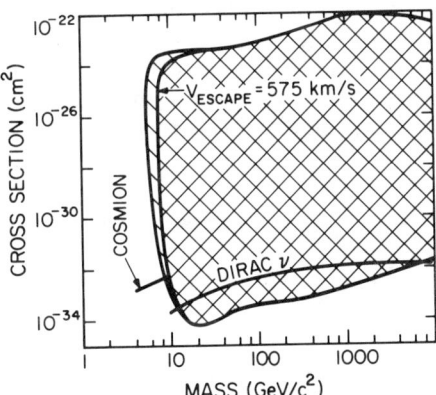

Fig. 3. Exclusion plot for the mass and interaction cross section with Ge of dark matter particles, using data from the UCSB/LBL/UCB experiment. All limits are 2σ, and the 575 km/s refers to an assumed particle velocity for escape from the galactic halo. The weak interaction cross section is indicated by "Dirac ν", and the "Cosmion" is a particle also solving the solar ν problem.

trinos, since these have spin-independent interactions which give coherent scattering in the mostly spin-zero nuclei of ionization detectors. Since the WIMPs would have thermal velocities with $v_{rms} \sim 270$ km/sec, and the earth moves through the galactic halo at ~ 230 km/sec, a WIMP scattering from a Ge nucleus could give a measurable recoil energy to that nucleus. This energy is small, and only a fraction of that energy goes into ionization, so one needs to measure down to keV energies, with sensitivity increasing exponentially as the threshold energy is decreased. Fortunately the ionization efficiency has been measured,[24] agreeing with theory[25] down to at least 2 keV, corresponding to 10 keV recoil energy.

Two such direct detection experiments have published[26,27] results. Both were designed to search for neutrinoless double beta decay in ^{76}Ge, which constitutes 7.8% of the ultra high purity Ge used in semiconductor ionization detectors. The Ge has nearly 0.1% energy resolution in the search for a 2.041 MeV line from the summed electron energies in the transition $^{76}\text{Ge} \rightarrow\, ^{76}\text{Se} + 2e^-$, and unprecedently low background rates (< 1 count/keV·kg·year) had to be achieved to reach a lifetime limit of 10^{24} years.[6]

The experiment[27] giving the more stringent limits will be described briefly here. While the UCSB/LBL double beta decay experiment[6] has used up to 8 Ge detectors, the data of concern here comes from two of those 160 cm^3 (0.9 kg) detectors which have been modified to measure keV energies. The Ge detectors were enclosed in a cavity formed by ten blocks of NaI scintillator of 15-cm thickness. Compton scattering, which provides the main source of radioactivity background at low energies, is suppressed not only by the vetoing of the scattered γ in the NaI or another Ge detector, but also because many of the γ's come from cascade decays, and the detection of one of those other γ's would also eliminate the event. The NaI was surrounded by passive shielding and the apparatus placed underground with a 600 m.w.e. overburden. Much effort had been devoted to finding materials of low radioactivity and developing fabrication techniques to minimize the material inside the NaI anticoincidence shield, but in the DM energy domain new background problems arose. While some sources have been eliminated, the main remaining activities are in the Ge itself, as can be seen from Fig. 2. One see below a relatively flat background of $\sim 1/2$ count/keV·kg·day, the rate rises due to ^{3}H, a spallation product in the Ge from cosmic ray bombardment. The Ga (10.4 keV) and Zn (9.7 keV) X-ray peaks are also a result of cosmogenic activities.[6] Below 3 keV the electronic noise dominates.

The results shown in Fig. 2 are used in Fig. 3, assuming a DM halo density of 0.3 GeV/cm^3 (5.3×10^{-25} g/cm^3), to produce an exclusion plot of DM particle mass vs. the particle's interaction cross section in Ge, with the weak interaction level indicated by the line labelled "Dirac ν". Note that over most of the Dirac ν mass range a particle-antiparticle asymmetry must exist. The level of the interaction of most Cosmions is also shown.

For the Cosmions the best limit comes from four Si detectors of 15 g each, which were substituted for two Ge detectors. Although this was supposed to be only an engineering run and the background level in the Si is high (~ 20 counts/keV·kg·day), the lower mass of the Si nucleus (giving more recoil energy), its greater ionization efficiency, and the lower threshold (~ 0.7 keV has been obtained) of Si detectors give this technique greater sensitivity for low DM masses, despite the small quantity of Si and the high background in this case. So far our Saclay collaborators have calibrated the ionization efficiency of Si down to 2 keV, again finding agreement with theory,[25] so data can be used only to that energy, corresponding to 7 keV recoil energy. Even with that limit, these data exclude Cosmions > 6 GeV/c^2, closing the window of allowed Cosmion masses to just 4 to 6 GeV/c^2.

The Ge data exclude a Dirac ν from 11 GeV/c^2 to 6 TeV/c^2. Since a Dirac ν above ~ 2 TeV/c^2 would make $\Omega > 1$, the limits above that value are superfluous for that particle, closing the allowed range to 4 to 11 GeV/c^2. The high mass limits are quite relevant for some other weakly interacting particles, however. In particular, technibaryons can be ruled out as DM. The technicolor idea, an extension of quantum chromodynamics, provides an alternative to the Higgs mechanism, and indeed it provides a better explanation of the W and Z masses. Like the proton, there should be a technibaryon which is essentially stable and would be an excellent DM candidate, but its mass must lie in the now excluded region.

The scattering results have been used for other purposes as well. Microcharged shadow matter[28] has also been eliminated[27] as DM. This would have arisen if there were any particle which connected our universe with that occupying the same space as ours but otherwise interacting with us only gravitationally, as suggested by string theory.[29] The measurements also can set limits on supersymmetric particles. While less than 8% of the Ge is an isotope with spin, which had been thought to be necessary for the spin-dependent interaction of a Majorana particle, an additional scalar interaction has been found.[30] An even stronger interaction would exist if there is a sufficiently light ($\lesssim 30$ GeV/c^2) Higgs particle, and earlier data have already been used[31] in this model to set limits on supersymmetric parameters competitive with, but complementary to, those from high energy colliders.

Thus existing data already provide powerful constraints on possible particles. The Ge and particularly the Si data will be improved, but it is difficult to get to very much lower cross sections, or to set good limits on many Majorana particles, such as the LSP. There is real need for cryogenic detectors with lower thresholds and possibly the extremely useful discriminants provided by simultaneous measurement of ionization and phonons and by directionality, utilizing the earth's motion through the galactic DM halo.

That motion could also provide an excellent check that one was observing a dark matter signal, since there would be an annual effect of $\sim 10\%$ on the signal rate above a fixed threshold as the earth's motion around the sun aided or opposed the motion of the solar system through the DM halo. A second check would be given by observing the signal in more than one material, since the signal from one nuclear type completely determines that for another, albeit with a different magnitude and energy range. Perhaps it will be some as yet undeveloped cryogenic detector which will have the happy task of proving that the mysterious matter that makes up most of the universe has at last been found.

Thanks are due to the author's collaborators in these experiments: R.M. Eisberg, B. Magnusson, and M.S. Witherell from UCSB; F.S. Goulding and A. Smith from LBL; B. Sadoulet from UCB; and G. Gerbier, J. Rich, and M. Spiro from Saclay. The work was partially supported by the U.S. Atomic Energy Commission.

REFERENCES

1. B. Cabrera, these Proceedings.
2. F. Zwicky, *Helv. Phys. Acta.* **6**, 110 (1933).
3. A.H. Guth, *Phys. Rev. D* **23**, 347 (1981); A.D. Linde, *Phys. Lett.* **108B**, 389 (1982); A. Albrecht and P.J. Steinhardt, *Phys. Rev. Lett.* **48**, 1220 (1982).
4. E.D. Loh and E.J. Spillar, *Ap. J.* **307**, L1 (1986).
5. D.J. Hegyi and K.A. Olive, *Ap. J.* **303**, 56 (1986).
6. D.O. Caldwell, R.M. Eisberg, D.M. Grumm, D.L. Hale, M.S. Witherell, F.S. Goulding, D.A. Landis, N.W. Madden, D.F. Malone, R.H. Pehl, and A.R. Smith, *Phys. Rev. Lett.* **54**, 281 (1985); *Phys. Rev. D* **33**, 2737 (1986); *Phys. Rev. Lett.* **59**, 419 (1987); D.O. Caldwell, *Int. J. of Mod. Phys.* **A4**, 1851 (1989).
7. D.N. Spergel, D.H. Weinberg, and J.R. Gott III, Inst. for Adv. Study preprint IASSNS–AST 88/35 (1988, to be published in Phys. Rev. D) and references therein.
8. P. Sikivie, these Proceedings.
9. R.L. Gilliland, J. Faulkner, W.H. Press, and D.N. Spergel, *Ap. J.* **306**, 703 (1986) and references therein.
10. E.W. Kolb and K.A. Olive, *Phys. Rev. D* **33**, 1202 (1986); K. Greist and D. Seckel, *Nucl. Phys.* **B283**, 681 (1987).
11. G.B. Gelmini, L.J. Hall, and M.J. Lin, *Nucl. Phys.* **B281**, 720 (1987).
12. S. Raby and G.B. West, *Nucl. Phys.* **B292**, 793 (1987); *Phys. Lett.* **B194**, 557 (1987); *Phys. Lett.* **B200**, 547 (1988).
13. G.B. Ross and G.C. Segrè, *Phys. Lett.* **B197**, 45 (1987).
14. S. Raby and G.B. West, *Phys. Lett.* **B202**, 47 (1988).
15. G.F. Giudice and E. Roulet, *Phys. Lett.* **B219**, 309 (1989).
16. J. Faulkner, "New and Exotic Phenomena", ed. O. Fackler and J. Tran Thanh Van, Editions Frontières (France, 1987), p. 387.
17. J. Faulkner and F.J. Swanson, *Ap. J.* **329**, L47 (1988).
18. The analysis used here is based on the work of K.A. Olive and M. Srednicki, *Phys. Lett.* **B205**, 553 (1988).
19. J. LoSecco et al., *Phys. Lett.* **B188**, 388 (1987).
20. E. Kuznik, "New and Exotic Phenomena", ed. O. Fackler and J. Tran Thanh Van, Editions Frontières (France, 1987), p. 215.
21. Y. Totsuka, University of Tokyo preprint UT-ICEPP-87-02 (1987, unpublished).
22. M. Srednicki, R. Watkins, and K.A. Olive. *Nucl. Phys.* **B310**, 693 (1988).
23. M.W. Goodman and E. Witten. *Phys. Rev. D* **31**, 3059 (1985).

24. C. Chasman, "Penetration of Charged Particles in Matter — A Symposium", Nat. Acad. of Sciences (1970), p. 16, and references therein.
25. J. Lindhard, M. Scharff, and H.E. Schiott, *Kgl. Dan. Vidensk. Selsk. Mat.-Fys. Medd.* **33**, 14 (1963); J. Lindhard, V. Nielsen, M. Scharff, and P.V. Thomsen, loc. cit. **33**, 10 (1963).
26. S.P. Ahlen, F.T. Avignone III, R.L. Brodzinski, A.K. Drukier, G. Gelmini, and D.N. Spergel, *Phys. Lett.* **195**, 603 (1987).
27. D.O. Caldwell, R.M. Eisberg, D.M. Grumm, M.S. Witherell, B. Sadoulet, F.S. Goulding, and A.R. Smith, *Phys. Rev. Lett.* **61**, 510 (1988).
28. H. Goldberg and L.J. Hall, *Phys. Lett.* **B174**, 151 (1986); R. Holdom, *Phys. Lett.* **B166**, 196 (1985).
29. D.J. Gross, J.A. Harvey, E. Martinec, and R. Rohm, *Phys. Rev. Lett.* **54**, 502 (1985).
30. K. Greist, Fermilab–Pub–88/52–A (1988).
31. R. Barbieri, M. Frigeni, and G.F. Giudice, *Nucl. Phys.* **B313**, 725 (1989).

SUPERCONDUCTING DETECTORS FOR LABORATORY DARK MATTER SEARCHES

B. Cabrera

Physics Department, Stanford University, Stanford, California 94305

Abstract

Convincing data exist showing that galaxies are surrounded by non-luminous matter. This "dark matter" shows itself through its gravitational interactions with luminous matter, and dominates the mass of galaxies by about a factor of ten. Our understanding of early universe nucleosynthesis together with the observational data on the relative abundance of the light elements set $\Omega_{baryon}<0.2$ (0.2 of the critical density). However, $\Omega=1$ is an unstable equilibrium, and inflationary models of the early universe predict that $\Omega=1$ to high precision. Thus, it seems likely that 90% of the matter of the universe is non-baryonic and weakly interacting, and that the dark matter halos may be largely non-baryonic. These particles would then be present in the cosmic rays and be observable with sensitive laboratory detectors.

In our laboratories at Stanford, we have undertaken several research programs to search for dark matter candidates in our galaxy with laboratory based detectors. The first effort is a search for magnetic monopoles in the cosmic rays. These would be supermassive (10^{16}-10^{19} GeV/c^2) and a density of only one per 10-10,000 km^3 would be sufficient to account for the local dark matter around our galaxy. We have been operating a 1.3 m^2 times 4π sr detector utilizing eight SQUIDs. It is the largest superconductive monopole detector. The second effort involves the development of large mass (~1 kg) elementary particle detectors capable of sensing weakly interacting particles. These utilize silicon crystals at temperatures below 1 K, have spatial resolution and would measure the total energy deposition. Such detectors will be used for direct dark matter searches and for neutrino experiments capable of setting better limits on the neutrino mass.

PART A:
SEARCH FOR MAGNETIC MONOPOLES IN THE COSMIC RAYS

Introduction

In 1931, P.A.M. Dirac proposed the existence of magnetically charged particles to explain the observed quantization of electric charge. He showed that only integer multiples of a fundamental magnetic charge g (Dirac charge) = hc/4πe are consistent with quantum mechanics. Many years of experimental searches produced no convincing candidates. In 1974 't Hooft and independently Polyakov showed that in all true unification theories (those based on simple or semi-simple compact groups) magnetically charged particles are necessarily present. The modern theory predicts the same long-range field and thus the same charge g as the Dirac solution; now, however, the near field is also specified leading to a calculable mass. The standard SU(5) model predicts a monopole mass of 10^{16} GeV/c^2, much heavier than had been considered in previous searches. More recent theories based on supersymmetry or Kaluza-Klein models yield even higher mass values up to the Planck mass of 10^{19} GeV/c^2.

Such supermassive magnetically charged particles would possess qualitatively different properties from those assumed in earlier searches. These include necessarily nonrelativistic velocities from which follow weak ionization and extreme penetration through matter. Thus such particles may very well have escaped detection in earlier searches based on heavy ionization of relativistic monopoles.

Although grand unificaiton theories (GUTs) are very clear in their prediction of the existence of monopoles, cosmological theories based on GUTs lead to impossibly high or unobservably low predictions for monopole particle flux limits with the latter results being exponentially model dependent. Thus, only astrophysical arguments provide guidance for the relevant detector sensing areas for experiments. An upper bound of 10^{-15} cm^{-2} sr^{-1} s^{-1} is obtained assuming an isotropic flux from arguments based on the existence of the 3 microgauss galactic magnetic field (Parker bound). The Parker bound becomes less severe linearly with monopole mass for masses above 10^{17} GeV/c^2. In addition, several authors have demonstrated that models incorporating monopole plasma oscillations would allow a much larger particle flux, approaching in some cases the local galactic dark mass limit. All of these bounds assume particle velocities in gravitational virial equilibrium, i.e., very near 10^{-3} c.

Superconducting technologies, many developed at Stanford University over the last decade, have led naturally to very sensitive detectors for magnetically charged particles. These superconducting detectors directly measure the magnetic charge independently of particle velocity, mass, electric charge and magnetic dipole moment [1]. In addition, the detector response is based on simple and fundamental theoretical arguments which are extremely convincing. By far, the most definitive positive identification of a magnetic charge would be made with a superconducting detector. Conventional particle detectors, though larger, are less satisfactory for positive identification of monopoles since the interaction of monopoles with matter through ionization processes is less well understood [2].

Operation of Three Loop Detector

Between February, 1983 and March, 1986, we operated a three loop coincident superconducting magnetic monopole detector [3]. The rms current noise levels in an effective noise bandwidth of ≈ 0.16 Hz were 0.02 ϕ_0/L in all three loops, less than 1% of the signal expected from the passage of a Dirac magnetic charge through one of the two turn loops (4 ϕ_0/L). The sensing area based on this low noise operation was 476 cm^2 (71 cm^2 loop area and 405 cm^2 near miss area) for events greater than 0.1 ϕ_0/L in at least two of the three loops. The data were extremely clean and no candidate events were seen. Based on the 1008 days of accumulated active running time, these data set an upper limit of 4.4×10^{-12} cm^{-2} sr^{-1} s^{-1} (90% C.L.) on any uniform flux of magnetic monopoles passing through the earth's surface at any velocity [3].

New Large Area Octagonal Detector

Our group at Stanford, consisting of B.Cabrera, M. Huber, M. Taber and R. Gardner, has designed, constructed, and brought into continuous operation an eight loop detector [4] with a cross section averaged over 4π sr of 1.3 m^2 for double coincident events and a signal-to-noise ratio of 30 for the passage of a single Dirac charge. This detector shown schematically in Fig. 1a, is composed of eight planar superconducting detection coils arranged around a cylinder with an octagonal cross section. Each coil is a gradiometer 16 cm wide and 6 m long and is connected to a high sensitivity rf SQUID current sensor. The entire assembly is surrounded by a superconducting lead shield and housed in a dewar which is enclosed in µ-metal. This detector has a sensing area 30 times larger than that of the three loop detector.

A very important design feature for the large scale use of superconducting coils as monopole detectors is the use of a gradiometer winding pattern [5,6]. The sensitivity

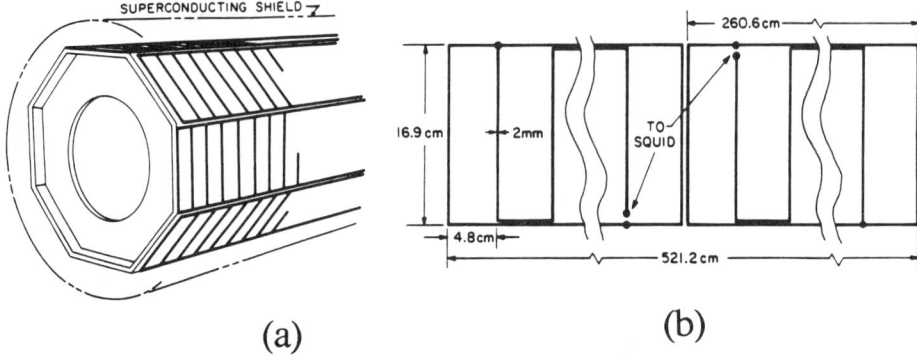

Fig. 1. (a) Schematic diagram of octagonal detector.
(b) Loop parameters for octagonal detector.

of a gradiometer to external magnetic field changes is substantially reduced over that from a simple coil whereas the sensitivity to the passage of a magnetic charge remains high since the particle passes through only one element of the gradiometer pattern. We have utilized computer calculations to optimize our coil design under constraints from dewar size and minimum acceptable signal to noise ratio. Our design is shown in Fig. 1b. The conducting elements are NbTi ribbon, 2mm wide and 50 μm thick. It is a repeating rectangular pattern with an aspect ratio of 0.6 to 1.0. A further improvement was achieved by breaking the loop up into a number of parallel elements which are connected together to one SQUID. This technique reduces the coupling losses to the SQUID from a $1/L$ proportionality to $1/\sqrt{L}$, where L is the inductance of the single series loop [7].

The new 1.3 m^2 monopole detector uses eight low-noise SQUID systems, one for each panel of the detector array. We use rf SQUIDs manufactured by Biomagnetic Technologies Inc. [B.T.i.], (formerly S.H.E. Corp.) together with rf electronics from Quantum Design. The external shield consists of 0.038 inch-thick sheets of μ-metal mounted on an aluminum frame. The shield provides an absolute field below 10 milligauss throughout most of its interior. We are utilizing a Model 1200 closed-cycle liquid helium refrigeration system from Koch with the new octagonal detector. There are two advantages from closed-cycle operation: lower long-term operating costs, and the virtual elimination of disturbances that are produced by liquid cryogen transfers.

The geometry of our detector provides greater confidence in identifying true monopole events than our earlier detectors. A monopole can intersect at most two loops. We exclude from consideration all signals appearing in one loop only or in three or more loops. The response of the detector is calculated from a numerical simulation of a uniform flux of monopoles and this calculation is used to obtain the sensing area.

We began obtaining low noise data on May 5, 1987 [8]. Figure 2 shows typical filtered data (one point every 10 sec) from the computer data acquisition system. Included is a calibration signal showing the signal to noise for a Dirac size magnetic charge. As of April 15, 1989, we have accumulated ~ 6,000 hours of live time. A preliminary analysis of these data [8] contain no candidate events and set a limit of ~ 7.8×10^{-13} cm^{-2} sr^{-1} s^{-1} (90% C.L.) for any particle flux of magnetically charged particles passing through the surface of the earth. We intend to run the detector continuously for at least three years.

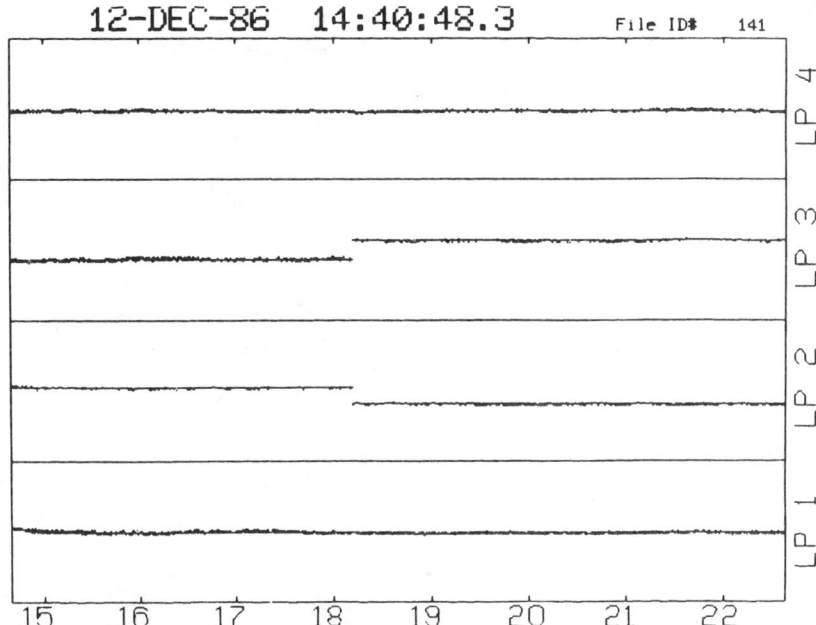

Fig. 2. Digitally filtered data from octagonal detector showing four SQUID readouts, including a Dirac-sized calibration signal through two coils.

Particle Flux Limits From Our Detector Exposures

In Fig. 3 we summarize the current status of our research efforts with respect to astrophysical bounds. The Parker bound as modified to include supermassive monopoles has a mass independent floor at ~ 10^{-15} cm^{-2} sr^{-1} s^{-1} and rises linearly with mass above ~ 10^{17} GeV/c^2 until it intersects the local dark matter bound around the Planck mass (10^{19} GeV/c^2). Since more recent unification theories suggest a monopole mass approaching the Planck mass, searches at a flux level of ~ 10^{-13} cm^{-2} sr^{-1} s^{-1} are particularly important. Also shown in Fig. 3 are possible monopole particle flux levels 2 to 3 orders of magnitude above the Parker bound. These models are based upon monopole plasma oscillations within the galaxy. Detailed computer simulations performed by the Cornell group confirm the stability of such solutions and they are not ruled out by any galactic observations.

The particle flux limit obtained from our three loop superconductive detector was 4.4 × 10^{-12} cm^{-2} sr^{-1} s^{-1} at 90% C. L. Four other groups (Chicago-FermiLab-Michigan; IBM, Yorktown Heights; Imperial College; and NBS, Boulder) had obtained similar limits, for a combined world limit of 1.4 × 10^{-12} cm^{-2} sr^{-1} s^{-1} at 90% C. L. In a preliminary analysis of ~ 15 months of data from our new octagonal detector, we find no candidate events for the passage of a monopole through the detector and set an upper limit on such a particle flux of ~ 7.8 × 10^{-13} cm^{-2} sr^{-1} s^{-1} (90% C.L.). In addition, the IBM, Yorktown Heights group is operating a 1 m^2 × 4π sr sensing area detector and has reported a similar preliminary result in slightly more run time. The combined world limit for velocity independent detectors is now ~ 3 × 10^{-13} cm^{-2} sr^{-1} s^{-1} (90% C.L.).

These data effectively rule out the plasma oscillation models and approach the mass-dependent Parker bound near the Planck mass.

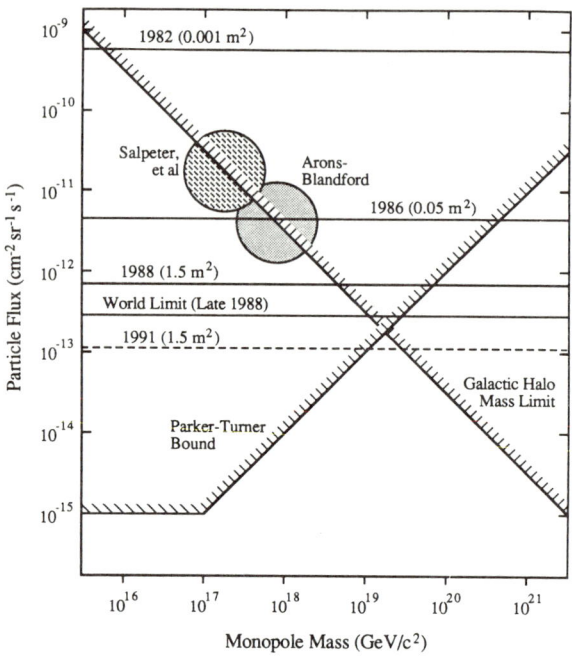

Fig. 3. Cosmic ray particle flux limits from our superconductive detectors compared with various astrophysical limits.

Acknowledgements

Our group at Stanford has included B. Cabrera, M. Huber, M. Taber and R. Gardner and J. Bourg. This research is funded in part by DOE Contract DE-AM03-76-SF00-326.

PART II:
SILICON CRYSTAL ACOUSTIC DETECTORS FOR NEUTRINOS

Introduction

Semiconductor diode particle detectors now provide the highest energy resolution (~ 3 keV FWHM for 1 kg) and the lowest thresholds available (~ 4 keV) for large mass detectors. In this energy range less than 30% of the deposition energy is converted directly into electron-hole pairs which produce the observed signal in the semiconductors, the rest forming phonons. The characteristic energies of these phonons is ~ 1 meV, 10^3 less than the excitation energy for an electron-hole pair in a semiconductor (~ 1 eV). Thus in principle, energy resolutions over an order of magnitude better are possible if the phonon signal is used. Recent work on bolometers, which sense the thermal phonons, has demonstrated an improvement in energy resolution (most recently 17 eV, FWHM, by Moseley, McCammon, et al [9]) by measuring the temperature rise in a small ultra-low heat capacity sensor (~ 10^{-5} g of silicon). B. Cabrera, L. Krauss and F. Wilczek [10] suggested scaling such bolometers up to a mass ~ 1 kg.

Motivated by this suggestion, our group, now composed of B. Cabrera, B. Young and A. Lee at Stanford and B. Neuhauser at San Francisco State University, has proposed direct sensing of the ballistic phonons produced by an event in a large insulating single crystal [11]. Since this wavefront carries information on the event total energy and location within these silicon crystal acoustic detectors (SiCADs), substantial improvements in background rejection are possible over detection schemes which measure total energy deposition only. A threshold of 1 keV or better is important for our interest in using SiCADs as neutrino detectorsand for dark matter searches. For example, at power reactors all nuclear recoil signals from elastic neutrino scattering are below 10 keV with 60% above 1 keV, and a 1 kg SiCAD with a 1 keV threshold would register an event rate of ~ 100 events/day.

Ballistic Phonons from Point Events in Crystal Cubes

For a deposition of energy in a well localized volume within a single crystal of silicon, a thermal-like spectral phonon distribution is generated with a characteristic temperature of 10-20 K. This spectral distribution arises from the rapid decay of electron-hole excitations to the band edge, first generating very short wavelength phonons, which quickly relax to longer wavelength phonons within less than ~10 nsec. The decay rates are very strongly dependent on phonon energy ($\propto v^5$) and for wavelengths of several hundred lattice spacings further decays are entirely negligible. These longer wavelength phonons propagate throughout the crystal with little scattering and no dispersion. This mode is called the the ballistic phonon mode [12]. An interesting and important aspect of ballistic phonon propagation is that strong focussing effects occur within the crystal, although the propagation is dispersionless. These experimentally verified focussing patterns permit reconstruction of event locations within the crystal and can resolve tracks or multiple scattering events.

Experiments with Titanium Transition Edge Sensors on Silicon Wafers

Our first generation of SiCADs utilize superconducting transition edge devices as phonon senors. These devices consist of thin superconducting lines deposited on the surfaces of the silicon crystals and biased with a constant current. For currents below a latching critical current, self-terminating voltage pulses are observed. These pulses are caused by the heating of line segments above the superconducting transition temperature. We have demonstrated this technique using aluminum films with alpha particle sources [13], and most recently we have obtained a factor of one hundred improvement in energy resolution and threshold using titanium films in x-ray experiments [14]. These most recent sensors are made by depositing 40 nm of titanium on crystalline silicon wafers which are 1 mm thick. The polished wafer faces are perpendicular to the [100] axis. The meander pattern, shown in Fig. 4a, was produced using conventional photolithographic techniques and consist of 299 parallel lines each 2 µm wide with 3 µm space between lines. The pattern is aligned parallel to the [110] crystalline axis of the wafer. The active area of the pattern is 4.5 mm long and 1.5 mm wide. The normal resistance just above their superconducting transition temperature ($T_c \approx 312$ mK) is ~ 18 kΩ per 5 mm line. These films have sharp transition widths of ≤ 3 mK (from 10% to 90% of the resistive transition), indicating homogeneous properties across the film. A typical superconducting to normal resistive transition for one of these patterns is shown in Fig. 4b.

Here we describe two recent x-ray experiments with a 24 µC source of ^{241}Am. The decay spectrum of ^{241}Am is dominated by two alpha particle energies at ≈ 5.5 MeV. In addition to these two alphas are a nuclear gamma at 60 keV and two atomic

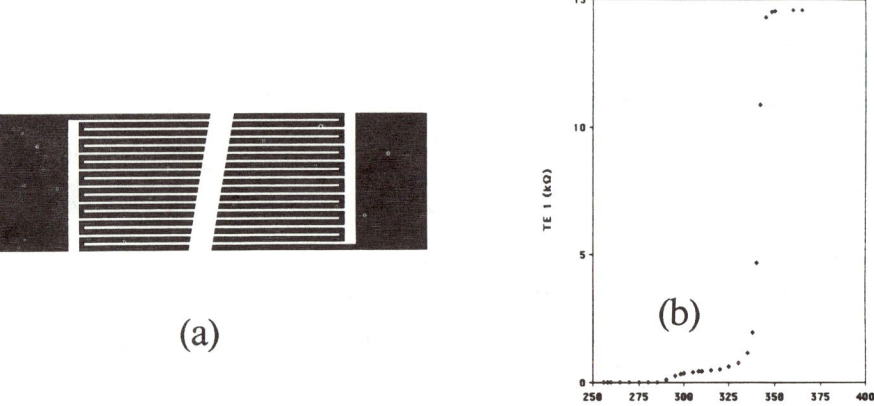

*Fig. 4. (a) Photolithographic patterning for titanium sensors.
(b) Resistive transition as a function of temperature.*

x-rays at 14 and 18 keV. An appropriate absorber placed in front of the source readily stops all of the alphas and essentially all of the 14 and 18 keV x-rays before they reach the sensor; however the 60 keV gamma rays, being much more penetrating, pass through the absorber attenuated in number by ~ 0.5. We used .005" (125 µm) thick foils of Sn and Pb as absorbers.

Using a cryopumped ^{3}He refrigerator at 0.3 K we biased the titanium/silicon devices at the foot of the resistive transition in temperature and below the latching critical current in bias current, and we observe self-terminating voltage pulses when the film is bombarded by phonons produced by the interaction of an incident photon within the Si substrate. These pulses occur because the phonon energy reaching the Ti film from the photoelectron in the silicon is sufficient to drive portions of the film normal. To estimate the threshold energy density E_ρ necessary for this superconducting to normal state transition, we integrate the heat capacity of the superconductor from the bias temperature T to T_c:

$$E_\rho = \int_T^{T_c} C_{es}(T) dT ,$$

where C_{es} is the electronic heat capacity below T_c as given by the BCS theory. We assume that the time constants are sufficiently long to allow thermal equilibrium. The asymptotic form near T_c is $E_\rho \approx 5N(0) \Delta_0^2 (1-T/T_c)$, where Δ_0 is the gap at T = 0 and N(0) is the density of states at the Fermi surface in the normal metal.

For our Ti films, $N(0) \approx 4 \times 10^{22}$ cm^{-3} eV^{-1} and $\Delta_0 \approx 47$ µeV, yielding $E_\rho \approx 22$ eV/µm^3 at $T/T_c = 0.95$. In terms of the energy density per unit area of film, E_σ, we get $E_\sigma \approx (22$ eV/µm$^3) \times (40$nm$) = 0.88$ eV/µm^2. We may also estimate the minimum detectable energy for a Ti sensor as follows. In our current experimental setup we use a cryogenic GaAs MESFET voltage-sensitive amplifier with an amplifier noise level of $\Delta V_{rms} \approx 1$ nV/$\sqrt{Hz}$ in a 1 MHz bandwidth. At $T/T_c \approx 0.95$ and a bias current (I_b) of 60 nA we can detect a minimum sensor resistance of $\Delta R_{rms} = \Delta V_{rms}/I_b = 17$ Ω. At $T \geq T_c$ the resistance of our sensor is ~ 8.5 Ω per square of the Ti film. Therefore the area of film which corresponds to a normal area resistance of $\Delta R_{rms} = 17$ Ω is (17 Ω/8.5 Ω) × (2 µm x 2 µm) = 8 µm^2. The minimum detectable energy deposited in the Ti sensor is then $\Delta E_{rms} < E_\sigma \times $Area$_{rms} = (0.88$ eV/µm$^2) \times (8$ µm$^2) = 7$ eV. The inequality holds because Joule heating contributes to the actual area driven normal.

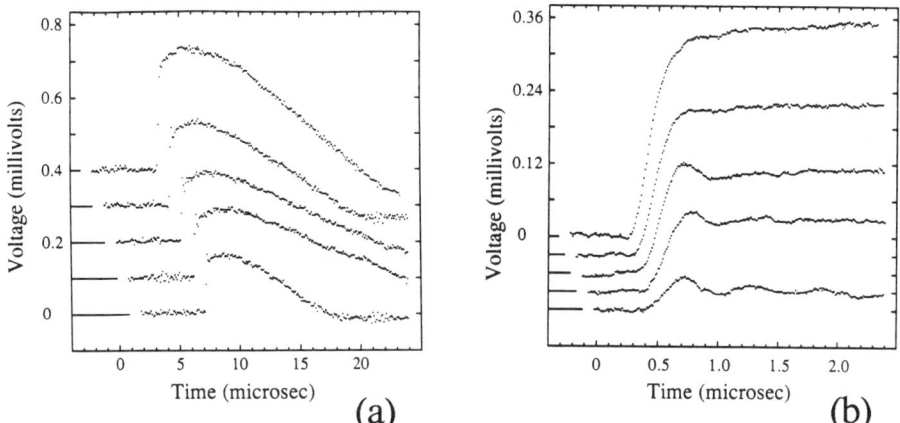

*Fig. 5. (a) Individual pulses from 60 and 25 keV x-rays.
(b) Leading edges of pulses with 140 ns rise time.*

Figure 5 shows (for I_b = 120 nA and T ≈ 286 mK) some typical single pulses, of ~ 2-6 μsec duration, resulting from the interaction of 60 keV gamma rays (from ^{241}Am) in the Si substrate of a Ti sensor. The leading edge of each pulse (Fig. 5b) shows an electronics-limited risetime of ~ 140 nsec.

Figure 6a is a pulse height spectrum obtained for the case of a .005" (125 μm) thick Pb absorber. The prominent peak towards the center of the spectrum is the photopeak due to interaction of 60 keV gamma rays in the Si substrate of the detector. The sharp feature at 11 keV is consistent with the position of the Compton edge for 60 keV gamma rays in silicon. In Fig. 6b, we show a pulse height spectrum under similar running conditions, but with a .005" Sn absorber. We note that the prominent features due to the 11 keV and 60 keV interactions appear, as expected, in the same positions as they did with the Pb absorber. Furthermore, in the Sn absorber spectrum, we observe an additional peak around 25 keV. This peak is readily identified as secondary emission of K_α x-rays from the Sn.

We can qualitatively reproduce these typical pulse height spectra by applying Monte Carlo methods to a simple model of the detector response. We assume that an incident photon, if it interacts at all, interacts in the silicon substrate only once, and that this interaction is point-like. We imagine that, to first order, the phonon energy generated by such a scattering event spreads spherically outward through the crystal, producing heated spots on the surfaces. The energy density of such a heated spot is then given by

$$E_\sigma(\rho, z) = z E_0 / [4\pi(z^2 + \rho^2)^{3/2}]$$

where ρ is the radius of the spot, z is the perpendicular distance from the Si surface to the location of the scattering event, and E_0 is the total energy. Defining E_{sc} as the critical energy density required to drive the detector film normal, we find that the area driven normal A is given by

$$A(z) = \pi \rho^2 = \pi [(zE_0/4\pi E_{sc})^{2/3} - z^2] = \pi z_0^2 [(z/z_0)^{2/3} - (z/z_0)^2]$$

where $z_0^2 \equiv E_0/(4\pi E_{sc})$. The amplitude of the voltage pulse resulting from an interaction in the crystal is simply proportional to the amount of line driven normal, and the number of interactions producing pulses is proportional to $(dA/dz)^{-1}$ for events uniformly distributed in z. For incident particles of given energies, we can then plot a theoretical pulse height spectrum for the Ti transition edge response -- as shown in Fig. 6c for the case of incident 30 and 60 keV photons. By plotting the calculated normal area A as a function of interaction depth z for various initial particle energies, one obtains a family of curves as shown in Fig. 6d. Assuming this simple model is qualitatively valid, we can extract important spatial information about events by comparing such curves with those obtained experimentally.

We are in the process of carrying out the more computer intensive calculations which include phonon focussing effects and we are working to fabricate 1 and 2 mm thick, double-sided (Ti meander patterns on both sides of the Si wafer) transition edge devices. These devices will enable us to do timing experiments and to obtain spatial information about each event occurring in the detector. We shall then be able to experimentally measure Fig. 6d and obtain an energy for each event.

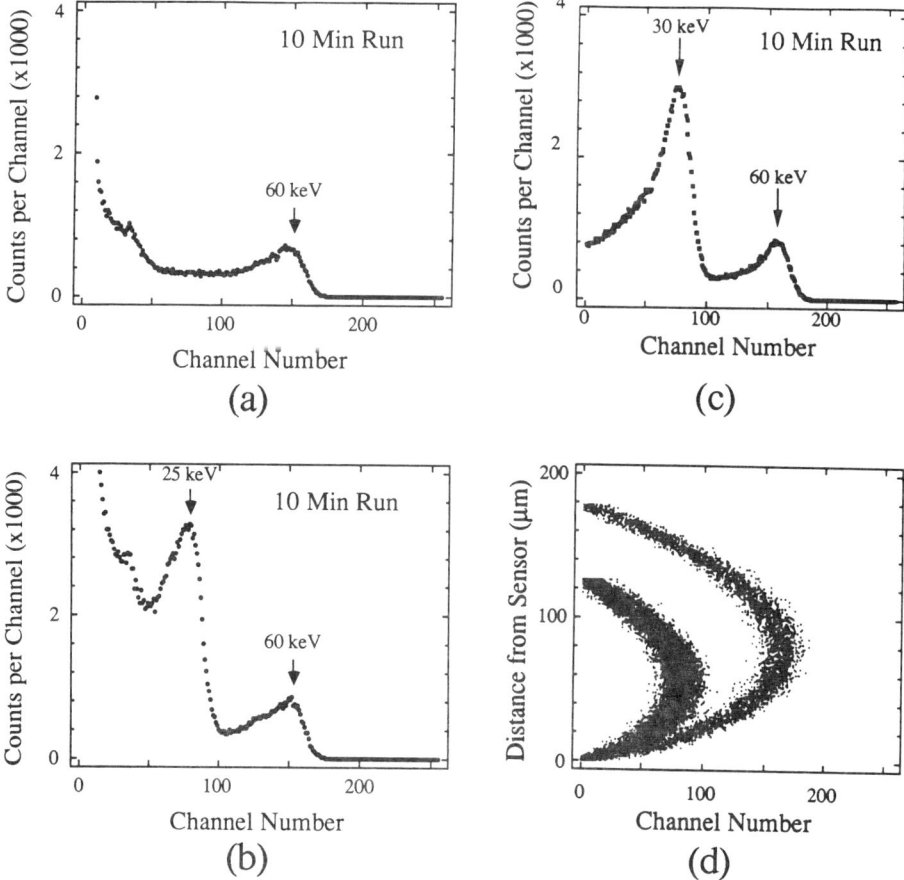

Fig. 6. (a) Pulse height spectrum for ^{241}Am source using Pb absorber.
(b) Pulse height spectrum for same geometry using Sn absorber.
(c) Monte Carlo generated pulse height spectrum from simple model.
(d) Pulse height versus distance to Ti sensor from Monte Carlo.

Conclusions

Utilizing titanium superconducting transition edge devices as phonon sensors on Si crystal surfaces, we believe that an energy threshold of ~ 1 keV can be achieved for a 1 kg SiCAD. Such a detector would be of great interest for a number of experiments including dark matter searches for weakly interacting neutral particle candidates and low energy neutrino experiments to set better limits on the neutrino mass.

Acknowledgements

The work at Stanford has been performed by B. Cabrera, C. J. Martoff, A. T. Lee, and B. A. Young. Also, B. Neuhauser now at San Francisco State University has participated extensively. This work has been funded in part by a Research Corporation Grant, a Lockheed Research Grant and DOE Contract DE-AM03-76-SF00-326.

References

[1] B. Cabrera, *Phys. Rev. Lett.* **48**, 1378 (1982).
[2] See the review article by D. E. Groom, *Phys. Rep.* **140**, 323 (1986).
[3] B. Cabrera, M. Taber, R. Gardner and J. Bourg, *Phys Rev Lett.* **51**, 1933 (1983); and R. Gardner, B. Cabrera, M. Taber and M. Huber, to be submitted to *Phys. Rev.* **D**.
[4] R. D. Gardner, Ph. D. Thesis, Stanford University (1987); and M. E. Huber, Ph. D. Thesis, Stanford University (1988).
[5] S. Bermon, P. Chaudhari, C. C. Chi, C. D. Tesche, and C. C. Tsuei, *Phys. Rev. Lett.* **55**, 1850 (1985).
[6] J. Incandela, M. Campbell, H. Frisch, S. Somalwar, M. Kuchnir, and H. R. Gustafson, *Phys. Rev. Lett.* **53**, 2067 (1984).
[7] S. Somalwar, H. Frisch, J. Incandela, and M. Kuchnir, *Nucl. Inst.* **A226**, 341 (1984).
[8] M.E. Huber, B. Cabrera, M.A. Taber and R.D. Gardner, *Jap. J. of Appl. Phys.* **26**, 1687 (1987); and M.E. Huber, B. Cabrera, M.A. Taber and R.D. Gardner, Proceedings of Applied Superconductivity Conference, San Francisco, Aug., 1988, in press.
[9] S.H. Moseley, J.C. Mather and D. McCammon, J. Appl. Phys. **56**, 1257 (1984); D. McCammon, S.H. Moseley, J.C. Mather and R.F. Mushotzky, J.Appl. Phys. **56**, 1263 (1984).
[10] B. Cabrera, L. Krauss and F. Wilzcek, Phys. Rev. Lett. **55**, 25 (1985).
[11] B. Cabrera, J. Martoff and B. Neuhauser, Stanford Preprint No. BC44-86.
[12] See for example: **Nonequilibrium Phonon Dynamics**, ed. W.E. Bron, NATO ASI Series **B124**, Plenum Press, N.Y., 1985. An excellent review of recent developments in phonon physics.
[13] B. Neuhauser, B. Cabrera, C.J. Martoff and B.A. Young, *Jap. J. of Appl. Phys.* **26**, 1671 (1987).
[14] B. A. Young, B. Cabrera, A. T. Lee and C. J. Martoff, Proceedings of Applied Superconductivity Conference, San Francisco, Aug., 1988, in press.
[15] A. T. Lee, Stanford Preprint No. BC73-88.

SEARCH FOR DARK MATTER AXIONS

P. Sikivie
Department of Physics, University of Florida,
Gainesville, Florida 32611

ABSTRACT

The axion is a hypothetical particle which was postulated to explain why the strong interactions conserve P and CP. If the axion mass is of order 10^{-5} eV, axions could constitute the dark matter which appears clustered in the halos of galaxies. Dark matter axions can be detected by stimulating their conversion to photons in a strong laboratory magnetic field. The general principle of such an experiment will be described as well as the status of ongoing searches.

INTRODUCTION

The axion[1,2,3] was postulated more than ten years ago to explain why the strong interactions conserve P and CP[4]. Consider the Lagrangian of QCD:

$$\mathcal{L}_{QCD} = -\frac{1}{4} G^a_{\mu\nu} G^{a\mu\nu} + \sum_{i=1}^{n} \left[\bar{q}_i \gamma^\mu D_\mu q_i - m_i q^\dagger_{Li} q_{Ri} - m^*_i q^\dagger_{Ri} q_{Li} \right]$$

$$+ \frac{\theta g^2}{32\pi^2} G^a_{\mu\nu} \tilde{G}^{a\mu\nu}. \tag{1.1}$$

The last term is a 4-divergence and hence does not contribute in perturbation theory. It does however produce non-perturbative effects associated with the existence of QCD instantons. As a consequence, the physics of QCD depends upon the value of the parameter θ. Using the Adler-Bell-Jackiw anomaly of the $U_A(1)$ current, one can readily show that the physics of QCD depends upon θ only through the combination of parameters

$$\bar{\theta} \equiv \theta - \arg \det m_q = \theta - \arg (m_1 m_2 \ldots m_n) \tag{1.2}$$

If $\bar{\theta} \neq 0$, QCD violates P and CP. The absence of P and CP violations in the strong interactions therefore places an upper limit upon $\bar{\theta}$.

The best constraint comes from the present experimental upper limit on the neutron electric dipole moment which yields[5]:

$$\bar{\theta} \lesssim 10^{-8} \qquad (1.3)$$

We then face the question: why is $\bar{\theta}$ so small? Recall that in the standard model of the strong and electroweak interactions, the quark masses originate in the electroweak sector of the theory. This sector must violate P and CP to produce the correct weak interaction phenomenology, in particular $K_L \to 2\pi$. There is no reason in the standard model to expect the overall phase of the quark mass matrix to exactly match the value of θ from the QCD sector in order to set $\bar{\theta} < 10^{-8}$. In particular, if CP violation is intro- duced in the manner of Kobayashi and Maskawa[6], the Yukawa couplings that give rise to the quark masses are arbitrary complex numbers and hence arg det m_q and $\bar{\theta}$ have no reason to take on any special value at all.

Peccei and Quinn[1] proposed a solution to this problem by postulating the existence of a global $U_{PQ}(1)$ quasi-symmetry. This $U_{PQ}(1)$ must have the following properties

1. it is a symmetry of the classical theory, i.e. a symmetry of the theory at the Lagrangian level,

2. it is broken explicitly by those non-perturbative QCD effects (instantons and the like) which make the physics of QCD depend upon the parameter θ,

3. it is broken spontaneously by the vacuum expectation value of some scalar field.

To see how the existence of a $U_{PQ}(1)$ quasi-symmetry yields $\bar{\theta} = 0$ (up to tiny corrections due to the CP violating interactions responsible for $K_L \to 2\pi$) consider the theory defined by

$$\mathcal{L}_{PQ} = -\frac{1}{4} G^a_{\mu\nu} G^{a\mu\nu} + \frac{1}{2} \partial_\mu \varphi^\dagger \partial^\mu \varphi + \sum_{i=1}^{n} \bar{q}_i \gamma^\mu D_\mu q_i$$

$$- (K_i q^\dagger_{Li} q_{Ri} \varphi + \text{h.c.}) + \frac{\theta g^2}{32\pi^2} G^a_{\mu\nu} \tilde{G}^{a\mu\nu} - V(\varphi^\dagger \varphi) . \qquad (1.4)$$

$\mathcal{L}_{PQ}$ has a classical $U_{PQ}(1)$ symmetry under which $\varphi \to e^{i\alpha}\varphi$ and $q_j \to e^{-i/2\,\alpha\gamma_5} q_j$ for $j=1\ldots n$. Assuming the potential V has the shape of a "Mexican hat", the $U_{PQ}(1)$ symmetry is spontaneously broken by the vacuum expectation value of the scalar field φ:

$$\langle\varphi(x)\rangle = v\, e^{i\alpha(x)} . \quad (1.5)$$

The quarks acquire masses

$$m_i = K_i\, v\, e^{i\alpha} \quad (1.6)$$

and hence

$$\bar{\theta} = \theta - \arg(m_1\ldots m_n)$$

$$= \theta - \arg(K_1\ldots K_n) - n\alpha . \quad (1.7)$$

The important difference between the theory defined by Eq. (1.1) and the theory defined by Eq. (1.4) is that in the former $\bar{\theta}$ is a function solely of the parameters in the theory, whereas in the latter $\bar{\theta}$ is a function also of the dynamical field α. As a result, those non-perturbative effects which make the physics of QCD depend upon the parameter $\bar{\theta}$ will, in the theory defined by Eq. (1.4), produce an effective potential $V_{eff}(\alpha)$ for the dynamical field α. It can be shown that the minimum of $V_{eff}(\alpha)$ occurs at a value of α such that

$$\bar{\theta} = 0 \pmod{\pi} , \quad (1.8)$$

i.e. such that the theory conserves P and CP. For pedagogical reasons, we did not include the electroweak interactions in our example of Eq. (1.4). Actually, the electroweak interactions only add a few mostly non-essential complications to the implementation of the Peccei-Quinn mechanism. The main difference from the description given above is that the CP violating interactions responsible for $K_L \to 2\pi$ will induce a small value for $\bar{\theta}$. This induced value of $\bar{\theta}$ is however always much smaller than 10^{-8}. For example, Georgi and Randall[7] found $\bar{\theta} \approx 10^{-17}$ for the Kobayashi-Maskawa model of CP violation.

Weinberg and Wilczek[2] pointed out that the Peccei-Quinn solution to the strong CP problem implies the existence of a light pseudo-scalar particle, called the axion. The axion is the Pseudo-Nambu-Goldstone boson associated with the spontaneous breaking of the Peccei-Quinn quasi-symmetry, i.e. the axion field is

$$a(x) = v\,\alpha(x) \qquad (1.9)$$

where v and $\alpha(x)$ are defined by Eq. (1.5). One can calculate the properties[2,8] of the axion using current algebra or chiral Lagrangian techniques. The axion mass is given by (in the limit $m_s \gg m_u, m_d$)

$$m_a = |N|\,\frac{f_\pi m_\pi}{v}\,\frac{\sqrt{m_u m_d}}{m_u + m_d} \qquad (1.10)$$

where

$$N = 2\sum_f t_f\,Q_f^{PQ}\,. \qquad (1.11)$$

In Eq. (1.11), the sum is over all colored left-handed Weyl fermions f, the Q_f^{PQ} are the Peccei-Quinn charges of these fermions and the t_f are given by $\mathrm{Tr}(T_f^\alpha T_f^\beta) = t_f\,\delta^{\alpha\beta}$ where the T_f^α ($\alpha = 1\ldots 8$) are the $SU^c(3)$ generators for the color representation to which f belongs ($t_3 = t_{\bar{3}} = 1/2$, $t_6 = t_{\bar{6}} = 5/2$, etc.). For example, $N = n$ in the model of Eq. (1.7). I will sometimes use $f_a \equiv \frac{v}{N}$. The coupling of the axion to two photons is given by

$$\mathcal{L}_{a\gamma\gamma} = \frac{\alpha}{8\pi}\,\frac{a}{f_a}\left[\frac{N_e}{N} - \left(\frac{5}{3} + \frac{m_d - m_u}{m_d + m_u}\right)\right] F_{\mu\nu}\tilde{F}^{\mu\nu} \qquad (1.12)$$

Here α is the fine structure constant and

$$N_e = 2 \sum_f (Q_f^\gamma)^2 Q_f^{PQ} , \qquad (1.13)$$

where Q_f^γ is the electric charge of fermion f in units of e. In many axion models $\frac{N_e}{N} = \frac{8}{3}$. In particular, this is true in all grand unified axion models which implement the successful Georgi-Quinn-Weinberg[9] prediction of $\sin^2\theta_w$. Comparing Eq. (1.10) with (1.12), one finds that in these models $\mathcal{L}_{a\gamma\gamma}$ is given uniquely in terms of the axion mass by

$$\mathcal{L}_{a\gamma\gamma} = - g_\gamma \frac{\alpha}{\pi} \frac{m_a}{0.6 \ 10^{16} \ eV^2} \ a \ \vec{E}\cdot\vec{B} \qquad (1.14)$$

with $g_\gamma = \frac{m_u}{m_d+m_u} \approx 0.36$. In the limit where CP is conserved, the coupling of the axion to fermion f has the form

$$\mathcal{L}_{a\bar{f}f} = i \ g_f \frac{m_f}{v} \ a \ \bar{f}\gamma_5 f . \qquad (1.15)$$

The coefficients g_f that appear in Eq. (1.15) are rather model-dependent. Formulas which give the g_f in terms of the parameters that define an axion model are given in Refs. (8), including those for the case where f is a proton or neutron.

When the axion was first proposed, it was thought that the breaking of $U_{PQ}(1)$ occurred at the electroweak scale, i.e. $v \sim 250$ GeV. The corresponding axion was searched for in various laboratory experiments but was not found[3]. Soon, however, it was discovered[10] how to construct axion models with arbitrarily large values of v. These were called "invisible" axion models because for $v \gg 250$ GeV, the axion is so weakly coupled that the event rates in the axion search experiments mentioned above are hopelessly small. However such axions are still constrained by astrophysical and cosmological considerations. These imply that the axion mass is most likely in the range 10^{-3} eV $\gtrsim m_a \gtrsim 10^{-6}$ eV. The $m_a \lesssim 10^{-3}$ eV constraint[11] follows from an analysis of the supernova SN 1987a. The $m_a \gtrsim 10^{-6}$ eV constraint[12] follows from a calculation of the cosmological energy density of axions produced during the QCD phase transition, which yields

$$\rho_a(t_o) \approx \rho_{crit}(t_o) \, (\frac{0.6 \, 10^{-5} \text{eV}}{m_a})^{7/6} \qquad (1.16)$$

where $\rho_{crit}(t_o)$ is the present critical energy density for closing the universe. These axions are cold dark matter[13] and, if m_a is of order 10^{-5} eV, could be the stuff that appears clustered in the halos of galaxies and clusters of galaxies.

THE CAVITY DETECTOR OF GALACTIC HALO AXIONS

As we saw in the preceding section, it is conceivable that the earth is bathed in a sea of galactic halo axions. If the galactic halo is made up exclusively of axions, their density in the solar neighborhood[14] would be of order $\frac{1}{2} 10^{-24}$ gr/cm^3 and their velocity dispersion would be of order 10^{-3} times the speed of light. A most likely mass range for such axions would include ($\hbar = c = 1$)

$$2\pi(242 \text{ MH}_z) = 10^{-6} \text{ eV} \lesssim m_a \lesssim 10^{-4} \text{ eV} = 2\pi(24.2 \text{ GH}_z). \qquad (2.1)$$

The possibility of detecting these dark matter axions constitutes an exciting prospect and one which, as we will see, is not entirely unrealistic.

a. <u>The general principle</u>

About five years ago, it was pointed out[15] that axions can be searched for by stimulating their conversion to photons in a strong magnetic field. The relevant coupling is given in Eq. (1.14). In particular, an electromagnetic cavity permeated by a strong magnetic field can be used to detect galactic halo axions. The latter have velocities β of order 10^{-3} and hence their energies

$$\epsilon_a = m_a + \frac{1}{2} m_a \beta^2 \qquad (2.2)$$

have a spread of order 10^{-6} above the axion mass. Consider a cylindrical electromagnetic cavity of arbitrary cross-sectional shape, permeated by a large static approximately homogeneous longitudinal magnetic field $\vec{B} = B_o \hat{z}$. When the frequency $\omega = 2\pi f$ of an appropriate cavity mode equals $m_a[1 + O(10^{-6})]$, galactic halo axions can convert to quanta of excitation (photons) of that cavity mode. Only the $TM_{n\ell o}$ modes couple in the limit where the cavity is much smaller than the de Broglie wavelength $\lambda_a = 2\pi(\beta m_a)^{-1} \approx 2\pi \, 10^3 \, m_a^{-1}$ of the galactic halo axions. The power on resonance ($\omega_{n\ell} = m_a[1 + O(10^{-6})]$) from axion $\rightarrow$ photon conversion into the $TM_{n\ell o}$ mode is[15,16,17]

$$P_{n\ell} = (\frac{\alpha}{\pi} g_\gamma \frac{N}{v})^2 \, V \, B_o^2 \, \rho_a \, C_{n\ell} \, \frac{1}{m_a} \, \min(Q_L, Q_a)$$

$$= 0.2 \, 10^{-19} \text{ Watt } (\frac{V}{500 \text{ liter}}) (\frac{B_o}{8 \text{ Tesla}})^2 \, C_{n\ell} \, \cdot$$

$$\cdot \, (\frac{\rho_a}{\frac{1}{2} 10^{-24} \text{ gr/cm}^3}) \, [\frac{m_a}{2\pi(3 GH_z)}] \, \text{Min} \, (\frac{Q_L}{Q_a}, 1) \quad (2.3)$$

where V is the volume of the cavity, ρ_a is the density of galactic halo axions on earth, Q_L is the quality factor of the cavity and $Q_a \equiv 10^6$ is the "quality factor" of the galactic halo axion signal, i.e. the ratio of their energy to their energy spread. $C_{n\ell}$ is a model dependent form factor defined as follows. For $TM_{n\ell o}$ modes, one has

$$\vec{E} = \hat{z} \, \psi_{n\ell}(x,y) \, e^{-i\omega_{n\ell} t}$$

$$\vec{B} = \frac{i}{\omega_{n\ell}} \, \hat{z} \times \vec{\nabla}_t \, \psi_{n\ell}(x,y) \, e^{-i\omega_{n\ell} t} \quad (2.4)$$

where x and y are the transverse coordinates and $\psi_{n\ell}$ satisfies

$$\left(\frac{\partial^2}{\partial x^2} + \frac{\partial^2}{\partial y^2}\right)\psi_{n\ell} + \omega_{n\ell}^2 \, \epsilon \, \psi_{n\ell} = 0 \qquad (2.5)$$

and the boundary condition

$$\psi_{n\ell}\Big|_{\Gamma} = 0 \qquad (2.6)$$

on the boundary Γ of the cross-sectional area S of the cavity. ϵ is the dielectric constant which may depend upon x and y. The magnetic permeability μ was assumed to be equal to one everywhere. The form factor $C_{n\ell}$ is defined by

$$C_{n\ell} = \frac{|\int_S d^2x \, \psi_{n\ell}|^2}{S \int_S d^2x \, \epsilon |\psi_{n\ell}|^2} . \qquad (2.7)$$

For a cavity of rectangular cross-section

$$C_{n\ell} = \frac{64}{\pi^4 n^2 \ell^2} \quad \text{for n and } \ell \text{ odd}$$

$$= 0 \qquad \text{otherwise} . \qquad (2.8)$$

For a circular cross-section

$$C_{nm} = \frac{4}{(X_{on})^2} \delta_{om} \qquad (2.9)$$

where X_{on} is the nth zero of the Bessel function $J_o(x)$. To detect the power $P_{n\ell}$, a hole must be made in the cavity wall through which the electromagnetic radiation can be brought to the front end of a microwave receiver. The quality factor Q_L which appears in Eq. (2.3) is the loaded quality factor given by

$$\frac{1}{Q_L} = \frac{1}{Q_w} + \frac{1}{Q_h} \qquad (2.10)$$

where $\frac{1}{Q_w}$ is the contribution due to absorption into the cavity walls and $\frac{1}{Q_h}$ is the contribution from the hole. The maximum power that can be brought to the front end of the microwave receiver is $\frac{Q_L}{Q_h} P$ where P is given by Eq. (2.3).

Because the axion mass is only known in order of magnitude at best, the cavity must be tunable and a large range of frequencies must be explored seeking the axion signal. The cavity can be tuned by inserting a dielectric rod into it[18], or by moving sideways inside the cavity a dielectric rod or metal post[19]. Using Eq. (2.3), one finds that to obtain a given signal to noise ratio s/n, the search rate is

$$\frac{df}{dt} = \frac{2.6 \text{ GHz}}{\text{year}} \left(\frac{3n}{s}\right)^2 \left(\frac{V}{500 \text{ liter}}\right)^2 \left(\frac{B_o}{8 \text{ Tesla}}\right)^4 c^2 \cdot$$

$$\cdot \frac{\rho_a}{\frac{1}{2} 10^{-24} \text{ gr/cm}^3} \left(\frac{20K}{T_n}\right)^2 \left(\frac{f}{3 \text{ GHz}}\right)^2 \cdot \begin{cases} Q_w/Q_a & \text{if } Q_w < 3Q_a \\ \frac{27}{4}\left(1 - \frac{Q_a}{Q_w}\right)^2 & \text{if } Q_w > 3Q_a \end{cases}$$

$$(2.11)$$

where T_n is the sum of the physical temperature of the cavity plus the noise temperature of the microwave receiver. Eq. (2.11) was derived assuming

1) that when $Q_L < Q_a$, i.e. when the cavity bandwidth is larger than the axion bandwidth, one uses the possibility of looking at Q_a/Q_L axion bandwidths simultaneously,

2) that Q_h has been adjusted so as to maximize the search rate. For $Q_w < 3Q_a$, the optimal $Q_h = \frac{1}{2} Q_w$ (and hence $Q_L = \frac{1}{3} Q_w$) whereas for $Q_w > 3Q_a$, the optimal Q_h is such that $Q_L = Q_a$.

Actually, the best possible quality factors attainable at present, using oxygen free copper, are only of order 10^5 in the GHz range. (One cannot make the cavity of ordinary superconducting material since the cavity must be permeated by a strong magnetic field. It is possible however that at some point in the future, high T_c superconductors will make available much larger B_o and/or Q_w then has been envisaged up till now.) The factor Q_w/Q_a on the RHS of Eq. (2.11) is therefore of order 10^{-1}. On the other hand, we have measured[20] noise temperatures as low as 3K (in the 1.2 - 1.6 GHz range) in commercially available microwave receivers. Provided $C = O(1)$, Eq. (2.11) suggests then that a galactic halo axion search is feasible with presently available technology. One of the problems such a search faces however is that a large volume empty cylindrical cavity has a low resonant frequency in its lowest TM mode: $f = 0.29$ GHz $(\frac{40\text{cm}}{R})$ where R is the radius of the cavity. A large cylindrical cavity is convenient for exploring the low frequency end of the range (2.1). However, one wishes to extend the search to frequencies well above 1 GHz. How is this to be done? Use of the higher $TM_{n\ell o}$ modes entails an unaffordable loss in sensitivity since their form factors, as shown by Eqs. (2.8) and (2.9), fall off like the square of the mode number. One can however increase the frequency of the lowest TM mode by inserting one or several metal posts into the cavity. In this way one can in fact obtain simultaneously large volume, high frequency and form factors C of order one. Alternatively one can adopt the brute force approach of filling up the available volume with many small cavities all resonant at the same frequency. These issues are discussed extensively in Ref. (19) which presents the results of computer simulations which optimize the cavity design to obtain the highest possible cosmic axion search rate for a given magnet size.

b. The status of ongoing experiments

There are at present two cavity detectors of galactic halo axions taking data, one at Brookhaven National Laboratory[18], the other at the University of Florida[20]. In addition, a group at KEK is working on the construction of a third detector[21].

The Brookhaven experiment has been running for about two years. It uses a 11 liter cavity placed in a 6 Tesla superconducting solenoid when searching near 1.2 GHz. To search at higher frequencies, an insert is placed inside the 6 Tesla coil which increases its magnetic field to 9 Tesla at some expense in available volume. The Brookhaven experiment system noise temperature is of order 15 K. They have searched over a relatively large frequency band ranging from about 1.1 GHz to about 4 GHz. The upper limit they place on the local axion density over this range is of order 10^3 times larger than the local halo density.

The assembly of the Florida detector has just been completed and we have started to take data. Our detector has a 7 liter cavity placed inside a 9 Tesla superconducting solenoid. During the next few months, we are planning to search in the 1.3 to 1.7 GHz range. In this bandwidth, our HEMT microwave receiver/amplifier has noise temperature as low as 3 K. When the detector is cooled to 2.2 K by pumping on the H_e bath, the system noise temperature is 3 K + 2.2 K = 5.2 K. In the 1.3 to 1.7 GHz range, we hope to place a limit on the local axion density which is somewhat more stringent than that obtained by the Brookhaven experiment.

ACKNOWLEDGEMENT

This work was supported in part by the U.S. Department of Energy under contract No. DE-FG05-86ER40272.

REFERENCES

1. R. D. Peccei and H. Quinn, Phys. Rev. Lett. $\underline{38}$ (1977) 1440 and Phys. Rev. $\underline{D16}$ (1977) 1791.
2. S. Weinberg, Phys. Rev. Lett. $\underline{40}$ (1978) 223; F. Wilczek, Phys. Rev. Lett. $\underline{40}$ (1978) 279.
3. Recent reviews include: J. E. Kim, Phys. Rep. $\underline{150}$ (1987) 1; H.-Y. Cheng, Phys. Rep. $\underline{158}$ (1988) 1; R. D. Peccei, "The strong CP problem", DESY preprint 88-109 (Aug. 1988).
4. G. 't Hooft, Phys. Rev. Lett. $\underline{37}$ (1976) 8 and Phys. Rev. $\underline{D14}$ (1976) 3432; R. Jackiw and C. Rebbi, Phys. Rev. Lett. $\underline{37}$ (1976) 172; C. G. Callan, R. F. Dashen and D. J. Gross, Phys. Lett. $\underline{63B}$ (1976) 334.
5. V. Baluni, Phys. Rev. $\underline{D19}$ (1979) 2227; R. J. Crewther, P. Di Vecchia, G. Veneziano and E. Witten, Phys. Lett. $\underline{88B}$ (1979) 123.
6. M. Kobayashi and K. Maskawa, Progr. Theor. Phys. $\underline{49}$ (1973) 652.
7. H. Georgi and L. Randall, Nucl. Phys. $\underline{B276}$ (1986) 241.
8. W. A. Bardeen and S.-H. H. Tye, Phys. Lett. $\underline{74B}$ (1978) 229; J. Ellis and M. K. Gaillard, Phys. Lett. $\underline{74B}$ (1978) 374; T. W. Donnelly et al., Phys. Rev. $\underline{D18}$ (1978) 1607. More recent treatments include: M. Srednicki, Nucl. Phys. $\underline{B260}$ (1985) 689; P. Sikivie, in "Cosmology and Particle Physics", Ed. by E. Alvarez et al., pp 143-169, World Scientific, 1987.

9. H. Georgi, H. R. Quinn and S. Weinberg, Phys. Rev. Lett. 33 (1974) 451.
10. J. Kim, Phys. Rev. Lett. 43 (1979) 103; M. A. Shifman, A. I. Vainshtein and V. I. Zakharov, Nucl. Phys. B166 (1980) 493; M. Dine, W. Fischler and M. Srednicki, Phys. Lett. 104B (1981) 199; A. P. Zhitnitskii, Sov. J. Nucl. Phys. 31 (1980) 260.
11. G. Raffelt and D. Seckel, Phys. Rev. Lett. 60 (1988) 1793; M. S. Turner, Phys. Rev. Lett. 60 (1988) 1797; R. Mayle et al., Phys. Lett. 203B (1988) 188; T. Hatsuda and M. Yoshimura, Phys. Lett. 203B (1988) 469; A. Burrows, M. S. Turner and R. P. Brinkmann, Phys. Rev. D39 (1989) 1020; N. Iwamoto, Univ. of Toledo preprint (1988).
12. L. Abbott and P. Sikivie, Phys. Lett. 120B (1983) 133; J. Preskill, M. Wise and F. Wilczek, Phys. Lett. 120B (1983) 127; M. Dine and W. Fischler, Phys. Lett. 120B (1983) 137. See also, R. Davis, Phys. Lett. 180B (1986) 225; D. Harari and P. Sikivie, Phys. Lett. 195B (1987) 361..
13. J. Ipser and P. Sikivie, Phys. Rev. Lett. 50 (1983) 925.
14. M. S. Turner, Phys. Rev. D33 (1986) 889.
15. P. Sikivie, Phys. Rev. Lett. 51 (1983) 1415.
16. P. Sikivie, Phys. Rev. D32 (1985) 2988.
17. L. Krauss, J. Moody, F. Wilczek and D. Morris, Phys. Rev. Lett. 55 (1985) 1797.
18. S. DePanfilis et al., Phys. Rev. Lett. 59 (1987) 839.
19. C. Hagmann et al., "Cavity design for a cosmic axion detector", Univ. of Florida preprint (Dec. 1988).
20. S.-I. Cho et al., Japanese J. of Appl. Phys. 26 (1987) Suppl. 26-3, 1705.
21. S. Inagaki et al., "Study of Galactic Axions by a Cavity Haloscope", KEK preprint (1988).

BOLOMETRIC DETECTION OF RARE EVENTS:
A CHALLENGE FOR LOW TEMPERATURE PHYSICS

M. Chapellier[†], G. Chardin[‡], H. Ji and J. Joffrin
Laboratoire de Physique des Solides — Bat. 510, Orsay, France

[†]also DPhG/SRM, IRF, CEN Saclay, Gif/Yvette, France
[‡]also DPhPE/SEPh, IRF, CEN Saclay, Gif/Yvette, France

INTRODUCTION

One of the outstanding problems of physics may be found in the so called missing mass of the Universe. Observations of the tangential speed of luminous matter in galaxies have revealed an anomalous relation between speed v and distance R to the galactic center: instead of $v \propto 1/R$ at large distances, nearly constant velocities are observed. Assuming the validity of Newton's law at such distances, this anomalous behavior must then be ascribed to the existence of otherwise invisible mass distributions extending far outside the luminous core of the galaxy. Further assuming that the mass density of the Universe is critical, as expected from inflation models and esthaetical prejudices, implies that at least part of this "missing mass" must be ascribed to the presence of as yet undiscovered Weakly Interacting Massive Particles (WIMP's), relics of the Big Bang. These particles would have escaped detection until now due to their unknown but a priori very small interaction cross section with ordinary matter. These non dissipative particles are expected to have velocities typical of the galactic virial velocity ($v/c \approx 10^{-3}$). At such low velocities, the main source of energy loss is due to elastic collisions with nuclei, resulting in recoil energies in the range 1-10 keV. Stronger interactions are already excluded by Germanium and Silicium detectors, which have been until now the most sensitive devices.

Therefore, the physicist is confronted to the following problems:
– the detection of a nucleus recoil of energy in the 1–10 keV range
– the discrimination of such a cosmic event from the background radioactivity inside the material (the expected number of cosmic events is $\approx 1000-5000$ kg^{-1} day^{-1} keV^{-1} for vector couplings, but is limited to $\approx 0.1-1$ event kg^{-1} day^{-1} keV^{-1} for couplings proportional to the spin of the target nuclei).

The latter problem —the contamination by internal radioactivity— which may fake a cosmic interaction, is in fact the *most challenging* problem. On the other hand, the first problem may be solved by calorimetry.

CALORIMETRIC DETECTION

A pure crystal has a specific heat expected to follow the Debye law $C \propto (T/\theta)^3$. An energy of 10 keV is $1.6 \; 10^{-15}$ J which, dissipated in 10^{-4} s is only $1.6 \; 10^{-11}$ W. This is known to be sufficient to heat a thermistor. As an example, a 10 keV dissipation in a crystal of 1 cm^3 of Al$_2$O$_3$ will result in raising the crystal temperature by $\Delta T \approx 10^{-8}/T^3$. At $T \leq 100$ mK, ΔT may be rather easily detected.

However, problems which remain to be solved include the following:
– the validity of the Debye law down to low temperatures (≈ 1 mK) is not extensively tested. Any kind of impurities, dislocations, defects, quadrupolar interactions may easily overwhelm the heat capacity predicted by the Debye law. It should be noted, however, that only anomalous heat capacities with short time constants should be considered, as they will be of no concern if they involve time constants much in excess of the characteristic time of the heat impulse.

© 1989 American Institute of Physics

– in addition, any non crystalline substance (metal junctions, amorphous materials, and the thermistor itself) will dramatically increase the specific heat.

– purity of material : nearly everything is radioactive (remember that you emit yourself around 8000 neutrinos per seconds !!!)

FIRST PRACTICAL DESIGN

A dilution fridge without He^4 pot has been built (now commercially available : P. Pari, DPhG/SRM, CEN Saclay, 91191 Gif/Yvette, France, phone 33-1-69 08 74 48), and has been operated in a He storage Dewar. With this dilution, temperatures of 70 mK, starting from room temperature, may be achieved in less than 4 hours. A massive composite bolometer comprising 1.6 g of Al_2O_3, on which a RuO_2 film (1 kΩ batch) was deposited, has been used for these preliminary measurements. These RuO_2 films have a variable hopping range conductivity $R = R_o \exp (T_o/T)^{1/4}$ and have very reproducible characteristics. Alpha particles are emitted by an Am^{241} radioactive source. Data acquisition involves a numerical oscilloscope coupled to a Macintosh II microcomputer. Figures below displays an averaged signal (on 180 signals) together with an histogram showing a 5% resolution .

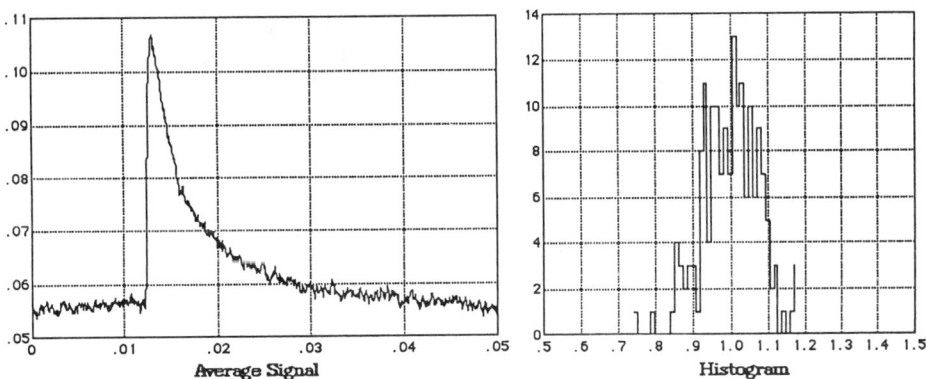

Average Signal   Histogram

FUTURE PROSPECTS

Although encouraging, our results are still preliminary. Individual alpha particles have been detected with a $\approx$5% event by event resolution. However, the heat impulse produced is much smaller (by a factor $\approx$20) than expected from a naive calculation. This may be due to extra heat capacity from the graphite pods. We are planning to replace these electrical contacts by Al superconducting wires. In addition, RuO_2 films may be replaced by other batches ($\approx$10kΩ) with a better sensitivity at temperatures $\approx$70 mK. A cooled FET amplifier at 4.2 K will be used in order to improve the signal/noise ratio and, possibly, the bandwidth. For larger samples, and in order to test the validity of the Debye law, lower temperatures (down to 10mK) will be achieved.

CONCLUSIONS

Interesting problems of low temperature physics are met in these new experiments; the ultimate aim of detecting interactions of dark matter particles are, however, still very distant. Other approaches will also be attempted as the signature of a calorimetric detection lacks the redundancy needed to ascertain the detection of dark matter particles.

A COSMIC AXION DETECTOR*

C. Hagmann, P. Sikivie, N.S. Sullivan and D.B. Tanner
University of Florida, Gainesville, FL 32611

AXION PROPERTIES

The axion[1] is a hypothetical particle which was introduced to solve the strong CP problem. Peccei and Quinn proposed a U(1)-symmetry to explain the absence of CP violations in QCD, manifested in the upper limit on the neutron electric dipole moment. At low energies the symmetry is spontaneously broken and the associated pseudo-Nambu-Goldstone boson is the axion. Massless at the classical level, it aquires a small mass due to instanton effects. The theory does not predict its mass and couplings to quarks, electrons and photons, though they are related to the symmetry breaking scale f_a :

$$m_a \simeq \frac{0.62 \times 10^{-5}\,\text{eV}}{(f_a/10^{12}\,\text{GeV})} = \frac{2\pi\,1.5\,\text{GHz}}{(f_a/10^{12}\,\text{GeV})} \tag{1}$$

The coupling of the axion field a to the electromagnetic field is given by

$$L_{a\gamma\gamma} \simeq -8.4 \times 10^{-4} \frac{a}{f_a} \vec{E} \cdot \vec{B} \tag{2}$$

Laboratory searches have eliminated axions which have $f_a < 10^3\,\text{GeV}$, while astrophysical and cosmological considerations have narrowed the possible mass range to $10^{-6}\,\text{eV} < m_a < 10^{-3}\,\text{eV}$. Axions of this mass have important cosmological consequences. In the early universe at $T \simeq 1\,\text{GeV}$, the axion mass 'turns on' due to quantum effects and the axion field starts to oscillate. These oscillations correspond to a highly degenerate zero-momentum Bose condensate. Relic axions are still around today and contribute to the present energy density of our universe :

$$\Omega_a \simeq 0.85 \cdot (m_a/10^{-5}\,\text{eV})^{-1.18} \tag{3}$$

If axions close the universe then they are likely to form the dark halo around our galaxy which is known to have a density of $\rho_{halo} \simeq 0.3\,\text{GeV} \cdot \text{cm}^{-3}$.

AXION DETECTION

Our detector makes use of the electromagnetic coupling of the axion, which allows for a decay of one axion into one photon in the presence of a static background field B_0. Axion to photon conversion is greatly enhanced in a microwave cavity resonating at the axion energy ($\omega = m_a \cdot (1 + O(10^{-6}))$), where the $O(10^{-6})$ spread comes from the axion kinetic energy.
The power in mode nl is given by ($\vec{B}_0 = B_0 \hat{z}$)

$$P_{nl} \simeq 2 \times 10^{-20}\,\text{W}\,\left(\frac{V}{500\,\ell}\right)\left(\frac{B_0}{8\text{T}}\right)^2 C_{nl}\left(\frac{\rho_a}{\rho_{halo}}\right)\left(\frac{m_a}{2\pi\,1.5\text{GHz}}\right) Q_{nl} \tag{4}$$

© 1989 American Institute of Physics

where V is the cavity volume, Q_{nl} is the loaded quality factor of the cavity and

$$C_{nl} = \frac{|\int_V d^3x\, E_{z,nl}|^2}{V \cdot \int_V d^3x\, \epsilon(x)\, |E_{z,nl}|^2} \quad (5)$$

Because the axion mass is only known in order of magnitude the cavity has to be tunable and a wide range of frequencies has to be scanned. The search rate for given signal to noise ratio s/n is determined by V, B_0, Q_{wall} and system noise temperature T_n.

RESULTS

$T_n = T_{bath} + T_{amp} = 2\,\text{K} + 3\,\text{K} = 5\,\text{K}$ @ $1.2 - 1.6$ GHz
$Q_{wall} \simeq 10^5$
$B_0 = 8.5\,\text{T}$
$V \simeq 6\,\ell$
Frequencies covered : $1.48 - 1.50$ GHz with $s/n = 1/300$

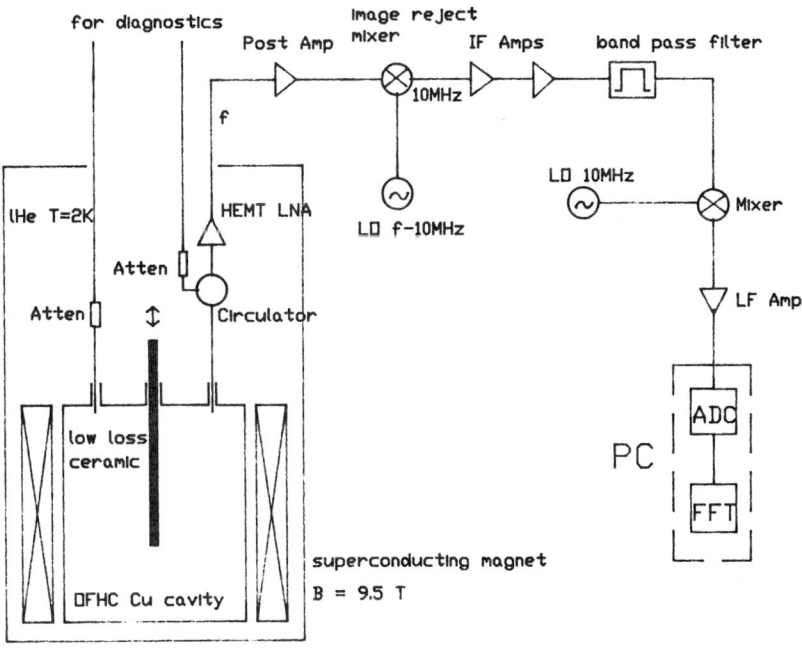

Fig.1: schematic diagram of axion detector.

* Research supported by DOE grant DOE-F605-86ER-40272.

REFERENCES

1. J.E. Kim Phys.Rep. **150** (1987) and references quoted therein

LOW TEMPERATURE TECHNIQUES

CHAIRMAN

Hidehiko Ishimoto
Institute for Solid State Physics
University of Tokyo

This session was made possible, in part, through
a generous donation provided by

6056 West Jefferson Boulevard / Los Angeles, California 90016 / (213) 870-9383

Scientific Instruments, Inc.
Telephone 407/881-8500 • Telex 51-3474 • FAX 407/881-8556
4400 W. Tiffany Drive • Mangonia Park • West Palm Beach, Florida 33407

QUASIPARTICLE SPECTROMETRY, THE ENGINEERING AND THE PHYSICS

A M Guénault and G R Pickett
Lancaster University, Lancaster, U.K. LA1 4YB

ABSTRACT

Despite the fact that the quasiparticle spectrometer works, there are several difficulties with understanding its operation. In an attempt to understand these problems we have discovered some very interesting behaviour of the dilute quasiparticle gas which responds at high velocity in an extremely nonlinear way. This nonlinearity has pointed us in the right direction for understanding the low velocity response, hitherto not understood. We have also observed fine-structure at the onset of pair-breaking which shows interesting "quantum" behaviour.

INTRODUCTION

It is four years since we suggested that it might be possible to devise a spectrometer operating with ballistic quasiparticles in superfluid ^{3}He-B at low temperatures[1]. Such a device is conceptually simple; a quasiparticle beam is generated by a source at a point A, monitored by a remote detector at B and in the intervening space the influence on the beam of whatever parameter we are interested in can be observed. Such devices should yield information both directly on quasiparticle dynamics and indirectly on the embedding condensate configuration.

When we made the original suggestion we had only very crude notions of how we could produce a source, based on a pulsed ohmic heater, which in the event turned out not to be practicable. However, the subsequent observation[2] of pair-breaking by wire resonators at a critical velocity immediately provided a highly controllable source and hence the key to making the spectrometer. In the intervening period we have succeeded in making a number of these devices which make use of vibrating wire resonators both as source and detector. We can readily produce ballistic quasiparticles and detect them at distances up to ~ 1 cm from the source (see for example references 3 and 4).

So far so good. However, to use such a spectrometer in any quantitative way we need to understand precisely how it works and and it is on these aspects that we have concentrated our recent efforts. Two things have come out of this work. First, the more we study the devices the more complex the behaviour appears, (one is tempted to say the more we study them the less we understand but that is overstating the case) but secondly we have thrown up a great deal of interesting new physics concerning the dilute quasiparticle gas.

We begin by looking at some of the problems. Already at the outset there are several difficulties concerning the wire resonators taken alone. First we consider low velocities. If we treat the wire resonators as approximating to an infinite cylinder moving transversely in the liquid then a naive kinetic model suggests that the force per unit length on a cylinder of radius a moving at velocity v through a gas of particles of density n exchanging momentum of order p_F with the cylinder should be

$$F \simeq a\, n\, p_F\, v. \tag{1}$$

This expression gives a good picture of the damping by ^{3}He quasiparticles in normal ^{3}He-^{4}He solutions. Unfortunately in ^{3}He-B this quantity turns out to be a factor of order $k_B T/E_F$ smaller than the experimentally observed force. Furthermore, we have implicitly assumed normal, rather than Andreev scattering processes and of course when we scatter a quasiparticle normally we expect a momentum transfer of approximately p_F, whereas a scattered quasihole will transfer $-p_F$. Therefore the momentum transfer in a gas with particle-hole symmetry should be zero for normal processes on this naivest of naive models.

The experimentally observed force on the wire is in fact consistent with the expression

$$F \simeq a\, \rho_n\, v_g\, v. \tag{2}$$

Despite the problems this expression leads to a temperature dependence of the damping of a resonator which is proportional to $\exp(-\Delta/k_B T)$. This at least yields us a very fine thermometer (at least down to temperatures where the damping becomes comparable to the natural damping of the resonator, but more of that below).

At higher velocities, i.e. where the wire velocity becomes comparable to the Landau velocity, Δ/p_F, pair breaking begins. We need to understand this process since we rely on the emitted quasiparticles to provide our beam source. Naively we expect pair breaking to begin when the relative velocity of the wire and condensate reaches the Landau velocity. For an infinite cylinder this occurs first along the two extremal lines on the surface when the cylinder velocity reaches half the Landau value. In fact, pair breaking is observed to begin at even lower velocities, roughly 0.25 – 0.3 times the Landau value, for all the wires we have examined. This is almost certainly due to the depression of the gap in the vicinity of the wire either by a boundary layer of liquid of different phase (planar phase for example) or because of distortion arising from the flow field. A reduced gap near the wire yields a lower critical velocity which would lead to the population of local quasiparticle states bound at the wire surface. Nevertheless we would still have to postulate some extra mechanism for pumping these local states up to energies where they can escape from the local potential well into the bulk liquid to provide an energy loss mechanism.

This latter is an important point as the final states of the quasiparticles after they have escaped from the wire constitute the beam. We would like to know the velocities of these particles, their angular distribution (i.e. how narrow a beam we get?) and their effective masses (are they quasiparticles or quasiholes?).

If we now put two wires together in a spectrometer configuration then we can indeed see the effect on one wire of the "quasiparticle wind" from the other when driven above the pair-breaking velocity. There are also some interesting problems here.

The force on the detector wire appears at double the frequency of that of the source wire. Since we expect the source wire to emit quasiparticles in the forward direction and quasiholes in the rearward direction we might expect a pulse of excitations to arrive at the detector every half cycle of the source motion, pulses alternating between bursts of quasiparticles and bursts of quasiholes. For normal scattering processes of the excitations at the detector wire surface this would give alternate 'pull' and 'push' forces and the detector would respond at the same frequency as the source. That the detector response is at **double** the source frequency may mean, that we have Andreev reflection processes occurring, or that the production process is not what we think, or that the detection process is not what we think.

One solid fact to emerge is that the force on the detector wire is proportional to the **power** dissipated in the source wire, suggesting that the energy goes into producing quasiparticles at or near the dispersion minimum. If we assume that the force on the detector depends linearly on the excitation flux then if all the particles are created with energy Δ, i.e. near the minimum in the dispersion curve, then the flux is simply given by $\dot{Q}/\Delta$ where $\dot{Q}$ is the source power.

If the force on the detector wire is a result of Andreev scattering processes then we would expect the flow field around a moving detector wire to modify the scattering considerably. The detector wire normally drives a SQUID detector. We can vary the amount that the detector wire is free to move under the influence of the "quasiparticle wind" by the rigidity of the feedback from the SQUID loop. Whether the wire is free to move or held virtually stationary there is no significant change in the force observed.

DAMPING BY THE QUASIPARTICLE GAS

Let us now look at some of our recent results on single wires.

The primary motivation for the work reported below was the wish to see if we could observe any change in critical velocity associated with the phase transition observed by the Moscow group[5] at around $0.4T_c$. In the event we found that too much heat is dissipated in running the vibrating wire resonators up to the critical velocities at such high temperatures. Fortunately, the attempt meant that we ran the resonators at high velocity at much higher temperatures than previously with initially surprising results.

The experimental details are briefly as follows: The experiments are performed on a series of vibrating wire resonators in superfluid ^{3}He-B. The experimental configuration is similar to what we have used in the past. The ^{3}He is cooled by contact to copper nuclei, themselves cooled by adiabatic demagnetization to a temperature of around 50 μK. The magnetic excitation field for the resonators is provided by the final demagnetization field, typically 32 mT in the present work.

The resonators used here have a loop diameter of about 3 mm (rather than 8 mm used in our previous work). The voltage generated by the movement of the wire in the excitation field is detected by a two-phase lock-in amplifier with a high quality transformer as an input preamplifier. We observe the damping of the resonator as a function of velocity is observed, as follows: The frequency is set to a value close to resonance, and the vector voltage measured as the drive current is slowly incremented upwards from zero. After small corrections are made for the self-inductance of the wire (negligible in practice) and for any zero error in the lock-in amplifier, the measured voltage gives the velocity of the wire as an average along the length perpendicular to the field. The driving current gives a measure of the similar average driving force on the resonator. This procedure leads to a velocity-force curve.

The measurement is then repeated for a series of temperatures, from base temperature (less than 100 μK at 0 bar pressure, see below) up to around $T_c/4$. The temperature is adjusted by a steady heat input provided by a second wire, driven rather violently, and the ^{3}He temperature is measured (and monitored as constant) from the resonance width of a third gently-driven wire. A set of curves[6] of velocity versus drive level taken in this way is shown in figure 1.

In the figure we see at low velocities the straightforward response of the resonator to the existing thermal quasiparticle distribution. The resistance is seen to increase with temperature as the thermal density increases. At a critical velocity (which in this temperature range is completely temperature-independent) pair breaking begins and the effective resistance of the liquid ^{3}He increases very rapidly. Two important features of the pattern are:

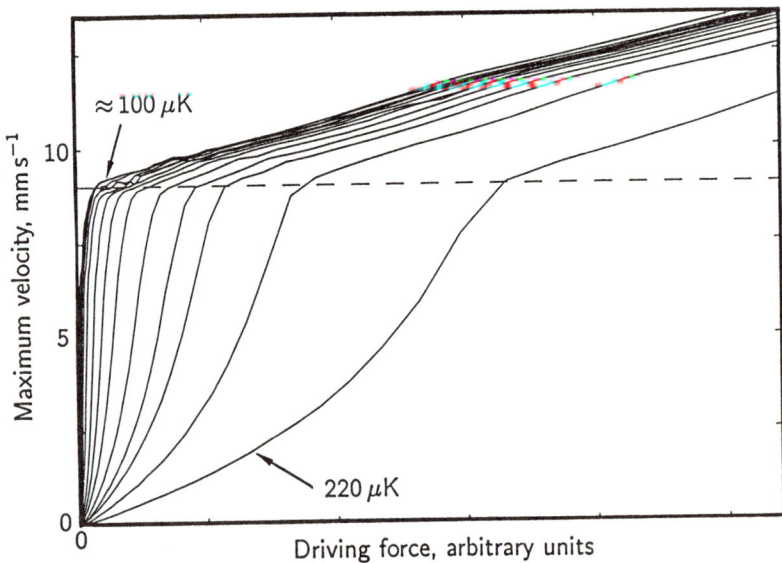

Figure 1. Velocity versus drive level curves for a 4.5 μm diameter wire in ^{3}He-B at zero bar and 32 mT for a series of temperatures from $\leq$100 to 220 μK. The velocity scale represents the maximum velocity reached by the resonator. The dashed line is drawn at 9 mm/s.

a the quasiparticle damping at the lower velocities is highly non-linear, the slope of the curve just below the critical velocity being almost three times greater than that a zero velocity, and

b at the pair-breaking edge all the curves are remarkably similar. Just above the critical velocity one can see structure on the lower temperature curves. This structure can be followed as the temperature is increased, always appearing at the same velocity, and it only disappears when at the high temperatures the drive increments become too large for the structure to be resolved.

Before this attempt to run at high temperature we had never seen this quasiparticle nonlinearity. At base temperatures presumably the curves below v_c have always been linear. At our base temperature the quasiparticle damping is still equally nonlinear, but, as we have always suspected, it has become much less than the linear nuisance damping of the (not so very high Q) micron-wire resonators.

To analyse the results we simply note that the effective damping coefficient (=drive force per unit velocity response) as a function of velocity is given by the inverse slope of the curves in figure 1. This effective damping coefficient we designate G. The values of G plotted against velocity for the data of figure 1 are shown in figure 2. G is made up of three terms; G_0, the residual nuisance damping of the resonator from internal friction, losses in the associated circuit etc.; G_1, the response of the thermal quasiparticle distribution; and G_2, the extra damping introduced at higher velocities by pair breaking, i.e.

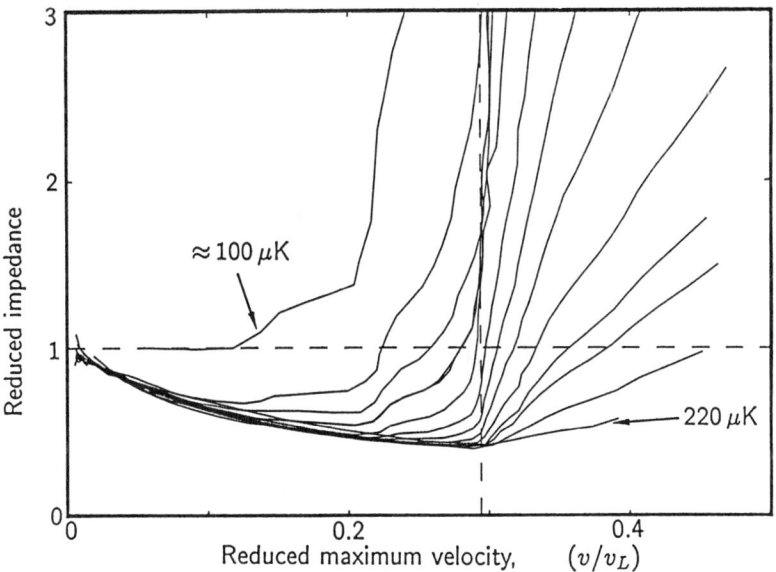

Figure 2. The data of figure 1 plotted as the effective damping coefficient, or impedance, as a function of maximum velocity, $v^* = v/v_L$, where the Landau velocity v_L is 30.6 mm s^{-1}. The curves are all normalised to unity at low velocities by dividing by the low velocity value of G.

$$G = G_0 + G_1 + G_2. \tag{3}$$

These terms can be separated if we make use of the assumptions that a) the nuisance damping of the resonator is constant and that b) the pair-breaking contribution depends only on the superfluid fraction (which is also constant over our temperature range since ρ/ρ_s does not deviate significantly from unity). At the lowest temperatures the quasiparticle damping G_1 is very much less than G_0. Hence at base temperature $G = G_0 + G_2$ only and thus if we subtract the base temperature G as a function of velocity from the measured values at the higher temperatures we are left with G_1, the quasiparticle damping coefficient alone. The result of this procedure for the data of figures 1 and 2 is shown in figure 3.

In the figure it is apparent that the nonlinearity in the quasiparticle damping is universal and independent of temperature. The nonlinearity is very large, the value of G falling at v_c to only a third of its low velocity value. Furthermore, despite the enormous increase in damping when pair breaking begins we can follow the quasiparticle damping as it continues through the critical velocity and continues smoothly to much higher velocities. (The noise on the curves above v_c arises from the subtraction process of two large quantities compounded with the fact that the data form a linear interpolation between data points and at the higher temperatures these points are quite widely spaced above v_c.)

Why is the quasiparticle damping so nonlinear? The nonlinearity is very

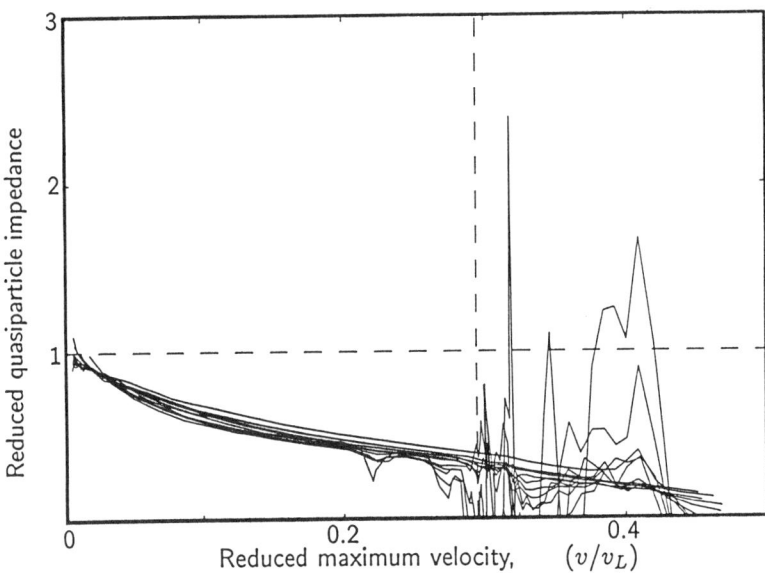

Figure 3. The normalized quasiparticle damping coefficient. for the data of figure 1, plotted on the same scale as figure 2. The form of curve is clearly universal and extends well above the pair-breaking region. The noise on the lower temperature curves is a consequence of the subtraction process.

surprising since the coefficient falls with increasing velocity. The increase in curvature of G_1 towards low velocity suggests perhaps that we should plot $1/G_1$ versus v. Such a plot yields a reasonable straight line, i.e. we can write $1/G_1 = a + bv$.

At this point we need to return to the simple one-dimensional kinetic argument touched on in the introduction above. Assume the wire to be a flat paddle of width a, moving with velocity v much lower than the group velocity v_g of the excitations. If the excitations have number density n then the leading side of the paddle intercepts of order $na(v + v_g)$ excitations per unit time and the trailing side $na(v_g - v)$. If each excitation exchanges of order p_F momentum with the paddle then the net force is:

$$p_F na(v_g + v) - p_F an(v_g - v) \approx p_F anv$$

which as we pointed out above is much less than the force observed, and in any case reduces to zero when we remember that there are equal numbers of quasiparticle and quasiholes which transfer opposite momenta to the paddle.

However, what we do not take into account with this argument is the change in the dispersion curve as observed in the frame of the paddle. For example a paddle travelling at v relative to the condensate would observe the dispersion curve to be as shown in figure 4. Excitations approaching the paddle from the front are to be found at A for quasiparticles and B for quasiholes. For excitations approaching from

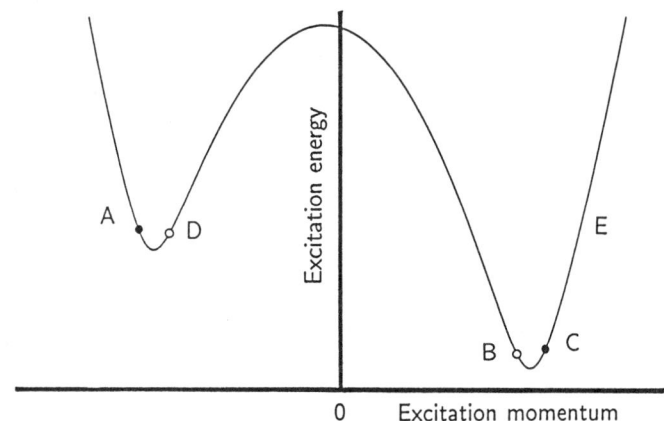

Figure 4. The dispersion curve for the excitations as seen from the reference frame of a paddle moving in the positive x-direction (see text). Excitations approaching from the front are found at A and B for quasiparticles and quasiholes, respectively, and quasiparticles and quasiholes approaching from the rear at C and D.

the rear, quasiparticles are found at C and quasiholes at D.

Now let us look again at the simple one-dimensional argument again. Per unit time the paddle intercepts $\sim (v_g + v)na$ particles approaching from the front. These particles can be normally scattered as there are empty states on the other side of the dispersion curve (at E in the figure). Therefore momentum $\sim p_F(v_g + v)na$ is transferred to the paddle per unit time. Quasiholes appearing from the forward direction cannot be normally scattered as there are no quasihole states on the other side of the dispersion curve, and these must undergo Andreev processes. This is also true for quasiparticles approaching from the rear, whereas quasiholes from the rear can be normally scattered. The number of quasiholes from the rear intercepted is $\sim (v_g - v)na$ and momentum transferred is $-p_F$ (both magnitude and direction are reversed compared with incoming quasiparticles). The total force is thus the sum of the forward side quasiparticle processes and the rearward side quasihole processes, i.e.

$$F = p_F(v_g + v)an + p_F(v_g - v)an \approx p_F v_g na.$$

Now the v terms cancel rather than the v_g terms giving us a much larger force than in the earlier picture since the group velocity v_g is of the order of metres per second whereas the wire velocity v is only millimetres per second. The force on this less naive argument is thus much more in line with that experienced in practice by a wire. However, the most interesting result is that the force is found to be independent of paddle velocity. Since the quasiparticles scattered from the front of the wire and the quasiholes scattered from the rear of the wire both transmit the same momentum on scattering, any increase in the wire velocity simply increases the quasiparticle collision rate and reduces the quasihole collision rate by the same amount, the total momentum change remaining the same.

At the low velocities when the difference in energy between the two minima in the dispersion curve of figure 4 is less than the thermal; spread of occupied states then the simple ideas above would break down. All classes of particles could then be normally scattered to some extent and the force would start to fall. In fact, a simple argument would suggest that only $\approx p_F v/k_B T$ of the $-p_F$ excitations are immobilised at low velocities, giving $F \approx anp_F v.(p_F v_g/k_B T)$, the simple expression with the $E_F/k_B T$ factor required to fit the experiment.

Given that a moving cylinder is a three dimensional object with a range of magnitudes and directions of the relative velocities between wire and condensate, a one-dimensional argument only gives a hint at the behaviour. Furthermore, we are looking at a resonator moving sinusoidally, not at a fixed constant velocity. Nevertheless, the rapidly increasing velocities with increasing drive seen in figure 1 and the approach of $1/G_1$ to proportionality with v at high v (damping force is proportional to Gv) suggests that this approach may be a step in the right direction..

The momentum change of the excitations undergoing Andreev scattering is only of order $(\Delta/E_F).p_F$, quite negligible, so that the wire will only detect those scattering processes in the normal channel. More work needs to be done on the scattering processes at the surface and the normal/Andreev scattering ratio.

Finally we can comment on the lowest liquid temperature we reach. The absence of nonlinearity at base temperature is, we now can see, a clear signature that $G = G_0$ with G_1 negligible. A careful search for nonlinearity convinces us that $G_1 < G_0/20$, and hence that the base temperature of the liquid cannot be higher than $100\,\mu$K. The nonlinearity in G_1 has been studied from about $130 - 220\,\mu$K.

In summary, we see that a vibrating wire in ^{3}He-B is a very unusual mechanical system even before the pair-breaking velocity is reached. The quasiparticle damping force rises with velocity to a saturation level independent of velocity. This potential runaway system is halted by the onset of pair-breaking, ensuring that the damping increases rapidly again at higher velocities.

STRUCTURE AT THE PAIR-BREAKING EDGE

Having seen how we can separate and examine the quasiparticle damping term what can we say about the pair-breaking behaviour, the term G_2 of equation 3? This term demonstrates three main features. With thick wires ($\phi{\geq}100\,\mu$m) there is a rapid and smooth increase in damping force which begins at a wire velocity of around $v_L/4$ (v_L being the Landau velocity, Δ/p_F) or around half the value we would expect for normal Landau pair breaking when the geometry of the wire is taken into account. This steady "background" increase is seen in all wires we have used. However, when we use much thinner wires $(6 - 12\,\mu$m) there appears a wealth of detailed structure superimposed on the smooth background and which is extremely wire-dependent, and varies with pressure.

The structure has two main features. First there are deviations from the background wich are highly reproducible and which we may ascribe to the particular shape of the wire. That is to say each excrescence on the wire (and the thin wires are **highly** non-circular) will provide a local pair-breaking centre and the particular distribution of these leads to the signature of a particular wire. These features we would expect to see at all pressures simply scaled by the appropriate Landau velocity. Broadly speaking, this is indeed what the experiments show.

However, secondly there is a further set of features which are hysteretic and take the form of very well-defined steps in the velocity-force curve. These steps are also highly reproducible but on sweeping the drive up and down the system does not always describe the same orbit jumping some of the steps in a sytematic manner. A typical example of such structure is shown in figure 5. The steps are very reminiscent of the early SQUID "staircase" pattern (which is generated in the same way with an oscillating system being limited by hitting a dissipative jump of some sort at an extreme of an AC cycle). Although these steps (and jumps) are wholly reproducible from run to run at the same pressure, the pattern changes entirely when the pressure is changed. This structure, therefore, does not scale with v_L. We do not yet have enough data to say whether the structure is associated with a constant velocity independent of pressure. Perhaps quantized vortex lines are being observed in some way, but at present this effect remains an unexplained curiosity.

THE QUASIPARTICLE GENERATOR

As we have seen, a strongly-driven resonator provides a quasiparticle source. In spite of the advances described above, we still cannot give a full characterisation of the properties of the source.

In the spectrometer, we need to drive the source wire quite strongly, typically at 2–4 times the critical velocity in order to get a measurable signal on the detector. It would be nice to know both the time of emission and the spectral content of the beam. Unfortunately we cannot measure these directly. As discussed by Stamp[7], we expect a wire driven east to generate easterly going quasiparticles and westerly going quasiholes. Since the excitations are created in a region of reduced energy they have to be further excited to escape into the bulk liquid and how these further processes may modify the hole/particle purity and direction is not clear.

We are currently studying the variation of amplitude and phase of the detected beam as a function of separation angle and pressure.

THE DETECTION PROCESS

The above topics are all associated with the oscillation of a single wire. The structure is an interesting bit of physics which may lead us into other topics and the nonlinearity in the quasiparticle damping has helped us to understand the influence of the quasiparticle gas on the wire. However, our most difficult problem is that of

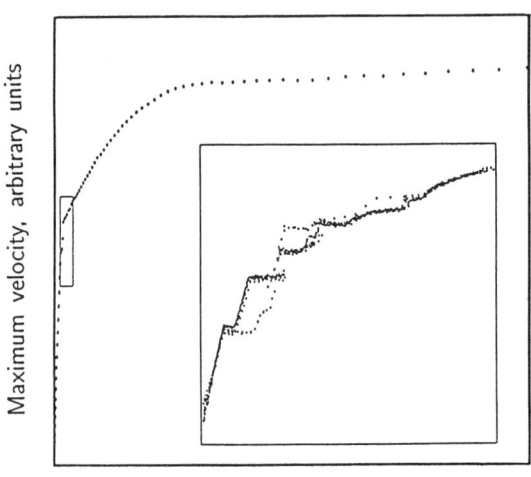

Figure 5. A velocity-drive curve taken for a 4.5 μm diameter wire at 6.8 bar. The inset picture is a magnified view of the box on the main curve. The system will traverse the steps quite reproducibly both up and down with the occasional different (but also reproducible) path.

how the detector wire detects when we have a two-wire spectrometer running.

If we follow the argument of Stamp above that the wire emits a beam of quasiparticles in the forward direction and a beam of quasiholes in the rearward direction, then, as mentioned in the introduction, the detector wire should be bombarded by alternate pulses of quasiparticles and quasiholes. If the reflection processes are normal then the wire should be alternately pulled once and pushed once each cycle of the source wire oscillation and should respond at the same frequency. However, we know the detector responds at double the source frequency. We can hold the detector wire virtually still by feedback to eliminate any flow field around the wire which would mean that we can detect the force while there is no relative motion between condensate a detector wire. It is possible that the pulsed beam of quasiparticles and quasiholes is dense enough that a counterbeam of superflow is set up around the wire which might contribute to frequency doubling under certaincircumstances.

CONCLUSION

The ballistic quasiparticle spectrometer is working but its operation is more complex than we first thought. The quasiparticle wind can be detected at distances up to 1 cm from source to detector. A source operated at frequency f produces a detector force at frequency $2f$, which is accurately proportional to $\dot{Q}$, the source power dissipation. Since $\dot{Q}$ is itself a pathological function of the source drive current this is a significant experimental test.

We are beginning to understand the physics of the source mechanism (although a regression to a pulsed ohmic heater might make life even simpler if this were technically feasible). However, the receiver mechanism remains an elusive mystery.

The signals we observe at present are rather delicate, so that the initial choice of scattering experiments is limited. We hope to observe the influence on the beam of phase boundaries, surfaces and flow fields. Other suggestions are welcome.

REFERENCES

1. A. M. Guénault and G. R. Pickett, Physica B+C, 126 [Proc. 17th Int. Conf. on Low Temp. Phys. (LT17), North Holland, Ed. U. Eckern, A. Schmid, W. Weber and H. Wuhl, Amsterdam] 260 (1984).
2. C. A. M. Castelijns, K. F. Coates, A. M. Guénault, S. G. Mussett, and G. R. Pickett, Phys. Rev. Lett. 56, 69 (1986).
3. K. F. Coates, A. M. Guénault, S. G. Mussett and G. R. Pickett, Europhysics Letters 2, 523 (1986).
4. J. P. Carney, K. F. Coates, A. M. Guénault, G. R. Pickett and G. F. Spencer, Japanese Journal of App. Phys. 26 (Suppl. 26-3, Proc LT18), 1823 (1987).
5. Yu. M. Bunkov, private communication.
6. J. P. Carney, A. M. Guénault, G. R. Pickett and G. F. Spencer, to be published.
7. P. C. E. Stamp, private communication.

SOME DESIGN FEATURES, NON-FEATURES, AND EX-NON-FEATURES OF THE CORNELL MICROKELVIN CRYOSTAT

E.N.Smith
Cornell University, Ithaca, N.Y. 14853

ABSTRACT

Cryostats intended for very low temperature use must meet very stringent demands in the areas of vibrational isolation, electrical isolation, and reliable thermometry. The efforts which have been made in these areas during the construction of the new microkelvin cryostat at Cornell have met with varying success. In this paper will be described both a number of our ideas which we feel have worked well (features, in the jargon of the American advertising industry), some which seemed like good ideas at the time, but which should not be repeated elsewhere (non-features, by logical extension). Also corrections to some of the less successful approaches will be discussed, which have led to the production of some ex-non-features.

INTRODUCTION

Slightly over four years ago, after an extended period of grant applications, site selection, and architectural design, ground was broken for the new Cornell Microkelvin Laboratory. Its siting was underground (for aesthetic considerations, as far as the university was concerned, but also good from a standpoint of vibrational isolation from the user viewpoint) immediately adjacent to the existing low temperature group laboratories. This location is in the long run very desirable because of the interactions between people working in the new laboratories and those in the pre-existing laboratory space, and for the sharing of apparatus between different experiments, although during the period of construction, the disturbance to ongoing experiments was rather more substantial than we had imagined it would be. The subsequent year and a half was devoted principally to the construction of the concrete shell of the new laboratory, improving the watertightness of the rooms–building 7 or 8 meters below the water table is to be avoided when possible, erection of the shielded rooms within, preparation of the support systems for the cryostats and installation of pumps, plumbing, electrical services, etc., and finally the installation and testing of the two dilution refrigerators purchased from Oxford Instruments[1]. With this phase of the building completed, during the last two and a half years the smaller cryostat has been in fairly continuous use by various experimenters with small experiments suitable for quick toploading. Meanwhile, construction and testing of the nuclear demagnetization stage and the initial round of experiments for it has been going on in the main cryostat, which is just now attaining temperatures in the range of 100 microkelvin with usably small heat leaks. There are, of course, a number of working cryostats in the world capable of achieving temperatures in the range under 100 microkelvin, and it was our intent principally to copy methods successfully used elsewhere, creating a conservatively designed cryostat. However, we have certainly not been able to resist all changes from what others have done, and thus the Cornell machine has a number of its own individual characteristics, some good and some bad, which are already at this point worth identifying. A more detailed and complete description of the cryostat must unfortunately wait for a few months, as only within the last two or three weeks has our thermometry been getting to a point where we can make reliable measurements to determine pre-cool times, ultimate temperatures, heat leaks, etc.

VIBRATIONAL ISOLATION

A very basic feature of any cryostat intended for achieving ultralow temperatures by adiabatic demagnetization is vibrational isolation. There were several provisions made from the start in constructing our cryostat which were intended to provide stable mounting, and these are enumerated here:

1) A stable footing. The floors of the rooms in which our two new cryostats are located are separated from the level of the rest of the physics building by about 3 meters in height. Transmission of vibrations between the two structures must pass through fiber expansion joints rather than directly through concrete (this is also true for the floors under our pumping systems), providing substantial isolation, while the area directly under the support pillars for the cryostat (to be described below) is concrete walls approximately 700 mm thick going clear to bedrock.

2) A massive, but dissipative, tripod support system. Concrete sewer pipes, approximately 1 m in diameter and 2.5 m tall and filled with sand, form a support system for the cryostat. These concrete pillars rest upon the shielded room floor, which rests directly on the underlying concrete floor. Steel plates are set into the top of these sewer pipes for purposes of anchoring the actual cryostat top plate. Additional 200 mm diameter aluminum pipes, imbedded in the sand, rise another 2 m above two of these pillars to provide a relatively dead support for a wooden beam to which all the pumping lines are mounted before being attached to the cryostat. Thus far, to the top of the concrete support pillars, the scheme has been very satisfactory. Accelerometers mounted on the top plates show vibration levels lower than any other location we have found in the physics building (we unfortunately do not have an absolute calibration on the accelerometers, and hence no numerical value for the vibration level). The single wooden beam with only two support posts is less satisfactory for support of the pumping lines. Whereas it is quite stiff in one direction, it is inadequately rigid perpendicular to the line of support. We had initially intended this beam only as a support which should not introduce additional vibrations. Unfortunately,

at this time there is no convenient way to provide this other axis of stiffness, so we have had to resort to other techniques to limit this vibrational coupling, described later.

3) Air legs. Commercial air legs from TMC[2] were placed on top of the sand-filled columns to act as a low-pass filter to remove any residual higher frequency vibrations coming through this part of the support system. These legs had proportional control valves regulating a continuous (very small) air leak to establish leveling of the platform independent of such things as the level of cryogenic fluids.

4) High rigidity support structure. The structure connecting the top plate of the cryostat to the support legs needed itself to have a high resonant frequency, to avoid internal vibrational modes. This was achieved with the use of triangular frames made with large rectangular cross-section aluminum TIG welded together. Filling the hollow sections of some of the aluminum beams with lead shot provided a structure which was acoustically very dead.

5) Decoupling of pumps. In an attempt to get away from oil-cracking problems reputed to be caused by oil diffusion pumps, causing potential blocking of refrigerator return lines, we had decided to use Roots type pumps instead (more on this later). The disadvantage is a greater level of vibration production, so the pumping lines had dual-gimbal vibration isolators of the type described by Kirk and Twerdochlib[3], both between the pumps and the shielded room wall, and also before going onto the cryostat platform.

6) Superinsulated dewar. To avoid unnecessary production of vibrations directly on the cryostat, we used a superinsulated stainless dewar rather than one with a liquid nitrogen jacket which could produce additional noise from the boiling of the nitrogen.

With these provisions, which were substantially more ambitious than those applied on most of the other cryostats used in our group, we cooled the demagnetization stage down for the first time, and were somewhat disappointed to discover heat leaks on the order of 100 nW, perhaps about an order of magnitude worse than on any other machine which we owned. It was at this point necessary to consider what additional problems we might have, and also to install some additional diagnostic instrumentation. We put a number of small piezoelectric transducers[4] onto our cryostat, the most useful of which seemed to be those glued onto the support rods for the copper stage, and used the output of these examined with a spectrum analyzer to help pin down our remaining problems. A number of the problems which we discovered, and the corrective measures taken are enumerated below:

1) Construction of the copper stage. This stage was intended to contain as little as possible of potentially glassy systems, which have been traditionally blamed for 'time-dependent' heat leaks of long duration. The stage was formed by rolling high purity copper bars of approximately 10 mm diameter into a roughly rectangular cross section of 5 mm by 13 mm. Twelve of these bars were arranged in pairs, radially oriented and e-beam welded into a 12 mm thick plate

which is to be used for mounting of the samples. Additionally, 6 bars of similar fabrication approximately 4 mm by 7 mm were situated between these other pairs, as indicated in figure 1. After welding, the entire copper structure was oxygen annealed in a vertical oven to boost the RRR to about 1000. Initially to keep the bars separated, quartz spacer rings 20 mm in diameter were placed at several levels along the 430 mm length of the bundle, with the bars held snugly against this central ring by a loop of copper wire around the outer perimeter. Thin spacers cut from a silicon wafer were used to keep the copper bars away from each other azimuthally. The total mass of the copper bars amounts to 58 moles; the effective amount at the maximum field (weighted by the average value of B^2 at each point) is 30 moles. We found at least two difficulties with this scheme of construction. There was a lack of rigidity of the whole assembly, as the bars were not sufficiently bound together, and there was the (postulated) ability of small vibrations to cause frictional heating between the spacers and the copper bars. After several attempts to bind the wires tighter and to attach the spacers more securely at low temperatures by the use of a small amount of vacuum grease, the spacer scheme was abandoned, and both the quartz rings and the silicon wafers were replaced by a number of small dots of epoxy between adjacent bars. With much less than 1 gram of epoxy used, the change in effective rigidity of the bundle was remarkable–tilting the cryostat would no longer cause the bundle to bend off to the side by the amount necessary to make it hang vertically.

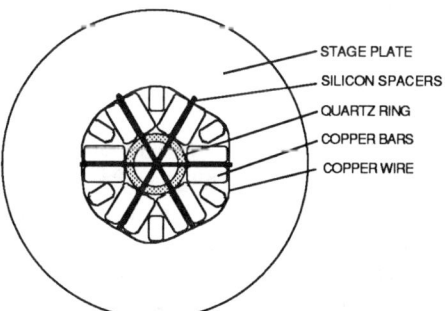

Figure 1. A schematic view of the geometrical arrangement of the copper demagnetization bundle as viewed from the bottom, indicating the initial use of quartz spacer rings and silicon wafer spacers.

2) Centering of the air legs. Initially, almost no difference was observed in either the behavior of the heating or the signals from our piezos upon pressurizing the air legs. Eventually we discovered that as the legs were not bolted to the platform, when pressure was released from the legs, their upper surface dropped away from the platform, which was supported by stops. Each time the pressure was cycled, contact between legs and platform was resumed at a

slightly different location, and the ratcheting effect slowly moved the platform off to the side until it rested against the stationary part of the air legs after repeated operations. Once the problem was recognized, new spacers were made, the legs were recentered, and there was an immediate improvement of a factor of ten in the amplitude of vibrations measured at several of the major peaks in the measured spectrum.

3) Support of the pumping lines. After several cooldowns with the goal of understanding heat leaks, we decided that a more thorough isolation of the cryostat at room temperature was desirable. The behavior of the piezoelectric transducers was monitored with essentially everything detached from the cryostat head. One item at a time was reconnected to see the effect. It became clear that reconnecting the pumping lines increased the vibration level immensely. This was partly due to the two-point mounting of the support bar just before the cryostat head. However, we found that much could be done to limit the vibrations being transmitted by the pipes by clamping lengths of thin (1 mm) lead sheet for lengths of 250 or 300 mm along the pipes with hose clamps, which appeared to introduce substantial dissipation into the flexing of the pipes. Additionally, the metal bellows used in the vibration isolators appeared to have some sort of internal resonance at about 100 Hz, which was very nearly situated at one of the resonant peaks associated with operation of the Roots pumps. This problem was cured by wrapping some foam rubber lightly around the bellows. A further complication is that the gas handling system is situated on a rather flexible intermediate floor which is separated from the cryostat, but which shakes considerably when people walk on it. This motion in turn flexes one end of some of the smaller pipes, and can rock the cryostat. The still pumping line penetrates the wall of the shielded room, which is also a relatively flexible, freestanding structure, and is therefore another route for potential coupling. In fact, the original mounting of the ventilation system was such that large amounts of fan vibrations were coupled directly into the shielded room walls, and in turn transmitted relatively efficiently to the cryostat. This source of coupling has since been broken, by the replacement of some sheet metal with fabric ducting.

4) Pumping system vibrations. The pumping system has both a rotary mechanical pump and two Roots pumps. The former has a substantial production of vibrations at 40 Hz, and the latter at about 100 Hz. We found that the noise from the rotary pump could be effectively eliminated by the use of sandboxes on the interconnecting lines between the primary pump and the Roots pumps, and the rearrangement of the valving for bypassing the Roots pumps so that it was all done outside the shielded room (initially there had been two connections of the rotary pump lines through the shielded room walls). Residual noise from the Roots pumps was largely reduced by bags of lead shot strapped to pumping lines and flexible bellows sections, and the use of lead sheet on the pumping lines. In any case, the latter case is ultimately less serious, as during demagnetizations, once the heat switch is off and the mixing chamber temperature is less critical, the Roots pumps may be shut off until the next precool.

5) Dewar and magnet system. The cryostat is really much longer than most demagnetization systems. This is because we had wished to have the possibility of installing a second magnet system below the one presently in use, either for a second stage of demagnetization, or for the performance of experiments in a controlled large field at low temperatures. As a consequence, the pendulum frequency for the cryostat is substantially lower than usual, and is rather close to that of the air suspension system. As this was thought to be a likely source of problems, we proceeded to try to stiffen the system as much as possible by putting rather rigid spacers between the layers of thermal shielding in the cryostat to limit as much as possible the relative motions in the high field region. The main effect of this seemed to be to increase the heat leaks due to direct thermal conductivity. When we looked at the piezoelectric transducers with the cryostat floating on air, we could see little difference at room temperature associated with the extra bracing. Finally, after achieving an amplitude of vibrations down by nearly two orders of magnitude from our starting point, with all of the cans mounted, the superconducting magnet installed on the cryostat, and all pumping lines connected, we raised the stainless steel dewar flask into place, and were rather distressed to see approximately a factor of ten increase in the vibrational level as we bolted the dewar in place. It soon became clear that the cryostat was very sensitive to the dewar being pressed hard against the superconducting magnet, which had a clearance of about 6 mm radially in a length of somewhat over 3 m. Eventually we made a tapered flange to insert between the dewar and the top plate of the cryostat, spaced with a neoprene gasket to allow some degree of adjustment by nonuniform tightness placed of different screws, and placed a number of touch sensors around the rim of the top and bottom of the magnet. While not an elegant solution, it seems to have effectively solved our problem with this touch. We still do not really understand why this contact should be so important in this cryostat. We have close tolerances between the magnet and the dewar walls in other systems in the laboratory, and have never experienced such sensitivity. Perhaps it is because the majority of our other dewars have fiberglass rather than stainless steel necks, and the lossier neck material eliminates the coupling of vibrations through this route.

AN ADDITIONAL MECHANICAL PROBLEM

One additional mechanical problem which we have experienced in the operation of our cryostat, and which we have heard by word of mouth to be a problem in many other cryostats elsewhere, is the progressive plugging of cold traps, and eventually of the return line of the dilution refrigerator, by light hydrocarbons. We had always supposed these to be the cracking products of diffusion pump oil, and had thought that replacing the diffusion pump by a Roots type pump would effectively eliminate the problem. We were disappointed to discover that we had more problems with our new refrigerators than with any of our existing cryostats. We have come to the tentative conclusion that the real problem lies at least in part with cracking of pump oil in the primary pump, associated in a very non-linear way with the circulation rate. As one of the advantages of the Roots pumps is that they can operate effectively with a higher backing pressure, there is a tendency in such a system to run the primary pump with a much higher than usual inlet pressure. Operating the system with a higher grade pump oil (Alcatel 300 rather than Alcatel 100) seems to have lessened the frequency with which we have to warm up the nitrogen temperature traps, but certainly has not completely solved the problem. One of the difficulties is that the material produced (mass-spectrometry has suggested propene as a potential candidate) seems to freeze well above liquid nitrogen temperature, and so blocks off the cold trap in the small diameter inlet tubes rather than coating the large surface area of adsorbent within the body of the traps. An attempt at putting a long length of tubing containing room temperature zeolite before the nitrogen trap seemed to help for some days, but eventually had no longer term effects, presumably having the delaying effect of a gas-chromatograph column, but not having sufficient binding energy to permanently retain the offending gases. Perhaps a trap at some intermediate temperature (a Freon refrigerator, or dry ice) would help solve the problem. We still do not have a complete solution.

THERMOMETRY

Probably the most useful thermometer available below dilution refrigerator temperatures is NMR thermometry, and the most used material is platinum. Some initial innovations in the design of our Pt NMR thermometer caused great problems that were solved by the substitution of a model patterned rather closely on the design used much earlier in Jülich[5]. Our first design had attempted to eliminate any heating produced in the H_1 coil from reaching the stage by physically separating it from the platinum wires of the thermometer by a significant distance, and thermally anchoring it to the mixing chamber. Additionally, the axis of the platinum wire bundle had been parallel to the H_0 field in this construction, so that the rf coil had been of a Helmholtz configuration rather than a solenoid, requiring an additional decrease in filling factor. The net effect was to reduce the expected signal to a few percent of what it might optimally be, which might still have been acceptable in view of the expected large signals, except that the physical mounting of the coils produced acoustical resonances coupled to the coil which looked very much like large NMR signals with a decay time of a millisecond or two, completely masking any real signal. We unfortunately spent quite some time trying to remedy the situation by replacing coaxial lines and stiffening mountings before finally just replacing the entire assembly.

In some cryostats, shielding is provided around the Pt NMR assembly to protect against interference from external fields, such as the fringe fields from the demagnetization solenoid. We have not used such a shield, having felt that the complications of computing the effect of the shield on the applied field were more of a nuisance that the small frequency shift produced by the energization of the main magnet (~1 mT for an 8T central field).

A feature which can be a nuisance during initial diagnostics, but which is thereafter a considerable convenience, is the shunting of such magnets as the H_0 field for the NMR, or the activation solenoid for the superconducting heat switch, by persistent current switches. These have worked quite conveniently in our cryostat, the switches being mounted at the still level in the vacuum can. An additional feature which exists in some of the other cryostats in our laboratory and which will probably soon be incorporated here is an additional parallel shunt of some resistance low enough to give a time constant of several seconds to changes of current in these coils, such that one can use much less care in ramping the current supplied to the coils.

The coaxial lines for taking rf signals in and out of the cryostat were a combination of commercial and home-made constructions. From room temperature to 4.2K, we have used 0.85 mm diameter BeCu hardline coax purchased from Uniform Tubes[6], while continuing down to lower temperatures we used a superconducting outer shield of 0.75 mm inner diameter niobium tube, with a teflon dielectric and a niobium-titanium inner conductor of suitable diameter to produce a 50 ohm characteristic impedance. These seem to have worked quite satisfactorily. For some of the audio-frequency instrumentation, such as capacitance measurement, we have used flexible stainless steel shielded cable with a cupro-nickel clad niobium-titanium inner conductor, available from Cooner[7]. It should be noted that furnished with the cryostat were several vacuum dielectric coaxial cables which were for all practical purposes useless except as convenient ports for the installation of our own lines. At low frequencies, the central wire of the line had numerous resonances coupling to the electrical signals, and when pulsed at rf for the NMR, they produced microphonic signals with long ringdown times. To add insult to injury, neither the inner or outer conductors made reliable contact with the installed connectors.

For calibration of the NMR thermometry, and as a general check of the cryostat performance at temperatures down to 1 mK, we have used three other thermometry systems, with varying success. First, resistance thermometers are useful at temperatures down to about 20 mK reliably, and occasionally down to rather lower temperatures, depending on electrical pickup. We have used some shaved-down Speer 100 ohm resistors (no longer manufactured), and also some Dale 1K ohm thick-film resistors[8], which as noted in the reference seem to have good reproducibility within a batch, although substantial variation from one batch to another. At higher temperatures, these resistors appear to follow a nice $\ln R \, \alpha \, T^{-.25}$ behavior, though at lower temperatures near 20 mK there seems to be some deviation, probably because of self-heating.

Next, we have used nuclear ordering (NO) thermometry down to the lowest dilution refrigerator temperatures. The ^{60}Co NO source is mounted on the mixing chamber of the dilution refrigerator rather than on the nuclear cooling stage, to avoid the steady state heating from the radioactive decays, but is in useful contact as long as the heat switch is activated and there is no large magnetic field on the stage, so that the heat capacity and thermal time constants are small. The NO thermometer is not useable when the main magnetic field is high in any case, as the fringe field from the solenoid saturates the magnetic shielding on the scintillator at room temperature. We find good agreement between the NO temperatures and the Pt temperatures based on T_1 measurements assuming a 30 msec-K Korringa constant.

Finally we have a ^{3}He melting curve thermometer which at the present time is mounted on the stage plate. This has good resolution above about 1 mK, and the provision of a fixed point at the superfluid transition for the ^{3}He. In our first runs on the cryostat, we were unable to use the instrument because of electrical problems, and on the last cooldown, the time constant for response seemed long and there was considerable hysteresis present. This has been traced to a contamination of about 1000 ppm ^{4}He, which seems to have been provided free of charge with the six-9's ^{3}He which we purchased. At the present time we are in the process of repurification. The moral of the story is that it is always desirable to check the purity of your ^{3}He upon initial receipt. Eventually, we might well remove this thermometer from the stage plate and put it on the mixing chamber with the NO thermometer, since it represents additional heat capacity, and doesn't function at the lowest temperatures.

Mounting of the experiments onto the experimental stage is done with conical joints of copper. A series of holes with a #0 Morse taper are bored in the top plate of the stage, and a mating male taper is machined onto copper mounts for experiments. The two pieces are pulled together either with a 6 mm thread below the tapered section, onto which a washer and nut are tightened against the plate from below, or with 6 3 mm screws on a flange above the plate. In most cases, we have bored out the center of the tapered section, and filled it with a short section of 3 mm diameter tungsten rod, which has a smaller expansion coefficient than copper, and which should the provide additional forces upon cooling. We observed electrical resistances of about 30 nano-ohms across such joints, about half of which could be attributed to the resistance of the bulk copper. It has been pointed out to us that the tungsten rod may be a bad idea, because with repeated thermal cycling there may well be flow of the copper and an eventual loosening of the joint with this differential thermal contraction going on. Thus far we have not observed a problem, but we have a very limited number of cooldowns. Such a joint has also been used to connect the tin heat switch to the stage.

ELECTRICAL ISOLATION

Several basic measures have been undertaken to keep unwanted electrical signals out of the cryostat. The entire cryostat is located in an electromagnetically shielded room, which in its original delivered state provided 120 dB of isolation from rf signals. All of the pumping lines going through the shielded room naturally must have electrical isolation, and both the 60 Hz power lines into the room, and instrumentation lines have been provided with filtration against electromagnetic interference. The cryostat within the shielded room is also electrically floating except for a single intentional grounding to the shielded room wall. All computers are kept outside of the shielded room, as at the moment are the rf equipment, and as much as possible of the analog electronics. Low frequency lines are, as mentioned above, coupled through EMI filters, which are very effective at taking out rf noise. These are effectively a pi-type filter with about 20 nF to ground on each side of a small series inductor. An unanticipated difficulty with these units has been that the capacitance to ground is so large that it constitutes a significant current path even at 60 Hz, and effectively has provided an additional source of ground loops to put additional line frequency noise into the cryostat. By judicious use of isolation transformers and disconnection of some grounds, it has been possible to work around the problem thus far, but eventually we probably need to work on a scheme which relies less on capacitance and more on inductance in the filtering. Presently we are not actually filtering the rf lines for the NMR experiments, but we have isolated the grounds, coupling the transmit and receive lines through small transformers.

The shielded room environment is powered by a motor-generator set, and has its own ground, separate from the rest of the building. One interesting feature of the motor-generator set is that it runs slightly asynchronously with the normal 60 Hz line. Thus it is possible to determine whether stray 60 Hz signals originate inside or outside the shielded room environment by seeing with which power source they are synchronous.

ACKNOWLEDGEMENTS

Over the last four years, a large number of people have contributed a tremendous amount of time, energy and ideas to the design, construction and operation of our laboratory. This has included faculty, postdocs, visitors, technicians, and graduate students. J. Amato, M. Freeman, R. Germain, P. Hakonen, D. McQueeney, R. Mihailovich, P. Moster, L. Opsahl, J. Parpia, L. Pollack, R. Richardson, J. Ross, W. Sprenger, and E. Varoquaux have all had major involvement with the main demagnetization cryostat, while almost everyone in the low temperature group has contributed suggestions, criticism, and interest. Finally, none of this work would have been possible without the support of the NSF through grant DMR 84-16040, contributions to the cost of construction from Cornell University, and a contribution to instrumentation costs from IBM.

REFERENCES

1. Oxford Instruments, 3A Alfred Circle, Bedford, MA 01730. (617) 275-4350.
2. Technical Manufacturing Corp., 15 Centennial Drive, Peabody, MA 01960. (617) 532-6330.

3. Kirk, W.P. and M. Twerdochlib, Rev. Sci. Instr. **49**, 1360 (1979).
4. Piezo Electric Products, Inc., 212 Durham Ave., Metuchen, NJ 08840. (201) 548-2800. We used model SG-3M.
5. Buchal,C., J. Hanssen, R.M. Mueller and F. Pobell, Rev. Sci. Instr. **49**, 1360 (1979).
6. Uniform Tubes Inc., 7th Avenue, Collegeville, PA 19426. (215) 539-0700.
7. Cooner Wire Co., 9186 Independence Ave., Chatsworth, CA 91311. width 0pt (818) 882-8311.
8. Li, Q, C.H. Watson, R.G. Goodrich, D.G. Haase, H. Lukefahr, Cryogenics **26**, 467 (1986).

NEW DESIGN FOR A COPPER DEMAGNETIZATION STAGE

J. D. Kilian[†], P. B. Chilson, G. G. Ihas and E. D. Adams
University of Florida, Gainesville, FL 32611

Copper is used almost exclusively as the refrigerant material in demagnetization stages for achieving temperatures of 1 mK or lower. Early copper bundles were constructed of fine wires to minimize eddy current heating during demagnetization. This construction presented some problems in achieving the necessary rigidity to avoid vibrational eddy–current heating without introducing time–dependent heat leaks through the use of epoxy or other bonding material. More recent constructions have used slitted solid copper bars to achieve rigidity without the use of epoxy. However, the slitting is expensive and allows a filling factor of only ~ 0.6.

We have developed a new design for a copper demagnetization stage which has the following objectives: maximize the filling factor within the high field region of the magnet, maximize the structural rigidity of the stage, minimize the eddy current and time–dependent heat leaks, and optimize the conductivity of the stage.

A bundle geometry of flat, closely–spaced plates was chosen to achieve a filling factor of nearly unity without compromising other design consideration (see Fig. 1). The plates were cut with appropriate widths to constitute a cylindrical cross–section when assembled. Using plates 3.0 mm thick with 0.05 mm spacing provided a filling factor of ~ 0.97.

Commercial "oxygen–free high–conductivity" (OFHC) copper was used. Values of the residual resistivity ratio [R (4.2 K)/R (300 K)], RRR, were all ~ 100 for raw samples taken from different sheets. However, for annealed samples (see below), RRR from 100 to 1000 were obtained. The copper used in the bundle was cut from stock which gave RRR $\simeq 1000$.

After annealing and welding the bundle to the end flange as described below, 0.025 mm – thick sheets of paper (cigarette paper stock) were inserted between plates for insulation. In the initial test of the bundle, it was tied with cotton cord at two places along its length. A minute amount of epoxy will be used to bond these plates together in the final construction. This construction resulted in the high structural rigidity necessary for low vibrational eddy–current heating.

In order to obtain high thermal conduction between the bundle plates and the end flange where experimental cells are mounted, diffusion welding was used. This process requires very flat surfaces held in close contact. A stainless steel clamp structure was used to hold the assembled plates together. While held by the clamp structure, the end of the assembled plate was finished using an end–mill fly cutter. On the final pass, the cutting depth was $2.5 \mu m$ and no cutting oil was used. Then this and all other surfaces were ground on a lapping table using straight strokes, without lapping oil. Greened stainless steel sheet 25 μm thick was placed between the copper plates and between the copper and stainless clamp to prevent sticking during welding. Besides affixing the copper plates to the experimental flange, various items such as the heat

[†]Current Address: HDQ. US Army Stat. Def. Comm. P. O. Box 15280 Arlington, VA 22215–0280

switch thermal length and a melting curve thermometer body were welded to the top of the flange at the same time.

The end flange was held tightly against the plate assembly by means of molybdenum rods to secure the end of the clamp. Molybdenum has a thermal expansion coefficient of 5.6×10^{-6} K^{-1} compared with 1.7×10^{-5} K^{-1} for copper. Diffusion welding was achieved with the oven at 920°C for 15 minutes, then at 870°C for 12 hours, using an atmosphere of 92% helium and 8% hydrogen.

The quality of the diffusion welds were tested by measuring their electrical conductivity at room temperature. No deviation from bulk resistivity (1.6 $\mu \Omega - cm$) could be observed for any plate–to–flange connection. At 6 mK thermal conductivity measurement showed all bulk copper characteristics.

This test stage consists of 117 moles of copper. The stage to be constructed for the new microkelvin laboratory will contain 250 moles and readily accommodates a two stage system.

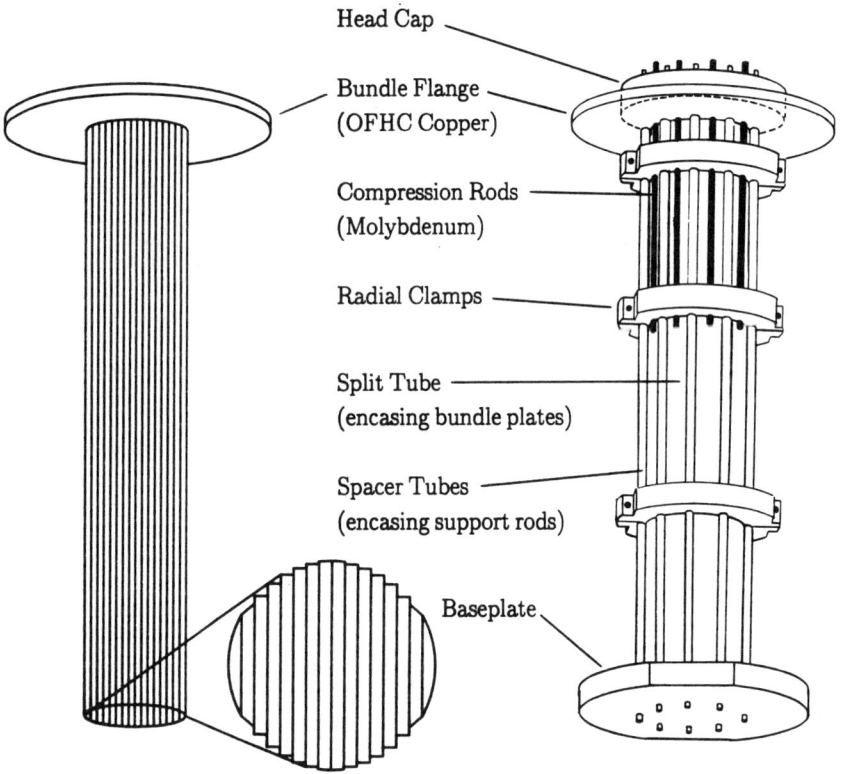

Fig. 1 Bundle geometry and welding apparatus.

This work has been supported by National Science Foundation grants no. DMR–8419267, DMR–8519007 and DMR–8615593.

VIBRATION ANALYSIS OF AN ULTRA–LOW TEMPERATURE CRYOSTAT*

G. J. Labbe, H. Royce, Y. Takano and G. G. Ihas
University of Florida, Department of Physics, Gainesville, FL 32611

Vibration elimination is one of the principal engineering concerns in constructing an ultra–low temperature cryostat. Since couplings, resonant frequencies, and dissipations are very difficult to calculate, one approach is to build a system whose components can be tuned. An important part of this approach is to know the vibration spectra of components and assemblies as construction progresses. Vibration spectra of the first University of Florida microkelvin cryostat are being compiled as the cryostat is built. A block diagram of the interconnected components of the cryostat with qualitative coupling strengths is shown in Fig. 1. A battery–powered Sundstrand Q–Flex accelerometer[1] is being used with a PAR 113 pre–amplifier and a Rapid Systems[2] PC based Fourier spectrometer. The acceleration resolution is typically 0.5 Hz and 2 μg for spectra from 1 to 100 Hz. The amplitude resolution (2 $\mu g/\omega^2$) is therefore 0.5 μm at 1 Hz.

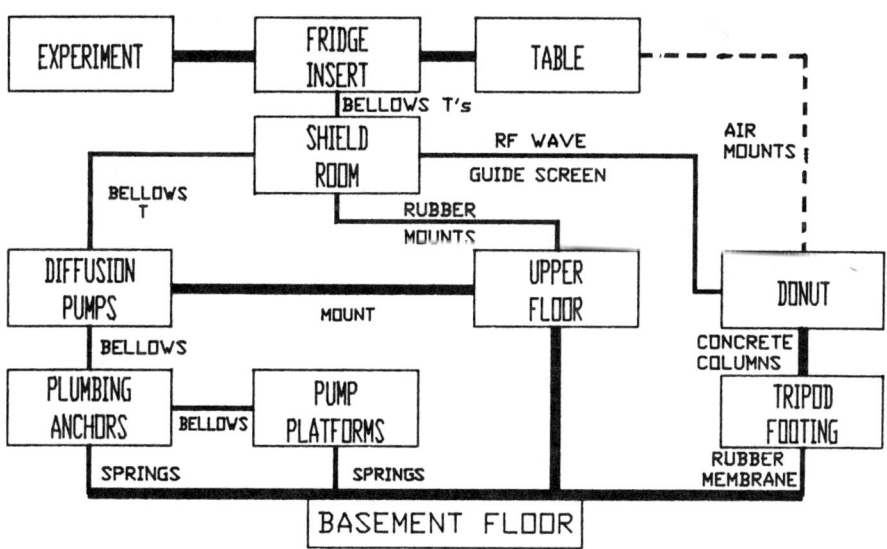

Fig. 1. Diagram of the interconnected components on the cryostat, with relative coupling strengths indicated by line thickness (thicker = stronger).

The concrete tripod is 13 feet tall, sits on three 2–ton concrete pads situated below the basement floor of the building with columns which pass through the basement floor with only soft rubber weather seals connecting them to the floor. The columns also go through the shield room floor without contact. The cryostat table inside the shield room floats on three air mounts[3],

*Supported in part by NSF Grants DMR 8419267 and DMR–8519007.

© 1989 American Institute of Physics

which are securely bolted to the column tops.

Vibration spectra have been taken at several locations in the building and on the cryostat support structure. Some salient features of these results are shown in Fig. 2. The spectra of the basement floor (a) and the tripod base (b) are not very different except for the 60 Hz peaks (spectrum 4 is amplified 5 times to show detail). With equipment and people operating in the building, the isolation aspects of this design are expected, however, to be apparent. The tripod and donut (c) act as a driven oscillator, responding to the 60 Hz acoustic noise present in the building due to the electrical power transformers and motors. The floating table (d and e) does not show this 60 Hz resonance. The upper level floor (f), which is a typical type of environment to house a cryostat, is one of the worst environments measured.

The shield rooms "float" on 2" cubes of rubber spaced in a 12" square array except near the edges, where they are spaced 6" apart, an appear to be more active (g) than the upper level building floor.

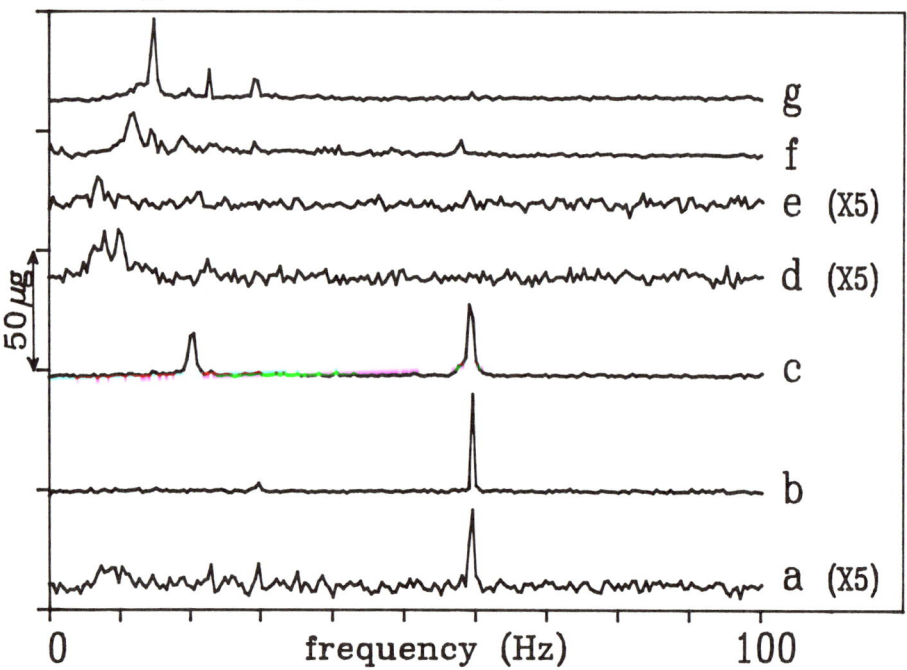

Figure 2. Vertical vibration spectra of: (a) basement floor; (b) tripod pad; (c) tripod donut; (d) floating table; (e) floating table weighted with 500 extra pounds; (f) building upper floor; and (g) shield room floor. (a), (d) and (e) have been amplified by a factor of 5.

REFERENCES

1. Sundstrand Data Control, Overlake Indust. Park, Redmond, WA 98052.
2. R1200/R35 by Rapid Systems, 433 N. 34th Street, Seattle, WA 98103.
3. Kinetic Systems, 20 Arboretum Road, Roslindale, MA 02131.

DILUTION REFRIGERATOR WITH INTERNAL CRYOGENIC ^{3}He CYCLE

Yu. M. Bunkov, D. A. Sergatskov, J. Nyéki*, I. N. Ivoilov
Institute for Physical Problems, Kosygin st. 2, Moscow 117334, USSR

A dilution refrigerator with an internal circulation of ^{3}He has been designed, constructed and tested. Two alternatively working pumping lines ensure a continuous cryogenic cycle of ^{3}He circulation. Each pumping line consists of a ^{3}He cold plate which is maintained at its working temperature with a charcoal adsorption pump. In such a construction the ^{3}He circulated in the dilution refrigerator does not warm up above the still temperature and no cryogenic valves are needed to operate the ^{3}He circulation.

The cryogenic circulation of ^{3}He in dilution refrigerators allows the experimenter to get rid of huge booster pumps, simplifies the vibroisolation of cryostats and is very advantageous for rotating cryostats. The main disadvantage of cryogenic circulation is that pumped ^{3}He does not immediately return to the dilution refrigerator, and some instabilities in his work may occur.

The first successful working dilution refrigerator with internal cryogenic cycle was designed by Edelman[1]. In his device ^{3}He was pumped out from the still by cold plate which formed the lower part of ^{3}He pot being pumped by an adsorption pump. ^{3}He condensed on the cold plate and then was dropping back to the incoming line of the dilution refrigerator. The height of liquid ^{3}He column in this line determined the circulation rate of ^{3}He. Only one adsorption pump with large volume was used in the device.

Another version of cryogenic cycle was described in ref. 2, where ^{3}He was pumped out from the still directly by two alternatively operating adsorption pumps placed in 4K bath. A special collector between 1K pot and the still had been designed to secure the stable ^{3}He circulation rate.

By using two independent and alternatively operating pumping lines we have improved the cryogenic cycle realized by Edelman. Fig. 1 shows schematic diagram of the adsorption pump [1] which is able to pump 1 mole of ^{3}He with rate up to 10^{-3} mole/s; 1K pot [2]; ^{3}He cold plate [3]; still [4]; continuous [5] and step [6] heat exchangers; mixing chamber [7] and cooling tubes for adsorption pumps [8].

The ^{3}He circulation occurs due to gravity. Since the density of concentrated phase is less than the density of the diluted one, the distance between the ^{3}He cold plate and the still must be much larger that the distance between the still and the mixing chamber. The flow impedance Z of heat exchangers (in our refrigerator $Z=4.10^7 cm^{-3}$) should be optimized as well to ensure sufficient rate of ^{3}He circulation. Condensation of the mixture and the initial

*Permanent address: Institute of Experimental Physics, Slovak Academy of Sciences, Solovjevova 47, 043 53 Košice, Czechoslovakia

cooling process are accomplished with an external rotary pump. After reaching 200 mK the refrigerator is switched to the internal cryogenic cycle. The adsorption pump is cooled and starts to pump ^{3}He from the ^{3}He cold plate. Gaseous ^{3}He from the still condenses on the lower part of ^{3}He cold plate and immediately flows down into the continuous heat exchanger. At the same time the second adsorption pump is warming up. Recovered ^{3}He is going through 1K pot, where it is condensing and filling the ^{3}He pot of the regenerating ^{3}He cold plate. The switching of the adsorption pumps must be made accurately to avoid instabilities of circulation rate. In this way we obtain a temperature $\sim$13 mK at circulation rate of 4.10^{-4} mole/s and for the external heat input to the mixing chamber about 3 μW.

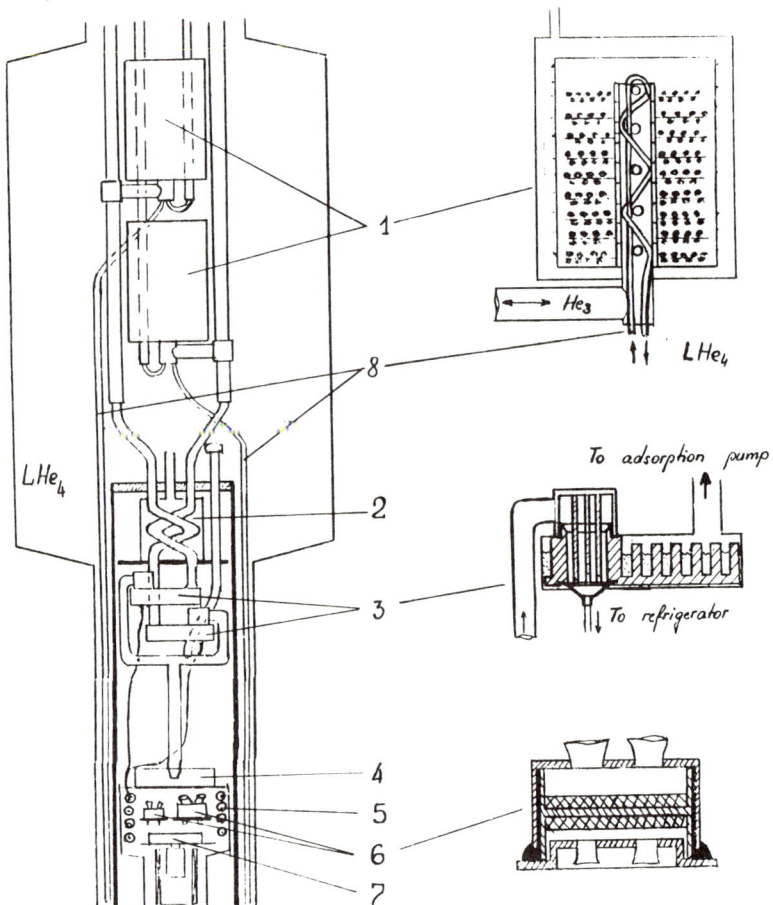

Fig.1. Schematic diagram of the dilution refrigerator

REFERENCES

1. V.S.Edelman, Cryogenics, **12**, 385 (1972)
2. V.A.Mikheev, V.A.Maidanov, N.P.Mikhin, Cryogenics, **24**, 190 (1984)

A HIGH-RESOLUTION, LOW-POWER ULTRASONIC SPECTROMETER

Norbert Mulders and John R. Beamish
Department of Physics and Astronomy
University of Delaware
Newark, DE 19716

We have developed an ultrasonic system (Figure 1) for high resolution sound velocity and attenuation measurements, using a pulse echo technique. The system consists of a CW signal generator, a gated amplifier to generate the RF pulses and a heterodyne receiver. It operates over a frequency range of 2 - 200 MHz. The excitation level can be varied from 1 nW to 1 mW pulse power (at 0.1% duty cycle). The system is completely computer controlled and can switch between frequencies, excitation levels and samples nearly instantaneously.

The velocity measurements are done by quadrature detection of the phase of the received RF pulses. To avoid the effect of slight imbalances in the quadrature detector on the determination of the attenuation, the amplitude was simultaneously but separately measured with a compensated diode detector. The phase and amplitude resolutions are better than 1 mRad and .005 dB, respectively, at a receiver input level of 50 pW. Both one and two echo measurements can be done with essentially the same resolution which makes the system attractive for experiments on high attenuation samples. By taking reference measurements the long term drift of the system can be reduced to within the random noise. The whole system operates at room temperature so that switching between experiments is easy.

A resolution of .001 Rad/.005 dB requires a signal to noise ratio (S/N) of 60 dB. To process short pulses (typically 2 microseconds long) without too much distortion, the receiver has to have a relatively large bandwidth of 2 MHz. The thermal noise in this bandwidth at the input of the receiver is -105 dBm (taking into account a 3 dB noise figure for the first amplifier and a doubling of the bandwidth because of down conversion in the heterodyne receiver). With a signal level of only -73 dBm a considerable amount of averaging has to be done to reach the desired S/N level. Most of this is done by boxcar gated integrators. By averaging over 300 pulses S/N is improved by 35 dB. The boxcars also eliminate most of the short term drift by active baseline subtraction. If required, further averaging can be done by computer.

The main sources of long term drift are changes in the DC offset of the boxcars, and in the gain (and phase) in the receiver and gated amplifier. Data are usually taken at a fixed experimental parameter, e.g. the temperature of the sample, while the system goes through a list of frequencies and amplitude levels. After all ultrasonic data have been taken, this parameter is changed to a new setpoint. At the beginning of each cycle the DC offsets are determined and during the time it takes to reach the new setpoint the ultrasonic system switches over to a reference sample to keep track of changes in the gain. A complete set of reference data are stored and can be used in later stages of the data processing to compensate for the drift.

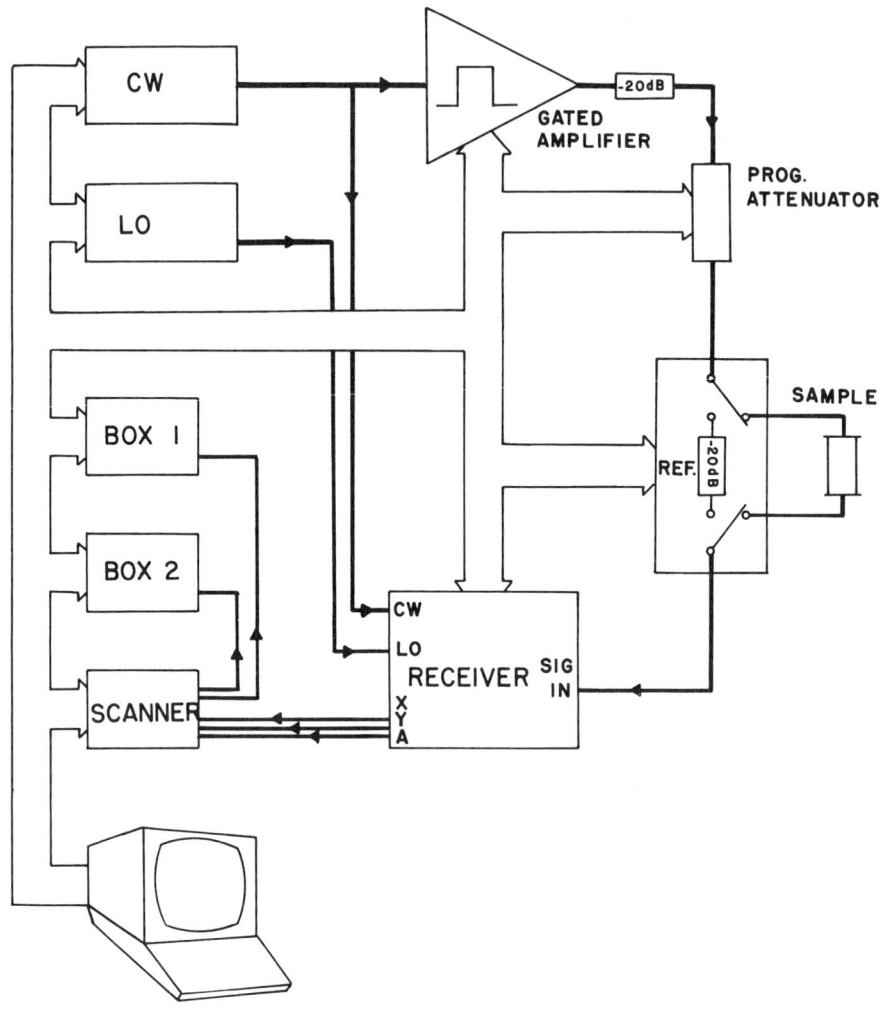

Figure 1 Ultrasonic system

        CW: RF signal generator (Marconi 2022A)
        LO: Local oscillator (Marconi 2018)
        BOX 1/2: Boxcar integrators (SRA 250) plus DMM (Keithley 195)
        SCANNER: RF scanner (Keithley 705/7062)
        GATED AMPLIFIER: Matec 310 HP
        PROG. ATTENUATOR: RF attenuator (Kay 4450)
        RECEIVER: preamplifier, mixers, splitters, amplifiers
            (Minicircuits); compensated diode detector (Matec 625)

ACOUSTIC STUDY IN LIQUID AND SOLID ^{3}He: USING NOVEL AND SOPHISTICATED ULTRASONIC TECHNIQUES

F. Guillon

University of Ottawa, Ottawa, Ont., Canada K1N 6N5

Non-linear effects in sound propagation in superfluid ^{3}He have been observed in several experiments in the past[1]. The effect was interpreted by Avenel et al.[2] as a saturation of the pair breaking mechanism a behaviour related to the superfluid state. On the other hand, the acoustic microscope work of Foster and Rugar [3,4] has shown that non-linear effects are different in superfluid and normal liquid ^{4}He. At very high power harmonic generation and frequency dependent coupling between phonon modes dominate the sound transmission in that superfluid state. There is an interest in distinguishing non-linear effect due to phonon from those arising from quasiparticles coupling in the superfluid state of ^{3}He by using acoustic microscope method of studying non-linear harmonic generation. In solid ^{3}He the suggested coupling[5] between the optical spin wave mode and a sound wave has not been observed experimentally owing to the technical difficulties in propagating a sound wave in a thin slab of solid ^{3}He below T=1 mK in the ordered uudd phase. Experiments are further complicated by the fact that solid helium is sensitive to large ultrasonic stresses[6] when high power acoustic waves are propagating through a thin crystal.

The purpose of this paper is to suggest briefly various new methods to investigate the properties of ^{3}He in the liquid and solid phase when power effects must be considered explicitly in the experimental design.

CRYOGENIC ACOUSTIC LENS FOR QUANTUM LIQUID

Recent developments in the use of acoustic lens for liquid study [7] and in the measurements of some acoustic properties of sintered materials[8] have led to the possibilities of implementing them in ultra-low temperature experiments.

Figure 1 shows the main features of the proposed design. Our previous study [7,8] has demonstrated that the surface of a sintered heat exchanger left unpolished can be used as an acoustic reflector for the sound waves coming from the lens. At Helium temperatures the low acoustic transmission at the interface between a conventional sapphire lens and the liquid helium used as the coupling fluid is usually resolved by evaporating a carbon matching layer in the lens cavity. It has been found[9] that this sapphire lens which is difficult to make could be replaced by an epoxy lens made of Stycast 1266 as shown in Fig.1. The low acoustic impedance and the very low acoustic attenuation below 1K[9,10] permit the use of this Stycast lens at high frequency without the use of a matching layer. Preliminary work[9]

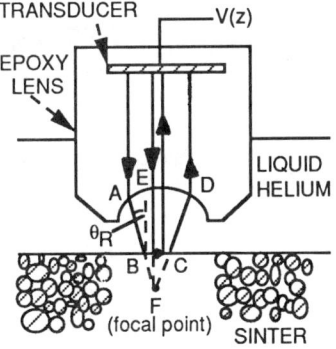

Fig.1. Acoustic lens configuration for quantum liquid. It shows the axial (EF) and leaky surface (ABCD) wave ray paths discussed extensively elsewhere.

© 1989 American Institute of Physics

showed that a 30 MHz sound beam from such a lens could be focalised on the flat surface of a ultrasonic transducer immersed in superfluid ^{4}He. This epoxy lens is made from a perfect Stycast rod free of any large bubbles. The ultrasonic transducer is immersed by pouring more Stycast at one end of the rod. The lens cavity can be made by two simple methods: grinding carefully using diamond paste and a ball bearing or inserting a ball bearing of the correct size inside the hardening Stycast rod during fabrication.

The arrangement shown in Fig.1 may be used in several ways to study in the high attenuation regime a thin layer of superfluid ^{3}He between the lens and the sintered surface. At low acoustic power small velocity changes can be measured by the acoustic material signature V(z) produced by the interference between the axial and leaky Rayleigh (surface) wave. This experiment can be performed by moving the lens or more appropriately by calibrating the fixed lens[7]. More interesting is the expected important contribution of the quasiparticles in the non-linear regime at high power. This kind of study should give valuable information consistent with the numerous results in the linear ultrasonic spectroscopy regime[1].

The solid ^{3}He phase might be studied through the leaky Rayleigh wave in the solid when the liquid-solid interface is moving away from the sinter towards the fixed lens. A great deal of experimental results[11] on either bulk or surface waves in various materials undergoing an antiferromagnetic transition showed that acoustic waves can probe efficiently the spin-phonon coupling. More recently using a second sound method Kagan and Kosevich[12] showed that surface waves in solid ^{4}He can be generated experimentally. The advantage of our design is that the propagation path along the surface can be much longer than across a thin crystal hence the enhanced sensitivity. However experiments should be done to show that high acoustic quality solid helium surface can be produced routinely in the proposed design.

ACKNOWLEDGEMENTS

This work was supported by NSERC.

REFERENCES

1. W.P.Halperin, Physica 109-110B, 1596(1982).
2. O.Avenel, M.Rouffard and E.Varoquaux, LT-17 Proceedings, 796(1984).
3. D.Rugar and J.S.Foster, Phys.Rev. B30, 259 (1984);
   D.Rugar, J.Appl.Phys, 56, 1338 (1984).
4. J.S.Foster, PhD dissertation, Stanford University (1984).
5. M.Roger, J.H.Hetherington and J.M.Delrieu, Rev.Mod.Phys. 55, 1(1983);
   M.C.Cross and D.S.Fisher, Rev.Mod.Phys. 57, 881 (1985).
6. A.Hikata, H.Kwan and C.Elbaum, Phys.Rev. B21, 3932 (1980);
   J.R.Beamish and J.P.Franck, Phys.Rev. B26, 6104 (1982).
7. F.Guillon, Ultrasonics, 27, 26 (1989).
8. F.Guillon, Can.J.Phys, 66, 963 (1988).
9. F.Guillon, unpublished results.
10. P.Doussineau and W.Schön, J.Physique, 44, 397 (1983).
11. E.Kawasaki and A.Ikushima, Phys.Rev. B1, 3143 (1970);
    A.K.Ganguly, K.L.Davis, D.C.Webb and G.Vittoria, J.Appl.Phys, 47, 2696 (1976); see also A.A.Maradudin, Non-Equilibrium Phonon Dynamics, ed. W.E.Bron, Plenum Press, New-York, p395, (1985).
12. M.Yu.Kagan and Yu.A.Kosevich, To appear Sov.J.Low.Temp.Phys,(1988).

FIELD–RESOLVED NQR SPECTROSCOPY FOR ULTRA–LOW TEMPERATURE THERMOMETRY.*

N. S. Sullivan, M. Rall and D. Zhou
University of Florida, Department of Physics, Gainesville, FL 32611

INTRODUCTION

The recent progress that has been made towards cooling samples to the microkelvin–millikelvin temperature range has led to the need for reliable thermometers with high sensitivity that are based on simple thermodynamic principles and thus unambiguous to interpret. The Zeeman splitting of nuclear quadrupole resonance lines can provide such a thermometer. Simple metals (with non–cubic crystal structures) such as Sb, Ga, Re and Mg are particularly promising; although many compounds (e.g. AuCl, $A\ell I_3$ and Sb_2O_3) could also be very useful. Metals with axially symmetric electric field gradients and reasonably short relaxation times (resulting from the coupling to conduction electrons) are the most practical. They can be welded to metallic surfaces including the nuclear refrigerants used, or even be an integral part of the refrigerant.

PRINCIPLE OF OPERATION

For nuclear spins $I > \frac{1}{2}$, possessing a nuclear electric quadrupole moment Q in an axially symmetric electric field gradient q, the energy levels of the states $I_z = \pm m$ are degenerate, and "pure" NQR is observed at the difference frequencies $\omega_m(Q) = (E_{\pm(m+1)} - E_{\pm(m)})/\hbar$. Application of an external B_0 lifts this degeneracy and for $m > \frac{1}{2}$,

$$E_{\pm m} = A[3m^2 - I(I+1)] \mp m\hbar\omega_L \cos\theta \tag{1a}$$

$A = e^2qQ/4I(2I-1)$, ω_L is the nuclear Larmor frequency, and θ is the polar angle giving the direction of B_0 with respect to the principal axis (O_z) of the electric field gradient. The case $m = \frac{1}{2}$ is special, because of the zeroth–order mixing of the states $m = \pm\frac{1}{2}$, and the energy eigenvalues are found to be

$$E_{\pm} = A[\tfrac{3}{4} - I(I+1)] \mp \tfrac{1}{2}mf\hbar\omega_L \cos\theta \tag{1b}$$

where $f = \{1+[\frac{1}{2}(2I+1)\tan\theta]^2\}$. The frequency splitting given by the last terms in equations (1a) and (1b) has been studied in detail[1], but what has not been pointed out in the literature is that the intensities of the two components are not equal at low temperatures. The line intensities are simply proportional to the difference in thermal population of the two energy levels $\pm m$. For $m > \frac{1}{2}$,

$$\frac{I[-(m+1) \to -m]}{I[(m+1) \to m]} = \exp[-\beta(2m+1)\hbar\Omega]\frac{\sinh[\frac{1}{2}\beta\hbar\omega_+]}{\sinh[\frac{1}{2}\beta\hbar\omega_-]} \tag{2}$$

*Research supported by the National Science Foundation–Low Temperature Physics–DMR8611620.

$\beta = 1/k_B T$, $\Omega = \omega_L \cos\theta$, and $\omega_\pm = \omega_m(Q) \pm \omega_L \cos\theta$.

Observation of the asymmetry of the intensities of the doublet and simultaneous measurements of their frequencies will provide a <u>direct measure</u> of the nuclear spin temperatures <u>without the need for additional calibration</u>. The advantage over conventional nuclear Curie–law thermometry ($I \propto T^{-1}$) is that there is no need for a careful calibration of spectrometer gain and stability control over a very extended temperature range. Careful measurements do need to be made to determine that the system is in thermal equilibrium, but this is true for all thermometers based on nuclear spin degrees of freedom in the sub–millikelvin regime. Only small RF fields need to be applied to detect the resonance absorption, particularly at low temperatures, thereby reducing the RF heating and saturation effects for long relaxation times.

PRACTICAL EXAMPLES

As an example we can cite Sb which has a trigonal arsenic type structure. There are two isotopes ^{121}Sb ($I = 5/2$), 57.25% abundance, $\nu_Q = 11.5289$ and 23.0611 MHz at 4K; and ^{123}Sb ($I = 7/2$), 42.75% abundance, $\nu_Q = 6.9996$, 13.9997 and 21.000 MHz at 4K. For Larmor frequencies $\simeq 0.3$ MHz ($B_0 \simeq 300$ gauss), Sb can be used in the 10–100 μK range. The relaxation times are relatively short $T_1 T = 0.092$ and 0.156 sec.K, for ^{121}Sb and ^{123}Sb, respectively. The line widths are narrow ($\lesssim 7.5$ kHz[2]), and single crystal spectra for pure NQR absorption have been observed in the 1–4K temperature range.[3][4] Re which has an hP2 structure is another possible candidate.

Although single crystals give the best resolution, powders can be used and are better suited for keeping the RF heating to a minimum. The powder spectrum obtained by summing over all values of $\cos\theta$ also has an asymmetry of the form of equation (2).

HIGH–TEMPERATURE METHODS

A variant of this method of determining nuclear spin temperatures from the asymmetries of nuclear resonance absorption has been used in studies of solid H_2 at low temperatures.[5] The perturbation by intramolecular dipolar interactions lifts the degeneracy of the $I_z = \pm 1 \to 0$ transitions and leads to an antisymmetric doublet because of the difference in thermal populations. This property has also been observed very recently for the proton resonance in the water of crystallization of gypsum.[6]

REFERENCES

1. T.P. Das and E.L. Hahn, Advance in Solid State Physics, Suppl. 1, (F. Seitz and D. Turnbull, Eds.), Academic Press, London, 1958
2. L.A. McLachlan and D.L. Williams, Proc. XIV Colloque Ampere, p. 462, Ljubljana, Sept. 1966, R. Blinc, ed. North Holland Pub. (1967).
3. J.J. Spocas and C.P. Slichter, Phys. Rev. <u>113</u>, 1463 (1959).
4. J. Buttet and P.K. Bailey, Phys. Rev. Lett. <u>24</u>, 1220 (1970).
5. N.S. Sullivan and R.V. Pound, Phys. Rev. <u>A6</u>, 1102 (1972); C.M. Edwards, D. Zhou and N.S. Sullivan, Phys. Rev. <u>B34</u>, 6540 (1986).
6. P. Kuhns, O. Gonen and J.S. Waugh, J. Mag. Res. <u>72</u>, 548 (1987).

MAGNETIC-FIELD-INDUCED METAL-INSULATOR TRANSITION IN DEGENERATELY DOPED n-TYPE GERMANIUM

M. J. Burns

Department of Physics, University of Florida, Gainesville FL 32611

INTRODUCTION

Many possible ground states have been proposed to describe degenerately doped semiconductors in large magnetic fields. Most such proposals speculate that the ground state is a collective state[1] and all proposed states are insulating. Degenerately doped Ge has recently been shown to undergo a metal-insulator transition in large magnetic fields at low temperatures[2-4]. Thermoelectric power measurements were undertaken on degenerately doped Ge in this insulating state in order to gain insight as to the nature of the density of states for the carriers.

EXPERIMENT

The material used was an uncompensated Sb-doped Ge crystal obtained from Eagle-Picher. Careful checks of dopant homogeneity were performed as described in references[2-4]. Samples were cut to sizes ~1.1x3.3x24mm with all edges ‖ to the <100> directions. The samples were etched and contacts made as decribed in references 2-4. Two contacts were placed on the longitudinal ends for the thermopower measurements and for current injection during resistivity and Hall measurements. Three small voltage contacts were placed on the sides to permit virtual-contact Hall measurements[5,6] and four-terminal resistance measurements. The sample was cantilevered into the vacuum with one end heat sunk to the crystadt. A small heater was epoxied to the free end of the sample and a differential thermocouple was used to measure the temperature gradient produced when the heater was switched on and off. Au wires were mounted to ends of the sample thus forming another differential thermocouple composed of Au-sample-Au. This is a modified version of the apparatus of Geballe, et al.[7]. All measurements were take with H along the <100> axis. Thermopower measurements were taken with the temperature gradient ‖ and ⊥ to H.

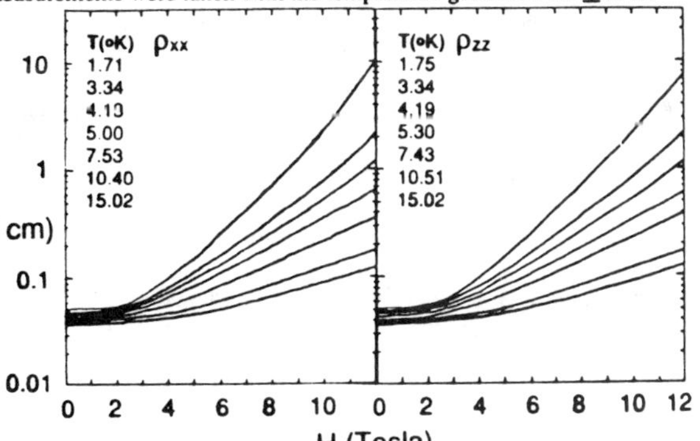

Figure 1

RESULTS

In figure 1 we show ρ_{xx}, ρ_{yy}, and in figure 2 the normalized Hall coefficient for a sample of donor density $N_D \sim 2.2 \times 10^{17} cm^{-3}$. Low field Hall measurements indicate $n \sim 1.7 \times 10^{17} cm^{-3}$ at 4.2°K. Figure 1 clearly shows a metal-insulator transition in ρ_{xx} and ρ_{yy} as H increases above ~4 Tesla, consistant with refs. 2-4. The Hall coefficient shows no sign of any transition, thus indicating that the apparent carrier concentration remains unchanged through the transition. Figure 3 shows the thermopower (S) plotted as 1/T. The thermopower was isotropic with respect to the field. The thermopower can often be naively considered as the entropy per carrier. For all systems σS --> 0 as T --> 0 is required by the third law of thermodynamics[8]. If σ stays nonzero as T-->0 then S-->0 as for any metal.

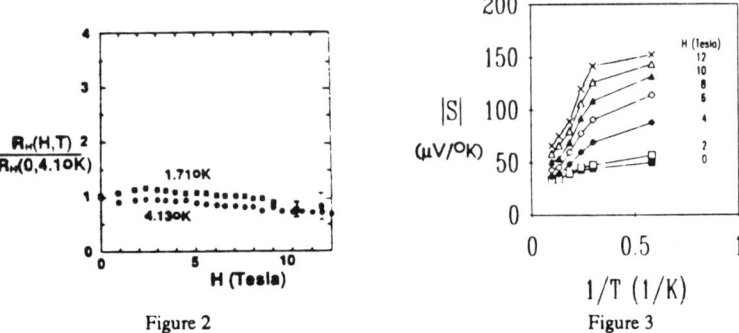

Figure 2   Figure 3

If σ -->0 as T-->0 then S can do anything. However, if S is nonzero at T=0, the density of states at the Fermi level must be zero[8,9]. In Mott type varaible range hopping, the denstiy of states, while localized, is finite, hence S-->0 as T-->0[10]. For Efros type interacting hopping, S may go to a nonzero constant as T-->0[11]. For systems with finite width gaps at the Fermi level such as intrinsic semiconductors, charge and spin density wave systems, the thermopower diverges as 1/T. If the Fermi level were in a energy gap (of finite width) in the density of states, S would display a 1/T dependence for $K_B T < E_{GAP}$[9].

Recent scaling theories which include both single-particle "localization" terms and the presence of disordered Coulomb interactions address the behavior of the conductivity near the metal-insulator transition in strong fields[12]. These theories indicate that at the field-induced metal-insulator transition the quasiparticle density of states remains non-zero while the quasiparticle diffusion constant vanishes. Unfortunately, neither the Hall effect or thermoelectric effects have been calculated in these models.

If the trend displayed in figure 3 does not change so that S is nonzero at T=0, the density of states for the carriers responsible for the transport must be vanishing as the system enters this high field insulating state.

ACKNOWLEDGEMENTS

We thank G.A. Thomas , P.F. Hopkins, R.M. Westervelt , B.I. Halperin and P.M. Chaikin for useful conversations. This work was supported by grants DSR-D-6 87-88 and MDA 972-88-J-1006.

REFERENCES

1. for reviews see B.I. Halperin, Jpn. J. Appl. Phys. 26, supp. 26-3, part 3 (1987) ; B.I. Halperin, in **Condensed Matter Theories, Vol.2**, ed. by. P. Vashista et al. (Plenum Press, 1987)
2. G.A. Thomas, M.J. Burns, P.F. Hopkins and R.M. Westervelt, Phil. Mag. **B56**, 687 (1987)
3. R.M. Westervelt, M.J. Burns, P.F. Hopkins, and A.J. Rimberg, in **Anderson Localization**, ed. T. Ando and H. Fukuyama, (Springer-Verlag, Berlin, 1988)
4. P.F. Hopkins, M.J. Burns, A.J. Rimberg, and R.M. Westervelt, Phys. Rev. B (in press)
5. H.J. Lippmann and U.F. Kuhrt, Z. Naturforschung 13a, 462 (1958)
6. H. Fritzche, in **Methods of Experimental Physics, Vol. 6B: Solid State Physics**, ed. by K. Lark-Horovitz and V.A. Johnson, (Academic Press, New York, 1959), Ch.8
7. T.H. Geballe and G.W. Hull, Phys. Rev. **94**, 1134 (1954)
8. A. Kohler, Abhandl. Brauns. Wiss. Ges. **3**, 49 (1951)
9. J. Tauc, **Photo and Thermoelectric Effects in Semiconductors**, (Pergamon, New York, 1962)
10. N.F. Mott and E.A. Davis, **Electronic Properties in Non-Crystalline Materials**, 2nd ed. (Oxford University Press, 1979)
11. M.J. Burns and P.M. Chaikin, J. Phys. C18, **L743** (1985)
12. C. Castellani, G. Kotliar, and P.A. Lee, Phys. Rev. Lett. **59**, 323 (1987)

EPILOGUE

CHAIRMAN

Mark W. Meisel
Department of Physics
University of Florida

QUANTUM FLUIDS AND SOLIDS: WHERE WE HAVE COME, WHERE WE ARE GOING

by D. M. Lee
Laboratory of Atomic and Solid State Physics,
Cornell University, Ithaca, NY 14853, USA

In 1908, eighty years ago, Professor Heiki Kammerlingh - Onnes became the first scientist to liquify helium. This truly formed the basis for our subject. Kammerlingh-Onnes was not only a great researcher, but a great organizer of research. He built a laboratory at Leiden which stayed at the forefront of low temperature physics for many years. Out of this laboratory came the important discoveries which have been the object of study by our profession for the remainder of the 20th century. Superconductivity and the observation of the λ transition in liquid ^{4}He constituted the great beginning of our subject.

One can trace through the history of quantum fluid research. The great schools of low temperature physics founded before World War II by Kapitza and Landau in the Soviet Union and by Simon, Kurti and Mendelsohn at the Clarendon Laboratory in Oxford came to rival and even surpass the achievements of the Leiden Laboratory particularly in the area of superfluid ^{4}He research. Around this time Fritz London introduced the grand idea of quantum mechanics on a macroscopic scale to explain superfluidity and superconductivity. Other laboratories around the world were also coming into the picture.

The development of the Kapitza helium liquifier into a commercially available research tool by Sam Collins of MIT greatly facilitated low temperature research throughout the world and especially in the USA. Another important development took place with the onset of nuclear technology during and shortly after World War II. As a byproduct of the hydrogen bomb effort, helium-3 became available in large quantities, thus providing a new quantum fluid and a new quantum solid for further study by low temperature physicists. Studies of ^{3}He-^{4}He mixtures were also made possible including the discovery of phase separation. A pioneer in these early studies on helium-3 was William Fairbank, who gave our banquet speech.

In 1957, Bardeen, Cooper and Schrieffer (BCS) announced their theory of superconductivity in which the "condensation" of Cooper pairs of electrons with opposite spin and $\ell = 0$ formed a coherent macroscopic ground state. This theory is one of the great constructs of the human mind and at long last provided an explanation for a phenomenon which had eluded human understanding for almost a half century. It has applicability far beyond low temperature physics. As an outgrowth of the BCS theory came the speculation that Cooper pairs with non zero relative orbital angular momentum could form. It was suggested that for $T \ll T_F$, liquid ^{3}He, which could not form s-wave pairs, could indeed become a superfluid with higher ℓ pairing.

An important technical development came upon the scene in the sixties. This was the ^{3}He-^{4}He dilution refrigerator which could provide continuous cooling below 10mk and could thus provide a stable temperature base for a nuclear cooling stage. This development as well as the development of Pomeranchuk cooling made studies down to a millikelvin far more convenient.

An interesting sidelight which made investigators in our field somewhat uncomfortable was the famous speech by A. B. Pippard at the dedication of the Thomas J. Watson Research Center of IBM in 1961. Pippard, a professor at the Rutherford Lab.in Cambridge, had been one of the pioneers in the application of microwaves to superconductivity. His studies of surface impedance led to the notion of coherence length. The central theme of Pippard's speech, entitled "the CAT and the CREAM", was that our physics, the application of quantum mechanics to condensed matter systems, was rapidly playing out as a result of the intensive efforts then underway and suggested that the stimulation would go out of our field just as for the case of classical physics in 1900. One cannot argue with the ultimate correctness of this thesis but I believe strongly that one can argue with the time scale which I am convinced was grossly underestimated. At the very time of this speech, Brian Josephson, one of Pippard's own graduate students, was discovering his famous effect. All the while the late John Wheatley and his research group were having an extremely productive period studying the various Fermi-liquid properties of ^{3}He. About a decade later the superfluid phases of liquid ^{3}He were discovered at Cornell. Shortly thereafter the magnetically ordered phases of solid ^{3}He were discovered at Cornell (low field phase) and here at the University of Florida (high field phase). Soon the theorists, led by Anthony Leggett were quickly working out the properties of superfluid ^{3}He in terms of the BCS theory for triplet pairs. Mark Twain, the famous American author and humorist probably gave the best answer to the Pippard thesis, namely "The reports of my death are greatly exaggerated". The idea is not to set up Pippard as a straw man. Rather we illustrate the lesson that almost everyone who looks into the future turns out to be dismally wrong. Therefore I shall not add my name to the long list of failed prophets but instead shall merely point out trends. Even classical physics is now having a recrudescence (a word recently popularized by Roland Combescot in considering the revival of interest in ^{3}He in 1972). Studies of turbulence and chaos have entered a whole new phase, for example. To illustrate, it was recently announced in the news that an asteroid narrowly missed colliding with the earth (by one half million miles). Events such as this can be studied by the new technique available from investigations of chaos as it pertains to planetary orbits.

We are now about to enter the last decade of the twentieth century. I fully expect that when we celebrate the millennium, the study of quantum fluids and solids will still be a lively field of endeavor (if we don't get hit by any large asteroids). At this conference we have seen hints as to how this might come about. I shall attempt to review some (not all) of the highlights of the conference. I apologize to you if your personal highlight is not one of those chosen to be cited herein. The only solace I can offer you is that you are in good company.

The first session included beautiful papers on two different versions of the Josephson effect, the spin current Josephson effect found by the group of Bunkov, Dimitriev, Borovik-Romanov et al. and the mass current Josephson effect seen by Avenel and Varoquaux in superfluid ^{3}He. In the latter study, a variety of fascinating Josephson phenomena have been observed with emphasis on observations in the A phase. Future experiments will study effects of the earth's rotation. Because the ideas are somewhat arcane, I shall spend a considerable amount of time (space) on the spin current experiments. The work of Bunkov, Dimitriev et al. involved setting up a phase difference in the order parameter for two connected cells of ^{3}He B by CW NMR techniques. The technique involved setting up a "homogeneously precessing domain" (HPD) in each cell. In the B phase, there is no frequency shift in pulsed NMR unless the tipping angle Θ is $\geq 104°$ which is large enough to "stretch" the dipolar interaction. By placing the ^{3}He sample in a field gradient and slowly reducing the d.c. field in a constant uniform large amplitude r.f. field, two domains form in the sample. As the d.c. field is reduced through the Larmor frequency, the spins in the low field region precess. This region of precessing spins grows in size as the field is reduced. In the high field region, the Larmor frequency is too high for precession to be induced by the applied r.f. In spite of the gradient, the low field domain precesses at a single well defined frequency. The reason for this is that the large continuous r.f. field tips the spins through an angle $\geq 104°$ The value of the tipping angle varies throughout the domain in such a fashion as to keep the spin precession in synchronism with the applied r.f. <u>in spite of</u> the presence of the field gradient. This spin locking is accomplished by longitudinal spin currents. Typical variations in tipping angle are $< 2°$ in these experiments.

If we have two interconnected chambers joined by a small tube, and the chambers are driven at two different NMR frequencies, the relative phases in the two HPD's will change. It is this change of phase which leads to excursions in the magnetization at particular phase differences in accordance with Josephson behavior. The phase slippage occurs at regular intervals in complete analogy with the Josephson effect.

The final talk in the session by Dana Browne explained the mysteries of macroscopic quantum tunneling to us.

In the following session, Frank Pobell told us about his fascinating experiments on helium confined to tiny bubbles in metals, followed by Roger Bowley's discussion of the theory of formation of a vortex ring about an ion either by thermal activation or by quantum tunneling. The final talk in the session was by Gregory Volovik on fractional statistics and analogs of the quantum hall effect in superfluid ^{3}He films. This talk was another example of Volovik's remarkable ability to relate phenomena in superfluid ^{3}He to other parts of physics. In his view, it seems as though ^{3}He is a microcosm of all physics, and therefore his talk provides inspiration for all of us working in the area of superfluid ^{3}He.

The last session on Monday consisted of two experimental talks, the first by Eric Smith who described the pitfalls and successes of the Cornell Microkelvin Laboratory now under construction. It was a triple feature talk with "design features, non-features and ex-non-features". The second talk was by George Pickett who gave a most entertaining account of his program to study quasi-particles in the range well below 500μK with vibrating loops of wire. Perhaps he will use these quasi particle slingshots to slay any unsuspecting errant boojum that happens by.

Tuesday started out with a session on low dimensional helium. The first paper by Machta was an account of theoretical studies of the superfluid transition (^{4}He) in porous materials. The emphasis in this paper was on the common topology of the many porous materials. Core vortices which can be shrunk to zero size and pore vortices which cannot be shrunk to zero because they contain some of the porous medium were introduced. Conditions where the superfluid transition is more Kosterlitz Thouless in character, or more of a bulk phenomenon, were considered.

Moses Chan then presented a beautiful account of the experimental situation for superfluid ^{4}He in porous glasses. One of the surprises in this talk was a small sharp specific heat signature of films with a superfluid transition near 0.1K. On theoretical grounds this signature was thought at the time to be unobservably small. The experiment was successfully repeated by John Reppy and Sheena Murphy at Cornell who measured the specific heat, but "with an added twist", they also measured the superfluid density via a torsional oscillator (sort of a "shake and bake" experiment). Chan also mentioned studies on more open porous media such as Xerogel and Aerogel. In this same session, Matthew Fisher studied (theoretically, of course) the superfluid transition at zero temperature in analogy with the metal insulator transition to predict critical exponents for very dilute films.

Beamish in his discussion introduced a clever new tool for investigating helium in porous media. He has used the velocity of transverse sound to probe the superfluid density. As the temperature is lowered, the superfluid density increases, resulting in a decoupling of the superfluid from the walls. Thus the walls no longer "feel" the superfluid mass and hence a corresponding increase in the transverse sound velocity occurs. Various porous media (aerogels and vycor) were investigated by this powerful new technique which complements observations with torsional oscillator studies.

The final talk in this session was given by Christopher Gould who gave a very nice account of studies at USC by Gould, Bozler and collaborators at

USC and Caltech. SQUID NMR was applied to a sample consisting of a grafoil substrate in contact with bulk liquid ^{3}He for $T \sim 1mK$. They found convincing evidence for magnetic ordering in the surface layer of ^{3}He adjacent to the grafoil. Peaks in the low field NMR spectra for H_o parallel to the graphite planes were an important manifestation of this ordering. One of the peaks persisted even at zero field!

The final invited session on Tuesday 25 April was devoted to studies of superfluid ^{3}He collective modes and related phenomena. The first contribution of the session was a theory paper by Nils Schopohl who treated the bound states and scattering states of quasi-particles at the interface between ^{3}He A and ^{3}He B. The roles of Andreev reflection and excited Cooper pair states were discussed.

The next four papers in this session were devoted to studies of the collective modes of superfluid ^{3}He B. Jukka Pekola presented results from Helsinki on the effect of rotation on the Zeeman split components of the real squashing mode. For example, Zeeman splitting can be observed for the sound propagation direction parallel to the magnetic field when the liquid is rotated. This is certainly not the case for non-rotating ^{3}He B. Textural effects in rotating ^{3}He A were also discussed. Roman Movshovich then discussed the five-fold Zeeman splitting of the squashing mode in a magnetic field. The effect of the broad evanescent regions associated with the strong coupling between zero sound and the squashing mode could only be surmounted by operating at temperatures well below 1mK, and at relatively high magnetic fields. Ross McKenzie then presented a theoretical study of the possibility of finding non-linear acoustic effects in ^{3}He B. He showed, for example, that it might be possible to generate new sound frequencies by stimulated Raman scattering from the real squashing mode. Finally Peter Fraenkel described the use of a newly developed broad-band transducer to make accurate measurements of the real squashing mode frequency over a wide range of temperatures and pressures by sweeping the frequency of the sound.

Following the formal sessions on Tuesday, there was a presentation of the posters. The posters presented were of the highest quality. A rich variety of new phenomena and new techniques was treated. Due to lack of time and space, the discussion of these posters will be left to the conference proceedings.

Wednesday, April 25 was a free day with a number of planned outings including a trip to the Kennedy Space Center and a canoe trip on the Santa Fe River. A good time was had by all who participated in these trips.

Thursday, April 26, began with a series of papers on solid helium. An heroic set of experiments on the hcp phase of solid ^{3}He was discussed by Mamiya, who gave strong evidence that the hcp phase orders ferromagnetically. Next we were treated to a description of computational wizardry by David Ceperly who described Monte Carlo calculations on the exchange frequencies of solid ^{3}He. It was necessary to include contributions of high order (up to six atom) exchange to obtain agreement between theory and experiments. A. F. Andreev from the Institute for Physical Problems in Moscow then gave a theoretical discussion of the motion of ions through solid helium. Beautiful geometrical representations of negative ions were presented. In liquid helium, we know that electrons form bubbles. In the solid, similar vacancies are created around the negative ions but the shapes of the vacancies are also determined by the crystal structure. With this model, calculations of drift velocity vs. applied electric field were performed which were able to explain the experiments.

The last three papers in the session on solid helium were by Dennis Greywall, Hiroshi Fukuyama, and Grégoire Bonfait. Greywall described precision measurements of the specific heat of layers of ^{3}He on grafoil for various coverages. (This contrasts with the work presented by Gould earlier in the conference in which the grafoil was in contact with bulk liquid ^{3}He). The specific heats observed by Greywall were very dependent on the coverage. The fluid or solid nature of the various layers as well as their magnetic properties were inferred from the results. Fukuyama described attempts to measure the upper critical field of the high field ordered phase of bcc solid ^{3}He. By measuring the pressure change with temperature in an isolated (constant density) cell at various magnetic fields, it is possible to determine the phase transition point between the high field ordered phase and the paramagnetic phase. Higher densities are required to keep critical fields within a manageable range. Preliminary data on the transition observed at 3.3 tesla and V = 22 cm^3/mole were presented. Finally Bonfait discussed an experimental search for the supersolid transition. In answering the rhetorical question "Why search for a supersolid?", he gave the brave answer "Why not?". The method used was to search for possible changes in plastic flow down to 4mK. So far no anomalous flow has been observed, but the search for the holy grail will go on.

Following the discussions of solid helium came the session on nuclear magnetic ordering. Nuclear demagnetization has given low temperature physicists the wherewithal to study directly the magnetic ordering in nuclear spin systems. Aarne Oja described neutron scattering experiments from specimens of copper whose nuclear spins were in ordered magnetic phases. Two ordered antiferromagnetic phases were observed. It is a tribute to the experimenters that two such difficult techniques, neutron scattering and nuclear demagnetization can be combined to give such impressive results. Yasu Takano then spoke about NMR studies on silver as well as an effort to attain nuclear ordering in this metal. The latter goal is quite difficult because of the tiny nuclear moment of silver. The work was performed at the Helsinki University of Technology. Particularly fascinating was the effect of coupling of the two silver isotopes, ^{107}Ag and ^{109}Ag via the nuclear dipolar interaction. This led to a coupled set of Bloch equations which in turn gave shifts of the resonance signal away from the respective Larmor frequencies of the two isotopes. The resonance experiments were necessarily performed at very low field ($H_o \leq 10$ Gauss) and the dipolar frequency was about 29.4 Hz. It was possible in these experiments to show that the exchange was anti ferromagnetic in nature. Finally, Georg Eska reported on NMR studies of the thallium isotopes ^{203}Tl and ^{205}Tl. His talk was mainly concerned with non-linear spin dynamics and nuclear order in this highly poisonous material. Thallium is especially interesting because it has a large exchange interaction.

The last session on Thursday contained three discussions of dark matter in the universe. The first, by David Caldwell gave a theoretical survey of the dark matter situation – why it is needed and what form it may take. For example it is known from rotational studies of galaxies that more matter is needed for a proper explanation. He also discussed Ge and Si detectors as possible candidates to observe dark matter particles. Blas Cabrera then gave an account of his ongoing search for the magnetic monopole. His original apparatus has been upgraded to allow more sophisticated coincidence measurements. Cabrera also discussed possible detection of weakly interacting massive particles (WIMPS) by means of Si or Ge bolometers and more exotic thermal detectors involving rotons in liquid helium and superconducting tunnel junctions as phonon detectors. Finally, P. Sikivie discussed progress on his experiment to detect axions, another hypothesized form of dark matter. The detection scheme here involves observing the conversion of axions into microwave photons in a strong applied magnetic field.

The detection of dark matter would be of fundamental importance to the understanding of our universe. Experimenters on the "dark" side face a most difficult task in finding particles which are hard to detect and whose properties are not well known. "May the force be with them!"

Following a delicious banquet dinner, Bill Fairbank gave an inspiring after dinner speech on Thursday night. On Friday morning, April 28th, the final working session of the conference was devoted to spin polarized quantum systems. The first paper by Alexander (Sasha) Meyerovich gave an account of the theoretical situation relating to transport phenomena in these systems. He gave a general theoretical framework for discussing them and also pointed out the importance of the role of boundaries.

Meyerovich's talk was followed by Isaac Silvera's discussion of the possibility of Bose-Einstein condensation in spin polarized hydrogen. There are two main approaches to achieving Bose condensation. One of these involves highly sophisticated atom traps to contain low density atomic hydrogen gas at very low temperatures. The other route is to work at high densities where Bose Einstein condensation will occur at higher temperatures. In this latter approach, the experiments must be carefully designed to minimize the effects of three body recombination. Toward this end, Silvera and his group have performed an experiment on high density spin polarized hydrogen at very high fields (~ 25 Tesla) and have shown that the three body recombination rate decreases with increasing field in this range.

Spin polarized hydrogen is a fascinating metastable system and will no doubt continue to be a rewarding field of investigation. Another metastable system, spin polarized liquid ^{3}He, provides an equally fertile area of study. The last technical paper of the conference was a contribution by Giorgio Frossati's Leiden group. The paper was given by Stefan Wiegers. A considerable enhancement of the superfluid transition temperature ($T_{A_1} = 3.25mK$) was observed in the highly polarized liquid, which was obtained by melting polarized solid ^{3}He. Vibrating wire viscometer measurements in partially polarized liquid ^{3}He yielded a small viscosity anomaly near T_{A_2}. It was suggested that this result could be related to a new phase transition in liquid ^{3}He. During the question period, it was pointed out that the putative phase transition was possibly of a textural nature, associated with dipole unlocking near T_{A_2}.

In summary, the main message of our conference is that the research field of quantum fluids and solids is alive and well, thank you. No longer must we heed the sneers and jeers of the particle physicists who question the relevancy of our work to the onward progression of fundamental science. During the 1960's, when Pippard made his famous speech, particle physicists were at the height of their productivity as well as their arrogance. They even referred to solid state physics as "squalid" state physics and considered our work to be a mere trivial consequence of the Schrodinger equation. In the meanwhile (bound) quarks have been discovered and the mysteries of quantum chromodynamics have been revealed. Nevertheless, calculations of the excited states of the upsilon particle (which consists of a b quark and an anti b quark bound to one another by the strong force) use the Schrödinger equation (the particles are heavy enough to keep relativistic corrections small). Thus even at very short length scales, the basic notions of quantum mechanics still seem to apply.

Especially compelling are the discussions by Volovik at our conference which directly relate quantum fluids to particle physics. The more modern approach of the 80's which should continue into the 90's and beyond is to emphasize the basic connectivity or unity, if you will, of all physics.

We may, with tongue in cheek, apply Hegel's philosophy to our present situation. You may recall the basic framework of this philosophy which is that each idea or system of ideas constitutes a thesis. Each thesis has an antithesis. The thesis and the antithesis interact to form a new synthesis. Let us define the happy world of quantum fluids and solids as the thesis. The antithesis would then correspond to the gloomy ideas of Pippard and the other doomsayers. What then is the new synthesis? In my view we must look to our beginnings. The centerpiece of our field is London's basic idea of quantum mechanics on a macroscopic scale. Of course we must continue our present line of research studying the existing quantum fluids and solids as well as the metastable fluids. But we must (at least some of us must) look beyond, toward ways that our ideas and methods might apply to new and unexplored systems. An example of this is the quantized Hall Effect. Another example is heavy Fermion superconductivity. And of course we shouldn't neglect high T_c superconductivity. Finally, we should be on the lookout for surprises which no one can predict. By pushing the accessible temperatures into the microkelvin range and the accessible magnetic fields up to 50 tesla and beyond, it may be that totally new states of matter will be discovered.

In closing, I wish to thank the conference organizers and all of the participants for making the conference a resounding success.

List of Participants

Adams, E. Dwight
Department of Physics
University of Florida
Gainesville, FL 32611

Adenwalla, Shireen
Physics Department
Northwestern University
Evanston, IL 60208

Alikacem, Nadir
P/G, P/H, MAPS II
University of Sussex
Falmer, Brighton BN1 9QH
England

Andreev, A. F.
Institute for Physical Problems
Academy of Sciences of the USSR
ul. Kosygina, 2
USSR 117334, Moscow

Avenel, Olivier
CEN Saclay DPhG/PSRM
91191 Gif–sur–Yvette Cedex
France

Barry, Jeremiah
Department of Physics
University of Florida
Gainesville, FL 32611

Beamish, John R.
Department of Physics and Astronomy
University of Delaware
Newark, DE 19716

Bedell, Kevin S.
Los Alamos National Laboratory
T–11, MS–B262
Los Alamos, NM 87545

Bernier, Michel E. R.
DPhG/PSRM
CEA–Orme des Merisiers
91191 Gif/Yvette Cedex
France

Bonfait, Grégoire
Departamento de Guimica
ICEN–LNETI
P2686 Sacavem Codex
Portugal

Borkowski, Lech S.
Department of Physics
Virginia Polytechnic Institute
  and State University
Blacksburg, VA 24061

Bowley, Roger M.
Department of Physics
Nottingham University
Nottingham NG7 2RD
England

Bozler, Hans M.
Physics Department
University of Southern California
Los Angeles, CA 90089–0484

Brewer, Douglas F.
University of Sussex
School of Mathematical
  and Physical Sciences
Brighton, BN1 9QH
United Kingdom

Brison, Jean–Pascal
Department of Physics
University of Florida
Gainesville, FL 32611

Brisson, John
Department of Physics
Harvard University
Cambridge, MA 02138

Browne, Dana A.
Department of Physics and Astronomy
202 Nicholson Hall
Louisiana State University
Baton Rouge, LA 70803–4001

Brusov, Peter N.
Department of Physics
 and Astronomy
Northwestern University
Evanston, IL 60208-3112

Bunkov, Yu. M.
Institute for Physical Problems
Academy of Sciences of the USSR
ul. Kosygina, 2
USSR 117334, Moscow

Burns, Michael J.
Department of Physics
University of Florida
Gainesville, FL 32611

Cabrera, Blas
Stanford University
Physics Department
Varian Bldg, Rm. 136, MC4060
Stanford, CA 94305

Caldwell, David O.
Department of Physics
University of California
Santa Barbara, CA 93106

Candela, Donald
Physics Department
Hasbrouck Laboratory
University of Massachusetts
Amherst, MA 01003

Carmi, Yoash
Physics Department
Case Western Reserve University
Cleveland, OH 44106

Ceperly, David
Department of Physics
1110 W. Green Street
Urbana, IL 61801

Chan, Moses H. W.
104 Davey Laboratory
Pennsylvania State University
University Park, PA 16802

Chapellier, Maurice
Laboratoire de Physique des Solides
Universite de Paris–Sud
91405 Orsay
France

Chilson, Phillip B.
Department of Physics
University of Florida
Gainesvill FL 32611

Close, John
Department of Physics
LeConte Hall
University of California
Berkeley, CA 94720

Corrada–Emmanuel, Andres
Department of Physics
Williams College
Williamstown, MA 01267

Davis, James C.
Physics Department
University of California
Berkeley, CA 94720

DeLong, Lance E.
National Science Foundation
1800 G Street, NW
Washington, DC 20550

Dmitriev, V. V.
Institute for Physical Problems
Academy of Sciences of the USSR
ul. Kosygina, 2
USSR 117334, Moscow

Dobbs, E. Roland
Royal Holloway and Bedford New College
Egham Hill
Egham, Surrey TW20 0EX
United Kingdom

Eska, Georg
am Lehrstuhl für Experimentalphysik V
der Universität Bayreuth
8580 Bayreuth,
Postfach 101251
West Germany

Fairbank, William M.
Department of Physics
Stanford University
Stanford, CA 94305–4060

Fisher, Matthew P. A.
IBM TJ Watson Research Center
P. O. Box 218
Yorktown Heights, NY 10598

Fraenkel, Peter N.
Physics Department
Cornell University
Clark Hall
Ithaca, NY 14853

Frossati, Giorgio
Kamerlingh Onnes Lab.
University of Leiden
Postbus 9506
2300 RA Leiden
The Netherlands

Fry, James N.
Department of Physics
University of Florida
Gainesville, FL 32611

Fukuyama, Hiroshi
Institute for Solid State Physics
University of Tokyo
Roppongi, Minato–ku
Tokyo 106 Japan

Geng, Qiquan
Chemistry Department
Massachusetts Institute of Technology
Room 6–133
Cambridge, MA 02139

Gould, Christopher M.
Department of Physics – SSC303
University of Southern California
Los Angeles, CA 90089–0484

Gramila, Tom J.
Physics Department
H–7 Clark Hall
Cornell University
Ithaca, NY 14853

Greywall, Dennis S.
AT & T Bell Laboratories
Room 1D152
Murray Hill, NJ 07974

Guénault, Anthony M.
Physics Department
University of Lancaster
Lancaster LA1 4YB
United Kingdom

Guillon, Francis
Department of Physics
University of Ottawa
Ottawa, Ontario K1N 6N5
Canada

Guyer, Robert A.
Department of Physics
Hasbrouck Laboratory
University of Massachusetts
Amherst, MA 01003

Hagmann, Christian
Department of Physics
University of Florida
Gainesville, FL 32611

Hallock, Robert B.
Department of Physics
University of Massachusetts
Amherst, MA 01003

Harrison, John P.
Physics Department
Queen's University
Kingston, Ontario K7L 3N6
Canada

Herr, Steven
Department of Physics
University of Florida
Gainesville, FL 32611

Hetherington, Jack H.
Physics Department
Michigan State University
E. Lansing, MI 48824

Hingerty, Arthur
Department of Physics
University of Florida
Gainesville, FL 32611

Hirai, Akira
Department of Physics
Kyoto University
Kyoto 606, Japan

Ihas, Gary G.
Department of Physics
University of Florida
Gainesville, FL 32611

Ishimoto, Hidehiko
Institute for Solid State Physics
University of Tokyo
7-22-1 Roppongi, Minato-ku
Tokyo 106, Japan

Jin, Chao
117 Clark Hall
Cornell University
Ithaca, NY 14853

Jochemsen, Reyer
Kamerlingh Onnes Laboratorium
P. O. Box 9506
2300 RA Leiden
The Netherlands

Joffrin, J.
Lab. Phys. Des Solides
Bat. 510, Université Orsay
91405 Orsay
France

Ketterson, John B.
Department of Physics
Northwestern University
Evanston, IL 60208

Kubota, Minoru
Institute for Solid State Physics
University of Tokyo
Roppongi 7-22-1, Minato-ku
Tokyo 106, Japan

Kumar, Pradeep
Department of Physics
University of Florida
Gainesville, FL 32611

Kurkijärvi, Juhani
Physics Department
Åbo Akademi
Porthansgatan 3
20500 Åbo
Finland

Labbe, Gregory J.
Department of Physics
University of Florida
Gainesville, FL 32611

Lee, David M.
Cornell University
Physics Department
Clark Hall
Ithaca, NY 14853

Lusher, Chris
Physics Department
Royal Holloway and Bedford
  New College
Egham, Surrey TW20 0EX
England

Machta, Jonathan
Department of Physics
Hasbrouck Laboratory
University of Massachusetts
Amherst, MA 01003

Mamiya, Takayoshi
Department of Physics
Nagoya University
Furocho, Chikusa-ku
Nagoya 464, Japan

Masuhara, Naoto
Department of Physics
Massachusetts Institute of Technology
Cambridge, MA 02139

McCall, K. R.
Department of Physics
Hasbrouck Laboratory
University of Massachusetts
Amherst, MA 01003

McKenzie, Ross H.
Department of Physics and Astronomy
Northwestern University
Evanston, IL 60208

Meisel, Mark W.
Department of Physics
University of Florida
Gainesville, FL 32611

Meyer, Horst
Department of Physics
Duke University
Durham, NC 27706

Meyerovich, Alexander E.
Department of Physics
Northwestern University
Evanston, IL 60208

Mishra, Suresh
Department of Physics
University of Florida
Gainesville, FL 32611

Miyamoto, Satoru
Department of Physics
University of Florida
Gainesville, FL 32611

Movshovich, Roman
Department of Physics
Cornell University
Clark Hall
Ithaca, NY 14853

Mulders, Norbert
Department of Physics and Astronomy
University of Delaware
Newark, DE 19716

Mullin, William J.
Department of Physics and Astronomy
Hasbrouck Laboratory
University of Massachusetts
Amherst, MA 01003

Muttalib, Khandaker
Department of Physics
University of Florida
Gainesville, FL 32611

Nunes, Geoff
Laboratory of Atomic
  and Solid State Physics
Clark Hall
Cornell University
Ithaca, NY 14853

Oja, Aarne
Department of Physics, Neutronhus
Risø National Laboratory
P. O. Box 49
DK–4000 Roskilde
Denmark

Osheroff, Douglas D.
Department of Physics
Stanford University
Stanford, CA 94305

Owers-Bradley, John R.
Department of Physics
Nottingham University
Nottingham NG7 2RD
England

Packard, Richard
Physics Department
University of California
Berkeley, CA 94720

Pekola, Jukka P.
Low Temperature Laboratory
Helsinki University of Technology
SF–02150 Espoo
Finland

Penfold, Peter
Oxford Instruments
3A Alfred Circle
Bedford, MA 01730

Pickett, George R.
Physics Department
University of Lancaster
Lancaster LA1 4YB
United Kingdom

Pobell, Frank
Lehrstuhl für Experimentalphysik V
Universität Bayreuth
Postfach 101251
D–8580 Bayreuth
West Gernmany

Puttika, William
Department of Physics
University of Florida
Gainesville, FL 32611

Qian, Yongjia
University of Wisconsin – Milwaukee
Department of Physics
1900 E. Kenwood Blvd.
Milwaukee, WI 53211

Rall, Markus
Department of Physics
University of Florida
Gainesville, FL 32611

Richardson, Robert C.
Department of Physics
Cornell University
Clark Hall
Ithaca, NY 14853

Royce, Howard
Department of Physics
University of Florida
Gainesville, FL 32611

Saarela, Mikko
Department of Theoretical Physics
University of Oulu
90570 Oulu
Finland

Salomaa, Martti M.
Low Temperature Laboratory
Helsinki University of Technology
SF–02150 Espoo
Finland

Sarma, Bimal K.
Physics Department
University of Wisconsin–Milwaukee
P. O. Box 413
Milwaukee, WI 53201

Saunders, John
Department of Physics
Royal Holloway and Bedford New College
University of London
Egham, Surrey TW20 0EX
United Kingdom

Schopohl, Nils
Institut Laue–Langevin
B.P. 156 X
38042 Grenoble Cedex
France

Shirahama, Keiya
Institute for Solid State Physics
University of Tokyo
Roppongi, Minato–ku
Tokyo 106, Japan

Sikivie, Pierre
Department of Physics
University of Florida
Gainesville, FL 32611

Silvera, Isaac F.
Department of Physics
Harvard University
Cambridge, MA 02138

Smith, Eric N.
Physics Department
H–12 Clark Hall
Cornell University
Ithaca, NY 14853

Stanton, Christopher J.
Department of Physics
University of Florida
Gainesville, FL 32611

Steel, Stephen C.
Department of Physics
Queen's University
Kingston, Ontario K7L 3N6
Canada

Sullivan, Neil S.
Department of Physics
University of Florida
Gainesville, FL 32611

Sun, Yuanshan
Department of Physics
University of Florida
Gainesville, FL 32611

Suzuki, Haruhiko
Tohoku University
Faculty of Science
Department of Physics
Sendai 980, Japan

Takano, Yasumasa
Department of Physics
University of Florida
Gainesville, FL 32611

Tanner, David B.
Department of Physics
University of Florida
Gainesville, FL 32611

Varoquaux, Eric
Lab. Physique des Solides
Bat. 510
91405 Orsay
France

Volovik, G. E.
Institute for Physical Problems
Academy of Sciences of the USSR
ul. Kosygina, 2
USSR 117334, Moscow

Warner, Kevin
Department of Physics and Astronomy
University of Delaware
Newark, DE 19716

Wiegers, S. A. J.
CRTBT
25 Avenue des Martyrs
166X — Centre de tri
38042 Grenoble Cedex
France

Williams, Gary A.
Physics Department
University of California
Los Angeles, CA 90024

Yip, S. K.
Departmetn of Physics
University of Maryland
College Park, MD 20742

Zhao, Zuyu
Physics Department
Northwestern University
Evanston, IL 60208

Zhu, Dawei
Department of Physics
University of Florida
Gainesville, FL 32611

Zieve, Rena
Physics Department
University of California
Berkeley, CA 94720

Zimmerman, George O.
Physics Department
Boston University
590 Commonwealth Avenue
Boston, MA 02215

Author Index

A

Abe, S., 267
Adams, E. D., 393
Adenwalla, S., 107, 109
Alikacem, N., 301
Ambegaokar, V., 39
Andreev, A. F., 273
Avenel, O., 3

B

Bassou, M., 290
Beamish, J. R., 182, 191, 399
Bernier, M. E. R., 290, 292
Bigelow, N. P., 257
Bonfait, G., 294
Borkowski, L. S., 219
Borovik-Romanov, A. S., 15, 27
Bowley, R. M., 149
Bozler, H. M., 201
Browne, D. A., 39
Brusov, P. N., 109, 111, 113
Bunkov, Yu. M., 15, 27, 147, 223, 397
Burns, M. J., 405

C

Cabrera, B., 342
Caldwell, D. O., 333
Candela, D., 241
Carmi, Y., 299
Castaing, B., 294
Chan, M. H. W., 170
Chapellier, M., 117, 245, 290, 364
Chardin, G., 364
Chilson, P. B., 393
Cho, H., 193
Chow, K., 39
Cross, M. C., 201

D

Dionne, R. J., 199
Dmitriev, V. V., 15, 27, 147, 223
Dobbs, E. R., 103, 105

E

Eguchi, K., 195
Epstein, J. L., 227
Eska, G., 316

F

Filatova-Novoselova, T. V., 111
Fisher, M. P. A., 179
Fraenkel, P. N., 96
Freed, J. H., 257
Friedman, L. J., 201
Fujii, Y., 123
Fukuda, T., 281
Fukuyama, H., 156, 281

G

Gallet, F., 193
Gebhardt, M., 123
Godfrin, H., 294
Gould, C. M., 201
Gramila, T. J., 217
Greywall, D. S., 213
Guénault, A. M., 371
Guillon, F., 401
Guyer, R. A., 161

H

Hagmann, C., 366
Hallock, R. B., 199, 225
Hara, Y., 286
Harań, G., 219
Harrison, J. P., 221
Hetherington, J. H., 288, 292
Higley, R. H., 225
Hirai, A., 286
Hu, Y., 217

I

Ihas, G. G., 393, 395
Inoue, S., 261
Ishimoto, H., 156, 281
Ivoilov, I. N., 397

J

Jacak, L., 219
Janú, Z., 147
Jeon, J. W., 243
Ji, H., 364
Joffrin, J., 117, 245, 364

K

Kato, T., 261
Ketterson, J. B., 107, 109
Kilian, J. D., 393
Kim, N., 75
Kondo, Y., 147
Kopnin, N. B., 51
Korhonen, J. S., 147
Krotscheck, E., 227
Krusius, M., 147
Kubota, M., 195
Kurikawa, I., 156
Kyynäräinen, J. M., 63

L

Labbe, G. J., 395
Lee, D. M., 75, 257, 409
Ling, R., 103, 105
Lipson, S., 299
Lomakov, M. V., 111

M

Machta, J., 161
Mamiya, T., 261, 267
Manninen, A. J., 63
McKenzie, R. H., 87
Meisel, M. W., 117
Meyer, H., 197
Meyerovich, A. E., 231
Minamide, Y., 261
Miura, Y., 261, 267
Mizusaki, T., 286
Mizutani, N., 328
Movshovich, R., 75
Mukharskiy, Yu. M., 15, 27, 147, 223
Mulders, N., 182, 191, 399
Mullin, W. J., 243

N

Nasten'ka, M. Y., 111, 113
Nyéki, J., 397

O

Ogawa, S., 156, 195, 281
Oja, A. S., 305
Okamoto, T., 281
Ono, M., 328

P

Pekola, J. P., 63
Pickett, G. R., 371
Pobell, F., 123
Polturak, E., 299
Popov, V. N., 111

R

Rall, M., 403
Rasmussen, F. B., 245
Richards, M., 301
Richardson, R. C., 217, 267
Roger, M., 288
Rotter, M., 290
Royce, H., 395

S

Saarela, M., 227
Sachrajda, A., 221
Sakayori, K., 281
Salmelin, R. H., 63, 119
Salomaa, M. M., 51, 115, 119
Sarma, B. K., 107, 109
Sasaki, Y., 286
Sauls, J. A., 87
Saunders, J., 103, 105
Schopohl, N., 55
Schuhl, A., 117, 245
Sergatskov, D. A., 27, 223, 397
Shirahama, K., 195
Sikivie, P., 352, 366
Silvera, I. F., 247
Smith, E. N., 382
Sprague, D. T., 225
Steel, S. C., 221
Sugiyama, S., 267
Sullivan, N. S., 366, 403
Suzuki, H., 328
Syskasis, E., 123

T

Takano, Y., 395
Tanner, D. B., 366
Tazaki, T., 281
Thomson, A. L., 201
Torizuka, K., 63
Tuttle, J., 197
Tvalashvili, G. K., 63

U

Uchiyama, T., 261

V

Van Keuls, F., 217
Varoquaux, E., 3, 75
Vermeulen, G. A., 245
Viertiö, H. E., 305
Volovik, G. E., 115, 136

W

Wada, N., 195
Warner, K., 191
Watanabe, T., 195
Waxman, D., 55
Weichman, P. B., 201
Williams, G. A., 193
Wojtanowski, W., 103, 105

Y

Yano, H., 261

Z

Zawadzki, P., 221
Zhao, Z., 107, 109
Zhong, F., 197
Zhou, D., 403

AIP Conference Proceedings

		L.C. Number	ISBN
No. 1	Feedback and Dynamic Control of Plasmas – 1970	70-141596	0-88318-100-2
No. 2	Particles and Fields – 1971 (Rochester)	71-184662	0-88318-101-0
No. 3	Thermal Expansion – 1971 (Corning)	72-76970	0-88318-102-9
No. 4	Superconductivity in d- and f-Band Metals (Rochester, 1971)	74-18879	0-88318-103-7
No. 5	Magnetism and Magnetic Materials – 1971 (2 parts) (Chicago)	59-2468	0-88318-104-5
No. 6	Particle Physics (Irvine, 1971)	72-81239	0-88318-105-3
No. 7	Exploring the History of Nuclear Physics – 1972	72-81883	0-88318-106-1
No. 8	Experimental Meson Spectroscopy –1972	72-88226	0-88318-107-X
No. 9	Cyclotrons – 1972 (Vancouver)	72-92798	0-88318-108-8
No. 10	Magnetism and Magnetic Materials – 1972	72-623469	0-88318-109-6
No. 11	Transport Phenomena – 1973 (Brown University Conference)	73-80682	0-88318-110-X
No. 12	Experiments on High Energy Particle Collisions – 1973 (Vanderbilt Conference)	73-81705	0-88318-111-8
No. 13	π-π Scattering – 1973 (Tallahassee Conference)	73-81704	0-88318-112-6
No. 14	Particles and Fields – 1973 (APS/DPF Berkeley)	73-91923	0-88318-113-4
No. 15	High Energy Collisions – 1973 (Stony Brook)	73-92324	0-88318-114-2
No. 16	Causality and Physical Theories (Wayne State University, 1973)	73-93420	0-88318-115-0
No. 17	Thermal Expansion – 1973 (Lake of the Ozarks)	73-94415	0-88318-116-9
No. 18	Magnetism and Magnetic Materials – 1973 (2 parts) (Boston)	59-2468	0-88318-117-7
No. 19	Physics and the Energy Problem – 1974 (APS Chicago)	73-94416	0-88318-118-5
No. 20	Tetrahedrally Bonded Amorphous Semiconductors (Yorktown Heights, 1974)	74-80145	0-88318-119-3
No. 21	Experimental Meson Spectroscopy – 1974 (Boston)	74-82628	0-88318-120-7
No. 22	Neutrinos – 1974 (Philadelphia)	74-82413	0-88318-121-5
No. 23	Particles and Fields – 1974 (APS/DPF Williamsburg)	74-27575	0-88318-122-3
No. 24	Magnetism and Magnetic Materials – 1974 (20th Annual Conference, San Francisco)	75-2647	0-88318-123-1
No. 25	Efficient Use of Energy (The APS Studies on the Technical Aspects of the More Efficient Use of Energy)	75-18227	0-88318-124-X

No. 26	High-Energy Physics and Nuclear Structure – 1975 (Santa Fe and Los Alamos)	75-26411	0-88318-125-8
No. 27	Topics in Statistical Mechanics and Biophysics: A Memorial to Julius L. Jackson (Wayne State University, 1975)	75-36309	0-88318-126-6
No. 28	Physics and Our World: A Symposium in Honor of Victor F. Weisskopf (M.I.T., 1974)	76-7207	0-88318-127-4
No. 29	Magnetism and Magnetic Materials – 1975 (21st Annual Conference, Philadelphia)	76-10931	0-88318-128-2
No. 30	Particle Searches and Discoveries – 1976 (Vanderbilt Conference)	76-19949	0-88318-129-0
No. 31	Structure and Excitations of Amorphous Solids (Williamsburg, VA, 1976)	76-22279	0-88318-130-4
No. 32	Materials Technology – 1976 (APS New York Meeting)	76-27967	0-88318-131-2
No. 33	Meson-Nuclear Physics – 1976 (Carnegie-Mellon Conference)	76-26811	0-88318-132-0
No. 34	Magnetism and Magnetic Materials – 1976 (Joint MMM-Intermag Conference, Pittsburgh)	76-47106	0-88318-133-9
No. 35	High Energy Physics with Polarized Beams and Targets (Argonne, 1976)	76-50181	0-88318-134-7
No. 36	Momentum Wave Functions – 1976 (Indiana University)	77-82145	0-88318-135-5
No. 37	Weak Interaction Physics – 1977 (Indiana University)	77-83344	0-88318-136-3
No. 38	Workshop on New Directions in Mossbauer Spectroscopy (Argonne, 1977)	77-90635	0-88318-137-1
No. 39	Physics Careers, Employment and Education (Penn State, 1977)	77-94053	0-88318-138-X
No. 40	Electrical Transport and Optical Properties of Inhomogeneous Media (Ohio State University, 1977)	78-54319	0-88318-139-8
No. 41	Nucleon-Nucleon Interactions – 1977 (Vancouver)	78-54249	0-88318-140-1
No. 42	Higher Energy Polarized Proton Beams (Ann Arbor, 1977)	78-55682	0-88318-141-X
No. 43	Particles and Fields – 1977 (APS/DPF, Argonne)	78-55683	0-88318-142-8
No. 44	Future Trends in Superconductive Electronics (Charlottesville, 1978)	77-9240	0-88318-143-6
No. 45	New Results in High Energy Physics – 1978 (Vanderbilt Conference)	78-67196	0-88318-144-4
No. 46	Topics in Nonlinear Dynamics (La Jolla Institute)	78-57870	0-88318-145-2
No. 47	Clustering Aspects of Nuclear Structure and Nuclear Reactions (Winnepeg, 1978)	78-64942	0-88318-146-0
No. 48	Current Trends in the Theory of Fields (Tallahassee, 1978)	78-72948	0-88318-147-9

No. 49	Cosmic Rays and Particle Physics – 1978 (Bartol Conference)	79-50489	0-88318-148-7
No. 50	Laser-Solid Interactions and Laser Processing – 1978 (Boston)	79-51564	0-88318-149-5
No. 51	High Energy Physics with Polarized Beams and Polarized Targets (Argonne, 1978)	79-64565	0-88318-150-9
No. 52	Long-Distance Neutrino Detection – 1978 (C.L. Cowan Memorial Symposium)	79-52078	0-88318-151-7
No. 53	Modulated Structures – 1979 (Kailua Kona, Hawaii)	79-53846	0-88318-152-5
No. 54	Meson-Nuclear Physics – 1979 (Houston)	79-53978	0-88318-153-3
No. 55	Quantum Chromodynamics (La Jolla, 1978)	79-54969	0-88318-154-1
No. 56	Particle Acceleration Mechanisms in Astrophysics (La Jolla, 1979)	79-55844	0-88318-155-X
No. 57	Nonlinear Dynamics and the Beam-Beam Interaction (Brookhaven, 1979)	79-57341	0-88318-156-8
No. 58	Inhomogeneous Superconductors – 1979 (Berkeley Springs, W.V.)	79-57620	0-88318-157-6
No. 59	Particles and Fields – 1979 (APS/DPF Montreal)	80-66631	0-88318-158-4
No. 60	History of the ZGS (Argonne, 1979)	80-67694	0-88318-159-2
No. 61	Aspects of the Kinetics and Dynamics of Surface Reactions (La Jolla Institute, 1979)	80-68004	0-88318-160-6
No. 62	High Energy e^+e^- Interactions (Vanderbilt, 1980)	80-53377	0-88318-161-4
No. 63	Supernovae Spectra (La Jolla, 1980)	80-70019	0-88318-162-2
No. 64	Laboratory EXAFS Facilities – 1980 (Univ. of Washington)	80-70579	0-88318-163-0
No. 65	Optics in Four Dimensions – 1980 (ICO, Ensenada)	80-70771	0-88318-164-9
No. 66	Physics in the Automotive Industry – 1980 (APS/AAPT Topical Conference)	80-70987	0-88318-165-7
No. 67	Experimental Meson Spectroscopy – 1980 (Sixth International Conference, Brookhaven)	80-71123	0-88318-166-5
No. 68	High Energy Physics – 1980 (XX International Conference, Madison)	81-65032	0-88318-167-3
No. 69	Polarization Phenomena in Nuclear Physics – 1980 (Fifth International Symposium, Santa Fe)	81-65107	0-88318-168-1
No. 70	Chemistry and Physics of Coal Utilization – 1980 (APS, Morgantown)	81-65106	0-88318-169-X
No. 71	Group Theory and its Applications in Physics – 1980 (Latin American School of Physics, Mexico City)	81-66132	0-88318-170-3
No. 72	Weak Interactions as a Probe of Unification (Virginia Polytechnic Institute – 1980)	81-67184	0-88318-171-1
No. 73	Tetrahedrally Bonded Amorphous Semiconductors (Carefree, Arizona, 1981)	81-67419	0-88318-172-X

No. 74	Perturbative Quantum Chromodynamics (Tallahassee, 1981)	81-70372	0-88318-173-8
No. 75	Low Energy X-Ray Diagnostics – 1981 (Monterey)	81-69841	0-88318-174-6
No. 76	Nonlinear Properties of Internal Waves (La Jolla Institute, 1981)	81-71062	0-88318-175-4
No. 77	Gamma Ray Transients and Related Astrophysical Phenomena (La Jolla Institute, 1981)	81-71543	0-88318-176-2
No. 78	Shock Waves in Condensed Matter – 1981 (Menlo Park)	82-70014	0-88318-177-0
No. 79	Pion Production and Absorption in Nuclei – 1981 (Indiana University Cyclotron Facility)	82-70678	0-88318-178-9
No. 80	Polarized Proton Ion Sources (Ann Arbor, 1981)	82-71025	0-88318-179-7
No. 81	Particles and Fields –1981: Testing the Standard Model (APS/DPF, Santa Cruz)	82-71156	0-88318-180-0
No. 82	Interpretation of Climate and Photochemical Models, Ozone and Temperature Measurements (La Jolla Institute, 1981)	82-71345	0-88318-181-9
No. 83	The Galactic Center (Cal. Inst. of Tech., 1982)	82-71635	0-88318-182-7
No. 84	Physics in the Steel Industry (APS/AISI, Lehigh University, 1981)	82-72033	0-88318-183-5
No. 85	Proton-Antiproton Collider Physics –1981 (Madison, Wisconsin)	82-72141	0-88318-184-3
No. 86	Momentum Wave Functions – 1982 (Adelaide, Australia)	82-72375	0-88318-185-1
No. 87	Physics of High Energy Particle Accelerators (Fermilab Summer School, 1981)	82-72421	0-88318-186-X
No. 88	Mathematical Methods in Hydrodynamics and Integrability in Dynamical Systems (La Jolla Institute, 1981)	82-72462	0-88318-187-8
No. 89	Neutron Scattering – 1981 (Argonne National Laboratory)	82-73094	0-88318-188-6
No. 90	Laser Techniques for Extreme Ultraviolt Spectroscopy (Boulder, 1982)	82-73205	0-88318-189-4
No. 91	Laser Acceleration of Particles (Los Alamos, 1982)	82-73361	0-88318-190-8
No. 92	The State of Particle Accelerators and High Energy Physics (Fermilab, 1981)	82-73861	0-88318-191-6
No. 93	Novel Results in Particle Physics (Vanderbilt, 1982)	82-73954	0-88318-192-4
No. 94	X-Ray and Atomic Inner-Shell Physics – 1982 (International Conference, U. of Oregon)	82-74075	0-88318-193-2
No. 95	High Energy Spin Physics – 1982 (Brookhaven National Laboratory)	83-70154	0-88318-194-0
No. 96	Science Underground (Los Alamos, 1982)	83-70377	0-88318-195-9

No. 97	The Interaction Between Medium Energy Nucleons in Nuclei – 1982 (Indiana University)	83-70649	0-88318-196-7
No. 98	Particles and Fields – 1982 (APS/DPF University of Maryland)	83-70807	0-88318-197-5
No. 99	Neutrino Mass and Gauge Structure of Weak Interactions (Telemark, 1982)	83-71072	0-88318-198-3
No. 100	Excimer Lasers – 1983 (OSA, Lake Tahoe, Nevada)	83-71437	0-88318-199-1
No. 101	Positron-Electron Pairs in Astrophysics (Goddard Space Flight Center, 1983)	83-71926	0-88318-200-9
No. 102	Intense Medium Energy Sources of Strangeness (UC-Sant Cruz, 1983)	83-72261	0-88318-201-7
No. 103	Quantum Fluids and Solids – 1983 (Sanibel Island, Florida)	83-72440	0-88318-202-5
No. 104	Physics, Technology and the Nuclear Arms Race (APS Baltimore –1983)	83-72533	0-88318-203-3
No. 105	Physics of High Energy Particle Accelerators (SLAC Summer School, 1982)	83-72986	0-88318-304-8
No. 106	Predictability of Fluid Motions (La Jolla Institute, 1983)	83-73641	0-88318-305-6
No. 107	Physics and Chemistry of Porous Media (Schlumberger-Doll Research, 1983)	83-73640	0-88318-306-4
No. 108	The Time Projection Chamber (TRIUMF, Vancouver, 1983)	83-83445	0-88318-307-2
No. 109	Random Walks and Their Applications in the Physical and Biological Sciences (NBS/La Jolla Institute, 1982)	84-70208	0-88318-308-0
No. 110	Hadron Substructure in Nuclear Physics (Indiana University, 1983)	84-70165	0-88318-309-9
No. 111	Production and Neutralization of Negative Ions and Beams (3rd Int'l Symposium, Brookhaven, 1983)	84-70379	0-88318-310-2
No. 112	Particles and Fields – 1983 (APS/DPF, Blacksburg, VA)	84-70378	0-88318-311-0
No. 113	Experimental Meson Spectroscopy – 1983 (Seventh International Conference, Brookhaven)	84-70910	0-88318-312-9
No. 114	Low Energy Tests of Conservation Laws in Particle Physics (Blacksburg, VA, 1983)	84-71157	0-88318-313-7
No. 115	High Energy Transients in Astrophysics (Santa Cruz, CA, 1983)	84-71205	0-88318-314-5
No. 116	Problems in Unification and Supergravity (La Jolla Institute, 1983)	84-71246	0-88318-315-3
No. 117	Polarized Proton Ion Sources (TRIUMF, Vancouver, 1983)	84-71235	0-88318-316-1

No. 118	Free Electron Generation of Extreme Ultraviolet Coherent Radiation (Brookhaven/OSA, 1983)	84-71539	0-88318-317-X
No. 119	Laser Techniques in the Extreme Ultraviolet (OSA, Boulder, Colorado, 1984)	84-72128	0-88318-318-8
No. 120	Optical Effects in Amorphous Semiconductors (Snowbird, Utah, 1984)	84-72419	0-88318-319-6
No. 121	High Energy e^+e^- Interactions (Vanderbilt, 1984)	84-72632	0-88318-320-X
No. 122	The Physics of VLSI (Xerox, Palo Alto, 1984)	84-72729	0-88318-321-8
No. 123	Intersections Between Particle and Nuclear Physics (Steamboat Springs, 1984)	84-72790	0-88318-322-6
No. 124	Neutron-Nucleus Collisions – A Probe of Nuclear Structure (Burr Oak State Park - 1984)	84-73216	0-88318-323-4
No. 125	Capture Gamma-Ray Spectroscopy and Related Topics – 1984 (Internat. Symposium, Knoxville)	84-73303	0-88318-324-2
No. 126	Solar Neutrinos and Neutrino Astronomy (Homestake, 1984)	84-63143	0-88318-325-0
No. 127	Physics of High Energy Particle Accelerators (BNL/SUNY Summer School, 1983)	85-70057	0-88318-326-9
No. 128	Nuclear Physics with Stored, Cooled Beams (McCormick's Creek State Park, Indiana, 1984)	85-71167	0-88318-327-7
No. 129	Radiofrequency Plasma Heating (Sixth Topical Conference, Callaway Gardens, GA, 1985)	85-48027	0-88318-328-5
No. 130	Laser Acceleration of Particles (Malibu, California, 1985)	85-48028	0-88318-329-3
No. 131	Workshop on Polarized ^{3}He Beams and Targets (Princeton, New Jersey, 1984)	85-48026	0-88318-330-7
No. 132	Hadron Spectroscopy–1985 (International Conference, Univ. of Maryland)	85-72537	0-88318-331-5
No. 133	Hadronic Probes and Nuclear Interactions (Arizona State University, 1985)	85-72638	0-88318-332-3
No. 134	The State of High Energy Physics (BNL/SUNY Summer School, 1983)	85-73170	0-88318-333-1
No. 135	Energy Sources: Conservation and Renewables (APS, Washington, DC, 1985)	85-73019	0-88318-334-X
No. 136	Atomic Theory Workshop on Relativistic and QED Effects in Heavy Atoms	85-73790	0-88318-335-8
No. 137	Polymer-Flow Interaction (La Jolla Institute, 1985)	85-73915	0-88318-336-6
No. 138	Frontiers in Electronic Materials and Processing (Houston, TX, 1985)	86-70108	0-88318-337-4
No. 139	High-Current, High-Brightness, and High-Duty Factor Ion Injectors (La Jolla Institute, 1985)	86-70245	0-88318-338-2

No. 140	Boron-Rich Solids (Albuquerque, NM, 1985)	86-70246	0-88318-339-0
No. 141	Gamma-Ray Bursts (Stanford, CA, 1984)	86-70761	0-88318-340-4
No. 142	Nuclear Structure at High Spin, Excitation, and Momentum Transfer (Indiana University, 1985)	86-70837	0-88318-341-2
No. 143	Mexican School of Particles and Fields (Oaxtepec, México, 1984)	86-81187	0-88318-342-0
No. 144	Magnetospheric Phenomena in Astrophysics (Los Alamos, 1984)	86-71149	0-88318-343-9
No. 145	Polarized Beams at SSC & Polarized Antiprotons (Ann Arbor, MI & Bodega Bay, CA, 1985)	86-71343	0-88318-344-7
No. 146	Advances in Laser Science–I (Dallas, TX, 1985)	86-71536	0-88318-345-5
No. 147	Short Wavelength Coherent Radiation: Generation and Applications (Monterey, CA, 1986)	86-71674	0-88318-346-3
No. 148	Space Colonization: Technology and The Liberal Arts (Geneva, NY, 1985)	86-71675	0-88318-347-1
No. 149	Physics and Chemistry of Protective Coatings (Universal City, CA, 1985)	86-72019	0-88318-348-X
No. 150	Intersections Between Particle and Nuclear Physics (Lake Louise, Canada, 1986)	86-72018	0-88318-349-8
No. 151	Neural Networks for Computing (Snowbird, UT, 1986)	86-72481	0-88318-351-X
No. 152	Heavy Ion Inertial Fusion (Washington, DC, 1986)	86-73185	0-88318-352-8
No. 153	Physics of Particle Accelerators (SLAC Summer School, 1985) (Fermilab Summer School, 1984)	87-70103	0-88318-353-6
No. 154	Physics and Chemistry of Porous Media—II (Ridge Field, CT, 1986)	83-73640	0-88318-354-4
No. 155	The Galactic Center: Proceedings of the Symposium Honoring C. H. Townes (Berkeley, CA, 1986)	86-73186	0-88318-355-2
No. 156	Advanced Accelerator Concepts (Madison, WI, 1986)	87-70635	0-88318-358-0
No. 157	Stability of Amorphous Silicon Alloy Materials and Devices (Palo Alto, CA, 1987)	87-70990	0-88318-359-9
No. 158	Production and Neutralization of Negative Ions and Beams (Brookhaven, NY, 1986)	87-71695	0-88318-358-7

No.	Title		
No. 159	Applications of Radio-Frequency Power to Plasma: Seventh Topical Conference (Kissimmee, FL, 1987)	87-71812	0-88318-359-5
No. 160	Advances in Laser Science–II (Seattle, WA, 1986)	87-71962	0-88318-360-9
No. 161	Electron Scattering in Nuclear and Particle Science: In Commemoration of the 35th Anniversary of the Lyman-Hanson-Scott Experiment (Urbana, IL, 1986)	87-72403	0-88318-361-7
No. 162	Few-Body Systems and Multiparticle Dynamics (Crystal City, VA, 1987)	87-72594	0-88318-362-5
No. 163	Pion–Nucleus Physics: Future Directions and New Facilities at LAMPF (Los Alamos, NM, 1987)	87-72961	0-88318-363-3
No. 164	Nuclei Far from Stability: Fifth International Conference (Rosseau Lake, ON, 1987)	87-73214	0-88318-364-1
No. 165	Thin Film Processing and Characterization of High-Temperature Superconductors	87-73420	0-88318-365-X
No. 166	Photovoltaic Safety (Denver, CO, 1988)	88-42854	0-88318-366-8
No. 167	Deposition and Growth: Limits for Microelectronics (Anaheim, CA, 1987)	88-71432	0-88318-367-6
No. 168	Atomic Processes in Plasmas (Santa Fe, NM, 1987)	88-71273	0-88318-368-4
No. 169	Modern Physics in America: A Michelson-Morley Centennial Symposium (Cleveland, OH, 1987)	88-71348	0-88318-369-2
No. 170	Nuclear Spectroscopy of Astrophysical Sources (Washington, D.C., 1987)	88-71625	0-88318-370-6
No. 171	Vacuum Design of Advanced and Compact Synchrotron Light Sources (Upton, NY, 1988)	88-71824	0-88318-371-4
No. 172	Advances in Laser Science–III: Proceedings of the International Laser Science Conference (Atlantic City, NJ, 1987)	88-71879	0-88318-372-2
No. 173	Cooperative Networks in Physics Education (Oaxtepec, Mexico 1987)	88-72091	0-88318-373-0
No. 174	Radio Wave Scattering in the Interstellar Medium (San Diego, CA 1988)	88-72092	0-88318-374-9
No. 175	Non-neutral Plasma Physics (Washington, DC 1988)	88-72275	0-88318-375-7

No.	Title	LCCN	ISBN
No. 176	Intersections Between Particle and Nuclear Physics (Third International Conference) (Rockport, ME 1988)	88-62535	0-88318-376-5
No. 177	Linear Accelerator and Beam Optics Codes (La Jolla, CA 1988)	88-46074	0-88318-377-3
No. 178	Nuclear Arms Technologies in the 1990s (Washington, DC 1988)	88-83262	0-88318-378-1
No. 179	The Michelson Era in American Science: 1870–1930 (Cleveland, OH 1987)	88-83369	0-88318-379-X
No. 180	Frontiers in Science: International Symposium (Urbana, IL 1987)	88-83526	0-88318-380-3
No. 181	Muon-Catalyzed Fusion (Sanibel Island, FL 1988)	88-83636	0-88318-381-1
No. 182	High T_c Superconducting Thin Films, Devices, and Application (Atlanta, GA 1988)	88-03947	0-88318-382-X
No. 183	Cosmic Abundances of Matter (Minneapolis, MN 1988)	89-80147	0-88318-383-8
No. 184	Physics of Particle Accelerators (Ithaca, NY 1988)	87-07208	0-88318-384-6
No. 185	Glueballs, Hybrids, and Exotic Hadrons (Upton, NY 1988)	89-83513	0-88318-385-4
No. 186	High-Energy Radiation Background in Space (Sanibel Island, FL 1987)	89-083833	0-88318-386-2
No. 187	High-Energy Spin Physics (Minneapolis, MN 1988)	89-083948	0-88318-387-0
No. 188	International Symposium on Electron Beam Ion Sources and their Applications (Upton, NY 1988)	89-084343	0-88318-388-9
No. 189	Relativistic, Quantum Electrodynamic, and Weak Interaction Effects in Atoms (Santa Barbara, CA 1988)	89-084431	0-88318-389-7
No. 190	Radio-frequency Power in Plasmas (Irvine, CA 1989)	89-045805	0-88318-397-8
No. 191	Advances in Laser Science–IV (Atlanta, GA 1988)	89-085595	0-88318-391-9
No. 192	Vacuum Mechatronics (First International Workshop) (Santa Barbara, CA 1989)	89-045905	0-88318-394-3
No. 193	Advanced Accelerator Concepts (Lake Arrowhead, CA 1989)	89-045914	0-88318-393-5